Advanced Structured Materials

Volume 225

Series Editors

Common engineering materials are reaching their limits in many applications, and new developments are required to meet the increasing demands on engineering materials. The performance of materials can be improved by combining different materials to achieve better properties than with a single constituent, or by shaping the material or constituents into a specific structure. The interaction between material and structure can occur at different length scales, such as the micro, meso, or macro scale, and offers potential applications in very different fields.

This book series addresses the fundamental relationships between materials and their structure on overall properties (e.g., mechanical, thermal, chemical, electrical, or magnetic properties, etc.). Experimental data and procedures are presented, as well as methods for modeling structures and materials using numerical and analytical approaches. In addition, the series shows how these materials engineering and design processes are implemented and how new technologies can be used to optimize materials and processes.

Advanced Structured Materials is indexed in Google Scholar and Scopus.

Vijay Kumar • Yogendra Kumar Mishra
Editors

Soft Materials for Functional Applications

Springer

Editors
Vijay Kumar
Department of Physics
National Institute of Technology Srinagar
Srinagar, Jammu and Kashmir, India

Yogendra Kumar Mishra
Smart Materials, NanoSYD, Mads Clausen Institute
University of Southern Denmark
Sønderborg, Denmark

ISSN 1869-8433 ISSN 1869-8441 (electronic)
Advanced Structured Materials
ISBN 978-981-97-9467-6 ISBN 978-981-97-9468-3 (eBook)
https://doi.org/10.1007/978-981-97-9468-3

This Springer imprint is published by the registered company Springer Nature Singapore Pte Ltd.
The registered company address is: 152 Beach Road, #21-01/04 Gateway East, Singapore 189721, Singapore

Preface

We are pleased to present this edited volume on the fascinating topic of soft materials for functional applications. This book brings together comprehensive contributions from well-known experts in the field, offering a broad overview of innovative advancements, research findings, and emerging trends in soft materials.

Soft materials play a crucial role in our everyday lives; they are ubiquitous, encompassing much of what we interact with daily. These materials occupy a unique phase between liquids and solids and are characterized by their distinct physical characteristics, often influenced by energies in their surrounding environments. The mesoscopic interactions between their components dictate the macroscopic behaviour of these materials. Their properties can be further engineered by incorporating nanostructures or additional components from foreign materials. The most significant examples include liquid metals, polymers, gels, elastomers, colloids, and biomaterials like proteins, enzymes, cells, etc.

Research into soft materials has seen a surge of interest due to their remarkable properties, including lightweight, low mechanical modulus, high stretchability, and a wide range of other functions, for example, sensing, actuating, insulating, and transporting; research on soft materials has recently increased interest. They are ideally suited for electronics, optoelectronic, electromechanical applications, intelligent textiles, biological and biomedical devices, flexible devices, robotics, and wearable electronics applications.

This book also delves into various soft functional materials' fundamental and applied properties. It explores critical advancements in the field, their functionalities in both pure and engineered forms, the challenges of integrating them into devices, and the performance of these devices in applications such as energy harvesting, soft grippers, robotics, biomimetic systems, drug delivery, surgical devices, innovative packaging, aerospace devices, etc., including self-reporting and self-healing materials. The underlying science and mechanism driving these properties are detailed to help readers comprehend the structure-property relationship in soft materials that exhibit such phenomena.

Each chapter in this book is a state-of-the-art review written by well-known specialists in the field. The chapters are self-contained with adequate cross-references and are presented clearly and concisely.

As we present this volume on soft materials, we sincerely thank the contributors, collaborators, and mentors whose expertise and insights have enriched this book. We hope this book will serve as a valuable resource for scholars, students, and enthusiasts, offering a deeper understanding of soft materials and their transformative impact on science and society.

In a nutshell, this book provides a wealth of information, offering scientists and engineers new insights for advancing their understanding of functional soft materials and inspiring the development of novel materials with unprecedented properties.

Srinagar, Jammu and Kashmir, India — Vijay Kumar
Sønderborg, Denmark — Yogendra Kumar Mishra

Contents

About the Editors

Vijay Kumar is working as an assistant professor in physics at the National Institute of Technology (NIT) in Srinagar (J&K). Before joining NIT Srinagar, he worked as an assistant professor in physics at Chandigarh University, Mohali. He was a postdoctoral researcher at the Phosphor Research Group, University of the Free State, South Africa. He holds a Ph.D. in physics from SLIET Longowal and an M.Sc. in physics from the National Institute of Technology Jalandhar, Punjab, India. His research experience and interests lie in the synthesis, processing, and characterization of functional materials, solid-state luminescent materials, bio-based/biodegradable polymers, and composites, drug delivery devices, wastewater treatment, polymer nanocomposites, and ion–solid interaction.

Yogendra Kumar Mishra is a professor MSO at Mads Clausen Institute, University of Southern Denmark (SDU), Sønderborg, Denmark. Prior to joining SDU, he was leading a scientifically independent group at functional nanomaterials chair, Institute for Material Science, Kiel University, Kiel Germany. He did habilitation (Dr. Habil.) in materials science from Kiel University in 2015 and Ph.D. in physics in 2008 from Inter University Accelerator Centre/Jawaharlal Nehru University, New Delhi, India. His main research focus is on 'Smart Materials'.

Synthesis of Pure and Engineered Soft Materials

Shamli Rajput, Rajinder Kaur, and Surbhi Sharma

Abstract In modern times, there are a number of soft materials in our daily lives, viz., plastic, hydrogels, and elastomers. Even the vast majority of human organs are composed of soft materials. Studies on soft materials are carried out mainly for two purposes: (a) developing a soft hand that mimics the human skin and can perform soft manipulations or an anthropomorphic/prosthetic hand and (b) generating soft actuators. Integrating chemistry and physics in the study of soft materials particularly focused on materials like colloids, polymers, surfactants, and liquid crystals, offers vital insight into biological soft material self-assembly mechanisms. The ultimate goal of exploiting soft materials is to manipulate/alter objects without damaging or leaving any marks behind. A steady hold resulting from an area contact would be the other benefit. Among the currently used soft materials, the most common are polymer composite, elastomer, natural rubber, synthetic rubber, and nanoparticulated polymer composite. Furthermore, many engineered and synthetic soft materials depend on self-organization, frequently predicated on systematic ordering generated by noncovalent interactions like hydrogen bonding or screened electrostatic forces. Presently, materials may be designed at the nanoscale level by humans using either bottom-up (self-organization) or top-down (atom by atom or molecule by molecule) approaches. Interestingly, soft nanotechnology is a component of the later approach, and soft materials' ability to self-organize can be used to produce a variety of nanostructures for a diversity of applications. Many engineered soft materials have been created and are essential to today's technology. Examples include soft nanotechnology, soft robotics, and soft lithography, which demonstrate the extensive importance of soft materials in biology and engineering. Understanding how biological materials self-organize as soft matter with flow

S. Rajput · R. Kaur · S. Sharma (✉)
Department of Chemistry, University Institute of Sciences, Chandigarh University, Mohali, Punjab, India

V. Kumar, Y. K. Mishra (eds.), *Soft Materials for Functional Applications*,
Advanced Structured Materials 225, https://doi.org/10.1007/978-981-97-9468-3_1

characteristics is a continuous and challenging task. It can act as an intermediate, i.e., between a crystalline solid and a liquid (therefore "soft") beyond the intramolecular covalent bond. Interestingly, atomic/molecular level manipulation can provide logical methods for producing functional materials that strongly connect with molecular system designs and are particularly helpful for creating innovative soft material functionalities. With the theme of all the exciting characteristics of soft materials developed by different researchers, the present chapter covers different synthetic routes and methods of some pure and engineered soft materials.

Keywords Soft materials · Pure soft materials · Engineered soft materials · Shape-memory polymer · Hydrogels · Micellar nanoparticles · Renewable resources

1 Introduction

Yao et al. [1] stated soft materials as materials that primarily exhibit entropic elastic restoring forces and have a low shear modulus compared to their bulk modulus. Soft materials comprise a wide variety of artificial materials made of polymers, such as gels, rubbers, self-adhesive materials, and a sizable number of materials made of molecules found naturally, like food or living tissues. Moreover, according to Xiao et al. [2], soft materials primarily exhibit entropic elastic restoring forces and have a low shear modulus compared to their bulk modulus. Hence, without using excessive force, their deformation can be felt with the hands or observed with the unaided eye. The creation and assembly of such soft materials and systems are central issues in the development of multiple functions and applications are highlighted by some of the researchers including sensing and separation by Lim et al. [3], environmental matters by Ariga et al. [4], energy conversion and storage by Gao et al. [5], advanced devices by Zeng et al. [6], and biological and biomedical applications by Samal et al. [7]. Due to the dramatic creation of useful materials and the advancement of nanotechnology, nanoscience has become more advanced and fascinating owing to the more contributions from the designs of nanoscale materials from molecular levels. Designing the material that can be placed in the category of soft material is highly crucial as one should have an idea about the possible interactions, which are noncovalent interactions that can perform reversible actions, demonstrating the nature of soft materials.

Interestingly, soft materials can react well to stimuli and the environment despite not being as highly durable as hard materials like metals, ceramics, and engineering plastics. Soft materials exhibit a high sensitivity to external stimuli and a low elastic modulus, which makes them particularly susceptible to surface instabilities caused by buckling. Baki et al. [8] explained some of the applications of the stimulus because of their sensitivity, soft materials are good choices for use as intelligent materials in biological scaffolds, sensors, microfluidic systems, nanoreactors, and therapies. Li et al. [9] conclude that the skin of humans and elastomer surfaces easily display a variety of regular or irregular wrinkles. Moreover, biological materials always inspire the exploration of noncovalent interactions, known as "soft," which combine flexible and fluid characteristics. The mechanical properties of soft materials are attributed to self-assembling supramolecular components that are stabilised via non-covalent interactions is based on experimental and theoretical frameworks and is a fascinating multidisciplinary research area some of which are studied by Frederix et al. [10] and Bisoyi et al. [11].

Considering the previously mentioned features of soft materials, Zhou et al. [12] and Pierini et al. [13] discovered that a variety of uses in the creation of composites from soft materials by combining other functional materials are increasingly popular to achieve complex and adjustable materials with improved properties for high-tech applications and one such new strategy is the addition of photothermal agents to the structure of soft materials in this regard.

The different types of pure and engineered soft materials known till now are represented in Fig. 1. Herein, in this chapter, we collectively present information on some of the essential pure and engineered soft materials and their methodology via various sources to get an idea for better design and applications of soft materials and the tabular description, which include important highlights—examples of different types of pure and engineered soft materials are represented in Table 1.

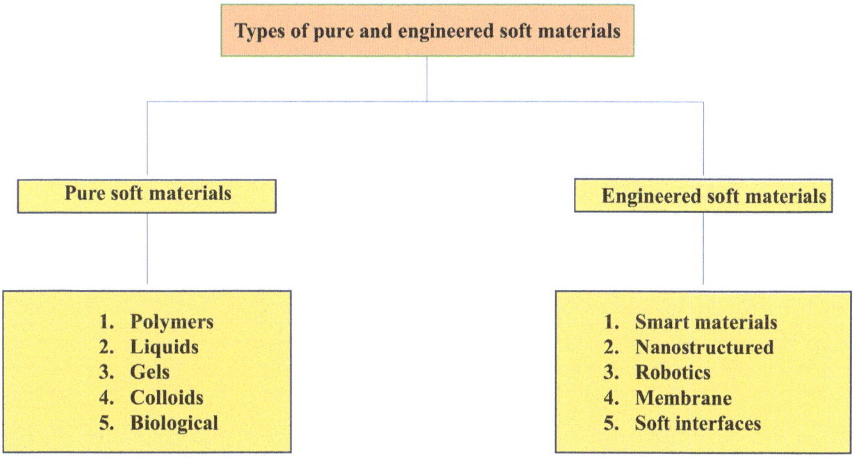

Fig. 1 Representing types of pure and engineered soft materials known

Table 1 Types of pure and engineered soft materials with examples

S. no.	Types of pure and engineered soft materials	Highlights	Examples	References
1.	Soft materials from renewable resources	We are developing sustainable energy solutions by providing practical possibilities for lowering the environmental impact and broadening the world's energy sources	Conversion of glucose obtained from biomass into alkanes Synthesis of isomaltulose from sucrose Synthesis of cardanol-based polymer	[14, 15]
2.	Polymer hydrogels	They can hold an immense amount of water in their cavities and continue connecting it while maintaining the network structure	Synthesis of poly(acrylic acid) hydrogel Synthesis of β-CD-*g*-(PNIPAM-*b*-PDMA) and β-CD-*g*-$(PNIPAM\text{-}b\text{-}PDMA)_3$ Synthesis of P(LE-IA-MPEG) hydrogel	[15, 16]
3.	Silicone elastomers	Its broad use and continuous research into new formulations and uses are encouraged by its stability, cleanliness of processing, and variety of applications	Formation of silicone elastomers with groups trifluoropropyl Synthesizing silicone elastomers with various stoichiometric ratios	[17, 18]
4.	Micellar nanoparticles	They offer stability, and their structural configuration enables surface property adjustments that improve biocompatibility and in vivo circulation time	Formation of cross-linked micellar nanoparticles	[19, 20]
5.	Shape-memory polymer	Soft robots, newly developed metamaterials, and programmable materials rely on shape-memory polymers	Poly(1,10-decylene adipate) diol synthesized by a polycondensation reaction	[21, 22]

2 Synthesis Procedures/Mythologies of Pure and Engineered Soft Materials

2.1 Soft Materials from Renewable Resources

A new technique has been recently developed that breaks down carbohydrates in the aqueous phase into liquid alkanes by using Scheme 1 as an outline for breaking down glucose generated from biomass which was synthesized by Serrano-Ruiz et al. [14]. Glucose is first dehydrated and furans are formed as a result of acid-catalyzed dehydration enzymes. Following that, these furans undergo hydrogenation and dehydration over a bifunctional catalyst. This catalyst has two functions: it efficiently promotes the hydrogenation and dehydration processes. These changes lead to the development of liquid alkanes with carbon chains that include 7–15 atoms. These alkanes distinguish themselves because of their molecular weight, which satisfies standards for gasoline components used in transportation. Their acceptability derives from their capacity to blend in with the current fuel system and engines built for conventional fuels generated from petroleum. This approach has potential in terms of the environment since it makes use of sugars obtained from biomass, which reduces dependency on limited fossil fuels and reduces corresponding greenhouse gas emissions. It resolves challenges with energy security and the environment, signifying an important development in sustainable fuel technology. In terms of technology, present work focuses on increasing production

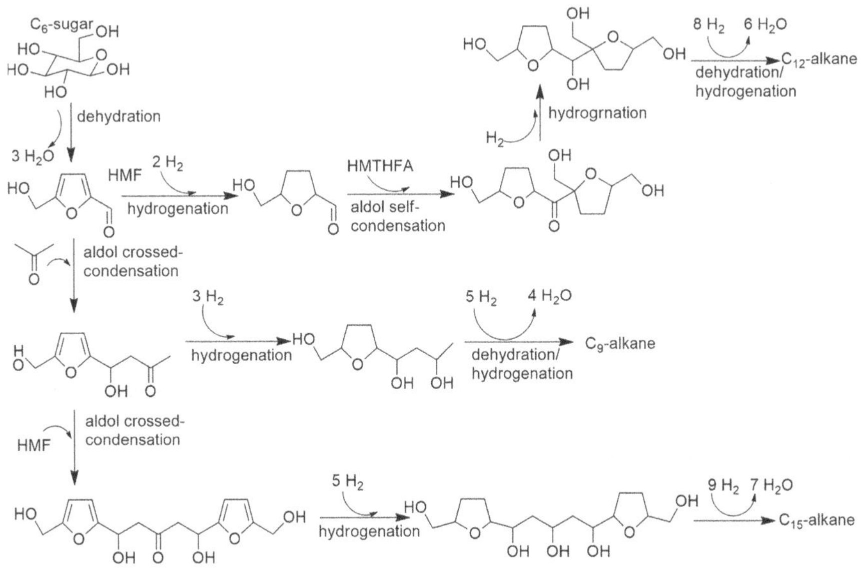

Scheme 1 Mechanism of the reaction for converting glucose obtained from biomass into alkanes (Reprinted with permission from [14])

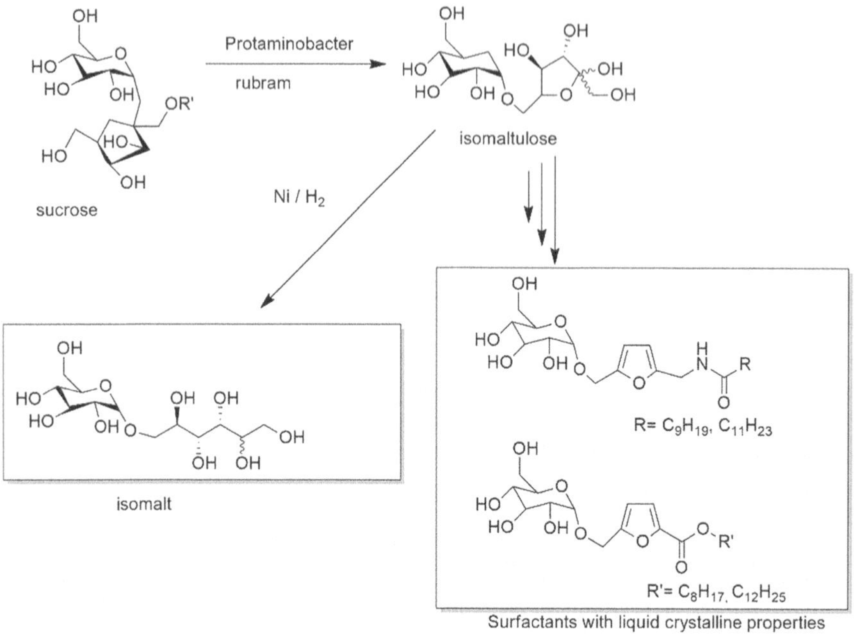

Scheme 2 Synthesizing isomaltulose from sucrose and confirming that the end products have commercial potential (Reprinted with permission from [25])

capacities, process economics, and catalyst efficiency optimization. The objective of these attempts is to enhance the feasibility and profitability of bio-based liquid alkanes as an economically feasible substitute for traditional fuels. Finally, Scheme 1 illustrates an important advancement in the transformation of glucose produced from biomass into liquid alkanes that may be used as fuel for transportation. Vemula et al. [23] and Lichtenthaler et al. [24] their invention highlights the ongoing development of sustainable energy solutions by providing practical possibilities for lowering the environmental impact and broadening the world's energy sources.

In addition to all of its applications, the synthesis of disaccharide intermediates, especially isomaltulose, has great industrial potential. Scheme 2 was synthesized by Heinze et al. [25] which describes how isomaltulose, a hydrogenated derivative of sucrose, is made by combining terminally *R*-glucosylated glucitol and mannitol in 1:1. This substance can be purchased commercially as a low-calorie sweetener because it has a flavor and texture similar to sucrose but has a lower glycemic index. Beyond its nutritional advantages, cellulose has industrial appeal. Isomaltulose derivatives have the potential to exhibit amphiphilic characteristics, which prepares them for use in liquid-crystalline materials. These materials provide special properties including reactivity to external stimuli and future research uses in advanced materials research because they possess ordered molecular configurations similar to those observed in liquid crystals. Furthermore, cellulose's economic significance is further increased by its adjustable nature. Biomass contains a lot of cellulose, which is a renewable energy source that helps to attain goals for sustainability by making isomaltulose and its derivatives.

In the final analysis, isomaltulose's synthesis and derivatives demonstrate its diverse industrial potential, extending from its use as a low-calorie sweetener to its incorporation into liquid-crystalline materials with amphiphilic characteristics. These developments highlight continued attempts to advance in materials science, sustainable manufacturing methods, and food science.

Synthetic surfactants have been produced using carbohydrates and vegetable oils in recent research. Understanding their interfacial characteristics, ecological significance, toxicological impacts, and dermatological characteristics has been the subject of research. In these investigations, genetically engineered *Escherichia coli* strains have been used to effectively and sustainably generate these surfactants. These surfactants' interfacial properties are essential in many industries, like the pharmaceutical, detergent, and cosmetics areas. Ecological evaluations check if they satisfy sustainability standards by analyzing their environmental effect and biodegradability. Studies on toxicology evaluate their safety characteristics, especially with reference to aquatic life and human health. Furthermore, dermatological analyses study their skin compatibility and possible allergenicity. Consecutively, these studies demonstrate the promise of bio-based surfactants made from vegetable oils and carbohydrates, emphasizing their adaptability and hope in ecological and health-related fields. Frost et al. and Deleu et al. [26] have shown the effective synthesis of catechol28 and phenol29 from D-glucose in Scheme 3.

Cardanyl acrylate was produced in Scheme 4 by Lochab et al. [27] via coupling an acrylate group to cardanol's reactive free hydroxyl group. A linear polymer was produced by using 0.8% azobisisobutyronitrile (AIBN) as a radical initiator, and

Scheme 3 Genetically modified *E. coli* to produce phenol and catechol via engineered pathways for aromatic amino acid metabolism (Reprinted with permission from [26])

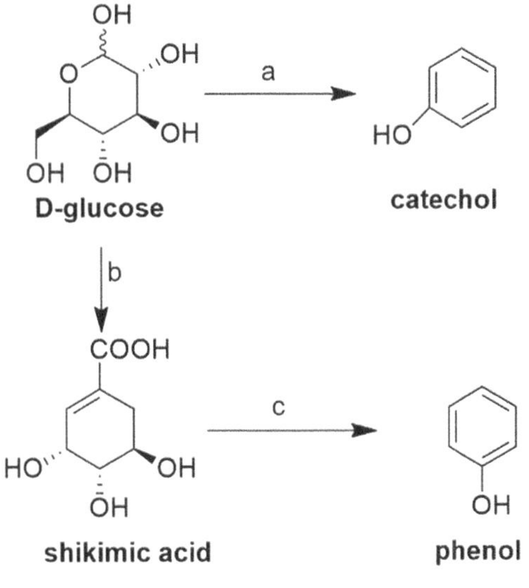

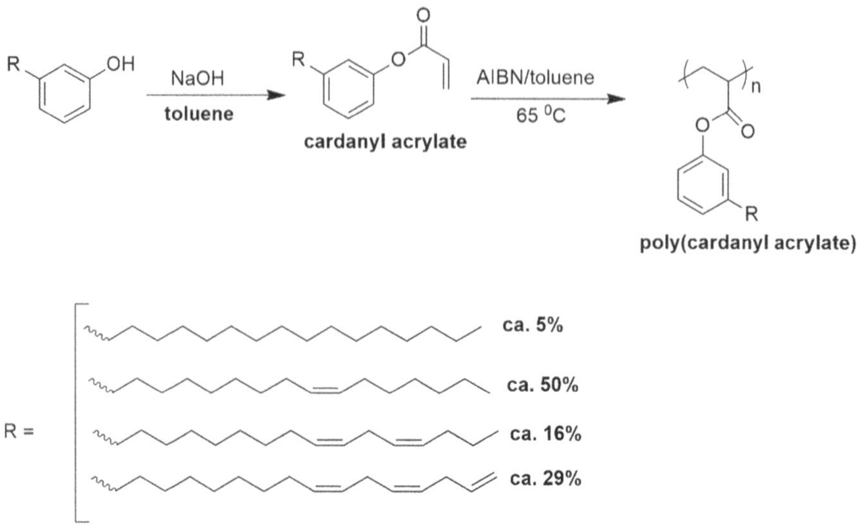

Scheme 4 Synthesis of cardanol-based polymer as soft material (Reprinted with permission from [27])

solution polymerization in toluene was performed. A chain growth polymerization mechanism was produced due to the polymerization process that AIBN initiated. This process enabled the cardanyl acrylate monomers to create covalent connections. Having characteristics that may be adjusted according to the polymer chain's molecular weight and composition, the resultant linear polymer has potential for use in adhesives, coatings, and other industrial applications. Cardanol's flexibility in polymer chemistry and its future potential use of environmentally friendly resources to create sustainable materials are highlighted by various modifications of the compound by functional group attachment. Cardanol and acryloyl chloride were reacted to create cardanyl acrylate. The resultant polymer developed cross-linking when exposed to air or UV light after removing the solvent. This procedure caused a transparent, thick coating to develop. The primary cross-linking method was oxidation, which was made possible by hydroperoxide intermediates produced by air or UV light exposure. By initiating free radical reactions, these hydroperoxides cause covalent bonds to form between polymer chains, which cross-link the chains. The amount of cross-linking and the conditions under which the polymerization occurs can be changed to modify the final film's characteristics, including transparency and thickness. Morales-Cerrada et al. [28] show how this technique applies cardanol-derived materials, which may be used to create resilient, eco-friendly films and coatings that are appropriate for a range of industrial and commercial uses.

Kasprzyk et al. [30] conclude bicyclic 2-pyridone derivatives require a slightly complex synthesis. Therefore, the nonaromatic structure of 5-oxo-2,3-dihydro-5H-(1,3) thiazolo (3,2-a) pyridine-3,7-dicarboxylic acid (referred to as TPA) was produced by the five steps that were thought to have taken place during the

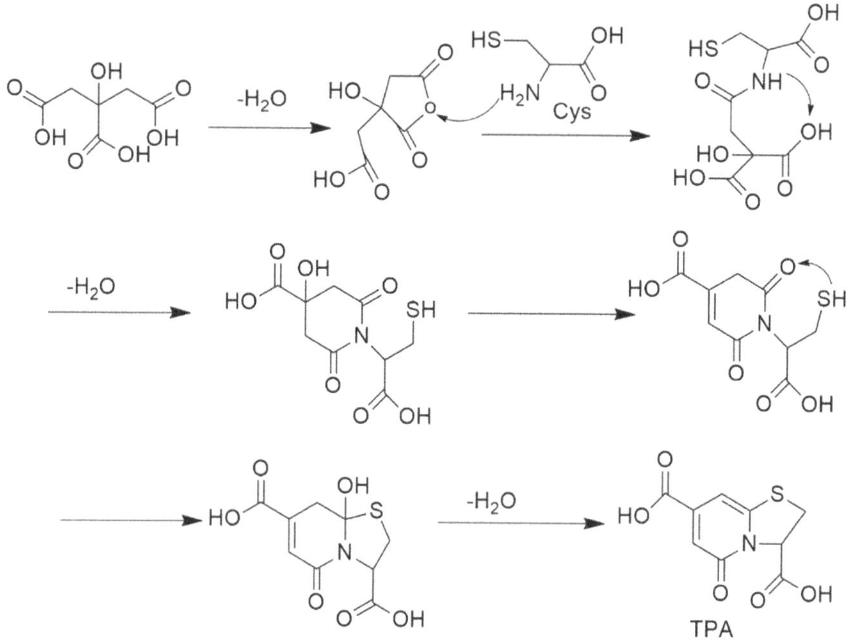

Scheme 5 Reaction mechanism for transforming nonaromatic fluorescent TPA, improving its fluorescence capabilities by chemical alteration (Reprinted with permission from [29])

nonaromatic molecule's creation in this study, as shown in Scheme 5 by Wang et al. [29]. These steps were comparable to the processes proposed by Kasprzyk and colleagues. The water solvent was extracted under vacuum from the CA + Cys aqueous combination in their innovative work, and after the combination was heated to a high temperature of 180.8 °C, a crude product containing TPA was produced. Considering the reaction conditions and preparation procedures, according to Li et al. [31], this study did not call for high-temperature operations or water evaporation; instead, the aqueous combination of CA + Cys was gently heated to produce. This simple method has various benefits, including, in contrast to previous suggestions, simple operation, an environmentally favorable reaction pathway, affordable cost, and global production.

Furthermore, the fresh product may have unique qualities and provide interesting functional applications because it is in a gel-like condition as compared to solid-state powders.

The high concentration of hydroxyl groups in cellulose limits its wide-ranging applications in natural conditions. Although these hydroxyl groups make the material more biocompatible and water-soluble, they make it more difficult to process and less useful in some situations. Because of its chemical structure, cellulose is more easily changed or processed than many synthetic polymers. However, cellulose's hydroxyl groups make it less amenable to conventional polymer processing methods. To get over this restriction, scientists have been looking at several methods, such as chemical modification or combining with other polymers, to improve the qualities of cellulose and increase its uses. These strategies aim to either lower the hydroxyl content or add properties that increase the material's adaptability to various substrates and processing techniques. Chang et al. [33], in addition, explain cellulose's hydroxyl groups create a few challenges that must be overcome to fully realize its potential in a variety of industries, from textiles and packaging to biomedical applications and the creation of sustainable materials. To overcome this limitation, Fu et al. [34] performed various chemical reactions to modify cellulose, yielding cellulose hydrogel. A type of polymer known as cellulose hydrogels is distinguished by their strong cross-linking and hydrophilic qualities, allowing them to expand and retain a lot of water. The cellulose chains in these hydrogels are interlinked, usually by physical or chemical methods, thereby generating three-dimensional networks. Because cellulose has a lot of hydroxyl groups, it is hydrophilic, which helps to explain why these hydrogels may absorb a lot of water. Because of this characteristic, they may be used in various biomedical and environmental uses, including water filtration technologies, substrates for wound dressings, drug delivery systems, and tissue engineering. Additionally, cellulose hydrogels are highly suitable for biomedical applications involving interactions with live tissues due to its biocompatibility and biodegradability. The goal of ongoing research by Singh et al. [35] is to maximize their porosity, mechanical strength, and environmental reactivity. The cross-linking method for preparing cellulose-based hydrogel is illustrated in Scheme 6 by Geng et al. [32]. Female hygiene products and infant diapers both employ these hydrogels, because of their incredible ability to absorb liquids without losing their original structure. Some hydrogel maintains its proper shape due to a combination of poor interactions, which include H-bonding, ionic interactions, and van der Waals interactions, as well as strong interactions, namely, covalent bonds, that collectively preserve the hydrogel when they're intact. Dasgupta et al. [36] demonstrated cellulose is more than just an industrial or

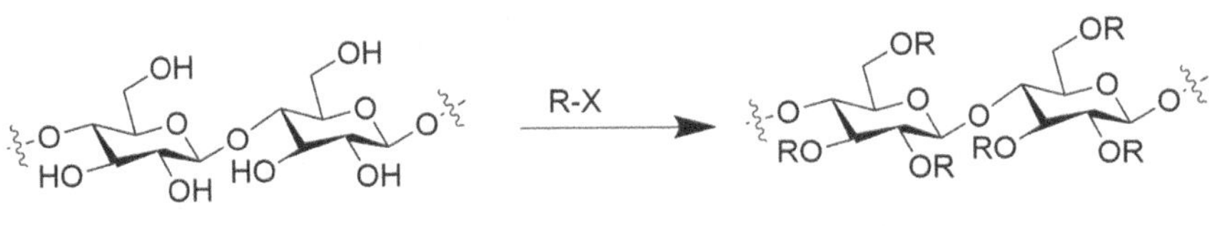

MC :R=H,CH_3
HPMC :R=H,CH_3,$CH_3CH(OH)CH_3$
EC :R=H,CH_2CH_3
HEC :R=H,CH_2CH_2OH
NaCMC :R=H,CH_2COONa

Scheme 6 Approach to cross-linking in cellulose-based hydrogel formation (Reprinted with permission from [32])

agricultural waste; rather, it has an important function in a considerable portion of our daily utility of cellulose hydrogel.

Traditional esterification process involves synthesis of cellulose hydrogels using carboxyl and hydroxyl groups. This process may occur at elevated temperature or in the presence of a condensing agent, such 1-ethyl-3-carbodiimide (3-dimethylaminopropyl) (EDC). EDC facilitates cross-linking without actively taking part in the coupling between carboxyl groups and hydroxyl or amine groups. Rather, it transforms into a urea derivative soluble in water, confirmed to have no harmful effects. Nasution et al. [38] method boasts several uses, particularly good conversion efficiency, low reaction temperature, readily separated by-products, and—most importantly—compatibility with a wide range of materials.

Cellulose or its derivatives, such as hydroxyethyl cellulose (HEC), hydroxypropyl cellulose (HPC), and carboxymethyl cellulose sodium (CMC), may contain the hydroxyl groups that are required for esterification. According to McMullen et al. [39] the carboxyl groups can come from synthetic polymers with multiple carboxyl groups, like polymethacrylic acid (PMAA) and polyacrylic acid (PAA), cellulose derivatives (e.g., CMC), dicarboxylic acid compounds, alginate, and hyaluronic acid (HA). In the synthesis of superabsorbent hydrogels, esterification cross-linking was employed to create HEC-CMC-HA hydrogels in the presence of the condensing agent EDC, as illustrated in Scheme 7. These resulting hydrogels exhibit biodegradability, sensitivity to changes in pH and ionic strength, and a high

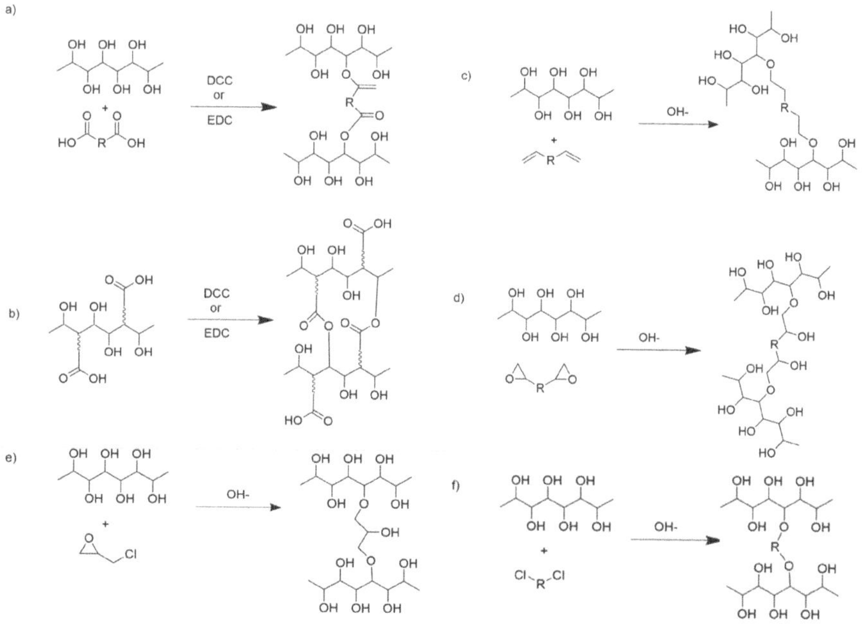

Scheme 7 Analyzing a variety of chemical reactions and cross-linking techniques used in the synthesis process (Reprinted with permission from [37])

capacity for holding water. Among cross-linking agents, divinylsulfone (DVS) is particularly favored. Its mechanism involves a Michael addition reaction between the carbon–carbon double bonds of DVS and the hydroxyl groups on cellulose, enabling the formation of polyelectrolyte hydrogels.

Numerous research papers have explored the creation of hydrogels based on cellulose through connecting methods involving compounds with both epoxy and halide groups, such as epichlorohydrin, as well as alkyl halides. Typically, these connecting processes take place under considering fundamental circumstances, i.e., at elevated temperatures. Diepoxy substances, mainly EGDE (ethylene glycol diglycidylether), known for their lower toxicity, have shown promise as agents that cross-link natural polymers. Capretti et al. [40] employed, in the creation of various hydrogels, β-cyclodextrin-CMC-based hydrogels, cellulose-based hydrogels for the adsorption of bisphenol A, and cationic cellulose hydrogels developed for drug delivery systems that are sensitive to pH or ions. In order to create cellulose-based gels, poly(ethyleneglycol diglycidyl ether) (PEGDE) has also been used as a macromolecular cross-linker by Ye et al. [41]. Hydroxypropyl cellulose hydrogels known as (HPC)-PEGDE, CMC-PEGDE, and CMC-*g*-poly(glycidyl methacrylate) were created by reacting epoxy and hydroxyl groups. The reaction of cross-linking involving strong alkaline conditions are required for halide functional groups and hydroxyl groups, which restricts their use in the creation of cellulose-based hydrogels. To overcome this limitation, some cross-linking procedure used to create cellulose hydrogels combines halide and epoxide functional groups, such as epichlorohydrin. This successful synthesis by Wu et al. [37] of hydrogels has been reported in a "one-step" procedure by the use of epichlorohydrin as a cross-linker, from cellulose in an aqueous solution of NaOH/urea.

The production of a superabsorbent cellulose/chitin hybrid hydrogel was undertaken by Kono et al. which is represented in Scheme 8 by Kundu et al. [42]. Cellulose and chitin emerge as primary candidates for producing biodegradable superabsorbent polymers (SAPs), being the first and second oldest organic materials on Earth, according to this order. Qureshi et al. [43] and Chen et al. [44] show polysaccharides have qualities that make them both biocompatible and biodegradable, making them particularly intriguing for producing biocompatible and ecologically favorable materials. Chitin, according to Tharanathan et al. [45], is a polymer that is structurally similar to cellulose and is made up of b-(1–4)-linked *N*-acetyl-D-glucosamine residues (GlcNAc), with some D-glucosamine residues (GlcN) substituting certain GlcNAc residues in the polymer chain. The production of hydrogels that are superabsorbent using chitin is illustrated in the scheme. In an Erlenmeyer flask containing LiCl/NMP (5 g of LiCl and 95 g of NMP), powdered chitin (1.0 g, 5.3 mmol for the monomer unit) is completely dissolved while being stirred with a Teflon impeller at 500 rpm at room temperature for 3 days. Subsequently, upon full dissolving, 0.64 g of DMAP (5.3 mmol) is added to the liquid followed by the addition of 1.1 g of BTCA (5.3 mmol), or a molar feed ratio of 1 for BTCA. Esterification is let to stir for a full day at room temperature. After that, the mixture is added to 500 mL of methanol/water (1:1) while being stirred to cause the product to precipitate. Using 10% (w/v) aqueous NaOH, the product

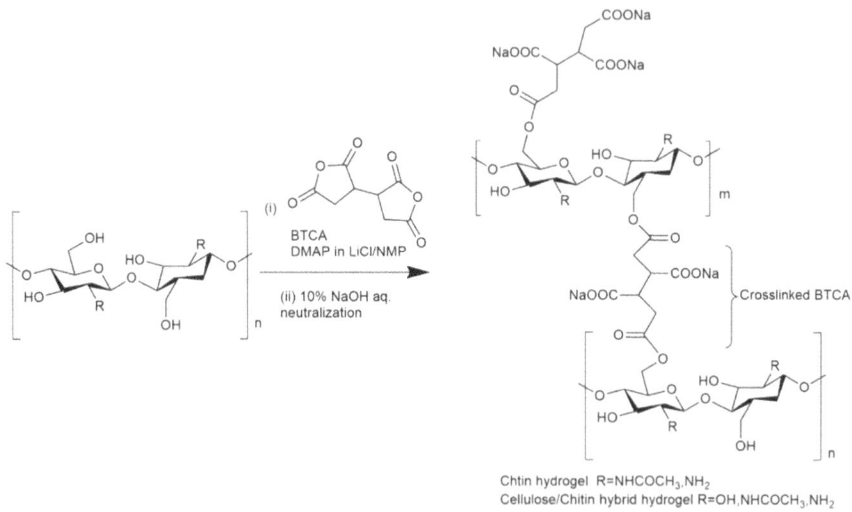

Scheme 8 Approach to creating a superabsorbent hydrogel made of chitin and cellulose (Reprinted with permission from [42])

is neutralized to pH 7.0 while being observed using a pH electrode. After filtering and again purifying with methanol and water by reprecipitation, the precipitate is dried under low pressure to get the finished product. Shen et al. [46] created a granular product—it is then broken down using a mixer and filtered using a 16-mesh sieve. Hybrid hydrogels are combined by dissolving cellulose (0.43 g, 2.65 mmol) in LiCl/NMP (2.5 g of LiCl and 47.5 g of NMP) and stirring at 500 rpm for 2 days—cellulose and chitin were produced. A chitin solution (0.50 g, 2.65 mmol) in LiCl/NMP (2.5 g of LiCl and 47.5 g of NMP) was added to this cellulose solution. Following 2 h of vigorous mixing, 0.64 g (5.3 mmol) of DMAP was added. Upon complete DMAP dissolution, the amount of BTCA supplied was 1.1 g (5.3 mmol), which means that each of the monomers of chitin and cellulose were given a 1:1 molar ratio of BTCA as explained by Kono et al. [47]. Purification, neutralization of unreacted BTCA, precipitation for hydrogel production, and esterification followed techniques comparable to those employed in the chitin hydrogel preparation by Kono et al. [48]. The chitin hydrogels and the hybrid hydrogels made of cellulose and chitin displayed notable degradability when exposed to chitinase, with the hybrid hydrogels further exhibiting degradation in the presence of cellulase. Khan et al. [49] indicate that both chitin and cellulose/chitin hydrogels hold promise for applications in agriculture and industry, and Llanes et al. [50] indicates these hydrogels could serve as viable alternatives to superabsorbent polymers (SAPs).

Gao et al. [51] fabricated pH-responsive, recyclable cellulose hydrogels for dietary injection of insulin by synthesizing carboxymethyl cellulose bonded with acrylate via a coupling reaction shown in Scheme 9. In 7.5 mL of deionized water, solutions of CMC-*g*-AA and AA at various CMC-*g*-AA/AA weight ratios

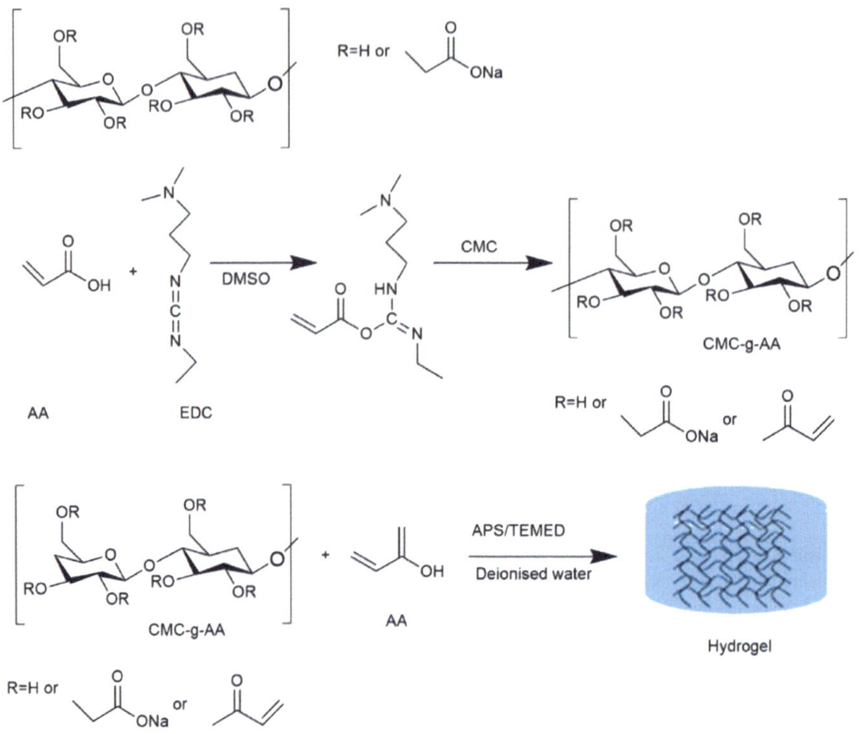

Scheme 9 Approach to fabricate hydrogel through the grafting of acrylate onto carboxymethyl cellulose (CMC-*g*-AA) (Reprinted with permission from [51])

were mixed. Following decomposition, the beginning for polymerization by free radicals took place with the introduction of TEMED as the catalyst and APS as the initiator, respectively. The process of polymerization continued at 25 °C for 12 h. The hydrogel samples were placed in a 200-mL beaker and submerged in deionized water following gelation. Over the course of week, the deionized water was replaced multiple times to eliminate unreacted agents and impurities. Biodegradable hybrid hydrogels made of carboxymethyl cellulose and poly(acrylic acid) react to pH, enlarging quickly under neutral environments and very little in acidic ones. Liu et al. [52] enable selective enzymatic degradation and accelerated insulin release. Administering these insulin-loaded hydrogels ingestion of food to diabetic rats by Sarangi et al. [53] causes a prolonged 6-h drop in fasting blood glucose levels, with over a tenfold increase in relative pharmacological availability compared to free insulin solution.

Mohammadbagheri et al. [54] and Priya et al. [55] produces a cost-effective, biodegradable hydrogel derived from cellulose was synthesized by Shahin Mohammadi-Khoo and colleague for the controlled release of fertilizer in Scheme 10. In this process, cellulose underwent modification and cross-linking via urea.

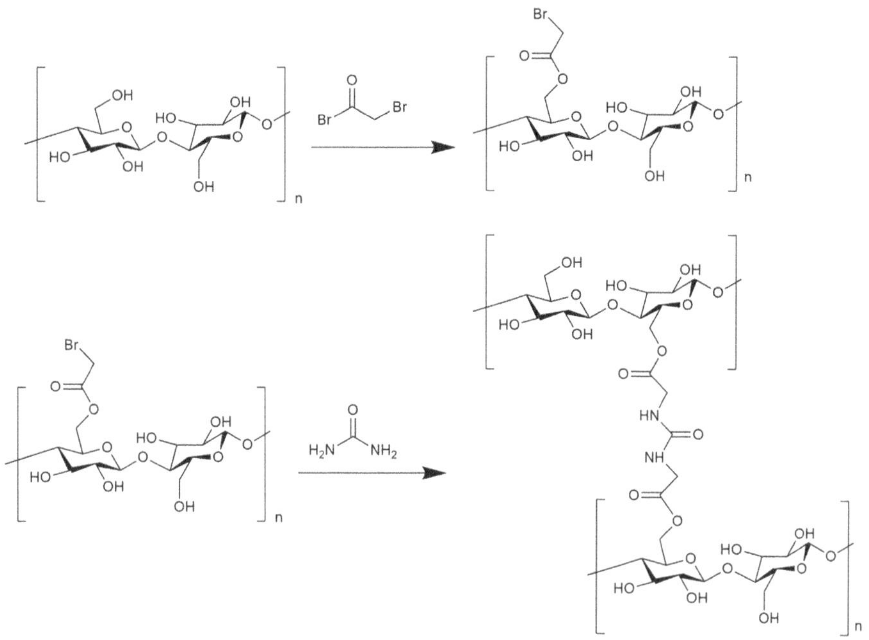

Scheme 10 Formation of biodegradable cellulose hydrogel cross-linked with urea (Reprinted with permission from [54, 55])

Subsequently, the hydrogel that was made was loaded using fertilizer to investigate its controlled release properties. The production of bromoacetylated cellulose (BA-Cell) and the resulting development of a hydrogel that has been treated with urea (CBA-Cell) involved a series of steps. Initially, 0.5 g of cellulose powder (9.259 mmol OH) was subjected to dry grinding in a mortar made of porcelain. During this process, after adding BAB (1.86 g, 9.259 mmol) in stages, the crushing process lasted for 2.5 h. The final composition underwent sequential washing using THF (tetrahydrofuran) and subsequent 5 h of vacuum drying.

For the synthesis of the urea-modified hydrogel (CBA-Cell), 1 g of BA-Cell, and 2.8 mmol of urea were combined in a 100-mL round-bottom flask. To this mixture, following the addition of TEA (0.58 g, 5.7 mmol) and DMA (30 mL), the reaction proceeded with stirring after 48 h at 85 °C. After the precipitate had formed, it was filtered, dissolved in methanol, and dried at 60 °C in a vacuum oven.

Sample characterization utilized thermal gravimetric analysis (TGA), elemental analysis, scanning electron microscopy (SEM), and Fourier transform infrared (FT-IR) spectroscopy. The gel-like substance displayed notable expanding capabilities in 0.9% NaCl solution, water from the tap and distilled drinking water. Further analyses included studying water holding and retention behaviors. Fertilizer flowing out from the packed hydrogel exhibited excellent control. Mohammadi-Khoo et al.

[56] suggest the hydrogel's potential as an advantageous addition to agricultural soil that retains moisture application [56].

2.2 *Synthetic or Engineered Soft Materials*

Polymer Hydrogels

Mahinroosta et al. [15] explain polymer hydrogels are linked, three-dimensional structure networks which include polymers that have the remarkable capacity to take in and retain huge amounts of water. These materials exhibit a gel-like consistency and can swell without losing their structural integrity. The production and formulation of hydrogels made of polymers continue to be an active area of research, exploring new materials and enhancing their functionalities for diverse practical uses.

The earliest hydrogel known to literature comes from 1960, when Wichterle and Lim created poly(2-hydroxethyl methacrylate) (PHEMA) and introduced it to the contact lens market. PHEMA indicated the modern-day hydrogel's capacity to retain its network structure even though absorbing moisture which was further studied by Bashir et al. [57]. Ahmed et al. [58] and Yang et al. [59] explain that because of their unique three-dimensional cross-linked polymer meshwork structure, hydrogels have a capacity to hold onto an immense amount of water in their cavities and continue to connect it while maintaining the network structure exaggerated. According to Manzoor et al. [60], the presence of polar hydrophilic moieties, such as SO_3H, OH, NH_2, COOH, $CONH_2$, etc., across the polymer network as branching groups, explains the appearance of such processes in hydrogels.

In 2020, Sun et al. [61] presented an interesting method for chemically engineering dynamic topological alterations in polymers in response to an individual chemical signal. Two controllable cross-linking reactions—reversible conjugate additions and thiol-disulfide exchange—were employed to transform the initial dimerised PEG and hydrophobic linear materials, a lightly cross-linked polymer, and a cross-linked hydrogel into an amphiphilic linear polymer, lightly cross-linked and linear random copolymers, a cross-linked polymer, and three distinct hydrogel matrixes as represented in Fig. 2.

Applications for hydrogels range from environmental technology to biomedical devices. They are a flexible family of materials. Natural polymers like cellulose, synthetic polymers like poly(acrylic acid) (PAA), polymerizable synthetic monomers, or mixtures of these can all be used to create them. Hydrogel synthesis generally involves the creation of chemical, physical, or hybrid linkages, which can be done by a variety of techniques including bulk polymerization, free radical processes, radiation, solution casting, mixing, or interpenetrating network development. *N,N′*-Methylenebisacrylamide (BIS) is used as a cross-linking agent and ammonium persulfate (APS) as an initiator. Scheme 11 shows how to synthesize poly(acrylic acid) hydrogel. In this procedure, APS, a free radical initiator, causes acrylic acid

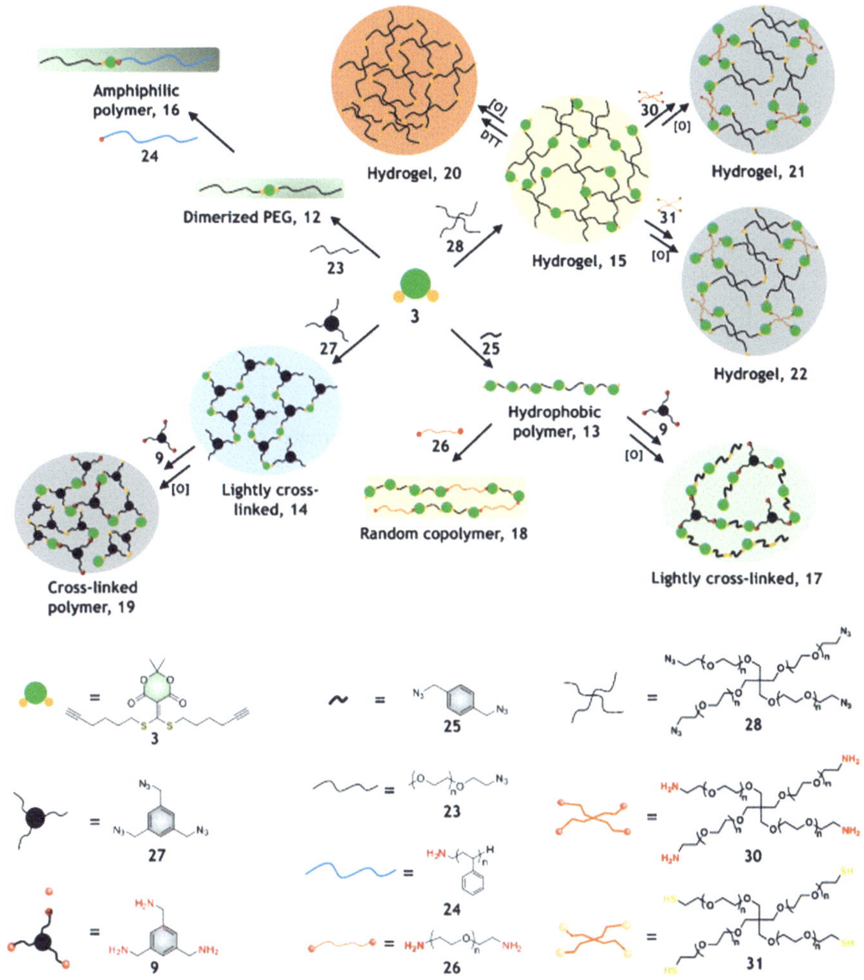

Fig. 2 Diagram illustrating the structural modifications generated by disulfide-thiol scrambling and click–declick chemistry in polymer synthesis. CuAAC was used to create dimerized PEG, hydrophobic linear polymer, hydrogel, and almost cross-linked polymer. By employing amine shifting using thiol-disulfide or the conjugate acceptor center of three processes, amphiphilic unique linear polymers, random copolymers, cross-linked polymers, and hydrogels were made. Note: within a fume hood, the reactions took place, handled with caution, stored, and used in safe locations due to the possible risks associated with poly(azide) compounds

to go through polymerization. When BIS is present, polymer chains may cross-link more easily, which results in the formation of a three-dimensional network structure which additionally is a characteristic of hydrogels. The synthesized PAA hydrogels have a number of significant characteristics that are essential to their wide range of uses. The cross-linked network imparts mechanical strength and may be modified by varying the BIS content and the polymerization conditions. PAA-based hydrogels

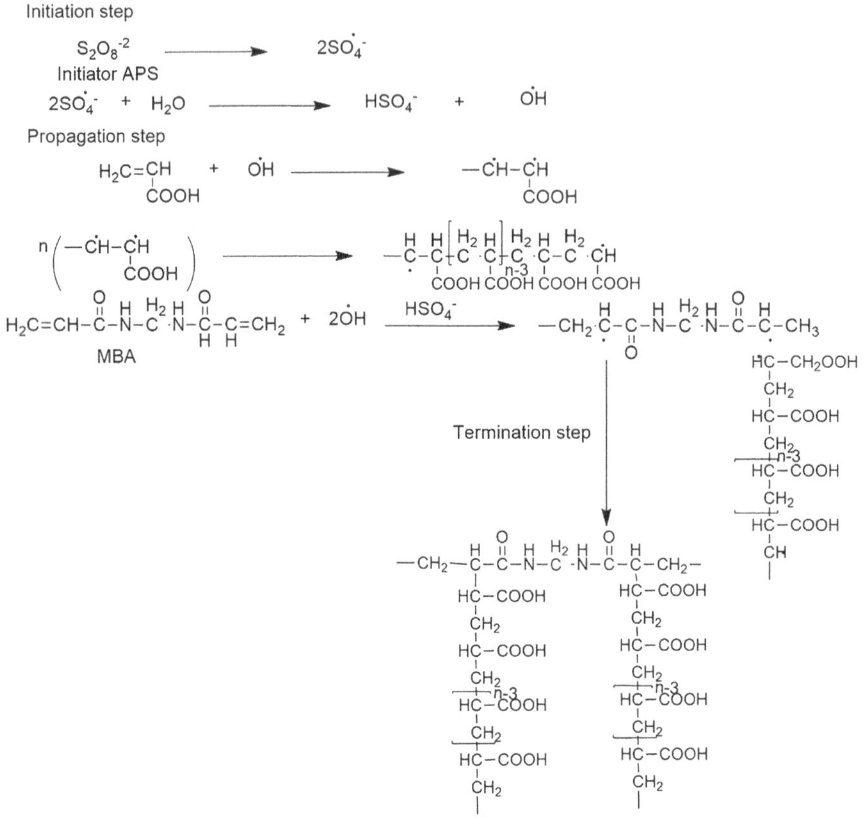

Scheme 11 Synthesis of poly(acrylic acid) hydrogel (Reprinted with permission from [62–64])

are naturally biocompatible and biodegradable, which makes them appropriate for use as frameworks in tissue engineering and other biomedical applications and drug delivery systems. Large volumes of biological fluids or water can be absorbed and kept in solution by them due to their swellability, which is useful for contact lens and wound dressing applications. Moreover, PAA hydrogels exhibit stimulus sensitivity, reacting to variations in temperature, pH, or ambient factors. For applications requiring the regulated release of medications or the detection of certain analytes in biological fluids, this adaptability is essential. The ionic conductivity and swelling behavior of PAA hydrogels are advantageous for electrochemical applications and may be utilized in sensors, actuators, and energy storage devices. Their application in bioelectronics and biosensors is further increased by their capacity to encapsulate and shield living biomolecules or cells. The goal of the research is still to improve PAA hydrogels' characteristics and functions by adjusting their chemical structure, cross-linking density, and adding additives or nanoparticles. These developments aim to maximize their efficacy in particular uses while ensuring durability and

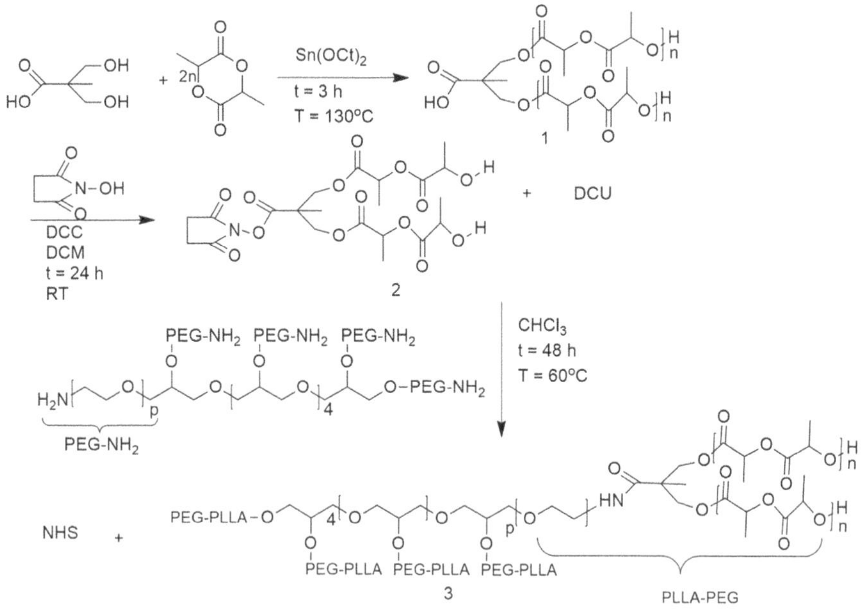

Scheme 12 The synthesis scheme for 8PEG-PLLA copolymer and poly(L-lactide) (PLLA) DCU is dicyclohexylurea; DCC is dicyclohexylcarbodiimide (Reprinted with permission from [7])

reducing ecological footprint. Bukhari et al. [62], Sharma et al. [63], and Zohuriaan-Mehr et al. [64] conclude that poly(acrylic acid) hydrogels are prepared by using techniques such as Scheme 11 provide a flexible framework for a variety of uses, including environmental sensors and biomedical devices.

In Scheme 12, Samae et al. [7] used trifunctional poly(L-lactide) (PLLA) and amine-functionalized polyethylene glycol (PEG) to synthesize a highly branched block copolymer called PEG-*b*-(PLLA)3. The ability of this copolymer to gel at a temperature that changes with temperature is an important feature for its use in injectable hydrogels. The synthesis employs ring-opening polymerization of L-lactide to couple trifunctional PLLA with amine-functionalized PEG. A central PEG core is joined to many PLLA branches to form the resultant copolymer structure. The resulting structure offers mechanical strength, adjustable thermoresponsive characteristics from PLLA, and biocompatibility and hydrophilicity from PEG. The copolymers exhibit thermoresponsive gelation behavior at less than 4 weight percent concentrations in the polymer solution. This indicates that the copolymer goes through a change in phase from a solution turning into a gel as the temperature increases. Interestingly, Ishida et al. found that changing the PLLA blocks' molecular weight and copolymer proportion precisely controlled the gelation's transition temperature. This tunability is essential for adjusting gelation to injection settings and particular biological applications.

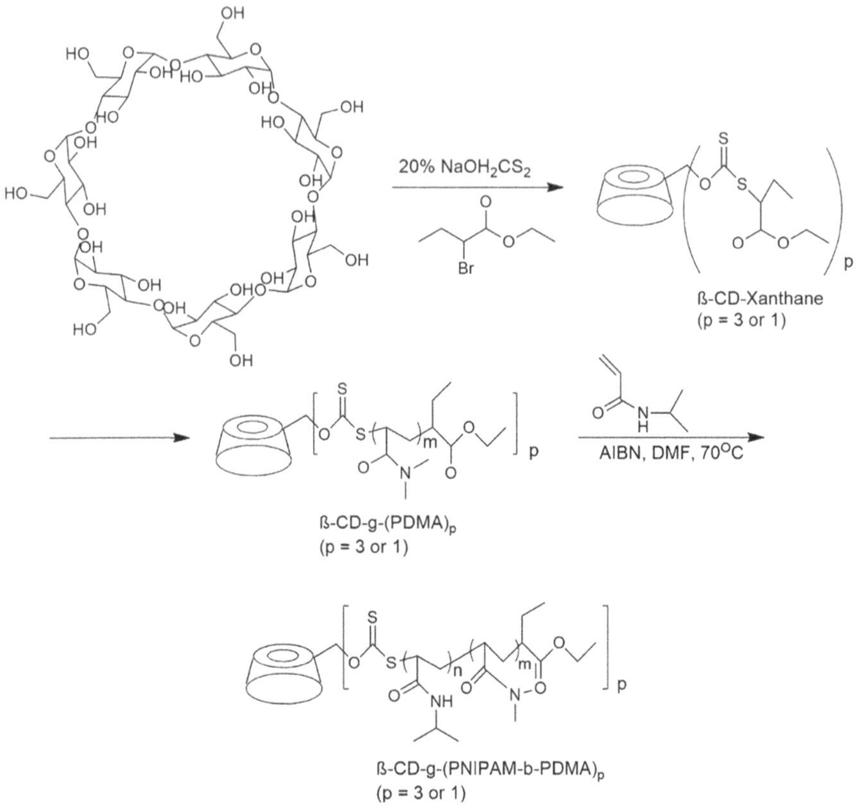

Scheme 13 Reaction route to synthesize β-CD-*g*-(PNIPAM-*b*-PDMA) and β-CD-*g*-(PNIPAM-*b*-PDMA)$_3$ (Reprinted with permission from [65])

After handling, the resultant hydrogels serve as injectable systems that may generate gels in situ. This feature is especially helpful in therapeutic situations (drug administration, tissue engineering, wound healing, etc.) where less invasive techniques are sought. Systemic hydrogels provide advantages such as ease of administration, conformability to irregular tissue shapes, and controlled release of therapeutic agents.

Zhang et al. (2012) synthesized a temperature-responsive hydrogel system by creating a copolymer with a three-arm star structure and a β-cyclodextrin (β-CD) core. The star-shaped copolymer was produced using the successive reversible addition-fragmentation polymerization (RAFT) technique, with β-CD xanthate serving as a chain transfer agent (Scheme 13). This copolymer was star-shaped—PDMA blocks are made of aqueous poly(*N*,*N*-dimethylacrylamide) that forms its arms and PNIPAAm blocks that react to temperature. Roka et al. [66] explain both blocks are soluble in water below the LCST of the PNIPAM segment, indicating the solubility of the copolymer. Above the LCST, however, the PNIPAM blocks

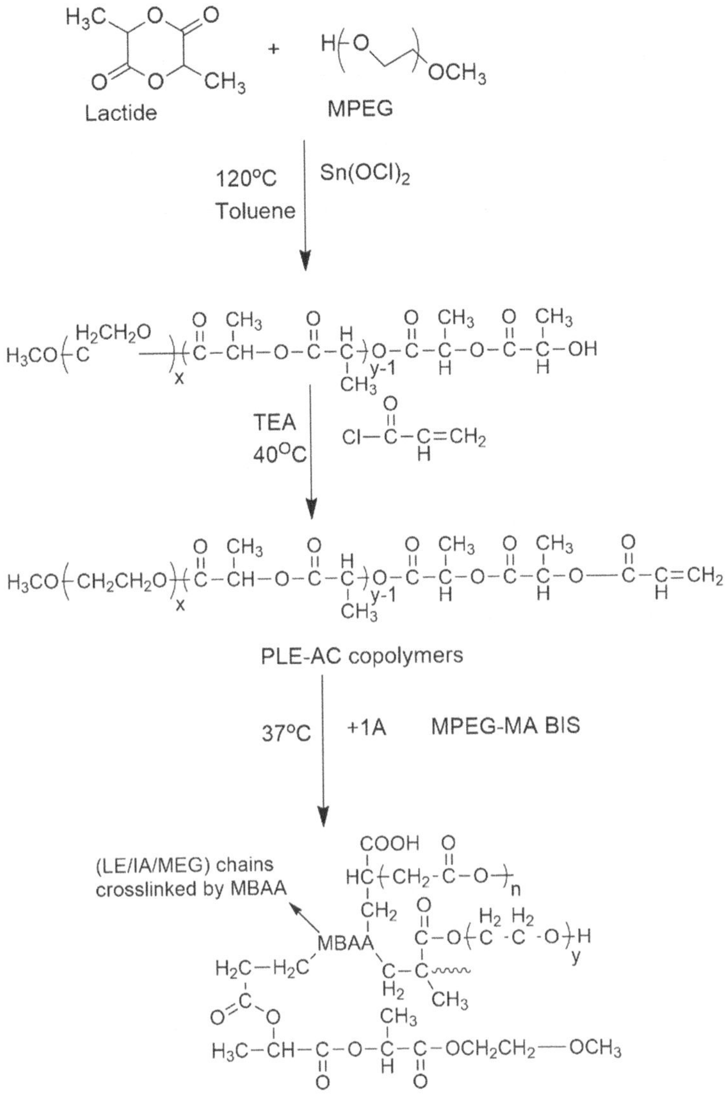

Scheme 14 Mechanism by which a P(LE-IA-MPEG) hydrogel is formed (Reprinted with permission from [67])

combine because they lose their solubility in water. Zhang et al. [65] suggest the unique gel formation behavior was controlled by the star-shaped structure and the thermally collapsed PNIPAAm chains, wherein gave the hydrogel that was formed interesting features.

Wang et al. and Popescu et al. (2012) [67] synthesized a pH-responsive smart hydrogel (Scheme 14) using free-radical polymerization catalyzed by heat, without employing organic solvents. The hydrogel was composed of methoxyl poly(ethylene glycol) (MPEG), poly(lactic acid) (PLA), and itaconic acid (IA), forming a copolymer referred to as P(LE-IA-MPEG). This synthesis approach resulted in an environmentally friendly process for creating a hydrogel with tunable pH-responsive properties, suitable for various applications. The influence of pH on the swelling ratio was seen in buffers with pH values ranging from 1.2 to 6.8. Many hydrogen bonds are present at low pH (1.2) because the hydrogel contains non-dissociated carboxylic acids. These hydrogen bond complexes thus restrict the existing network chains' ability to move or relax. Hydrogel swelling is caused by the partial ionization of carboxylic acid moieties at pH 6.8, which breaks hydrogen bonds and produces electrostatic repulsions between the chains of polymers I.

Mocanu et al. (2012) synthesized a multi-responsive cross-linked carboxymethyl pullulan hydrogel with side chains formed by amide bonds, incorporating polyoxyalkyleneamines, also known as Jeffamine molecules (M-600 and M-2005) (Scheme 15).

Interestingly, the pH-responsive characteristics of these hydrogels were caused by the presence of the Jeffamine moieties explained by Gupta et al. [68]; there are anionic carboxymethyl functional groups and thermo-associative characteristics. Proteins and antioxidants interact with one another to show how potentially useful they are as regulated medication release systems.

Ion-sensitive smart hydrogels, specifically P(NIPAM-*co*-CE), were synthesized with a notable feature—the incorporation of crown ether units by Liu et al. [16] in Scheme 16. One interesting characteristic of synthesizing ion-sensitive smart hydrogels, P(NIPAM-*co*-CE), was the addition of crown ether units. This special constituent was essential in raising the hydrogels' ionic reactivity. When the surrounding ionic environment varies, the resultant material responds with dynamic and intelligent activity. Its sensitivity corresponds to the crown ether units in the hydrogel structure. This makes it a good fit for uses requiring exact ion control and manipulation, including sensing devices or drug administration systems. This method of synthesis creates possibilities for the creation of advanced and adaptable biomaterials [16].

Silicone Elastomers

A subclass of silicone polymers known as silicone elastomers have become highly useful and adaptable materials with many uses in many different sectors, as suggested by Dvornic et al. [17]. Herzberger et al. [69] explain these elastomers are beneficial for various uses, including consumer goods, industrial settings, and medical equipment, due to their distinctive blend of temperature stability, flexibility, and biocompatibility. Usually, silicone elastomers are created by polymerizing monomers with a silicone basis. Mixing a silicone based polymer with a cross-linking and curing agent is common. After that, heat is applied to this mixture, starting a curing reaction that creates a three-dimensional network and gives the material elastomeric qualities, which was further explained by Madsen et al. [70].

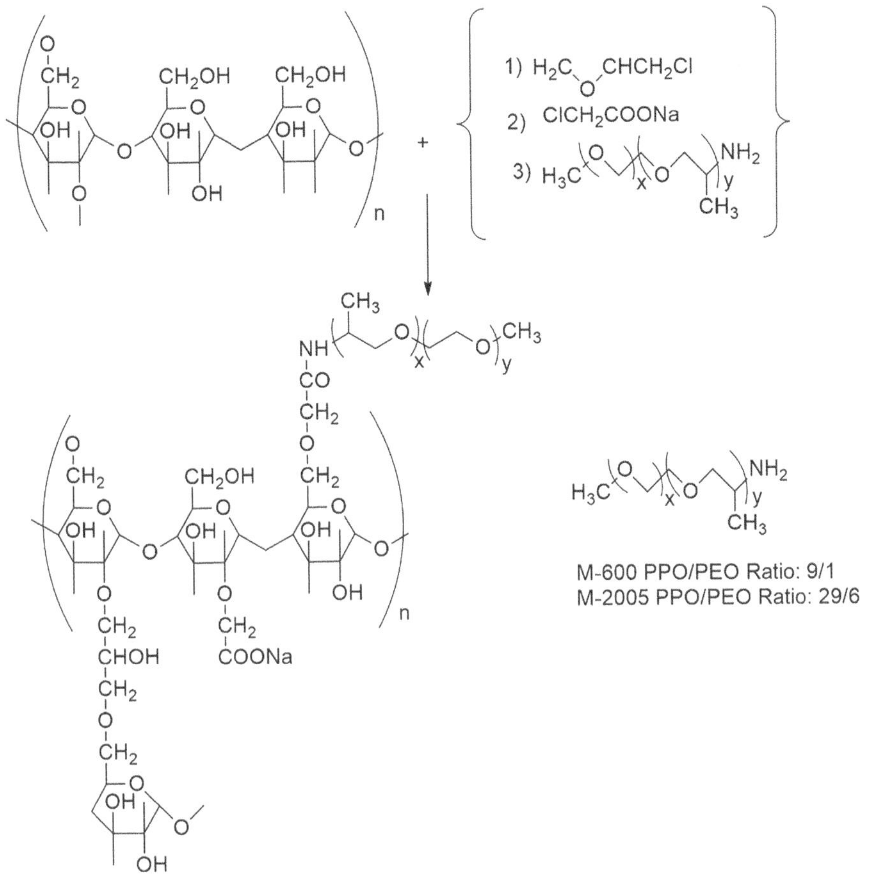

Scheme 15 Formation of a multi-responsive Jeffamine hydrogel using carboxymethyl pullulan (Reprinted with permission from [68])

Moreover, Bartholomew et al. [71] suggest if long-term device performance is required, silicone elastomers used in high-temperature applications must meet stringent reliability and durability requirements. Enhancing silicone elastomers' heat stability is a significant task that the scientific and industrial communities are addressing. However, Eduok et al. [72] explain conventional techniques like using heat-resistant fillers or modifying the chemical composition continue to have problems from significant flaws. Polydimethylsiloxane (PDMS), the most widely used silicone polymer, is well known for its exceptional characteristics and adaptability in various applications. Dimethylsiloxane (–[$Si(CH_3)_2O$]–) repeating units combine to produce a linear and flexible polymer chain that is known as PDMS.

This molecular structure is responsible for several important features that make PDMS very appealing, including flexibility and elasticity: PDMS maintains its

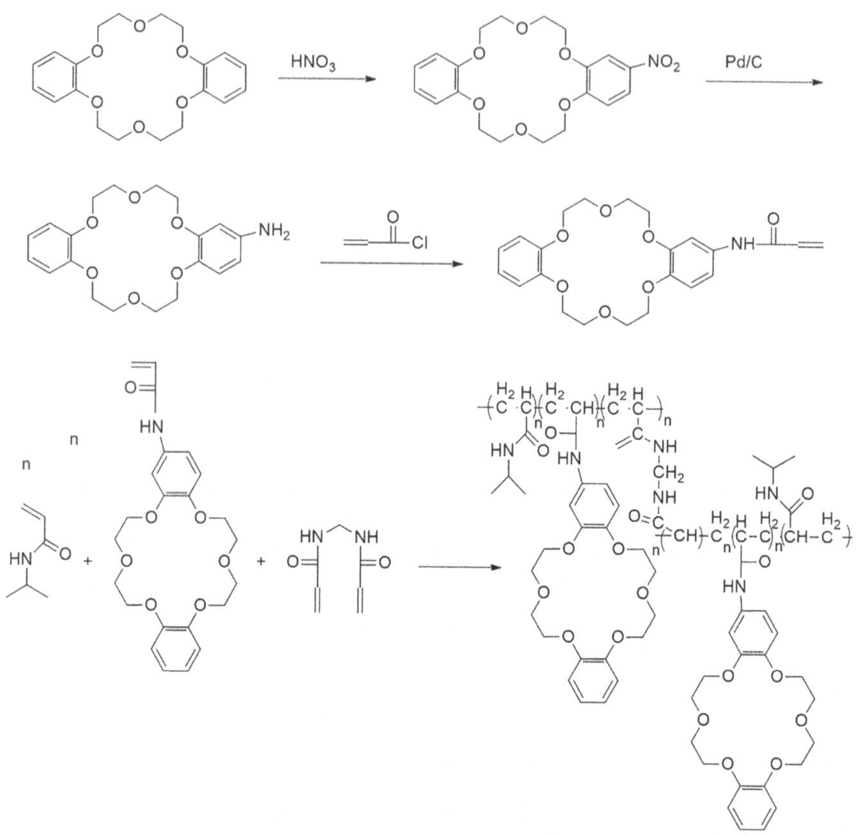

Scheme 16 Synthesis of p(NIPAM-*co*-CE) hydrogels (Reprinted with permission from [16])

mechanical qualities at temperatures between −50 and 200 °C while being elastic and flexible. Because of its durability, it may be used in applications like medical tubing, gaskets, and seals where flexibility is crucial. Chemical Inertness: water, acids, bases, and most organic molecules are among the substances to which PDMS is inert. Its resilience to chemical degradation assures its stability and durability in abrasive settings, such as industrial coatings and sealants. Low surface tension: PDMS has outstanding water repellency and nonstick properties because of the small surface tension. This characteristic makes it perfect for lubricants, mold-releasing agents, and surface coatings. Biocompatibility: PDMS is widely used in pharmaceutical and biomedical applications due to its biological inertness and lack of toxicity. Because of its compatibility with human tissues and fluids, it is employed in medical devices such as implants, prostheses, and catheters. Electrical insulation: PDMS is a good electrical insulator due to its superior dielectric qualities. This feature is essential for electronics applications, where PDMS is used as a protective coating, insulation, and encapsulation. Due to the combination of these qualities,

PDMS is now essential in a wide range of sectors, including electronics, aerospace, automotive, healthcare, and cosmetics. Ariati et al. [73] and Miranda et al. [74] explain its broad use and continuous research into new formulations and uses are encouraged by its stability, cleanliness of processing, and variety of applications.

Mazurek et al. [76] indicate the dielectric within DET can theoretically be any elastomer, although silicones are most frequently utilized because of their quick reaction, little viscoelastic loss, and low Tg. That enables it to use at various temperatures and frequencies. Due to its high gas permeability, hydrophobicity, and dielectric strength, silicones are resistant to moisture absorption and can withstand high humidity levels. Large actuation stresses and high energy densities are possible with silicone actuators, but they need the use of extremely high electric fields as explained by Pelrine et al. [77]. Zhu et al. [78] work improved silicones with lower electric field actuation efficiency may open the door for this technology's use in medical applications, such as implantable devices with operating voltage limits under 24. When the dielectric film's thickness is decreased and soft, high-permittivity elastomers are employed as the dielectric, the actuation strain can be raised. An additional option to work with heterogeneous materials according to Cai et al. [79] that are appropriately engineered and have alternating soft, isotropic, and anisotropic layers will increase the actuation strain. Vinyl end-functionalized polysiloxanes (Px) with varying mol% of trifluoropropyl groups (x) were synthesized using an anionic polymerization approach as described by Dascalu et al. (Scheme 17) [75]. The process involved the polymerization of 1,3,5-tris(3,3,3-trifluoropropyl)-1,3,5-trimethylcyclotrisiloxane (F3) and octamethylcyclotetrasiloxane (D4) in the presence of tetramethylammonium hydroxide (TMAH) as the catalyst. Since TMAH can break down into inactive components at high temperatures, it was chosen as the initiator. We were able to functionalize Px with vinyl end groups by using the end-capping reagent, and this should be used going ahead for thin-film cross-linking. Different feed ratios of D4 to F3 were used to adjust the quantity of CF3 groups in Px, and how much end-capping reagent is used may regulate the molecular weight.

A series of PDMS elastomers were synthesized (Scheme 18) by Meyers et al. [80] through an addition curing reaction, varying the stoichiometric ratio (r) between telechelic vinyl-functional PDMS and an eight-functional hydride oligomer. The stoichiometric imbalance values (r) were set at 1, 1.5, and 2, leading to differences in network structures. These ranged from networks with a high content of dangling PDMS chains and a significant sol fraction (unreacted species) to highly cross-linked structures with minimal dangling chains and sol fraction. When $r > 1$, an excess of cross-linker ensures that most PDMS chains are theoretically linked to the cross-linker at both ends. However, due to steric hindrance, the reaction does not proceed to completion, resulting in a two-phase system comprising a network and a sol fraction. As r increases from 1 to 1.5 and 2, the network becomes more cross-linked, with fewer dangling chains and reduced sol fraction, as predicted by traditional network theory. The sol fractions for $r = 1$, $r = 1$, and $r = 2$ were determined to be 29.3%, 8%, and 6.9%, respectively. The mechanical properties of the elastomers, analyzed through viscoelastic behaviour, showed predominantly

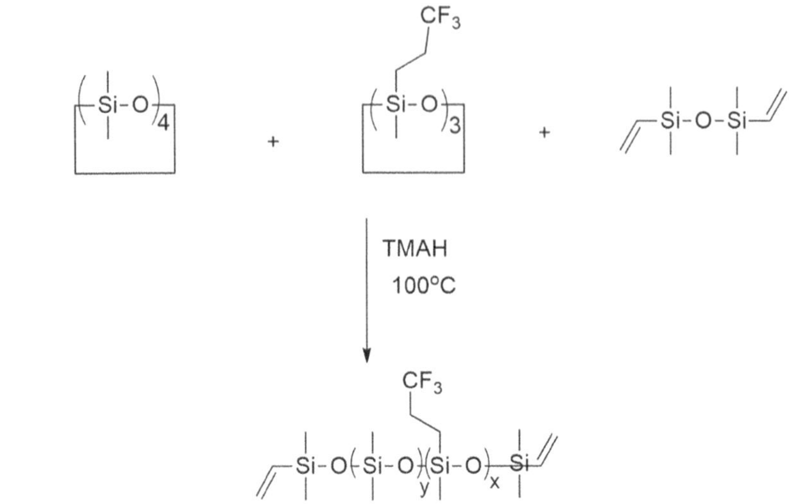

Scheme 17 Formation of silicone elastomers with groups trifluoropropyl (Reprinted with permission from [75])

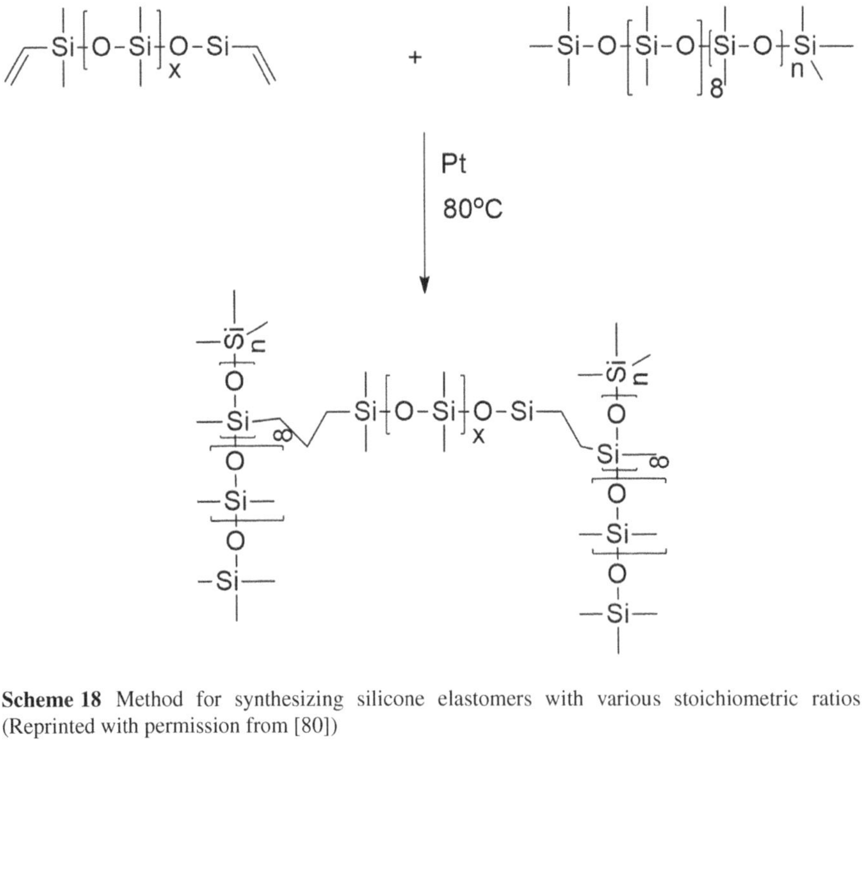

Scheme 18 Method for synthesizing silicone elastomers with various stoichiometric ratios (Reprinted with permission from [80])

elastic characteristics. Higher *r* values resulted in increased storage moduli (G') Swelling the elastomers in heptane allowed separation of the sol fraction from the network, with the removal of the sol fraction leading to a notable increase in G' across all samples.

The objective of this study is to understand how network structure influences the heat degradation of silicone elastomers and whether mobility plays a role in the degradation of PDMS. To modify the relative proportions of elastic, hanging, and sol fractions, different combinations of elastomers were produced by Liu et al. [81], where the sol fraction is the most mobile component and the elastic fraction is the least mobile. Goff et al. [82] study employed thermogravimetric analysis to examine the heat degradation of silicone networks that had varying concentrations of cross-linking. Analysis was also done by Wolf et al. [83] and Pouget et al. [84] on degradation products and modifications to the thermal characteristics of silicone elastomers brought on by the high-temperature treatment. However, the product becomes more expensive due to a chemical alteration of the silicone network, making it less desirable at an industrial level. Fang et al. [85] and Hegemann et al. [86] explain it is extremely difficult to find low-cost, simple ways to increase silicone elastomer thermal stability, particularly when trying to decrease silicone component thermal aging in order to extend the device's lifespan and operational range.

The special qualities of functionally terminated silicone oligomers—such as flexibility, thermal stability, and water repellency—make them essential in various industrial settings. These include monofunctional polydimethylsiloxane (PDMS) oligomers, or "macromonomers," which are especially useful because of how simple and flexible they are to synthesize and use to create customized materials. Usually, live anionic polymerization—a regulated polymerization process that provides exact control over molecular weight and structure—is used to synthesize monofunctional PDMS oligomers (Scheme 19) by Yilgör et al. [87]. Hexamethyl-

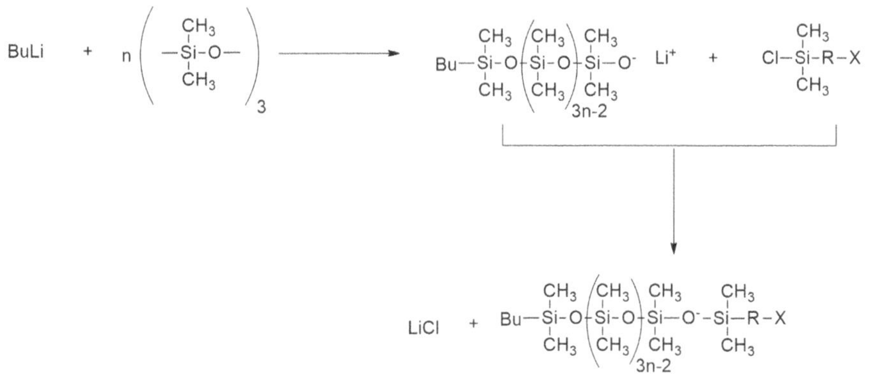

Scheme 19 Synthesis of anionic polymerization-based monofunctional PDMS oligomers (Reprinted with permission from [87])

cyclotrisiloxane (D3) is essential to the study of modern materials design since it is a necessary precursor for the manufacture of monofunctional polydimethylsiloxane (PDMS) oligomers. Xu et al. [88] indicate anionic initiators, including amides or alkali metal alkoxides, are essential for accurately controlling the polymer chain's growth during the live anionic polymerization of D3, which paves the way for customized characteristics and regulated molecular topologies. Maintaining monofunctionality throughout the polymerization process becomes necessary, which motivates careful control of the silicone oligomers' reactive end groups. This degree of control is essential to customizing the features of the resultant macromonomers by determining their compatibility with various materials and their reactivity in subsequent processes. With careful manipulation, the functional end group can be a flexible tool for other changes, including grafting, cross-linking, or mixing with other polymers. This adaptability makes the production of hybrid materials with improved qualities possible, which broadens the range of sectors in which PDMS oligomers may find use. These macromonomers' well-defined structures make it easier to create customized materials with regulated viscoelasticity, rheological behavior, and surface characteristics. Sanchez et al. [89] explain this degree of accuracy makes it possible to create materials that fulfill particular performance requirements, which promotes improvements in the fields of electronics, adhesives, coatings, and biomedical engineering.

A variety of techniques can be used to synthesise telechelic, organofunctionally terminated silicone oligomers with number average molecular weights ranging from roughly 500 to 50,000 g/mol. The ring-chain equilibration or redistribution reactions of low molecular weight α,ω-organofunctionally terminated silicones, especially a disiloxane as indicated in the reaction scheme, are a highly common easy approach (Scheme 20). The primary explanation for the equilibration reactions is the (Si-O) bond's somewhat ionic nature. In these reactions of ring-chain equilibration, the typical molecular proportion of the reaction mixture's cyclic monomer to end-blocker determines the weight of the oligomer that has to be formed by Köhler et al. [90]. Strong bases, such as sodium and potassium hydroxides, as well as strong acids like sulfuric acid, trichloroacetic acid, and sulfonic acids, can be used as catalysts in the equilibration processes for combustion. For these equilibration processes, reaction temperatures between 50 and 100 °C often provide

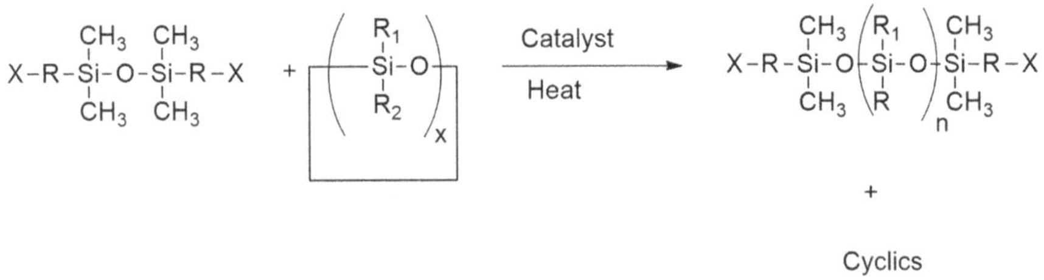

Scheme 20 Using equilibration processes, telechelic silicone oligomers are synthesized ($x = 3$–5) (Reprinted with permission from [90])

sufficient rates according to Sivasamy et al. [91]. The silicone oligomers' backbone architectures that were produced by equilibration reactions can be readily altered by combining D4 with different cyclic siloxane comonomers, including methylphenylsiloxane, methyltrifluoropropylsiloxane, and cyclic diphenlysiloxane. The nature and reactivity of the organofunctional end group most often influence the choice of acid or base catalyst in equilibration procedures. Strong bases are employed to prepare oligomers with basic end groups, like amines, and strong acids are used to prepare oligomers similar to carboxylic acids that have acidic end groups. When selecting a catalyst, extra caution needs to be used for end groups like vinyl, epoxy, and amine groups since they may react at high temperatures at acidic environments which include strong acids and bases. Catalysts in equilibrium processes are limited to cleaving the (Si O) bonds present in linear and cyclic species, such as expanding chains and end-blockers. Nevertheless, (Si R) and (R X) bonds are stable in these conditions. As a result, Bezlepkina et al. [92] explain cyclical by-products that make up roughly 10–15% of the total weight based on the oligomer's average molecular weight formed are nonfunctional and, at the conclusion of linear oligomers, they functionally terminate the processes.

There are two methods that were used to synthesize elastomers: either vinyl-Pt(0) hydrosilylates terminated polydimethylsiloxane (PDMS(vinyl)$_2$), or Sn(II) catalyzes hydroxyl-terminated polydimethylsiloxane (PDMS(OH)$_2$). The first step in the manufacture of Sn(II)-cured elastomers was treating PDMS(OH)$_2$ with a tetraethoxysilane (TEOS) cross-linker and a suitable quantity of PHMS-Pyx, a polymer that contains pyrene. To create homogenous, transparent elastomers in both situations, meticulously and cautiously combining a predecessor component was necessary since gelation happened quite quickly (around 30 s). Therefore, using a speed mixer, all components were thoroughly mixed right away once the catalyst was added as explained by Thoma et al. [93]. Using Perspex molds, circular elastomers with a consistent thickness of about 0.2 cm were created. To make sure that the copolymer dissolved completely in the mixture, THF had to be added. The second technique, involving Sn(II)-catalyzed cross-linking of PDMS(OH)2 with Py-TEOS, was demonstrated by Rossi et al. [18], (Scheme 21). A crucial phase in the formation of functional elastomeric materials has been made with the synthesis of a tin(II)-cured elastomer that incorporates Py-TEOS (pyridine-modified tetraethylorthosilicate). Through the use of Py-TEOS's adaptability and the special qualities of tin(II) catalysts, this novel technique creates elastomers with improved mechanical and thermal properties. Py-TEOS, a modified precursor created by adding pyridine functionality to tetraethylorthosilicate (TEOS), is the first step in the synthesis. This alteration gives the resultant elastomer particular chemical characteristics that affect its reactivity and compatibility with other materials. The final stage involves the introduction of tin(II) catalysts into the elastomer matrix. Tin(II) compounds serve as curing agents during the polymerization process, such as tin(II) octoate. The cross-linking events that result in the development of a three-dimensional elastomeric network are started by the tin(II) species in coordination with the accessible reactive sites on the Py-TEOS. The material gains mechanical strength and structural integrity from this cross-linking. The

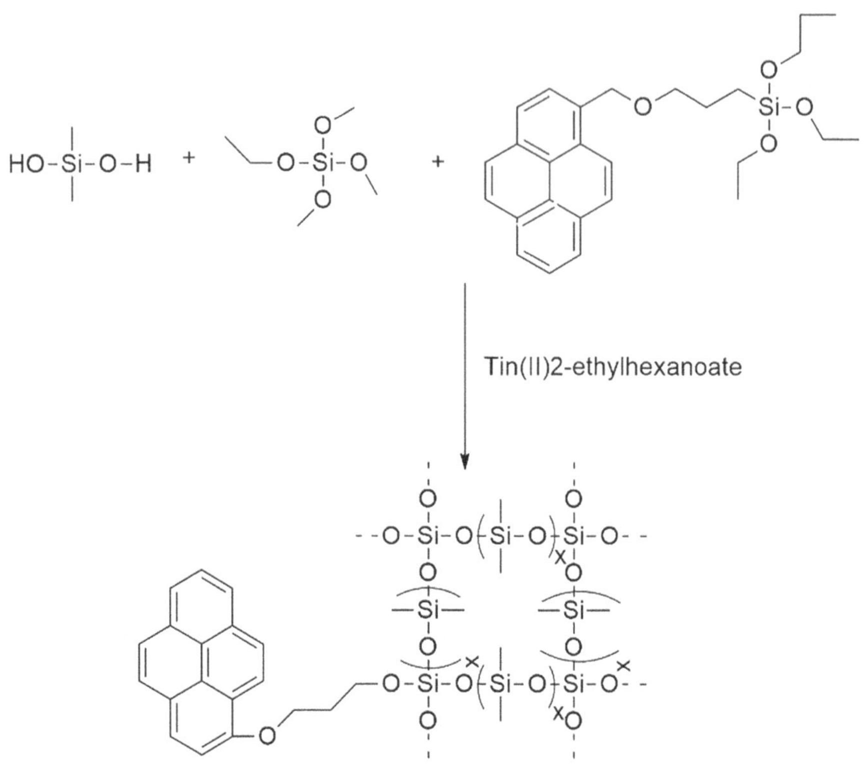

Scheme 21 Synthesis of pyrene labeled PDMS elastomer (Reprinted with permission from [18])

pyridine moieties included in the Py-TEOS-containing elastomer give it special characteristics. The elastomer functionalized with pyridine offers greater chemical resistance and thermal stability in addition to better adherence to specific substrates. As a result, the elastomer can be used in settings with harsh circumstances, like exposure to caustic materials or high temperatures. Moreover, the tin(II)-cured Py-TEOS elastomer has controllable features that enable the material's qualities to be customized to fulfill certain application needs. Precise control over the mechanical characteristics, flexibility, and curing kinetics of the elastomer can be achieved by varying the Py-TEOS-to-catalyst ratio or adding extra functional groups during synthesis. This tin(II)-cured Py-TEOS elastomer finds use in a variety of industries, such as electronics, automotive, and aerospace. Because of its strength and specially designed qualities, it is a great option for gaskets, seals, and coatings where toughness, flexibility, and resilience to extreme conditions are critical. The synthesis of customized materials, such as tin(II)-cured Py-TEOS elastomers, advances the continued evolution of high-performance materials with a variety of industrial uses as elastomer chemistry research advances.

Micellar Nanoparticles

Amphiphilic molecules in aqueous solutions self-assemble to produce tiny structures known as micellar nanoparticles. These molecules spontaneously form spherical micelles in the presence of water because they have both hydrophobic (which repels water) and hydrophilic (which attracts water) regions. While the hydrophilic portions of the amphiphilic molecules face the aqueous environment, the hydrophobic parts group closer together in the middle of the micelle to protect themselves from the surrounding water molecules. This configuration creates stable micellar nanoparticles by reducing the system's free energy. Micellar nanoparticles usually have a diameter that can vary, ranging in size from a few to tens of nanometers. They can be used in a variety of medicine delivery, cosmetic, and nanotechnology applications because of their tiny size and solid structure. For example, hydrophobic medicines can be enclosed within micellar nanoparticles to protect them from degradation and increase their solubility in aqueous solutions when used to deliver drugs. Furthermore, they are adaptable carriers for focused delivery systems due to their capacity to change surface characteristics and enclose various compounds. Calandra et al. [19] provide new micellar nanoparticle formulations, and uses remain under study to improve the stability, biocompatibility, and usefulness of these particles in industrial and biomedical environments. Moreover, Evans et al. [94] indicate that the word "micelle" describes the spherical configuration of these molecules, where the hydrophilic heads that interact with the surrounding water protect the hydrophobic tails from one another at the core. A specific type of micelle known as "soft micellar nanoparticles" is flexible or elastic. Compared with standard rigid nanoparticles, soft micellar nanoparticles have a high degree of structural flexibility and can change form in response to biological system interactions or environmental stimuli.

The word "soft" refers to these nanoparticles' capacity to modify or change their form, rather than being tightly packed into an established structure. This feature is highly helpful in biological applications, especially in drug delivery systems. Because soft micellar nanoparticles may dynamically change their size and form, they are better able to move through complicated biological environments like cellular walls or blood arteries. Additionally, soft micellar nanoparticles' deformability enables them to incorporate therapeutic substances, such as drugs or genetic material, more efficiently. This encapsulation shields the material being transported from degradation, making it easier to release it gradually at certain points throughout the body. These nanoparticles' flexibility also makes it possible for them to stick to cell membranes or pass through cellular barriers, which improves their capacity to carry payloads straight into cells.

To create structures where the hydrophobic core may encapsulate hydrophobic pharmaceuticals or imaging agents and the hydrophilic corona stabilizes the particle in the aqueous environment, soft micellar nanoparticles are constructed utilizing amphiphilic compounds that self-assemble in aqueous solutions. Bhattacharya et al. [95] suggest, in addition to offering stability, this structural configuration enables

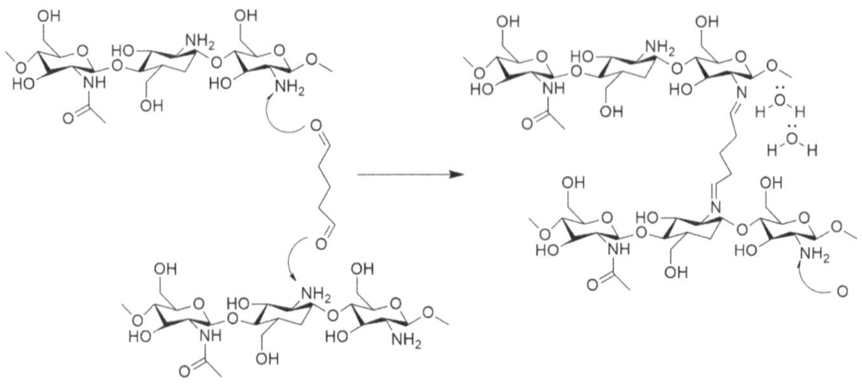

Scheme 22 Formation of cross-linked micellar nanoparticles (Reprinted with permission from [96])

surface property adjustments that improve biocompatibility and in vivo circulation time.

The cross-linker agent used in the Ch cross-linking procedure to produce Ch-NPs was glutaraldehyde (G). Huo et al. [96] in Scheme 22 illustrates how each aldehyde moiety of G combines with the main amino groups of Ch to form two imine linkages (Schiff bases). G has shown to be the most dependable and efficient covalent cross-linker agent for creating cross-linked Ch materials despite testing many other compounds. Ch and G react quickly, requiring a relatively small amount of G to produce extremely stable compounds that can withstand significant pH changes. Thus, we carried out the interaction between Ch and G in AOT and BHDC RMs under identical experimental circumstances to obtain insight into the impact of the micellar interface on the size and shape of Ch-NPs produced by the reverse micellar approach by Orellano et al. [97]. Ch-NPs have a wide range of uses because of their capacity to capture insoluble compounds and transport them in hydrophilic environments. This allows for regulating drug release, modifying bioavailability, and maintaining the same effects at lower concentrations.

Shape-Memory Polymer

Both soft robots and newly developed metamaterials and programmable materials rely on shape-memory polymers (SMPs), and flexible activator materials are perfect for this use. SMPs, responsive to stimuli, can retain a programmed temporary shape until triggered by the impact of single-direction shape memory (1 W SME), returning the polymer virtually completely to its final form. The process of form recovery is entropic, whereas heat causes the shape-memory effect, which was further studied by Chi et al. [98].

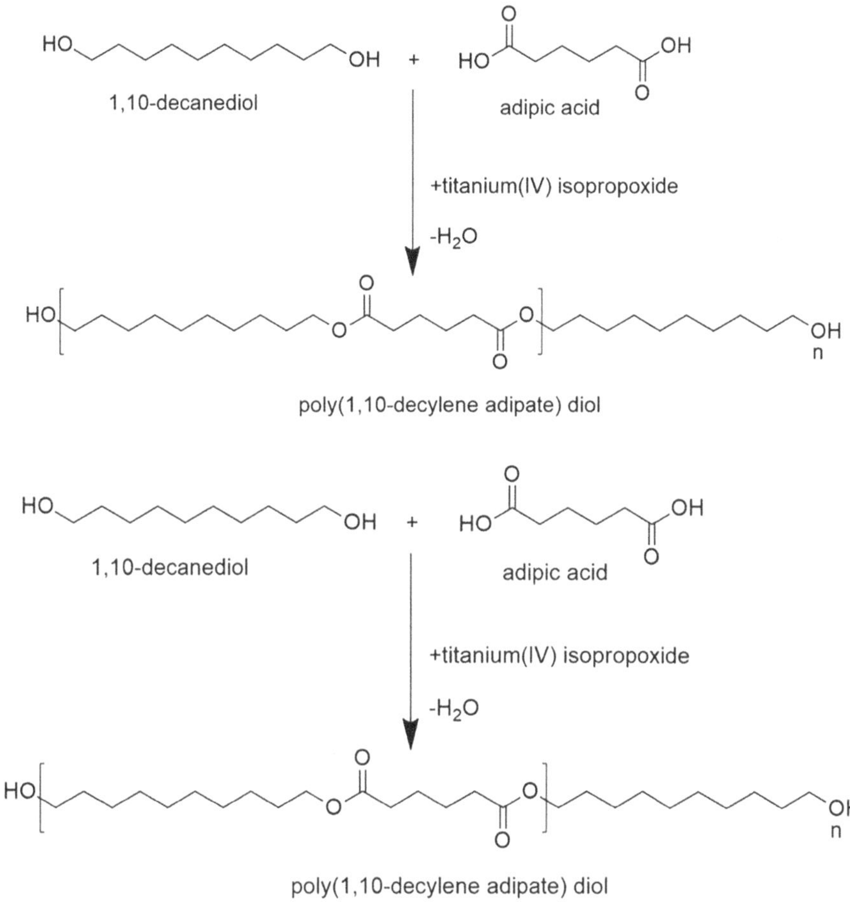

Scheme 23 Poly(1,10-decylene adipate) diol synthesized by polycondensation reaction (Reprinted with permission from [22])

Poly(1,10-decane adipate) diol (PDA), with a molecular weight of 3290 g/mol, is produced by mixing 1,10-decane diol and adipic acid, as demonstrated by Schönfeld et al. [22], and is shown in Scheme 23. This PDA is then further reacted by combining 1,4-butanediol with 4,4′-diphenylmethane diisocyanate to create polyester urethane (PEU). The PEU is transformed into a filament, and fused filament production is applied to create samples additively. Following thermomechanical processing, the PEU exhibits reliable actuation—with a maximum thermoreversible strain of 16.1%, it expands with freezing and contracts upon heating. After 100 heating-cooling cycles, the actuator's efficiency settles at 12.2%. A component of an actuator made from PEU is integrated into a gripper system, demonstrating the capability to pick up, transport, and deposit a hen's egg. Additionally, actuator elements incorporate customizable metamaterials into two kinds of unit cells that allow the creation of behaviors that change with temperature. Opportunities in soft robotics and shape morphing applications are expected due to this technique.

3 Summary and Outlook

In summary, the chapter describes a meticulous report on the synthetic route and mechanism of different pure and engineered materials. Natural soft materials, shape-memory polymers, hydrogels, and nanomaterials demonstrating soft nature are discussed in various research reports. The soft material field is broad and covers the physical and chemical behavior of materials to strengthen the theoretical and experimental concepts illustrated by different research groups. The most interesting fact about being deformed in the presence of any external force or stimulus is that soft material can be exploited for various applications, viz., multiple biological and biomedical applications, environmental issues, energy conversion and storage, specialized gadgets, and sensing and separation. Nature is always presented as a decent example of self-assembly that clearly demonstrates the soft nature of the building materials. Hence, researchers are getting enough inspiration to synthesize soft materials at molecular and nano levels to control the shape of soft nature materials by self-assembly. The manufacture of nanoparticles, the tagging of nanostructures, the design of nanomotors, the utilization of biomineralization, and the creation of functionalized delivery vectors are some of the nanotechnology applications where the power of self-organization may be most effectively applied. Although soft material is the most recent and exploited interest of scientists, the introduction of chiral groups, mixed chirality, different solvents, and environmental effects on the chiral molecules has to be exploited more so that the design of soft material would reach the next level.

Furthermore, if the synthesis idea of chiral induction, transfer, and amplification in the soft materials is taken care of, one could see a new route of smart soft material with robust control on the formation and decay. These kinds of novel ideas can be further exploited in various smart and efficient applications of sensors, supramolecular polymerizations, and other drug delivery vehicles. Hence, now is the right moment to reap the rewards of the last few decades' worth of research on soft materials and to use self-assembly to create customized molecules.

Acknowledgments The authors sincerely thank the Department of Chemistry for their support, UIS, Chandigarh University.

References

1. Yao, H., Liu, T., Jia, Y.Q., Du, Y.J., Yao, B.W., Xu, J.H., et al.: Water-insensitive self-healing materials: from network structure design to advanced soft electronics. Adv Funct Mater. **33**(48), 2307455 (2023)
2. Xiao, Z., Zhao, Q., Niu, Y., Zhao, D.: Adhesion advances: from nanomaterials to biomimetic adhesion and applications. Soft Matter. **18**(18), 3447–3464 (2022)
3. Lim, H.R., Kim, H.S., Qazi, R., Kwon, Y.T., Jeong, J.W., Yeo, W.H.: Advanced soft materials, sensor integrations, and applications of wearable flexible hybrid electronics in healthcare, energy, and environment. Adv Mater. **32**(15), 1901924 (2020)
4. Ariga, K., Jia, X., Shrestha, L.K.: Soft material nanoarchitectonics at interfaces: molecular assembly, nanomaterial synthesis, and life control. Mol. Syst. Des. Eng. **4**, 49–64 (2019)
5. Gao, M.R., Xu, Y.F., Jiang, J., Yu, S.H.: Nanostructured metal chalcogenides: synthesis, modification, and applications in energy conversion and storage devices. Chem. Soc. Rev. **42**, 2986–3017 (2013)
6. Zeng, W., Shu, L., Li, Q., Chen, S., Wang, F., Tao, X.M.: Fiber-based wearable electronics: a review of materials, fabrication, devices, and applications. Adv. Mater. **26**, 5310–5336 (2014)
7. Samal, S.K., Dash, M., Dubruel, P., Van Vlierberghe, S.: Smart polymer hydrogels: properties, synthesis and applications. In: Smart polymers and their applications, pp. 237–270. Woodhead Publishing (2014)
8. Baki, A., Wiekhorst, F., Bleul, R.: Advances in magnetic nanoparticles engineering for biomedical applications—a review. Bioengineering. **8**(10), 134 (2021)
9. Li, B., Cao, Y.P., Feng, X.Q., Gao, H.: Mechanics of morphological instabilities and surface wrinkling in soft materials: a review. Soft Matter. **8**, 5728–5745 (2012)
10. Frederix, P.W.J.M., Patmanidis, I., Marrink, S.J.: Molecular simulations of self-assembling bio-inspired supramolecular systems and their connection to experiments. Chem. Soc. Rev. **47**, 3470–3489 (2018)
11. Bisoyi, H.K., Urbas, A.M., Li, Q.: Soft materials driven by photothermal effect and their applications. In: Photoactive functional soft materials: preparation, properties, and applications, pp. 1–44. Wiley (2018)
12. Zhou, L.Y., Fu, J., He, Y.: A review of 3D printing technologies for soft polymer materials. Adv. Funct. Mater. **30**, 2000187 (2020)
13. Pierini, F., Nakielski, P., Urbanek, O., Pawlowska, S., Lanzi, M., De Sio, L., et al.: Polymer-based nanomaterials for photothermal therapy: from light-responsive to multifunctional nanoplatforms for synergistically combined technologies. Biomacromolecules. **19**, 4147–4167 (2018)
14. Serrano-Ruiz, J.C., Luque, R., Sepúlveda-Escribano, A.: Transformations of biomass-derived platform molecules: from high added-value chemicals to fuels via aqueous-phase processing. Chem. Soc. Rev. **40**, 5266–5281 (2011)
15. Mahinroosta, M., Jomeh Farsangi, Z., Allahverdi, A., Shakoori, Z.: Hydrogels as intelligent materials: a brief review of synthesis, properties and applications. Mater. Today Chem. **8**, 42–55 (2018)
16. Liu, X.W., Zhu, S., Wu, S.R., Wang, P., Han, G.Z.: Response behavior of ion-sensitive hydrogel based on crown ether. Colloids Surfaces A Physicochem. Eng. Asp. **417**, 140–145 (2013)
17. Dvornic, P.R.: Some recent advances in the silicon containing polymers. Mater. Sci. Forum. **214**, 131–138 (1996)
18. Rossi, N.A.A., Duplock, E.J., Meegan, J., Roberts, D.R.T., Murphy, J.J., Patel, M., et al.: Synthesis and characterisation of pyrene-labelled polydimethylsiloxane networks: towards the in situ detection of strain in silicone elastomers. J. Mater. Chem. **19**, 7674–7686 (2009)
19. Calandra, P., Caschera, D., Turco Liveri, V., Lombardo, D.: How self-assembly of amphiphilic molecules can generate complexity in the nanoscale. Colloids Surf Physicochem Eng. Asp. **484**, 164–183 (2015)

20. Razi, M.A., Wakabayashi, R., Tahara, Y., Goto, M., Kamiya, N.: Genipin-stabilized caseinate-hitosan nanoparticles for enhanced stability and anti-cancer activity of curcumin. Colloids Surf B Biointerfaces. **164**, 308–315 (2018)
21. Scalet, G.: Two-way and multiple-way shape memory polymers for soft robotics: an overview. Actuators. **9**, 10 (2020)
22. Schönfeld, D., Chalissery, D., Wenz, F., Specht, M., Eberl, C., Pretsch, T.: Actuating shape memory polymer for thermoresponsive soft robotic gripper and programmable materials. Molecules. **26**, 522 (2021)
23. Vemula, P.K., John, G.: Crops: a green approach toward self-assembled soft materials. Acc. Chem. Res. **41**, 769–782 (2008)
24. Lichtenthaler, F.W., Peters, S.: Carbohydrates as green raw materials for the chemical industry. Comptes Rendus Chim. **7**, 65–90 (2004)
25. Heinze, T., Liebert, T.: Unconventional methods in cellulose functionalization. Prog. Polym. Sci. **26**, 1689–1762 (2001)
26. Deleu, M., Paquot, M.: From renewable vegetables resources to microorganisms: new trends in surfactants. Comptes Rendus Chim. **7**, 641–646 (2004)
27. Lochab, B., Shukla, S., Varma, I.K.: Naturally occurring phenolic sources: monomers and polymers. RSC Adv. **4**, 21712–21752 (2014)
28. Morales-Cerrada, R., Molina-Gutierrez, S., Lacroix-Desmazes, P., Caillol, S.: Eugenol, a promising building block for biobased polymers with cutting-edge properties. Biomacromolecules. **22**, 3625–3648 (2021)
29. Wang, H.X., Yang, Z., Liu, Z.G., Wan, J.Y., Xiao, J., Zhang, H.L.: Facile preparation of bright-fluorescent soft materials from small organic molecules. Chem. A Eur. J. **22**, 8096–8104 (2016)
30. Kasprzyk, W., Bednarz, S., Bogdał, D.: Luminescence phenomena of biodegradable photoluminescent poly(diol citrates). Chem. Commun. **49**, 6445–6447 (2013)
31. Li, X., Du, X.: Surface enhanced electrochemiluminescence of the Ru(bpy)32+/tripropylamine system by au@SiO_2 nanoparticles for highly sensitive and selective detection of dopamine. Microchem. J. **176**, 107224 (2022)
32. Geng, H.: A one-step approach to make cellulose-based hydrogels of various transparency and swelling degrees. Carbohydr. Polym. **186**, 208–216 (2018)
33. Chang, C., Zhang, L.: Cellulose-based hydrogels: present status and application prospects. Carbohydr. Polym. **84**, 40–53 (2011)
34. Fu, L.H., Qi, C., Ma, M.G., Wan, P.: Multifunctional cellulose-based hydrogels for biomedical applications. J. Mater. Chem. B. **7**, 1541–1562 (2019)
35. Singh, N., Agarwal, S., Jain, A., Khan, S.: 3-Dimensional cross linked hydrophilic polymeric network "hydrogels": an agriculture boom. Agric. Water Manag. **253**, 106939 (2021)
36. Dasgupta, A., Mondal, J.H., Das, D.: Peptide hydrogels. RSC Adv. **3**, 9117–9149 (2013)
37. Wu, C., Li, J., Zhang, Y.Q., Li, X., Wang, S.Y., Li, D.Q.: Cellulose dissolution, modification, and the derived hydrogel: a review. ChemSusChem. **16**, e202300518 (2023)
38. Nasution, H., Harahap, H., Dalimunthe, N.F., Ginting, M.H.S., Jaafar, M., Tan, O.O.H., et al.: Hydrogel and effects of crosslinking agent on cellulose-based hydrogels: a review. Gels. **8**, 568 (2022)
39. McMullen, R.L., Ozkan, S., Gillece, T.: Physicochemical properties of cellulose ethers. Cosmetics. **9**, 52 (2022)
40. Capretti, M., Giammaria, V., Santulli, C., Boria, S., Del Bianco, G.: Use of bio-epoxies and their effect on the performance of polymer composites: a critical review. Polymers (Basel). **15**, 4733 (2023)
41. Ye, D., Chang, C., Zhang, L.: High-strength and tough cellulose hydrogels chemically dual cross-linked by using low- and high-molecular-weight cross-linkers. Biomacromolecules. **20**, 1989–1995 (2019)
42. Kundu, R., Mahada, P., Chhirang, B., Das, B.: Cellulose hydrogels: green and sustainable soft biomaterials. Curr. Res. Green Sustain. Chem. **5**, 100252 (2022)

43. Qureshi, M.A., Nishat, N., Jadoun, S., Ansari, M.Z.: Polysaccharide based superabsorbent hydrogels and their methods of synthesis: a review. Carbohydr. Polym. Technol. Appl. **1**, 100014 (2020)
44. Chen, J., Wu, J., Raffa, P., Picchioni, F., Koning, C.E.: Superabsorbent polymers: from long-established, microplastics generating systems, to sustainable, biodegradable and future proof alternatives. Prog. Polym. Sci. **125**, 101475 (2022)
45. Tharanathan, R.N., Kittur, F.S.: Chitin—the undisputed biomolecule of great potential. Crit. Rev. Food Sci. Nutr. **43**, 61–87 (2003)
46. Shen, X., Shamshina, J.L., Berton, P., Gurau, G., Rogers, R.D.: Hydrogels based on cellulose and chitin: fabrication, properties, and applications. Green Chem. **18**, 53–75 (2015)
47. Kono, H., Zakimi, M.: Preparation, water absorbency, and enzyme degradability of novel chitin- and cellulose/chitin-based superabsorbent hydrogels. J. Appl. Polym. Sci. **128**, 572–581 (2013)
48. Kono, H., Fujita, S.: Biodegradable superabsorbent hydrogels derived from cellulose by esterification crosslinking with 1,2,3,4-butanetetracarboxylic dianhydride. Carbohydr. Polym. **87**, 2582–2588 (2012)
49. Khan, F., Atif, M., Haseen, M., Kamal, S., Khan, M.S., Shahid, S., et al.: Synthesis, classification and properties of hydrogels: their applications in drug delivery and agriculture. J. Mater. Chem. B. **10**, 170–203 (2022)
50. Llanes, L., Dubessay, P., Pierre, G., Delattre, C., Michaud, P.: Biosourced polysaccharide-based superabsorbents. Polysaccharides. **1**, 51–79 (2020)
51. Gao, X., Cao, Y., Song, X., Zhang, Z., Zhuang, X., He, C., et al.: Biodegradable, pH-responsive carboxymethyl cellulose/poly(acrylic acid) hydrogels for oral insulin delivery. Macromol. Biosci. **14**, 565–575 (2014)
52. Liu, L., Yao, W.D., Rao, Y.F., Lu, X.Y., Gao, J.Q.: pH-responsive carriers for oral drug delivery: challenges and opportunities of current platforms. Drug Deliv. **24**, 569–581 (2017)
53. Sarangi, M., Padhi, S., Rath, G.: Non-invasive delivery of insulin for breaching hindrances against diabetes. Crit. Rev. Ther. Drug Carrier Syst. **41**(5), 1–64 (2023)
54. Mohammadbagheri, Z., Rahmati, A., Hoshyarmanesh, P.: Synthesis of a novel superabsorbent with slow-release urea fertilizer using modified cellulose as a grafting agent and flexible copolymer. Int. J. Biol. Macromol. **182**, 1893–1905 (2021)
55. Priya, E., Jha, A., Sarkar, S., Maji, P.K.: A urea-loaded hydrogel comprising of cellulose nanofibers and carboxymethyl cellulose: an effective slow-release fertilizer. J. Clean. Prod. **434**, 140215 (2024)
56. Mohammadi-Khoo, S., Moghadam, P.N., Fareghi, A.R., Movagharnezhad, N.: Synthesis of a cellulose-based hydrogel network: characterization and study of urea fertilizer slow release. J. Appl. Polym. Sci. **133**, 42935 (2016)
57. Bashir, S., Hina, M., Iqbal, J., Rajpar, A.H., Mujtaba, M.A., Alghamdi, N.A., et al.: Fundamental concepts of hydrogels: synthesis, properties, and their applications. Polymers (Basel). **12**, 1–60 (2020)
58. Ahmed, R., Afreen, A., Tariq, M., Zahid, A.A., Masoud, M.S., Ahmed, M., et al.: Bone marrow mesenchymal stem cells preconditioned with nitric-oxide-releasing chitosan/PVA hydrogel accelerate diabetic wound healing in rabbits. Biomed. Mater. **16**, 035014 (2021)
59. Yang, X.Y., Chen, L.H., Li, Y., Rooke, J.C., Sanchez, C., Su, B.L.: Hierarchically porous materials: synthesis strategies and structure design. Chem. Soc. Rev. **46**, 481–558 (2017)
60. Manzoor, A., Dar, A.H., Pandey, V.K., Shams, R., Khan, S., Panesar, P.S., et al.: Recent insights into polysaccharide-based hydrogels and their potential applications in food sector: a review. Int. J. Biol. Macromol. **213**, 987–1006 (2022)
61. Sun, X., Chwatko, M., Lee, D.H., Bachman, J.L., Reuther, J.F., Lynd, N.A., et al.: Chemically triggered synthesis, remodeling, and degradation of soft materials. J. Am. Chem. Soc. **142**, 3913–3922 (2020)
62. Bukhari, S.M.H., Khan, S., Rehanullah, M., Ranjha, N.M.: Synthesis and characterization of chemically cross-linked acrylic acid/gelatin hydrogels: effect of pH and composition on swelling and drug release. Int. J. Polym. Sci. **2015**, 187961 (2015)

63. Sharma, K., Kaith, B.S., Kumar, V., Kalia, S., Kumar, V., Swart, H.C.: Water retention and dye adsorption behavior of Gg-cl-poly(acrylic acid-aniline) based conductive hydrogels. Geoderma. **232**, 45–55 (2014)
64. Zohuriaan-Mehr, M.J., Pourjavadi, A., Salimi, H., Kurdtabar, M.: Protein- and homo poly(amino acid)-based hydrogels with super-swelling properties. Polym. Adv. Technol. **20**, 655–671 (2009)
65. Zhang, H., Yan, Q., Kang, Y., Zhou, L., Zhou, H., Yuan, J., et al.: Fabrication of thermo-responsive hydrogels from star-shaped copolymer with a biocompatible β-cyclodextrin core. Polymer (Guildf). **53**, 3719–3725 (2012)
66. Roka, N., Kokkorogianni, O., Kontoes-Georgoudakis, P., Choinopoulos, I., Pitsikalis, M.: Recent advances in the synthesis of complex macromolecular architectures based on poly(*N*-vinyl pyrrolidone) and the RAFT polymerization technique. Polymers (Basel). **14**, 701 (2022)
67. Popescu, I., Constantin, M., Pelin, I.M., Suflet, D.M., Ichim, D.L., Daraba, O.M., et al.: Eco-friendly synthesized PVA/chitosan/oxalic acid nanocomposite hydrogels embedding silver nanoparticles as antibacterial materials. Gels. **8**, 268 (2022)
68. Gupta, N.R., Arun Torris, A.T., Wadgaonkar, P.P., Rajamohanan, P.R., Ducouret, G., Hourdet, D., et al.: Synthesis and characterization of PEPO grafted carboxymethyl guar and carboxymethyl tamarind as new thermo-associating polymers. Carbohydr. Polym. **117**, 331–338 (2015)
69. Herzberger, J., Sirrine, J.M., Williams, C.B., Long, T.E.: Polymer design for 3D printing elastomers: recent advances in structure, properties, and printing. Prog. Polym. Sci. **97**, 101144 (2019)
70. Madsen, F.B., Daugaard, A.E., Hvilsted, S., Skov, A.L.: The current state of silicone-based dielectric elastomer transducers. Macromol. Rapid Commun. **37**, 378–413 (2016)
71. Bartholomew, E.R., Eykamp, G.R., Gibbs, W.E.: High temperature elastomeric compounds and polymers. Rubber Chem. Technol. **32**, 1587–1626 (1959)
72. Eduok, U., Faye, O., Szpunar, J.: Recent developments and applications of protective silicone coatings: a review of PDMS functional materials. Prog. Org. Coat. **111**, 124–163 (2017)
73. Ariati, R., Sales, F., Souza, A., Lima, R.A., Ribeiro, J.: Polydimethylsiloxane composites characterization and its applications: a review. Polymers (Basel). **13**, 4258 (2021)
74. Miranda, I., Souza, A., Sousa, P., Ribeiro, J., Castanheira, E.M.S., Lima, R., et al.: Properties and applications of PDMS for biomedical engineering: a review. J. Funct. Biomater. **13**, 2 (2022)
75. Dascalu, M., Dünki, S.J., Quinsaat, J.E.Q., Ko, Y.S., Opris, D.M.: Synthesis of silicone elastomers containing trifluoropropyl groups and their use in dielectric elastomer transducers. RSC Adv. **5**, 104516–104523 (2015)
76. Mazurek, P., Vudayagiri, S., Skov, A.L.: How to tailor flexible silicone elastomers with mechanical integrity: a tutorial review. Chem. Soc. Rev. **48**, 1448–1464 (2019)
77. Pelrine, R., Kornbluh, R., Joseph, J., Heydt, R., Pei, Q., Chiba, S.: High-field deformation of elastomeric dielectrics for actuators. Mater. Sci. Eng. C. **11**, 89–100 (2000)
78. Zhu, B., Wang, H., Leow, W.R., Cai, Y., Loh, X.J., Han, M.Y., et al.: Silk fibroin for flexible electronic devices. Adv. Mater. **28**, 4250–4265 (2016)
79. Cai, P., Wang, C., Gao, H., Chen, X.: Mechanomaterials: a rational deployment of forces and geometries in programming functional materials. Adv. Mater. **33**, 2007977 (2021)
80. Meyers, K.O., Bye, M.L., Merrill, E.W.: Model silicone elastomer networks of high junction functionality: synthesis, tensile behavior, swelling behavior, and comparison with molecular theories of rubber elasticity. Macromolecules. **13**, 1045–1053 (1980)
81. Liu, J., Yao, Y., Li, X., Zhang, Z.: Fabrication of advanced polydimethylsiloxane-based functional materials: bulk modifications and surface functionalizations. Chem. Eng. J. **408**, 127262 (2021)
82. Goff, J., Sulaiman, S., Arkles, B.: Applications of hybrid polymers generated from living anionic ring opening polymerization. Molecules. **26**, 2755 (2021)
83. Wolf, M.P., Salieb-Beugelaar, G.B., Hunziker, P.: PDMS with designer functionalities—properties, modifications strategies, and applications. Prog. Polym. Sci. **83**, 97–134 (2018)

84. Pouget, E., Tonnar, J., Lucas, P., Lacroix-Desmazes, P., Ganachaud, F., Boutevin, B.: Well-architectured poly(dimethylsiloxane)-containing copolymers obtained by radical chemistry. Chem. Rev. **110**, 1233–1277 (2010)
85. Fang, L., Yan, W., Chen, S., Duan, Q., Herath, M., Epaarachchi, J., et al.: Light and shape-memory polymers: characterization, preparation, stimulation, and application. Macromol. Mater. Eng. **308**, 2300158 (2023)
86. Hegemann, D., Gaiser, S.: Plasma surface engineering for manmade soft materials: a review. J. Phys. D. Appl. Phys. **55**, 173002 (2022)
87. Yilgör, E., Yilgör, I.: Silicone containing copolymers: synthesis, properties and applications. Prog. Polym. Sci. **39**, 1165–1195 (2014)
88. Xu, Z., McCarrol, J., Stenzel, M.H.: Flow-based assembly of nucleic acid-loaded polymer nanoparticles. Aust. J. Chem. **76**, 731–745 (2023)
89. Sanchez, C., Belleville, P., Popall, M., Nicole, L.: Applications of advanced hybrid organic–inorganic nanomaterials: from laboratory to market. Chem. Soc. Rev. **40**, 696–753 (2011)
90. Köhler, T., Gutacker, A., Mejiá, E.: Industrial synthesis of reactive silicones: reaction mechanisms and processes. Org. Chem. Front. **7**, 4108–4120 (2020)
91. Sivasamy, A., Cheah, K.Y., Fornasiero, P., Kemausuor, F., Zinoviev, S., Miertus, S.: Catalytic applications in the production of biodiesel from vegetable oils. ChemSusChem. **2**, 278–300 (2009)
92. Bezlepkina, K.A., Milenin, S.A., Vasilenko, N.G., Muzafarov, A.M.: Ring-opening polymerization (ROP) and catalytic rearrangement as a way to obtain siloxane mono-and telechelics, as well as well-organized branching centers: history and prospects. Polymers (Basel). **14**, 2408 (2022)
93. Thoma, J.L., Duhamel, J., Bertocchi, M.J., Weiss, R.G.: Long range polymer chain dynamics of highly flexible polysiloxane in solution probed by pyrene excimer fluorescence. Polymers (Basel). **10**, 345 (2018)
94. Evans, D.F., Ninham, B.W.: Molecular forces in the self-organization of amphiphiles. J. Phys. Chem. **90**, 226–234 (1986)
95. Bhattacharya, S., Samanta, S.K.: Soft-nanocomposites of nanoparticles and nanocarbons with supramolecular and polymer gels and their applications. Chem. Rev. **116**, 11967–12028 (2016)
96. Huo, Q., Liu, J., Wang, L.Q., Jiang, Y., Lambert, T.N., Fang, E.: A new class of silica cross-linked micellar core-shell nanoparticles. J. Am. Chem. Soc. **128**, 6447–6453 (2006)
97. Orellano, M.S., Longo, G.S., Porporatto, C., Correa, N.M., Falcone, R.D.: Role of micellar interface in the synthesis of chitosan nanoparticles formulated by reverse micellar method. Colloids Surf A Physicochem Eng Asp. **599**, 124876 (2020)
98. Chi, Y., Li, Y., Zhao, Y., Hong, Y., Tang, Y., Yin, J.: Bistable and multistable actuators for soft robots: structures, materials, and functionalities. Adv. Mater. **34**, 2110384 (2022)

Soft-Material-Based Devices and Technologies

Rahul Karmakar and J. Chakrabarti

Abstract Soft matter systems have length scales ranging from several nanometres to a few microns and slow time scales due to large sizes. These systems are termed soft due to their sensitivity to external perturbations as compared to pure molecular materials. Colloidal dispersion, polymers, membranes, and biological macromolecules are typical soft matter systems. Soft matter systems in equilibrium have been studied in great details. These systems can be easily driven out of equilibrium by external perturbations, like shear, temperature gradient, tuning electric and magnetic fields and applying pressure, and so on. The simplest non-equilibrium state of a system is steady state. This means that there is a time-independent current in the system. Example of typical non-equilibrium steady-state phenomenon includes phase transitions like shear-banding, lane formation, glass and gel formation, homogeneous and heterogeneous crystal nucleation and cluster formation, tracer diffusion in polymeric membrane under external force. We have investigated a couple of classes of soft matter systems in non-equilibrium situation: (1) steady-state behaviour of systems with temperature-sensitive interactions in the presence of temperature gradient and (2) motion of particles driven through a polymeric network. These systems in non-equilibrium situations may be helpful for various technological purposes. We observe long-ranged crystalline order in cold region in thermo-sensitive ligand capped noble metal nanoparticles in the presence of temperature difference. The aggregated order particles may be useful for

R. Karmakar
Department of Physics of Complex Systems, S. N. Bose National Centre for Basic Sciences, Salt Lake, Kolkata, India

Department of Chemical Engineering, Indian Institute of Technology Madras, Chennai, Tamil Nadu, India

J. Chakrabarti (✉)
Department of Physics of Complex Systems, S. N. Bose National Centre for Basic Sciences, Salt Lake, Kolkata, India

Department of Chemical and Biological Sciences, S. N. Bose National Centre for Basic Sciences, Salt Lake, Kolkata, India
e-mail: jaydeb@bose.res.in

V. Kumar, Y. K. Mishra (eds.), *Soft Materials for Functional Applications*, Advanced Structured Materials 225, https://doi.org/10.1007/978-981-97-9468-3_2

sensory applications. We also study thermo-responsive colloids that show change in diameter with temperature. We observe long-range order in hot region as well as in cold region in the presence of temperature gradient in this system beyond a certain diameter ratio. It can be helpful in designing structures at high temperature. In other class of problem, we explore different parameters of polymer network to reduce respiratory droplet permeation through a face mask. We observe that the efficiency of the face mask depends on composition of polymer, interaction strength of tracer and the network, the polymeric network rigidity, and the thickness of the confinement. Our results may be helpful in designing breathable face mask with better efficiency. Finally, we study a model system to control binary mixture separation through membrane using reverse osmosis technique. We observe performance of RO membrane in terms of permeation, solute rejection, and fouling can be controlled with changing relative interaction of membrane material with mixture. Our results may be useful for designing reverse osmosis membrane with better permeation, selectivity, and less fouling for longer use.

Keywords Colloid · Polymer · Non-equilibrium situation · Molecular simulation

1 Introduction

Soft matter systems have length scales ranging from several nanometres to a few microns and slow time scales [1]. These systems are termed soft due to their sensitivity to external perturbations as compared to pure molecular materials. Colloidal dispersion, polymers, membranes, and biological macromolecules are typical soft matter systems. Many different materials, like polymers, ceramics, semiconductors, and metals, may form soft matter systems. These systems have long been used in a broad spectrum of applications such as inks, cosmetics, food, paints, and rheological fluids [2]. It also shows potential application in industries [3] of drug delivery [4], biosensor supports [5], catalytic supports [6], antifouling coatings [7], and additives to paints/coatings [8].

Soft matter systems in equilibrium have been studied in great details [9]. Colloids can selectively be tailored in a static external field like laser optical fields [1]. A broad-ranging idea is to use optical fields to trap single particles a particle at a given fixed position. This type of set-up is known as an "optical tweezer". The another set-up is to impose a spatially oscillating external potential on a bulk colloidal suspension by means of crossed laser beams. The freezing transition in a such a periodically modulated external potential is different from the bulk freezing [10].

These systems can be easily driven out of equilibrium by external perturbations, like shear, temperature gradient, tuning electric and magnetic fields and applying pressure, and so on. The simplest non-equilibrium state of a system is steady state. This means that there is a time-independent current in the system. Example of typical non-equilibrium steady-state phenomenon includes phase transitions like shear-banding [11–13], lane formation [14], glass and gel formation, homogeneous and heterogeneous crystal nucleation [15] and cluster formation, long-ranged attraction due to temperature gradient [16], tracer diffusion in polymeric membrane [17] under external force. Soft matter systems show various steady-state structures [16, 18–20] by forming aggregates, like active matters aggregation where motion is assisted by energy source [21–25], ordering objects of micrometre size in low-Reynolds number flows [26], polymeric systems [27], aggregation of colloids in toggling magnetic field [28], electric field [14, 29, 30], periodic forcing [31], chemical reaction fuels [32, 33] and thermal gradients [34], and so on. Manipulation of these types of systems under non-equilibrium steady-state conditions [21, 22, 26, 28, 31–34] is relevant in different technological applications, like separation of components in organic solvents [35], sensors [34], biomolecules trapping [36], microfluidic applications and colloidal confinement [37], etc. It is also to be noted that structure formation for example aggregation state of ligand capped nanoparticles in equilibrium by changing parameters, like ligand coating and solvent conditions, is different from those in non-equilibrium situation [38, 39].

Here we discuss a couple of classes of soft matter systems of technological importance in non-equilibrium situations:

(i) **Systems with temperature-sensitive interactions in the presence of temperature gradient:** Huge potential application as Surface enhance Raman spectroscopy platform [34] lead us to study microscopic models to highlight the steady-state structure in systems in presence of temperature gradient with interaction parameters as per the local temperatures [40]. One class of thermo-responsive system consists of ligand-coated metal nanoparticles. Their interaction is temperature sensitive and forms aggregates at low temperature region in the presence of temperature difference. The aggregates are useful for high-resolution SERS measurements [34]. The interaction parameters are modelled by a repulsive electrostatic interaction and hydrophobic attraction due to the ligands depending on local temperature [40].

    PNIPAM hydrogels are another class of thermo-responsive system whose size depends on temperature which makes them suitable for a variety of applications ranging from fabrication of photonic crystals to drug delivery medium [41, 42]. The PNIPAM systems are modelled via the Lennard-Jones (LJ) interaction potential [43] with temperature-dependent van der Waals diameter σ [44]. The more compact particles at higher temperature are taken to have a smaller σ than the swollen ones at a lower temperature.

    We observe long-ranged crystalline order in cold region in thermo-sensitive ligand capped noble metal nanoparticles in the presence of temperature difference. For PNIPAM model we observe long-ranged order in hot region as well as in cold region in the presence of temperature gradient in this system beyond a certain diameter ratio.

(ii) **Motion of particles driven through a polymeric network:** Transport process of penetrating molecules (e.g., gas, ligands, reactants, etc.) to collectively permeate and flow through a soft matter medium (like polymer network) under the action of an external field or chemical gradient is another interesting topic in current days [45]. The application ranges from gas separation or recovery [46–48], desalination [49, 50], nano-filtration [51], medical treatments by dialysis or selective drug transport [52], intra-cellular transport [53] to respiratory droplet movement through face mask [54]. Naturally, manipulation of transport processes in external drive gains interest from both fundamental physical sciences and material engineering. Here, we study via stochastic dynamics different parameters of polymer network to reduce respiratory droplet permeation through a face mask in the presence of pressure difference required for normal breathing [55]. We observe that the efficiency of the face mask depends on composition of polymer, interaction strength of tracer and the network, the polymeric network rigidity, and the thickness of the confinement.

Finally, we study a model system to understand binary mixture separation through membrane using reverse osmosis technique through molecular simulations [56]. We observe performance of RO membrane in terms of permeation, solute rejection, fouling can be controlled with changing relative interaction of membrane material with mixture.

We study the model systems by molecular simulations. The structures are characterised in terms of bond orientation order parameters computed from the steady-state trajectories. The organisation of the rest of the chapter is as follows. In Sect. 2, we describe the steady-state structure of a fluid with temperature-sensitive interaction in temperature difference. In Sect. 3, we report on a model where temperature sensitivity of interaction is taken through change in diameter. Section 4 describes model FM where respiratory droplets movement takes place through polymeric network in the presence of external driving. Section 5 depicts model of RO. The simulation algorithms are described in Appendix 1 and the details of calculation of quantities of interest in Appendix 2.

2 Ordering in Suspension of Thermally Driven Ligand Capped Noble Metal Nanoparticles

Soret effect or thermophoresis is particle movement in the presence of temperature gradient [57–60]. It has been widely studied in the past [61–63] in colloids of micrometre size and nanometre-sized particles [57, 64–68]. Recent experimental studies [34] show formation of big clusters with high mono-dispersity index in colder region in an aqueous solution of noble metal (gold and silver) nanoparticles in the presence of temperature gradient. The nanoparticles in the experiment are capped by citrate ligand. This aggregated large cluster used as a base platform for Surface-Enhanced Raman Scattering (SERS) to ascertain chemical species of a part in billion. This leads to real-time, fast, and highly sensitive detection of discrete molecules from the solid-vapour and liquid-vapour interface. Further, the surface zeta potential of the metal nanoparticles falls in the colder region. These observations suggest that the electrostatic repulsion reduced in the colder region due to decrease in temperature. With decrease in electrostatic repulsion, ligand capped nanoparticle aggregation suggests higher attractive interaction in the colder region due to hydrophobic nature of ligands. Particle aggregation here [34] is driven by non-equilibrium situation in the presence of temperature difference in the system- and temperature-sensitive inter-particle interaction which we address here using a model system.

We model this situation with temperature-dependent electrostatic repulsion and hydrophobicity mediated attraction. We use widely celebrated Derjaguin-Landau-Verwey-Overbeek (DLVO) potential to mimic repulsive electrostatic potential between a pair of ligand-coated metal nanoparticles. The form of the potential is as follows:

$$V_R(r) = V_0 \frac{e^{-\kappa r}}{r}. \tag{1}$$

Here $r(= |\vec{r}_i - \vec{r}_j|)$ is the separation between a pair of nanoparticles i and j. V_0 is the pre-factor/amplitude of the repulsive potential given by $\frac{Z^2 e^2 e^{\kappa\sigma}}{\epsilon(1+\frac{\kappa\sigma}{2})^2}$, where ϵ is the dielectric constant of the medium and Ze is the effective surface charge on the nanoparticle. κ is the inverse Debye screening length given by

$$\kappa^2 = \frac{4\pi e^2}{\epsilon k_B T} n_p Ze. \tag{2}$$

n_p is the density of nanoparticle in salt-free solution. We take mean Debye screening length in the pre-factor V_0 and the smaller of the two Debye screening lengths for the damping term for the cross interaction between two particles at different temperatures.

The introduction of temperature dependence of the electrostatic repulsion in Eq. 1 is as follows. V_0 and κ both depend on the surface charge Z that is related to the surface zeta potential ϕ_0:

$$\frac{2Z\lambda_B}{\sigma} = (1 + \kappa\sigma/2)\phi_0. \tag{3}$$

This is valid for low values of ϕ_0 and η [69]. Here $\lambda_B(= \frac{e^2}{\epsilon k_B T})$ is the Bjerrum length. The temperature dependence is taken via ϕ_0. As mentioned in experiment, we also fix $\phi_0 = 1.2$ and 1.0 for hot and cold regions, respectively [34]. We get $\kappa\sigma = 3.0$ and $Z = 160$ in the hot region and $\kappa\sigma = 2.5$ and $Z = 106$ in the cold region solving the coupled equation Eqs. 2 and 3 for a given ϕ_0. We do not consider temperature dependence of the dielectric constant as dielectric constant in both the screening length and the pre-factor comes as product with $k_B T$ in equation which makes it temperature independent.

Ligands, coated on the metal nanoparticle surface, are solvophobic in character. A solvophobic nanoparticle gives repulsive force on solvent molecules by which stabilising solvent depleted vapour region of radius Λ_0 with the nanoparticle at the core. It creates a vapour-liquid interface where the Laplace pressure takes place [70]. We use $2\Lambda_0/\sigma \approx 1.27$ that is observed for Lennard-Jones systems in a sub-critical liquid [70]. When two such solute particles come to a close promixity, the solvent particles get depleted from the overlapping region. This is due to repulsion of both particles so that the vapour-liquid interface disappears. The Laplace pressure drops to zero in this region. The effective attraction comes from the imbalance of the Laplace pressure over the particles that is shown in Ref. [70]. We get the effective force between the particles is harmonic in form for small separation of the particles which is described in Ref. [70]. The harmonic attraction can be easily appreciated as follows: As the separation between the particles increases, more solvent particles come between two particles, which is not favourable energetically. We model this attraction between two large solvophobic solute particles through a harmonic potential of spring constant βf, as in Ref. [70]:

$$V_A(r) = \beta f (r - \sigma)^2, \tag{4}$$

where $(r - \sigma)[= s] \leq 2\Lambda_0$ is the surface-to-surface distance. $\beta = 8\pi\gamma_s[\frac{2\Lambda_0}{\sigma} - 1]$, where γ_s is the surface tension at the liquid-vapour interface and f the fraction of ligands attached to a nanoparticle surface in solvent. As ligands bound onto the nanoparticle surface lead to hydrophobicity, we take into account the dependence on f.

The spring constant βf shows temperature dependence as well. We use experimental value [71, 72] of γ_s in cold and hot temperature and determine β. We find that β is 1.29 times larger in the cold region than that in the hot region. f is dependent on temperature via the Boltzmann factor $\exp(-E_b/k_B T)$. E_b is the binding energy which we take $E_b \approx -11\text{kJ/mol}$ as reported for citrate binding to gold nanoparticles, $E_b \approx -11\text{kJ/mol}$ [73]. We get βf to be 50 for hot region and $\beta f = 300$ for cold region multiplying both β and f in corresponding temperature. We use geometric mean of the respective βf for cross interaction at two different temperatures.

We study the model thermo-sensitive particles in the presence of temperature difference defined as x-axis. We perform Brownian dynamics (BD) simulations (see Appendix 1 for details) on $N(= 6400)$ colloidal particles of diameter $\sigma = (114\,\text{nm})$ in an implicit aqueous solvent. The rectangular parallelepiped box lengths are $L_x = 39.5\sigma$ and $L_y = L_z = 18\,\sigma$ with the periodic boundary conditions in all three directions. We perform simulation at packing fraction $\eta = \frac{\pi}{6}\frac{N}{V}\sigma^3 = 0.26$. We show a schematic of our system in Fig. 1. The system is in equilibrium at hot temperature $T_H(= 373\,\text{K})$ (Fig. 1a). Then we create cold temperature $T_c(= 273\,\text{K})$ regions as Fig. 1b along x-axis. The interaction forces are computed from the potential forms described in Eqs. 1 to 4.

We use $\frac{e^2}{\epsilon\sigma}$ and σ as energy and length units, respectively, in our simulation, while $\tau_H(= \sigma^2/D_H)$ the unit of time. The random displacements $\vec{R}_{i,\alpha}(\delta t)$ have mean zero and variances $2D_\alpha \delta t$ due to fluctuating force during time δt. Here α

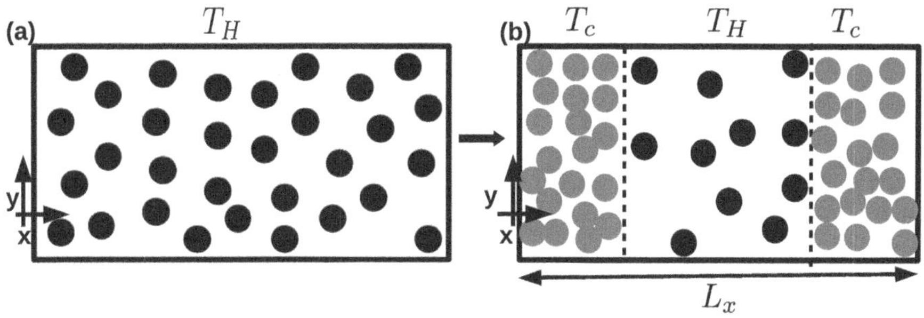

Fig. 1 (**a**) The system is equilibrated at hot temperature T_H. (**b**) We cool two ends at T_C. The middle region remains at hot T_H. The black particles are at hot region and grey particles are at cold region. (The z-axis not shown is pointing into the page for panel(**a**–**b**))

stands for hot (H) and cold (C) regions. δt is the integration time and D_α the Stokes-Einstein diffusion coefficient, given by $k_B T_\alpha / \Gamma_\alpha$ at temperature T_α, k_B the Boltzmann constant, and $\Gamma_\alpha = 3\pi\eta_\alpha\sigma$ is the viscous damping. The temperature dependence of viscosity is given in an empirical form, $2.761 \times 10^{-6} exp(\frac{1317}{T})$ Pa.s [74]. The discretised equation of motion is integrated with time step $0.0001\tau_H$. We average over five different trajectories for all time-dependent quantities.

First, we observe that the increase in density with time in cold region and that decreases with time in hot region after creating temperature difference, which is also evident from the schematic in Fig. 1b. We observe in Fig. 2a density $\rho(t)$ increases in cold region and decreases in cold region. The increased density and slower diffusion lead to long-range order formation in the cold region. Using the Unsupervised Machine Learning (UML) (see details in Appendix 2) algorithm, we find the presence of two clusters of different structures as revealed through the bond order parameters $\bar{q}_l$ (Appendix 2). We find that the cold region has predominantly body centred cubic (BCC) structure at steady-state coexisting with liquid phase [75] from bond order parameters $\bar{q}_4$ and $\bar{q}_6$ (Fig. 2b) (see Appendix 2 for details). The structural growth kinetics in the cold regions is calculated through the crystal fraction θ_c. θ_c denotes the ratio of solid-like particles and the number of particles present at a given time in the cold region. We observe that the crystallisation takes place in a two-step manner from an initial slow to rapid crystal formation and finally to saturation (Fig. 2c).

3 Long-Ranged Order in Thermo-Responsive Hydrogels in Hot Region

Poly-N-iso-propylacrylamide (PNIPAM) [76] hydrogels are thermo-responsive. The volume of the microgel particles undergoes reversible changes above lower critical solution temperature (LCST) [77, 78]. The cross-linked entangled polymer

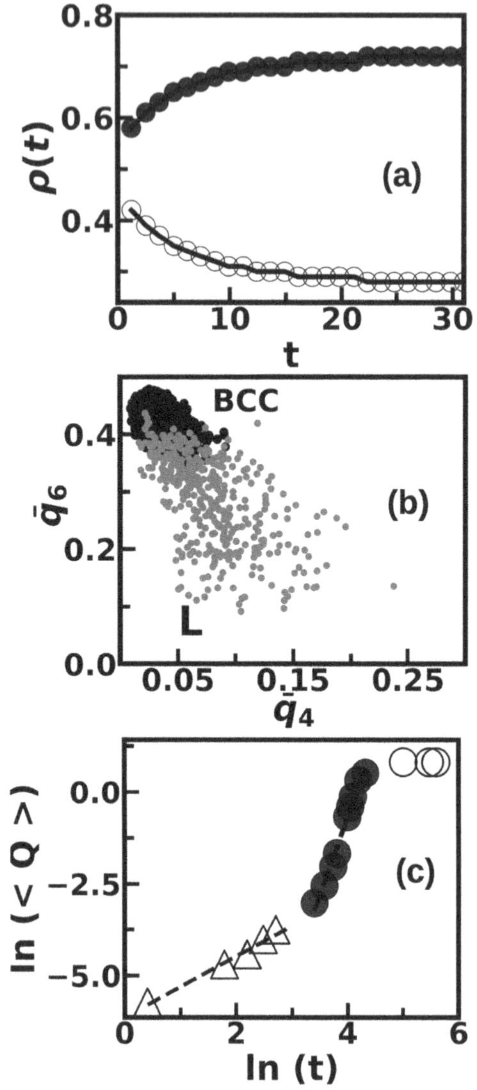

Fig. 2 (**a**) Density $\rho(t)$ in hot (open circle) region and cold (solid circle). (**b**) Scatter plot in $\bar{q}_4$ and $\bar{q}_6$ planes at large time in cold region, those Liquid (grey) and BCC (black) particles. (**c**) $\ln\langle Q\rangle$ vs. t plot. Open triangles are shown for lower time. Solid circles denote middle time values, and open circles show saturated portion. The dashed line is the fitted result

network in PNIPAM traps a large amount of water. The diameter of the PNIPAM particles increases due to trapped water at low temperature below the LCST. When heated above the LCST, they release water resulting in shrinkage of the particles in temperature window between 27 °Cand 34 °C [76, 79]. The tuneability of the LCST and the reversibility of the volume change of these classes of thermo-responsive particles are used in a variety of applications [80] in drug delivery [42], tissue engineering, cell culture, and photonic crystal [41, 81]. The formation of long-ranged crystalline order lies at the heart of synthesising photonic crystals via colloidal route. PNIPAM particles form ordered structures at low temperature [82] in

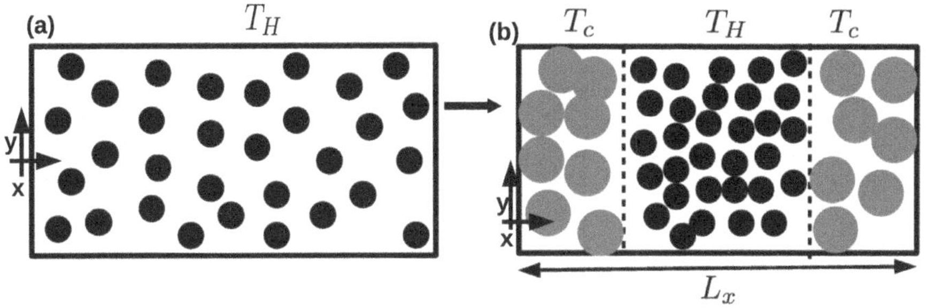

Fig. 3 (**a**) The system is equilibrated at hot temperature T_H. (**b**) We cool two ends at T_C. The middle region remains at hot T_H. The black particles are at hot region and grey particles are at cold region. Grey particles are larger in size than black particles. (The z-axis not shown is pointing into the page for panel (**a**–**b**))

the presence of temperature gradient. Moreover, PNIPAM has inherent advantages due to the reversibility of volume changes which make them excellent candidate for designing photonic crystals with tuneable properties [41].

We study thermo-responsive particles with temperature-dependent size to investigate order formation in due to temperature gradient. We treat thermo-responsive hydrogel as Lennard-Jones (LJ) ($4\epsilon[(\frac{\sigma}{r})^{12} - (\frac{\sigma}{r})^{6}$) fluid whose diameter is sensitive to temperature, σ_C in cold temperature T_C, and σ_H at hot temperature T_H, where $\sigma_C > \sigma_H$. For cross interaction between two different-sized particles, we use $(\sigma_H + \sigma_C)/2$. The particle dynamics are computed via the under-damped Langevin equation (Appendix 1) simulations with the PBC in three directions of the box.

Here we equilibrate the system at a higher temperature $\frac{k_B T_H}{\epsilon} = 1.2$(in LJ unit) in a liquid state (Fig. 3a) and then cool two ends at $\frac{k_B T_C}{\epsilon} = 1$ leaving the middle region at the centre of the box at hot temperature (Fig. 3b) as in the previous section. The schematic diagram in Fig. 3b shows that the diameter is larger in cold temperature T_C than that in hot temperature T_H. We ignore the temperature dependence of friction coefficient ζ in our calculation. This simplified assumption follows from the work [40] where it is observed temperature dependence of viscosity has almost no importance in structure formation in the presence of temperature difference.

In our simulations, ϵ strength of LJ potential is the unit of energy, σ_H the length unit, and $\tau(=\sqrt{\frac{m\sigma_H^2}{\epsilon}})$ the time unit. We set $\gamma = 100$. The discretised equations of motion are integrated with time step 0.001τ. We perform simulation on $N(= 4000)$ colloidal particles in a volume (V) of rectangular parallelepiped box of length $L_x = 57.1$ and $L_y = L_z = 10$ with the periodic boundary conditions in all three directions at the packing fraction $\eta = \frac{\pi}{6}\frac{N}{V}\sigma_H^3 = 0.36$. We average over three different trajectories for all time-dependent quantities. We take both cold regions as shown in Fig. 3b for considering data pertaining to the cold region.

We vary parameter $\sigma^* = \frac{\sigma_c}{\sigma_H}$ for a given $T^* = \frac{T_H}{T_C}$. We generate steady-state structural diagram in $\sigma^* - T^*$ plane for a given $T^*(= 1.2)$. We identify steady-state structure based on $(\bar{q}_4, \bar{q}_6)$ plot. We observe both regions remain liquid for a lower

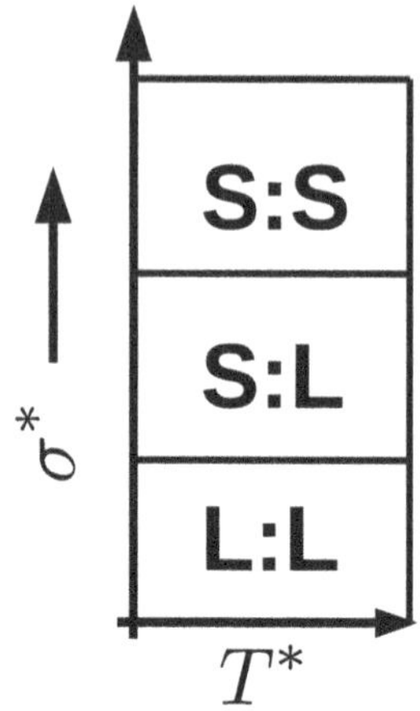

Fig. 4 Schematic of steady-state diagram in $\sigma^* - T^*$ plane. Symbols L and S denote liquid and ordered states, respectively. Left side denotes the structure of cold region, and right side for the hot region

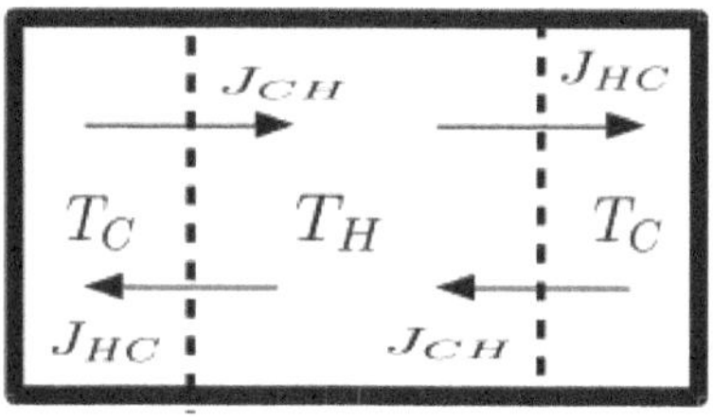

Fig. 5 Schematic of flux calculation. Arrows in two interfaces denote particle flux from hot region to cold region J_{HC} and cold region to hot region J_{CH}

value of σ^*. At first the cold region starts to emerge order formation, whereas hot region remains liquid with increasing σ^* value. Both hot and cold regions show long-range order for sufficiently large σ^*. This is shown schematically in Fig. 4.

The intriguing phenomenon of crystallisation in the hot region can be understood as follows. We calculate the net flux ΔJ at the two interfaces $\Delta J = J_{HC} - J_{CH}$ (Fig. 5), where J_{HC} is flux from hot region to cold region and J_{CH} is flux from cold region to hot region. We observe that J goes from negative value to zero and fluctuates around zero. This means that particles movement from cold region to hot region dominates over that from hot region to cold region and saturates. As particles diameter increases in cold region, they are pushed out from the cold region to the hot region due to excluded volume effect. As a result, the packing fraction increases also in the hot region due to increase in particle. Packing fraction increases in cold region as well due to large diameter in spite of reduction of particle number. Increase in packing fraction in both regions leads to the crystallisation in steady state. This is further reflected in overall pressure P in the system. We observe P suddenly increases in cold region initially due to a large number of particles with large diameter. Subsequently, P decreases with time as particles push out from the cold to hot region and also due to structural rearrangement. Simultaneously, P increases in hot region due to increases in particle number density. At steady state, pressure becomes flat in both hot and cold regions.

4 Motion of Fluid Droplets Driven Through a Face Mask

Face mask (FM) has become important in recent times after COVID-19 pandemic hit the globe. FM prevents spread of airborne diseases by capturing respiratory droplets emitted by an affected person while speaking, coughing, and sneezing [83]. These droplets consist of different salts, proteins in small amounts, and microorganisms all floating in water. According to the WHO recommendation, FM should be made up of three layers [84, 85]: hydrophobic material in the middle and the outermost layers and a polar material like cotton layer inside toward mouth [85, 86].

Commercial available N95 FMs with fibres of micrometre range filter 95% of the micro-droplets through electrostatic capture and mechanical interception [87]. But they show reduction in efficiency due to charge loss [88]. Several studies mainly after post-pandemic situation are performed to search for stable FM performances like tribo-electric generators [86, 89, 90], mixture of fabrics with different compositions [91], and multiple layers [84]. Nano-fibres [87] and metal-organic framework filters [92, 93] are also studied. But they show breathing difficulty due to nanometre size pores. FMs need to exchange essential molecules between the human body and the atmosphere so that normal breathing takes place [91]. For comfortable breathing, pressure drop inside FM is calculated for different types of FM materials, which is found to be around 2.5 $\pm$ 0.4 Pa [91]. Thus, designing a FM with better stability and breathability is still quite pertinent. Such design can be facilitated by microscopic understanding of droplet movement inside FM.

We model droplet movement through FM under pressure difference as required for normal breathing. We consider explicitly the middle layer of the tri-layer FM, composed of polymeric strands, each being composed of randomly distributed two kinds of beads (h-bead and p-bead) of mass m and the same diameter σ (1 μm) in a given ratio, and distributed randomly over the strand. A schematic diagram is shown in Fig. 6. We take the interpenetrating polymer network confined by asymmetric walls (h-wall at $z = 0$ and p-wall at $z = z_P$) favouring h-beads and p-beads representing the outermost and the innermost layer, respectively, following WHO recommendation [85]. Then we study tracer colloidal particle (TCP) permeation from the air side (h-wall) of the FM towards the side facing the mouth (p-wall) shown schematically in Fig. 6.

Let us consider the interaction in the polymer strands. The non-bonded interaction between two monomers i and j with separation r_{ij} is taken through the Lennard-Jones (L-J) 12-6 potential:

$$V_{\alpha\beta}(r_{ij}) = 4\epsilon_{\alpha\beta}\left[\left(\frac{\sigma}{r_{ij}}\right)^{12} - \left(\frac{\sigma}{r_{ij}}\right)^{6}\right], r_{ij} < 3\sigma. \tag{5}$$

Here $\alpha(= h, p)$ and $\beta(= h, p)$ stand for the type of beads, and r_{ij} is separation between two beads. The bonded interaction corresponding to stretching between

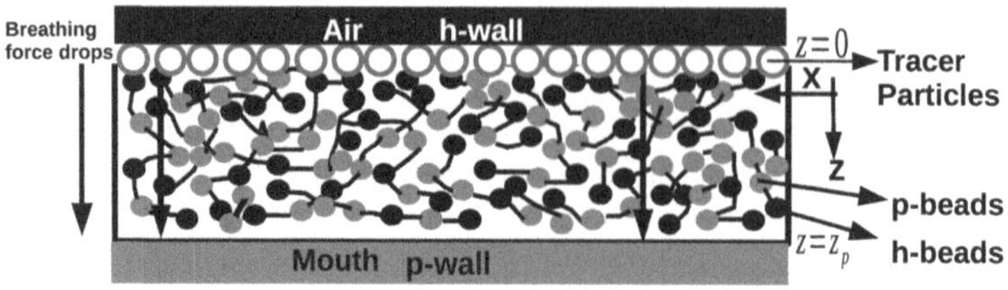

Fig. 6 A model schematic of polymeric network confined in asymmetric p(grey) and h(black) wall. Polymer consists of two beads: p-beads are shown in solid grey circle, and h-beads are shown in solid black circle. p-beads prefer p-wall and h-beads prefer h-wall. Tracer particles (open grey circle) are added close to h-wall. Tracers in the simulation prefer p-beads and p-wall. Force is added on each tracer particle (mimic breathing) from h-wall to p-wall. Force is indicated by black arrows from h-wall to p-wall. Force is maximum at $z = 0$ and zero at $z = z_P$. (The y-axis is not shown and is pointing into the page)

two neighbouring beads at separation r_{ij}:

$$V_{bond}(r_{ij}) = \frac{1}{2}k_b(r_{ij} - r_0)^2, \tag{6}$$

where $r_0 = 1.5\sigma$ is the equilibrium bond distance between connected monomers and k_b the force constant. The bond angle energy change is of harmonic form:

$$V_{angle}(\theta) = \frac{1}{2}k_\theta(\theta - \theta_0)^2, \tag{7}$$

where k_θ the force constant. $\theta = \cos^{-1}(\frac{\vec{r}_{ij}\cdot\vec{r}_{jk}}{|\vec{r}_{ij}||\vec{r}_{jk}|})$ is the angle generated by three successive monomers i,j,k in a polymer and equilibrium angle $\theta_0 = 114$ degrees [94]. We take $k_b = k_\theta = k$ to reduce the number of variables which is fair consideration for a rigid network where elastic deformation of any kind costs considerable energy values.

We place the polymeric network within two walls along z-direction: The p-wall interacts with a bead via the L-J 9-3 potential:

$$V_{f,wp}(z_i) = \epsilon_{f,wp}\left[\frac{2}{15}\left(\frac{\sigma}{z_P - z_i}\right)^9 - \left(\frac{\sigma}{z_P - z_i}\right)^3\right]. \tag{8}$$

Similarly, h-wall interacts with a bead:

$$V_{f,wh}(z_i) = \epsilon_{f,wh}\left[\frac{2}{15}\left(\frac{\sigma}{z_i}\right)^9 - \left(\frac{\sigma}{z_i}\right)^3\right]. \tag{9}$$

Here $f = (p, h)$ stands for the bead type. z_i is the z-coordinate of ith particle. We take same σ values for all the interactions, but $\epsilon_{h,wh} > \epsilon_{p,wh}$ and $\epsilon_{p,wp} > \epsilon_{h,wp}$.

Next, we define interaction of TCPs. They interact with each other with strength $\epsilon_{tr,tr}$. The strengths are $\epsilon_{tr,p}$ and $\epsilon_{tr,h}$ for interaction of TCPs with p- and h-beads. They also interact with two walls through similar L-J 9-3 potential with parameters $\epsilon_{tr,wp}$ and $\epsilon_{tr,wh}$ for p- and h-walls, respectively. The TCPs strongly interact more with the p-beads and the p-wall than the h-beads and the h-wall, respectively, so that $\epsilon_{tr,wp} > \epsilon_{tr,wh}$. We further put force over a TCP:

$$F(z) = F_0(1 - z/z_P). \tag{10}$$

The unit of energy is taken ϵ_{hh} in the simulation. We take the mass of water of density 1 gm/cm^3 in a sphere of diameter 1 μm as a unit of mass. The unit of length is the bead diameter (σ =1 μm) [95–98]. We fix dimensionless $\epsilon_{pp} = 0.33(1/3$ of $\epsilon_{hh})$. The cross interaction between h and p monomers follows the standard Berthelot mixing rule. We fix dimensionless wall particle interactions, $\epsilon_{h,wh} = 1.0$, $\epsilon_{p,wh} = 0.033$, $\epsilon_{p,wp} = 0.33$, and $\epsilon_{h,wp} = 0.033$, which ensures the polymer network is evenly spread throughout gap between two walls as shown in Fig. 6. The dimensionless L-J energy parameters for TCPs are defined as: $\epsilon_{tr,tr} = 1.0$, $\epsilon_{tr,wh} = 0.033$, $\epsilon_{tr,wp} = 1.0$. Masses of both beads and tracers are taken as density of water with corresponding diameters σ and σ_{tr} for beads and tracer, respectively. We vary several parameters: the network rigidity k, $\bar{\epsilon} = \epsilon_{tr,p}/\epsilon_{tr,h}$, and the confining length z_P. We also vary the reduce temperature $T^* = \frac{k_B T}{\epsilon_{hh}}$ and perform simulations.

The network consists of 27 identical polymeric strands, each having randomly distributed 50 monomers in a number ratio h:p. We set the packing fraction $\eta = \frac{\pi}{6}\frac{N}{V}\sigma^3 = 0.21$ of total polymeric beads in our model. It ensures almost 80% is voids in the system to mimic a reasonable real scenario [95]. We exert external force $F(z)$ on TCPs, which is taken to be maximum at h-wall and linearly vanishes at p-wall to ensure breathing. We use the under-damped Langevin equation of motion of the ith particle with mass m_i at position $\vec{r_i}(t)$ at time t for both the polymeric bead and the TCP (Appendix 1) for particle dynamics: We consider bonded and non-bonded interactions with other network beads, TCPs, and the walls for the network beads. We set $\vec{F}_i^{\alpha} = 0$. Similarly, TCPs interact with other TCPs, network beads, the wall, and in addition, we exert an additional external force, $\vec{F_i}$ on them. We also take the Gaussian white noise with zero mean and auto-correlation function, $6\zeta_i^{\alpha} k_B T \delta(t^{'} - t^{''})$, where T the temperature and k_B is the Boltzmann constant. Here $\zeta_i^{\alpha} = m_i^{\alpha}\gamma$, γ being the damping term that is due to friction of the network. We use the LAMMPS package [99] for simulations taking all the interactions as given by Eq. 5 to Eq. 10. We estimate a time scale, $\tau (= \sqrt{\frac{m\sigma^2}{\epsilon_{hh}}}) \sim 0.35$ millisecond using the values of mass m, σ, and ϵ_{hh}. The time step for integration is used $\delta t = 0.001\tau$, and damping term is set to $\frac{1}{\gamma} = 10\tau$. We averaged over five different independent trajectories for all thermodynamic quantities each 6500τ long. We use periodic boundary conditions (PBC) along x ($L_x = 30\sigma$) and y ($L_y = 20\sigma$) directions and no PBC along z-direction. We study the motion of the TCPs through the network from h-wall to p-wall. The ratios $\bar{\epsilon} = \frac{\epsilon_{tr,p}}{\epsilon_{tr,h}}$, k, and confinement length z_P are varied in the simulations.

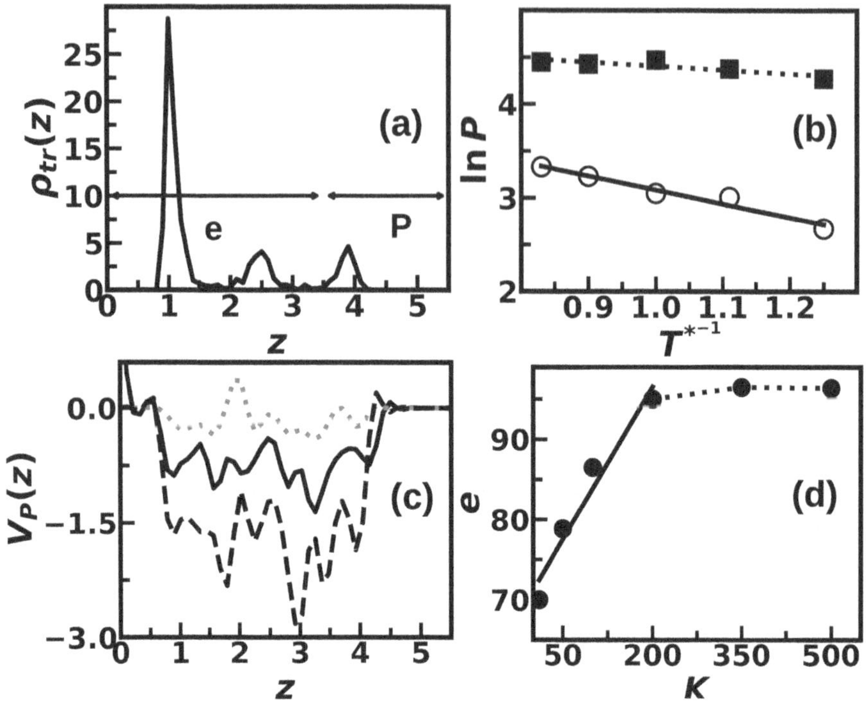

Fig. 7 (**a**) Tracer density profile $\rho_{tr}(z)$ versus z plot for $\sigma_{tr} = 1.5$, $T^* = 1.0$ with h:p=50:50, $F_0 = 0$, $\bar{\epsilon} = 3.3$, $z_P = 5.5$, $k = 50$. (**b**) ln P versus $\frac{1}{T^*}$ plot for $\sigma_{tr} = 1.5$ with $\bar{\epsilon} = 3.3$, h:p=50:50, $z_P = 5.5$, $k = 50$ for $F_0 = 1$ (open circle), $F = 10$ (close square). The best fitted lines are also shown. (**c**) $V_P(z)$ versus z plot with $T^* = 1.0$ for $z_p = 5.5$, $k = 50$, $F_0 = 1.0$ for $\bar{\epsilon} = 3.3$(black line); $\bar{\epsilon} = 2$ (grey dotted line) and $\bar{\epsilon} = 5$ (black dashed line). (**d**) e versus k plot (solid circles) for $\sigma_{tr} = 1.5$ with h:p=50:50, $T^* = 1.0$, $\bar{\epsilon} = 3.3$, $F_0 = 1$, $z_P = 5.5$. Solid line indicates the best fit, and dotted line is a guide to the eyes

We create an interpenetrating polymer network confined by h and p-wall without external force at first. We estimate mean pore size of the polymer network from tracer density profile $\rho_{tr}(z)$ along confinement z-direction without external drive. We observe in Fig. 7a distinct peaks along z-direction. The peak is strongest near h-wall. The tracers are started from h-wall, and the peak near h-wall denotes particles stopped (cannot pass through) mainly due to geometrical effect. We identify mean pore size $\sim$2 micrometre where the tracer density profile is sharply peaked near h-wall. As larger tracer will be stopped geometrically, our focus is on tracer size slightly smaller than the mean pore size. We restrict further analysis to droplet size $<$2 micrometre. The peak close to p-wall is permeation P (Fig. 7a) and efficiency of mask is $e = 100 - P$.

Next we apply driving force along +z-direction on the TCPs. We explore the temperature dependence of P. We observe that tracer permeation is an activated Arrhenius process [100] with barrier as shown in Fig. 7b. We observe the slope of the curve get flat with increasing driving force (Fig. 7b), which indicates the barrier can be tuneable with different parameters. We calculate the potential energy profile along the z-direction per tracer particle. We average these profiles over steady - state configurations. We identify $V_p(z)$ (the potential energy profile for interaction of the TCPs with p-beads) playing the dominant role in the profile traces prefers p-beads over h-beads. We observe in Fig. 7c that minimum of the energy of the profile gets deeper with increasing $\bar{\epsilon}$, where TCP interaction with p-beads increases. We explore other parameters responsible for tuning barrier as well and observe that the potential energy barrier responsible for the activation process increases with rigidity of network k, tracer bead interaction $\bar{\epsilon}$, tracer size, and decreases with confinement length z_P, driving force. WE specifically shows how efficiency increases with increases in network rigidity in Fig. 7d. Finally, we observe that the efficiency $e > 90\%$ of the mask can be achieved by composition of polymer (h:p ratio = 50:50), interaction strength of tracer and the network $\bar{\epsilon} > 3.3$, the network rigidity $k > 200 k_B T/\sigma^2$, and the thickness of the confinement $z_P = 5.5\,\mu\text{m}$.

5 Reverse Osmosis of Binary Mixture Through Polymeric Membrane

We study another application of polymer-based membrane which is Reverse Osmosis (RO) process for separation of components in a solution. A solution is characterised by the presence of solute of a given concentration. If a solution of lower solute concentration and that of higher solute concentration are connected with a semipermeable membrane, solvent from low solute concentration region flows to that higher solute concentration by the process of osmosis [101]. Reverse osmosis is a pressure-driven technique to overcome the natural osmotic pressure and enable solvent flow from higher to lower solute concentration in steady-state conditions [102]. RO membrane separation processes are widely used nowadays in both industry and household purposes [103, 104] like water purification.

The semipermeable membranes are thin film composite(TFC) membranes which are entangled polymers prepared from interfacial polymerisation process [105], widely used in RO process for separation of both polar [106] and non-polar [107] mixtures. Three quantities indicate RO membrane efficiency: Permeation P, the amount of solvent flow from one side to other side; Solute Rejection R, rejection of solute from solution; and Fouling F: the amount of solute gather inside membrane during permeation. Experiments show that changing relative interaction of membrane matrix [103, 106] with solute and solvent changes membrane efficiency as permeation of solvent, solute rejection, and fouling inside the membrane. Despite atomistically detailed simulation for transport processes [17], it is yet a challenge to design RO membrane with better permeation, solute rejection, and less fouling. A coarse-grain model addressing solute and solvent transport may aid designing a better RO membrane.

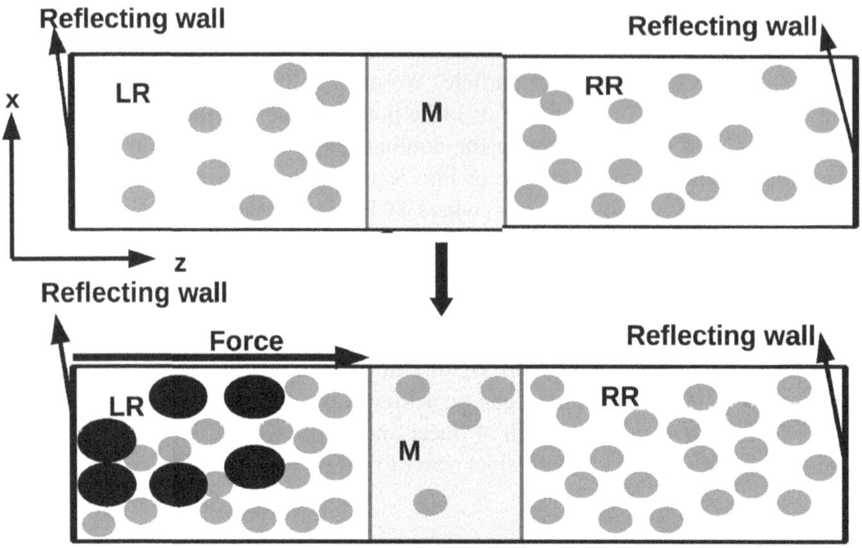

Fig. 8 Schematic of the simulation cell. The y-axis (not shown) is pointing into the page. Upper panel: Two reservoirs with S particles (grey) are added in two sides of membrane. Lower panel: After solvation of membrane, L particles (black) are added in LR

In our model the RO membrane is created as a polymeric network in an asymmetric confinement situation to the experimental IP process [105]. The strand composition and interactions are similar as described in the previous section. We take $k_b = k_\theta = k = 100$ for simplicity.

We attach two reservoirs (LR and RR) on left and right sides of the RO membrane (M). Two reflecting walls are placed at two extreme ends of two reservoirs. The wall interactions are similar to those in the previous section. Both reservoirs consist of solvent (S) particles. We solvate M without applying force as shown in Fig. 8 (upper panel). After equilibration, both LR and RR come to same density of S. Then we add solute (L) particles (1.8 times larger than S) in the LR removing the same number of S particles randomly from the LR so that the total number density of particles (S+L) remains the same in the box (Fig. 8, lower panel). The partial density of S becomes lower in LR compared to RR.

There are several interactions in the system with the strength parameters as follows: the bead-bead interaction ($\epsilon_{\alpha,\beta}$), bead-wall interaction ($\epsilon_{\alpha,wh}$ and $\epsilon_{\alpha,wp}$), solvent-solvent interaction ($\epsilon_{S,S}$), solute-solute interaction ($\epsilon_{L,L}$), solvent-solute interaction ($\epsilon_{S,L}$), solvent h-bead interaction ($\epsilon_{S,h}$), solvent p-bead interaction ($\epsilon_{S,p}$), solute h-bead interaction ($\epsilon_{L,h}$), and solute p-bead interaction ($\epsilon_{L,p}$). Then we add external force $F(z)$ along z to study the motion of both S and L from LR to RR through M.

We take unit of energy $\epsilon_{h,h}$ as unit of energy. The unit of length is diameter of S particles $\sigma_S = 0.5$ nm. We take the mass of water of density 1 gm/cm^3 in a sphere

Table 1 Diameter of particles

Diameter of different particles	Dimensionless value
σ_S	1.0
σ_L	1.8
σ_h	2.0
σ_p	2.0

Table 2 Table for ϵ

Nature of interaction	Dimensionless variable	Dimensionless value
h,h	$\epsilon_{h,h}$	1.0
p,p	$\epsilon_{p,p}$	0.33
h,wh	$\epsilon_{h,wh}$	1.0
p,wh	$\epsilon_{p,wh}$	0.033
p,wp	$\epsilon_{p,wp}$	0.33
h,wp	$\epsilon_{h,wp}$	0.033
S,S	$\epsilon_{S,S}$	1.0
L,L	$\epsilon_{L,L}$	1.0

Table 3 Mass of particles

Mass of different particles	Dimensionless value
m_S	1.0
m_L	5.832
m_h	8.0
m_p	8.0

of diameter 0.5 nm as a unit of mass. The polymeric bead diameter is taken ($\sigma_b = 2\sigma_S = 1$ nm). The solute particle diameter is taken $\sigma_L = 1.8\sigma_S$. The diameters of all different particles are given in Table 1. The other energy interaction parameters are given in Table 2. For cross interaction, diameters follow Berthelot rule. When preparing RO membrane, we use interaction of the wall with h- and p-beads. The masses of all different particles are given in Table 3.

We perform simulation using the LAMMPS package [99] using Nose-Hover thermostat. Using the values of mass m, σ_S, and ϵ_{hh}, we estimate a time scale, $\tau(=\sqrt{\frac{m\sigma_S^2}{\epsilon_{hh}}}) \sim 11$ picoseconds. The time step for integration is taken 0.001τ. All the quantities are averaged over three different independent trajectories each 600000τ long. We take $L_x = 30$ and $L_y = 20$, in x- and y-directions, respectively, with the periodic boundary conditions (PBC) and no PBC in z-direction. We vary: $\bar{\epsilon}_S = \epsilon_{S,h}/\epsilon_{S,p}$ and $\bar{\epsilon}_L = \epsilon_{L,h}/\epsilon_{L,p}$ at reduced temperature $T^* = \frac{k_B T}{\epsilon_{h,h}} = 1$. We study two types of solutions: (1) case of solution particles S,L preferring the h-beads and (2) case of solution particles S,L preferring the p-beads.

(1) Case of solution particles S,L preferring h-beads – We increase $\bar{\epsilon}_L = \frac{\epsilon_{L,h}}{\epsilon_{L,p}}$, keeping $\bar{\epsilon}_L > 1$. We measure solvent recovery or permeation P in RR, solute rejection R in RR when system reaches steady state and fouling F. We observe that both P and R increase with $\bar{\epsilon}_L$ but F also increases. Then, we tune $\bar{\epsilon}_S =$

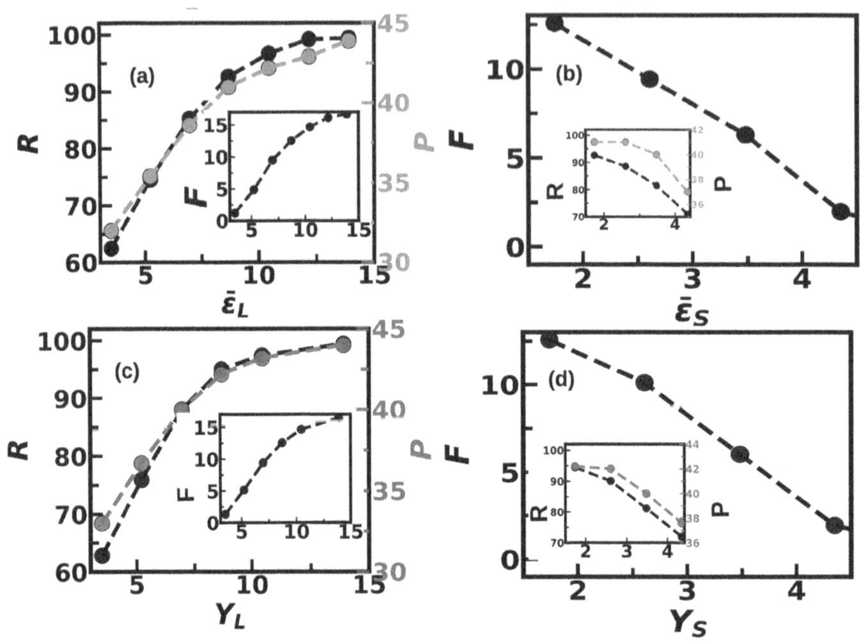

Fig. 9 RO performance parameters: (**a**) R vs. $\bar{\epsilon}_L$(left axis), P vs. $\bar{\epsilon}_L$(right axis, grey line). Inset: F vs. $\bar{\epsilon}_L$. (**b**) F vs. $\bar{\epsilon}_S$. Inset: R vs. $\bar{\epsilon}_S$(left axis) and P vs. $\bar{\epsilon}_S$(right axis grey line). (**c**) R vs. Y_L(left axis), P vs. Y_L(right axis, grey line). Inset: F vs. Y_L. (**d**) F vs. Y_S. Inset: R vs. Y_S(left axis) and P vs. Y_S(right axis grey line)

$\frac{\epsilon_{S,h}}{\epsilon_{S,p}}$ keeping $\bar{\epsilon}_S > 1$ and $\bar{\epsilon}_L = 8.71$. We observe F decreases linearly close to zero with increasing $\bar{\epsilon}_S$. However R, P decreases with $\bar{\epsilon}_S$. The results are shown in Fig. 9a–b. Qualitatively, the observations can be rationalised as follows: As $\bar{\epsilon}_S$ increases, S particles find more favourable to stay inside the M. Consequently, the L particles do not find favourable to stay inside M. They move towards PR, which leads to decrease in R and mixed phase in PR also which is undesirable for the RO performance. Thus, the increase in $\epsilon_{S,h}$ relative to $\epsilon_{S,p}$ impedes the performance of RO, although the fouling can be reduced to almost zero. We do not show data larger value for $\bar{\epsilon}_S$ where we do not reach steady state in our calculations.

(2) Case of solution particles S,L preferring p-beads – Here, we study the cases where solvent and solute particles favour p-beads, namely, $\bar{\epsilon}_S$, $\bar{\epsilon}_L < 1$. Here we fix $\bar{\epsilon}_S(= 0.57)$ and vary $\bar{\epsilon}_L$. As lower $\bar{\epsilon}_L$ denotes a strong L- and p-bead interaction, it is convenient to denote the data in terms of $Y_L = \bar{\epsilon}_L^{-1}$. Similarly we define $Y_S = \bar{\epsilon}_S^{-1}$. We observe R increases with increasing Y_L and finally comes to saturation for large Y_L. Simultaneously, F increases rapidly with increasing Y_L. We also observe that P also increases with increasing Y_L. We further vary $\bar{\epsilon}_S$ in the regime $\bar{\epsilon}_S < 1$ fixing $\bar{\epsilon}_L(= 0.114)$. We observe F

sharply decreases linearly close to zero with increasing Y_S. However, R, P also decreases with Y_S. The results are summarised in Fig. 9c–d.

Note that the relevant parameters in the study involve only the energy depth parameters of LJ interactions. This is sufficiently robust and can be defined for any inter-molecular interaction having a well-defined minimum. Hence, it may be interesting to point out the implication of our model for realistic systems. h-beads mimic non-polar moieties of the polymer network. We set higher interaction strength among h-beads as it has tendency to show hydrophobic collapse. Similarly, p-beads represent polar moieties. The solution consisting of S and L particles both preferring the h-beads can be thought of the case as model for a solution of non-polar components. Our calculation shows that the interaction of the non-polar solute needs stronger interaction with the non-polar moieties of the polymer membrane so that the non-polar solvent can easily flow through the membrane from mixture to the pure solvent. On the other hand, solution consisting of S and L particles both preferring the p-beads is a model for polar solution. Our calculation shows that polar solute interaction with polar moieties of the membrane should be stronger than the solvent interaction for easy passage of solvent from mixture to pure solvent. Our results at least qualitatively explain different experimental observations on how tuning the membrane properties affects RO performance [106]. However, stronger solute membrane interaction results in larger solvent permeation and solute rejection but more membrane fouling. Despite many theoretical and simulation studies on driven permeation, it is not clear how fouling as well as solute rejection and solvent permeation can be controlled with tuning different solute and solvent interactions with the polymer membrane matrix. We explore a microscopic understanding of these factors in this study by considering the explicit particle motions.

6 Conclusion

We consider different types of problems to study behaviour of soft matter systems in non-equilibrium steady-state conditions pertinent to technological applications. Our observation of long-ranged order formation may pave way for microscopic understanding of long-range order formation of temperature-dependent interaction in temperature difference. The presence of the long-ranged order in our system suggests that the surface area may be macroscopic and thus useful for sensory applications. Long-range order in hot region which we observe may also be interesting in practical scenario for designing ordered materials at high temperature. Our observation can be observed with more careful experiments in future. Our model studies on face mask may be helpful in designing mask with better efficiency while retaining normal breathability conditions at room temperature. Finally, our studies on RO performance could be useful to develop reverse osmosis membrane with better permeation, selectivity, and less fouling for longer use.

We focus on two aspects. In one class of problem, we explore structure formation in steady state in the presence of temperature gradient. As exposed surface area depends on its structure, our work of guided structure formation may help to design of large surfaces for surface-dependent applications as well. Moreover, the observation of hot crystal formation may guide an experimentally verifiable case of formation of long-ranged order in extreme conditions in steady state. It may be important to carry out simulations on more atomistic models to make direct comparison with experimental results. It may also be interesting to understand transport properties of these systems in detail to aid their applications. In other class of problem, we study permeation through a polymeric medium in steady state in the presence of external force. Our results on model face mask and RO can be verified through further experimentation and may be helpful in designing breathable face mask and RO membranes with better efficiency at room temperature.

Appendix 1

BD Simulations

The Brownian particle (i) position is updated in the simulation according to Langevin's equation [108] in the over-damped limit with random force $\vec{f}_{i,\alpha}$ with white noise (see details in Appendix A1):

$$\Gamma_\alpha \frac{d}{dt}\vec{r}_{i,\alpha} = \vec{\nabla}_i \sum_{j=1}^{N} V_i^{(\alpha)}(r_{ij}) + \vec{f}_{i,\alpha}(t). \tag{11}$$

Equation 11 for ith particle could be discretised to describe the position as follows [108]:

$$\vec{r}_{i,\alpha}(t+\delta t) = \vec{r}_{i,\alpha}(t) + \frac{1}{\Gamma_\alpha}\vec{\nabla}_i \sum_{j=1}^{N} V_i^{(\alpha)}(r_{ij})\delta t + \vec{R}_{i,\alpha}(\delta t). \tag{12}$$

Here α =(hot, H) and (cold, C) denotes the temperature regions in the simulation box. $V_i^{(\alpha)}$ is the total potential for ith particle in the αth region with other particles j with distance $r_{ij} = (|\vec{r}_i - \vec{r}_j|)$, $V_i^{(\alpha)} = \sum_{j=1}^{N} V_A^{(\alpha)}(r_{ij}) + V_R^{(\alpha)}(r_{ij})$. The derivative $\vec{\nabla}_i$ is calculated with respect to the coordinate of the ith particle. $\vec{R}_{i,\alpha}(\delta t)$ is corresponding random displacement with mean zero and variances $2D_\alpha \delta t$ due to fluctuating force during time δt. Here D_α the Stokes-Einstein diffusion coefficient, given by $k_B T_\alpha / \Gamma_\alpha$ at temperature T_α, k_B the Boltzmann constant, and $\Gamma_\alpha = 3\pi\eta_\alpha\sigma$ is the viscous damping and δt is the integration time.

The particle dynamics in the under-damped Langevin equation of motion is given by

$$m_i \frac{d^2 \vec{r}_{i,\alpha}}{dt^2} = -\zeta \frac{d\vec{r}_{i,\alpha}}{dt} - \nabla \sum_{j=1}^{N} V_i^{(\alpha)}(r_{ij}) + \vec{f}_{i,\alpha}(t). \tag{13}$$

Here α =(hot(H) and cold(C)) denotes the temperature regions in the simulation box. N is the total number of particles in the system. ζ is the friction coefficient. $V_i^{(\alpha)}$ is the total potential energy of the ith particle in the α region with other particles j at a distance $r_{ij}(= |\vec{r}_i - \vec{r}_j|)$. The components of $\vec{f}_{i,\alpha}(t)$ are the Gaussian white noise with zero mean and auto-correlation function, $6\zeta k_B T_H \delta(t^{'} - t^{''})$ at hot temperature T_H and $6\zeta k_B T_C \delta(t^{'} - t^{''})$ in cold temperature T_C. k_B is the Boltzmann constant. Here $\zeta = m\gamma$, where γ is the damping term due to friction.

Appendix 2

Structural Quantities

Bond Order Parameter

Structural parameters are calculated in the steady state as functions of time. We take both cold regions in our simulation set-up for considering data of the cold region. We calculate bond orientation order parameter distribution of order l which is defined as [109],

$$q_l q_l(i) = \frac{1}{N_b(i)} \sum_{j \in N_b(i)} q_l(i).q_l(j), \tag{14}$$

where $q_l(i).q_l(j) = \sum_{m=-l}^{m=+l} (\frac{1}{(\sum_{-l}^{l} |q_{lm}(i)|^2)^{1/2}}) \cdot (\frac{1}{(\sum_{-l}^{l} |q_{lm}(j)|^2)^{1/2}}) \cdot q_{lm}(i) q_{lm}^*(j)$. Here $q_{lm}(i) = \frac{1}{N_b(i)} \sum_j^{N_b(i)} Y_{lm}(\theta_{ij}, \phi_{ij})$. Y_{lm} are the spherical harmonics. l is an integer, and m goes from $-l$ to $+l$. (θ_{ij}, ϕ_{ij}) are orientation of bond between particle i. $N_b(i)$ is neighbouring particles (j) within cutoff distance. Cutoff distance is taken up to the first minimum of the pair correlation function in the hot bulk system with respect to direction (x-axis) of temperature difference.

To measure structural type, we also measure an average quantity $\bar{q}_{lm}(i)$, defined as follows:

$$\bar{q}_{lm}(i) = \frac{1}{\tilde{N}_b(i)} \sum_{j \in \tilde{N}_b(i)} q_{lm}(j), \tag{15}$$

where $\tilde{N}_b(i)$ is summed over all neighbours of particle i within cutoff distance up to the first minimum of the pair correlation function in the hot bulk system and the particle i itself. Finally, we calculate global rotational invariant quantity,

$$\bar{q}_l(i) = \sqrt{\frac{4\pi}{2l+1} \sum_{m=l}^{+l} |\bar{q}_{lm}(i)|^2}. \tag{16}$$

As a standard practice, we consider l= 4 and 6 in Eqs. 15 and 16 for each particle in a given configuration and plot $\bar{q}_4 - \bar{q}_6$ as reported in the literature [75].

Machine-Learning-Based Autoencoder

We briefly describe machine learning method. We construct the unsupervised machine learning (UML) algorithm based on global rotational invariant bond orientation order parameters as input for a fully connected neural-network-based autoencoders. We use this to detect the optimum number of clusters of different structures in the aggregate in lower-dimensional space as in Ref. [110]. We use $\bar{q}_l(i)$ as input to train encoder, where we use l = 1,. . . ,8. Autoencoders are widely used to project higher-dimensional input in lower-dimensional layer (called bottleneck) and finally reconstruct the input at decoder output. We train the network through minimising MSE (mean square error) between input and output. Bottleneck captures important features of the input in low dimension after training. Then, we identify optimum dimension of bottleneck layer, which comes out c = 2 in our case. We find $Y_1(i)$, $Y_2(i)$ for each ith particle in the bottleneck layer. Then, we use a probabilistic method called Gaussian mixture models (GMMs) [111, 112] to cluster data in this two-dimensional space using scikit-learn [113]. In GMM each particle is associated with a particular cluster in which cluster it shows highest membership percentage.

References

1. Löwen, H.: Colloidal soft matter under external control. J. Phys.: Condensed Matter **13**(24), R415 (2001)
2. Vogel, N., Retsch, M., Fustin, C.-A., Del Campo, A., Jonas, U.: Advances in colloidal assembly: the design of structure and hierarchy in two and three dimensions. Chem. Rev. **115**(13), 6265–6311 (2015)
3. Hyde, E.D.E.R., Seyfaee, A., Neville, F., Moreno-Atanasio, R.: Colloidal silica particle synthesis and future industrial manufacturing pathways: a review. Ind. Eng. Chem. Res. **55**(33), 8891–8913 (2016)
4. Geszke-Moritz, M., Moritz, M.: APTES-modified mesoporous silicas as the carriers for poorly water-soluble drug. Modeling of diflunisal adsorption and release. Appl. Surf. Sci. **368**, 348–359 (2016)
5. Bitar, A., Ahmad, N.M., Fessi, H., Elaissari, A.: Silica-based nanoparticles for biomedical applications. Drug Discov. Today **17**(19–20), 1147–1154 (2012)

6. Xie, W., Hu, L., Yang, X.: Basic ionic liquid supported on mesoporous SBA-15 silica as an efficient heterogeneous catalyst for biodiesel production. Ind. Eng. Chem. Res. **54**(5), 1505–1512 (2015)
7. Wang, J., Shah, Z.H., Zhang, S., Lu, R.: Silica-based nanocomposites via reverse microemulsions: classifications, preparations, and applications. Nanoscale **6**(9), 4418–4437 (2014)
8. Jo, K., Ishizuka, M., Shimabayashi, K., Ando, T.: Development of new mineral oil-based antifoams containing size-controlled hydrophobic silica particles for gloss paints. J. Oleo Sci. **63**(12), 1303–1308 (2014)
9. Löwen, H.: Melting, freezing and colloidal suspensions. Phys. Rep. **237**(5), 249–324 (1994)
10. Chakrabarti, J., Krishnamurthy, H.R., Sood, A.K.: Density functional theory of laser-induced freezing in colloidal suspensions. Phys. Rev. Lett. **73**(21), 2923 (1994)
11. Imhof, A., Van Blaaderen, A., Dhont, J.K.G.: Shear melting of colloidal crystals of charged spheres studied with rheology and polarizing microscopy. Langmuir **10**(10), 3477–3484 (1994)
12. Palberg, T., Würth, M.: Multiphase coexistence and non-linear rheology of colloidal. dispersions as observed in a model capillary viscosimeter. Journal de Physique I **6**(2), 237–244 (1996)
13. Vermant, J., Raynaud, L., Mewis, J., Ernst, B., Fuller, G.G.: Large-scale bundle ordering in sterically stabilized latices. J. Colloid Interface Sci. **211**(2), 221–229 (1999)
14. Dutta, S., Chakrabarti, J.: Length-scales of dynamic heterogeneity in a driven binary colloid. Phys. Chem. Chem. Phys. **22**(31), 17731–17737 (2020)
15. Pusztai, T., Tegze, G., Tóth, G.I., Környei, L., Bansel, G., Fan, Z., Gránásy, L.: Phase-field approach to polycrystalline solidification including heterogeneous and homogeneous nucleation. J. Phys.: Condensed Matter **20**(40), 404205 (2008)
16. Di Leonardo, R., Ianni, F., Ruocco, G.: Colloidal attraction induced by a temperature gradient. Langmuir **25**(8), 4247–4250 (2009)
17. Shen, M., Keten, S., Lueptow, R.M.: Dynamics of water and solute transport in polymeric reverse osmosis membranes via molecular dynamics simulations. J. Membrane Sci. **506**, 95–108 (2016)
18. Lin, X., Si, T., Wu, Z., He, Q.: Self-thermophoretic motion of controlled assembled micro-/nanomotors. Phys. Chem. Chem. Phys. **19**, 23606–23613 (2017)
19. Wu, Z., Lin, X., Wu, Y., Si, T., Sun, J., He, Q.: Near-infrared light-triggered "on/off" motion of polymer multilayer rockets. ACS Nano **8**(6), 6097–6105 (2014). PMID: 24806430
20. Duhr, S., Braun, D.: Two-dimensional colloidal crystals formed by thermophoresis and convection. Appl. Phys. Lett. **86**(13), 131921 (2005)
21. Lin, Z., Gao, C., Chen, M., Lin, X., He, Q.: Collective motion and dynamic self-assembly of colloid motors. Curr. Opin. Colloid Interface Sci. **35**, 51–58 (2018)
22. Singh, D.P., Choudhury, U., Fischer, P., Mark, A.G.: Non-equilibrium assembly of light-activated colloidal mixtures. Adv. Mater. **29**(32), 1701328 (2017)
23. Chari, S.S.N., Dasgupta, C., Maiti, P.K.: Scalar activity induced phase separation and liquid–solid transition in a Lennard-Jones system. Soft matter **15**(36), 7275–7285 (2019)
24. Weber, S.N., Weber, C.A., Frey, E.: Binary mixtures of particles with different diffusivities demix. Phys. Rev. Lett. **116**(5), 058301 (2016)
25. Stenhammar, J., Wittkowski, R., Marenduzzo, D., Cates, M.E.: Activity-induced phase separation and self-assembly in mixtures of active and passive particles. Phys. Rev. Lett. **114**(1), 018301 (2015)
26. Abe, Y., Zhang, B., Gordillo, L., Karim, A.M., Francis, L.F., Cheng, X.: Dynamic self-assembly of charged colloidal strings and walls in simple fluid flows. Soft Matter **13**(8), 1681–1692 (2017)
27. Smrek, J., Kremer, K.: Small activity differences drive phase separation in active-passive polymer mixtures. Phys. Rev. Lett. 118(9):098002 (2017)
28. Sherman, Z.M., Swan, J.W.: Dynamic, directed self-assembly of nanoparticles via toggled interactions. ACS Nano **10**(5), 5260–5271 (2016)

29. Dutta, S., Chakrabarti, J.: Anomalous dynamical responses in a driven system. EPL (Europhys. Lett.) **116**(3), 38001 (2016)
30. Dutta, S., Chakrabarti, J.: Transient dynamical responses of a charged binary colloid in an electric field. Soft Matter **14**(22), 4477–4482 (2018)
31. Jha, P.K., Kuzovkov, V., Grzybowski, B.A., De La Cruz, M.O.: Dynamic self-assembly of photo-switchable nanoparticles. Soft Matter **8**(1), 227–234 (2012)
32. Sawczyk, M., Klajn, R.: Out-of-equilibrium aggregates and coatings during seeded growth of metallic nanoparticles. J. Am. Chem. Soc. **139**(49), 17973–17978 (2017)
33. Grötsch, R.K., Wanzke, C., Speckbacher, M., Angı, A., Rieger, B., Boekhoven, J.: Pathway dependence in the fuel-driven dissipative self-assembly of nanoparticles. J. Am. Chem. Soc. **141**(25), 9872–9878 (2019)
34. Moronshing, M., Subramaniam, C.: Room temperature, multiphasic detection of explosives, and volatile organic compounds using thermodiffusion driven Soret colloids. ACS Sustain. Chem. Eng. **6**(7), 9470–9479 (2018)
35. Giddings, J.C.: Field-flow fractionation: analysis of macromolecular, colloidal, and particulate materials. Science **260**(5113), 1456–1465 (1993)
36. Duhr, S., Braun, D.: Optothermal molecule trapping by opposing fluid flow with thermophoretic drift. Phys. Rev. Lett. **97**(3), 038103 (2006)
37. Nikoubashman, A.: Self-assembly of colloidal micelles in microfluidic channels. Soft Matter **13**(1), 222–229 (2017)
38. Mai, Y., Eisenberg, A.: Self-assembly of block copolymers. Chem. Soc. Rev. **41**(18), 5969–5985 (2012)
39. Dutta, S., Patra, P., Chakrabarti, J.: Self-assembly in amphiphilic macromolecules with solvent exposed hydrophobic moieties. Biopolymers **110**(12), e23330 (2019)
40. Karmakar, R., Chakrabarti, J.: A long-range order in a thermally driven system with temperature-dependent interactions. Soft Matter **18**, 867–876 (2022)
41. Iwata, N., Koike, T., Tokuhiro, K., Sato, R., Furumi, S.: Colloidal photonic crystals of reusable hydrogel microparticles for sensor and laser applications. ACS Appl. Mater. Interfaces **13**(48), 57893–57907 (2021)
42. Ashraf, S., Park, H.-K., Park, H., Lee, S.-H.: Snapshot of phase transition in thermoresponsive hydrogel PNIPAM: role in drug delivery and tissue engineering. Macromol. Res. **24**(4), 297–304 (2016)
43. Smit, B.J.T.J.: Phase diagrams of Lennard-Jones fluids. J. Chem. Phys. **96**(11), 8639–8640 (1992)
44. Karmakar, R., Chakrabarti, J.: Hot crystals of thermo-responsive particles with temperature dependent diameter in the presence of a temperature gradient. J. Chem. Phys. **159**(3), 034904-1–034904-7 (2023)
45. Kim, W.K., Kanduč, M., Roa, R., Dzubiella, J.: Tuning the permeability of dense membranes by shaping nanoscale potentials. Phys. Rev. Lett. **122**(10), 108001 (2019)
46. Robeson, L.M.: Correlation of separation factor versus permeability for polymeric membranes. J. Membrane Sci. **62**(2), 165–185 (1991)
47. Park, T.J., Reznick, J., Peterson, B.L., Blass, G., Omerbašić, D., Bennett, N.C., Kuich, P.H.J.L., Zasada, C., Browe, B.M., Hamann, W. et al.: Fructose-driven glycolysis supports anoxia resistance in the naked mole-rat. Science **356**(6335), 307–311 (2017)
48. Obliger, A., Pellenq, R., Ulm, F.-J., Coasne, B.: Free volume theory of hydrocarbon mixture transport in nanoporous materials. J. Phys. Chem. Lett. **7**(19), 3712–3717 (2016)
49. Shannon, M.A., Bohn, P.W., Elimelech, M., Georgiadis, J.G., Mariñas, B.J., Mayes, A.M.: Science and technology for water purification in the coming decades. Nature **452**(7185), 301–310 (2008)
50. Geise, G.M., Park, H.B., Sagle, A.C., Freeman, B.D., McGrath, J.E.: Water permeability and water/salt selectivity tradeoff in polymers for desalination. J. Membrane Sci. **369**(1–2), 130–138 (2011)

51. Tansel, B., Sager, J., Rector, T., Garland, J., Strayer, R.F., Levine, L., Roberts, M., Hummerick, M., Bauer, J.: Significance of hydrated radius and hydration shells on ionic permeability during nanofiltration in dead end and cross flow modes. Separation Purification Technol. **51**(1), 40–47 (2006)
52. Stamatialis, D.F., Papenburg, B.J., Gironés, M., Saiful, S., Bettahalli, S.N.M., Schmitmeier, S., Wessling, M.: Medical applications of membranes: drug delivery, artificial organs and tissue engineering. J. Membrane Sci. **308**(1–2), 1–34 (2008)
53. Tabei, S.M.A., Burov, S., Kim, H.Y., Kuznetsov, A., Huynh, T., Jureller, J., Philipson, L.H., Dinner, A.R., Scherer, N.F.: Intracellular transport of insulin granules is a subordinated random walk. Proc. Natl. Acad. Sci. **110**(13), 4911–4916 (2013)
54. Yi, L., Fengzhi, L., Qingyong, Z.: Numerical simulation of virus diffusion in facemask during breathing cycles. Int. J. Heat Mass Transf. **48**(19–20), 4229–4242 (2005)
55. Karmakar, R., Ghosal, A., Chakrabarti, J.: Model studies on motion of respiratory droplets driven through a face mask. Europhys. Lett. **141**(2), 27001 (2023)
56. Karmakar, R., Chakrabarti, J.: Transport of particles through RO membrane in steady state condition. ChemistrySelect **9**(6), e202304343 (2024)
57. Duhr, S., Braun, D.: Why molecules move along a temperature gradient. Proc. Natl. Acad. Sci. **103**(52), 19678–19682 (2006)
58. Pérez-Madrid, A., Rubi, J.M., Mazur, P.: Brownian motion in the presence of a temperature gradient. Phys. A: Stat. Mech. Appl. **212**(3–4), 231–238 (1994)
59. Rubı, J.M., Mazur, P.: Simultaneous Brownian motion of n particles in a temperature gradient. Phys. A: Stat. Mech. Appl. **250**(1–4), 253–264 (1998)
60. Burelbach, J., Brückner, D.B., Frenkel, D., Eiser, E.: Thermophoretic forces on a mesoscopic scale. Soft Matter **14**(36), 7446–7454 (2018)
61. Michaelides, E.E.: Brownian movement and thermophoresis of nanoparticles in liquids. Int. J. Heat Mass Transf. **81**, 179–187 (2015)
62. Debuschewitz, C., Köhler, W.: Molecular origin of thermal diffusion in b e n z e n e+ c y c l o h e x a n e mixtures. Phys. Rev. Lett. **87**(5), 055901 (2001)
63. Putnam, S.A., Cahill, D.G.: Transport of nanoscale latex spheres in a temperature gradient. Langmuir **21**(12), 5317–5323 (2005)
64. Jiang, H.-R., Wada, H., Yoshinaga, N., Sano, M.: Manipulation of colloids by a nonequilibrium depletion force in a temperature gradient. Phys. Rev. Lett. **102**(20), 208301 (2009)
65. Wei, N., Wang, H.-Q., Zheng, J.-C.: Nanoparticle manipulation by thermal gradient. Nanoscale Res. Lett. **7**(1), 154 (2012)
66. Tsuji, T., Sasai, Y., Kawano, S.: Thermophoretic manipulation of micro-and nanoparticle flow through a sudden contraction in a microchannel with near-infrared laser irradiation. Phys. Rev. Appl. **10**(4), 044005 (2018)
67. Würger, A.: Thermal non-equilibrium transport in colloids. Rep. Progress Phys. **73**(12), 126601 (2010)
68. Tsuji, T., Iseki, H., Hanasaki, I., Kawano, S.: Negative thermophoresis of nanoparticles interacting with fluids through a purely-repulsive potential. J. Phys.: Condensed Matter **29**(47), 475101 (2017)
69. Smallenburg, F., Boon, N., Kater, M., Dijkstra, M., van Roij, R.: Phase diagrams of colloidal spheres with a constant zeta-potential. J. Chem. Phys. **134**(7), 074505 (2011)
70. Chakrabarti, J., Dutta, S.: Analytical form of forces in hydrophobic collapse. Chem. Phys. Lett. **620**, 109–113 (2015)
71. Pellicer, J., García-Morales, V., Guanter, L., Hernández, M.J., Dolz, M.: On the experimental values of the water surface tension used in some textbooks. Am. J. Phys. **70**(7), 705–709 (2002)
72. Kell, G.S.: Density, thermal expansivity, and compressibility of liquid water from 0. deg. to 150. deg.. correlations and tables for atmospheric pressure and saturation reviewed and expressed on 1968 temperature scale. J. Chem. Eng. Data **20**(1), 97–105 (1975)

73. Beurton, J., Lavalle, P., Pallotta, A., Chaigneau, T., Clarot, I., Boudier, A.: Design of surface ligands for blood compatible gold nanoparticles: effect of charge and binding energy. Int. J. Pharmaceut. **580**, 119244 (2020)
74. Das, P.K.: Effect of temperature on zeta potential of functionalized gold nanorod. Microfluidics Nanofluidics **21**(5), 95 (2017)
75. Lechner, W., Dellago, C.: Accurate determination of crystal structures based on averaged local bond order parameters. J. Chem. Phys. **129**(11), 114707 (2008)
76. Karthickeyan, D., Joshi, R.G., Tata, B.V.R.: FCC-HCP coexistence in dense thermo-responsive microgel crystals. J. Chem. Phys. **146**(22), 224503 (2017)
77. Du, H., Wickramasinghe, R., Qian, X.: Effects of salt on the lower critical solution temperature of poly (n-isopropylacrylamide). J. Phys. Chem. B **114**(49), 16594–16604 (2010)
78. de Oliveira, T.E., Mukherji, D., Kremer, K., Netz, P.A.: Effects of stereochemistry and copolymerization on the LCST of PNIPAm. J. Chem. Phys. **146**(3), 034904 (2017)
79. Kolker, J., Harrer, J., Ciarella, S., Rey, M., Ickler, M., Janssen, L.M.C., Vogel, N., Löwen, H.: Interface-induced hysteretic volume phase transition of microgels: simulation and experiment. Soft Matter **17**(22), 5581–5589 (2021)
80. Consiglio, G., Forte, G.: Molecular dynamics study of coil-to-globule transition in a thermo-responsive oligomer bound to various surfaces: hydrophilic surfaces stabilize the coil form. Phys. Chem. Chem. Phys. **20**(47), 29754–29763 (2018)
81. Cui, K., Gong, J.P.: Aggregated structures and their functionalities in hydrogels. Aggregate **2**(2), e33 (2021)
82. Hu, Z., Schall, P. et al.: Single crystal growth and anisotropic crystal-fluid interfacial free energy in soft colloidal systems. Phys. Rev. E **84**(1), 011607 (2011)
83. Poon, W.C.K., Brown, A.T., Direito, S.O.L., Hodgson, D.J.M., Le Nagard, L., Lips, A., MacPhee, C.E., Marenduzzo, D., Royer, J.R., Silva, A.F. et al.: Soft matter science and the covid-19 pandemic. Soft Matter **16**(36), 8310–8324 (2020)
84. Sharma, S., Pinto, R., Saha, A., Chaudhuri, S., Basu, S.: On secondary atomization and blockage of surrogate cough droplets in single-and multilayer face masks. Sci. Adv. **7**(10), eabf0452 (2021)
85. World Health Organization: Advice on the use of masks in the context of covid-19: interim guidance, 6 Apr 2020. Technical documents (2020)
86. Karmacharya, M., Kumar, S., Gulenko, O., Cho, Y.-K.: Advances in facemasks during the covid-19 pandemic era. ACS Appl. Bio Mater. **4**(5), 3891–3908 (2021)
87. Ullah, S., Ullah, A., Lee, J., Jeong, Y., Hashmi, M., Zhu, C., Joo, K.I., Cha, H.J., Kim, I.S.: Reusability comparison of melt-blown vs nanofiber face mask filters for use in the coronavirus pandemic. ACS Appl. Nano Mater. **3**(7), 7231–7241 (2020)
88. Karabulut, F.N.H., Höfler, G., Ashok Chand, N., Beckermann, G.W.: Electrospun nanofibre filtration media to protect against biological or nonbiological airborne particles. Polymers **13**(19), 3257 (2021)
89. Liu, G., Nie, J., Han, C., Jiang, T., Yang, Z., Pang, Y., Xu, L., Guo, T., Bu, T., Zhang, C. et al.: Self-powered electrostatic adsorption face mask based on a triboelectric nanogenerator. ACS Appl. Mater. Interfaces **10**(8), 7126–7133 (2018)
90. Ghatak, B., Banerjee, S., Ali, S.B., Bandyopadhyay, R., Das, N., Mandal, D., Tudu, B.: Design of a self-powered triboelectric face mask. Nano Energy **79**, 105387 (2021)
91. Konda, A., Prakash, A., Moss, G.A., Schmoldt, M., Grant, G.D., Guha, S.: Aerosol filtration efficiency of common fabrics used in respiratory cloth masks. ACS Nano **14**(5), 6339–6347 (2020)
92. Hao, Z., Wu, J., Wang, C., Liu, J.: Electrospun polyimide/metal-organic framework nanofibrous membrane with superior thermal stability for efficient pm2. 5 capture. ACS Appl. Mater. Interfaces **11**(12), 11904–11909 (2019)
93. Tang, Z., Kong, N., Zhang, X., Liu, Y., Hu, P., Mou, S., Liljeström, P., Shi, J., Tan, W., Kim, J.S. et al.: A materials-science perspective on tackling covid-19. Nat. Rev. Mater. **5**(11), 847–860 (2020)

94. Opaskar, C.G., Krimm, S.: Tetrahedral approximation in normal coordinate calculations of high polymers. Spectrochimica Acta Part A: Mol. Biomol. Spectrosc. **21**, 1165–1166 (1965)
95. Du, W., Iacoviello, F., Fernandez, T., Loureiro, R., Brett, D.J.L., Shearing, P.R.: Microstructure analysis and image-based modelling of face masks for covid-19 virus protection. Commun. Mater. **2**(1), 1–10 (2021)
96. Essa, W.K., Yasin, S.A., Saeed, I.A., Ali, G.A.M.: Nanofiber-based face masks and respirators as covid-19 protection: a review. Membranes **11**(4), 250 (2021)
97. Yim, W., Cheng, D., Patel, S.H., Kou, R., Meng, Y.S., Jokerst, J.V.: Kn95 and n95 respirators retain filtration efficiency despite a loss of dipole charge during decontamination. ACS Appl. Mater. Interfaces **12**(49), 54473–54480 (2020)
98. Choi, S., Jeon, H., Jang, M., Kim, H., Shin, G., Koo, J.M., Lee, M., Sung, H.K., Eom, Y., Yang, H.-S. et al.: Biodegradable, efficient, and breathable multi-use face mask filter. Adv. Sci. **8**(6), 2003155 (2021)
99. Thompson, A.P., Aktulga, H.M., Berger, R., Bolintineanu, D.S., Brown, W.M., Crozier, P.S., in 't Veld, P.J., Kohlmeyer, A., Moore, S.G., Nguyen, T.D., Shan, R., Stevens, M.J., Tranchida, J., Trott, C., Plimpton, S.J.: LAMMPS – a flexible simulation tool for particle-based materials modeling at the atomic, meso, and continuum scales. Comp. Phys. Commun. **271**, 108171 (2022)
100. Petrowsky, M., Frech, R.: Temperature dependence of ion transport: the compensated Arrhenius equation. J. Phys. Chem. B **113**(17), 5996–6000 (2009)
101. Lion, T.W., Allen, R.J.: Osmosis in a minimal model system. J. Chem. Phys. **137**(24), 244911 (2012)
102. Song, Y., Xu, F., Wei, M., Wang, Y.: Water flow inside polamide reverse osmosis membranes: a non-equilibrium molecular dynamics study. J. Phys. Chem. B **121**(7), 1715–1722 (2017)
103. Ridgway, H.F., Orbell, J., Gray, S.: Molecular simulations of polyamide membrane materials used in desalination and water reuse applications: recent developments and future prospects. J. Membrane Sci. **524**, 436–448 (2017)
104. Geise, G.M., Lee, H.-S., Miller, D.J., Freeman, B.D., McGrath, J.E., Paul, D.R.: Water purification by membranes: the role of polymer science. J. Polymer Sci. Part B: Polymer Phys. **48**(15), 1685–1718 (2010)
105. Lau, W.J., Ismail, A.F., Misdan, N., Kassim, M.A.: A recent progress in thin film composite membrane: a review. Desalination **287**, 190–199 (2012)
106. Mehta, R., Brahmbhatt, H., Bhojani, G., Mukherjee, M., Bhattacharya, A.: Poly (piperizinamide) with copper ion composite membranes: application for mitigation of Hexaconazole from water and combat microbial contamination. J. Hazard. Mater. **376**, 102–111 (2019)
107. Kushida, W., Gonzales, R.R., Shintani, T., Matsuoka, A., Nakagawa, K., Yoshioka, T., Matsuyama, H.: Organic solvent mixture separation using fluorine-incorporated thin film composite reverse osmosis membrane. J. Mater. Chem. A **10**(8), 4146–4156 (2022)
108. Ermak, D.L.: A computer simulation of charged particles in solution. I. Technique and equilibrium properties. J. Chem. Phys. **62**(10), 4189–4196 (1975)
109. Eslami, H., Khanjari, N., Muller-Plathe, F.: A local order parameter-based method for simulation of free energy barriers in crystal nucleation. J. Chem. Theory Comput. **13**(3), 1307–1316 (2017) PMID: 28195473.
110. Boattini, E., Dijkstra, M., Filion, L.: Unsupervised learning for local structure detection in colloidal systems. J. Chem. Phys. **151**(15), 154901 (2019)
111. Schwarz, G.: Estimating the dimension of a model. Ann. Stat. **6**, 461–464 (1978)
112. Baudry, J.-P., Raftery, A.E., Celeux, G., Lo, K., Gottardo, R.: Combining mixture components for clustering. J. Comput. Graph. Stat. **19**(2), 332–353 (2010)
113. Pedregosa, F., Varoquaux, G., Gramfort, A., Michel, V., Thirion, B., Grisel, O., Blondel, M., Prettenhofer, P., Weiss, R., Dubourg, V. et al.: Scikit-learn: machine learning in python. J. Mach. Learn. Res. **12**, 2825–2830 (2011)

Soft Materials for Sensor Applications

Amit Kumar Sharma, Garima Agrawal, Rohit, and Balbir Singh Kaith

Abstract Hydrogels are highly cross-linked, hydrophilic, and water-swellable polymeric networks which can be synthesized by the chemical or physical cross-linking of hydrophilic molecules. They can recognize the impact of external stimuli and can produce electrical, thermal, and optical signals to form sensors and diagnosis tools. Depending upon their physical or chemical structure, flexibility, biocompatibility, durability, and sensitivity, they can be integrated into different systems. This chapter mainly highlights the different types of physiochemical sensing mechanisms and recent advances in different types of sensors, such as temperature sensors, pressure/strain sensors, gas sensors, pH/ionic strength sensors, and humidity sensors. Moreover, some important biomedical applications of hydrogel-based sensors are highlighted. In addition to the recent advances and developments in hydrogel-based sensors, some drawbacks of conventional hydrogel sensors and their future perspectives are also disclosed in this chapter.

Keywords Hydrogels · Strain sensor · Biocompatible · Stimuli-responsive · Cross-linking

---

A. K. Sharma (✉)
Department of Chemistry, University Institute of Sciences, Chandigarh University, Mohali, Punjab, India

University Centre for Research and Development, Chandigarh University, Mohali, Punjab, India

G. Agrawal
School of Chemical Sciences, Indian Institute of Technology Mandi, Mandi, Himachal Pradesh, India

Indian Knowledge System and Mental Health Applications Centre, IIT Mandi, Mandi, Himachal Pradesh, India

Technology Innovation Hub in Human-Computer Interaction (iHub), Mandi, Himachal Pradesh, India

Rohit · B. S. Kaith
Department of Chemistry, Dr B R Ambedkar National Institute of Technology Jalandhar, Jalandhar, Punjab, India

V. Kumar, Y. K. Mishra (eds.), *Soft Materials for Functional Applications*, Advanced Structured Materials 225, https://doi.org/10.1007/978-981-97-9468-3_3

1 Introduction

In our daily lives, we come across in touch with numerous devices that are extremely important to sense and accordingly respond to the distinct inputs from their external environment. Such materials are termed as "sensors." It is derived from the Latin word *sensus* meaning "to sense something" [1]. The idea of sensing originated from the traditional five human body senses, i.e., touch, smell, hear, taste, and sight. All these human senses first receive signals from the external environment and then respond accordingly. The classification of sensors is also inspired by the human senses, as touch, hearing, and sight represent physical stimuli, whereas taste and smell respond to chemical stimuli [2]. Physical sensors can be defined as devices that can analyze and deliver information about the physical properties of the materials, whereas chemical sensors can transform the chemical information of the system, which may be analyzed into an analytically important signal [3]. On this basis, the sensors can be categorized as physical or chemical sensors. They can also be classified as electrochemical sensors, optical sensors, electric sensors, radiation-sensitive sensors, mass-sensitive sensors, and magnetic sensors.

Over the last two decades, tremendous research has been carried out to design hydrogel-based soft materials that can be applied in different types of sensing applications [4, 5]. Hydrogels are highly cross-linked polymer structures, capable of imbibing a large amount of water molecules and releasing it slowly under the impact of external stimuli such as pH of the swelling medium, temperature, concentration of reactants, light, humidity, and applied electric or magnetic field [6] (Fig. 1). The important characteristics of hydrogels such as porosity, flexibility, consistency, flexibility, and biocompatibility make them a suitable choice of materials to be used in different biomedical applications. Such types of materials can be classified as physically cross-linked hydrogels or chemically cross-linked hydrogels depending upon the type of cross-linking method followed during their synthesis [7]. The network structure of chemically cross-linked hydrogels is formed by covalent bonds, and as a result, they don't dissolve easily in water, whereas there exists a transient junction with dynamic cross-linking in the case of physically cross-linked hydrogels. Physically cross-linked hydrogels can show non-covalent interactions such as hydrogen bonding and hydrophobic or electrostatic interactions [7]. Hydrogels can be classified as natural, synthetic, or hybrid, depending on their source of origin. Nowadays, hydrogels are gaining increasing attention in different fields for various applications that may require different dimensions and configurations of the synthesized hydrogel matrix [8]. For example, fiber-based gels or microfibers depict a valuable solution to soft robotics, whereas sheet-shaped hydrogels demand proper adhesion to the surface of the skin [9, 10]. The properties of hydrogels can be easily tuned during their synthesis. They are also environmental stimuli respondents, and this characteristic is mainly governed by their specific mess size, framework ionic charge, and cross-linking degree [11]. All these specific features of the hydrogels are used to make specific sensors that can be used in different fields such as horticulture, environmental, agriculture, and biomedical sectors.

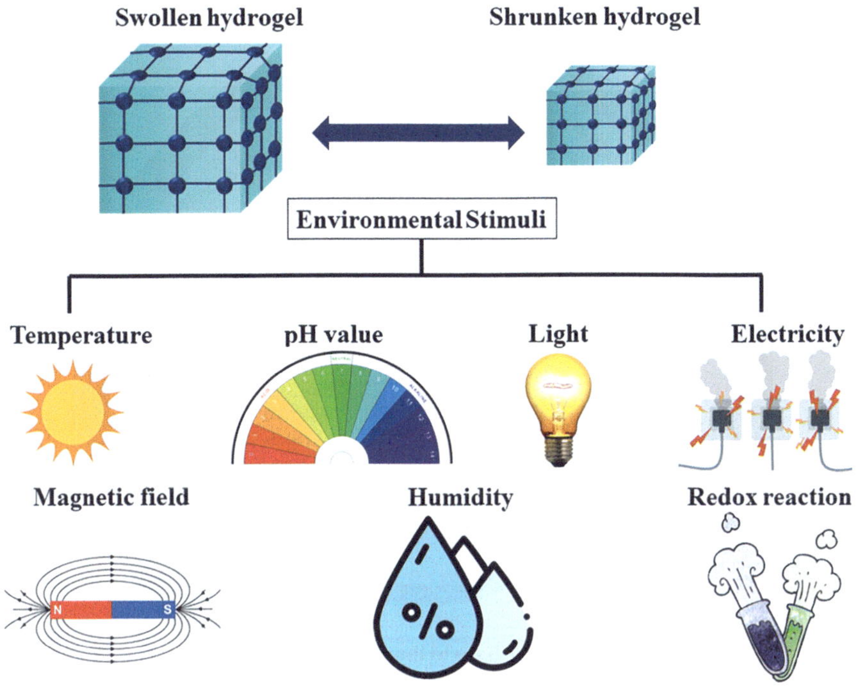

Fig. 1 Different types of environmental factors governing the stimuli-responsive behavior of hydrogels

2 Classification of Physicochemical Sensing Mechanism

The sensing mechanism of any sensor usually depends upon the specific recognition of the target analyte that can induce some specific changes in its properties such as volume change, color change, change in shape/size, etc. Hydrogels being stimuli-responsive can be used in different types of sensing applications, and different sensing mechanisms may be operational in such devices. Nowadays, self-healing hydrogel-based sensors are being used in different biomedical applications (Fig. 2).

2.1 *Temperature Sensors*

Temperature-responsive sensors are extremely important in industries during the food and chemical production processes. Temperature-responsive hydrogel sensors are being utilized widely because of their unique and special characteristic features [12]. They can tune their properties as per the desired requirement of the system, and they are also biocompatible as a result of which they can be used inside the body

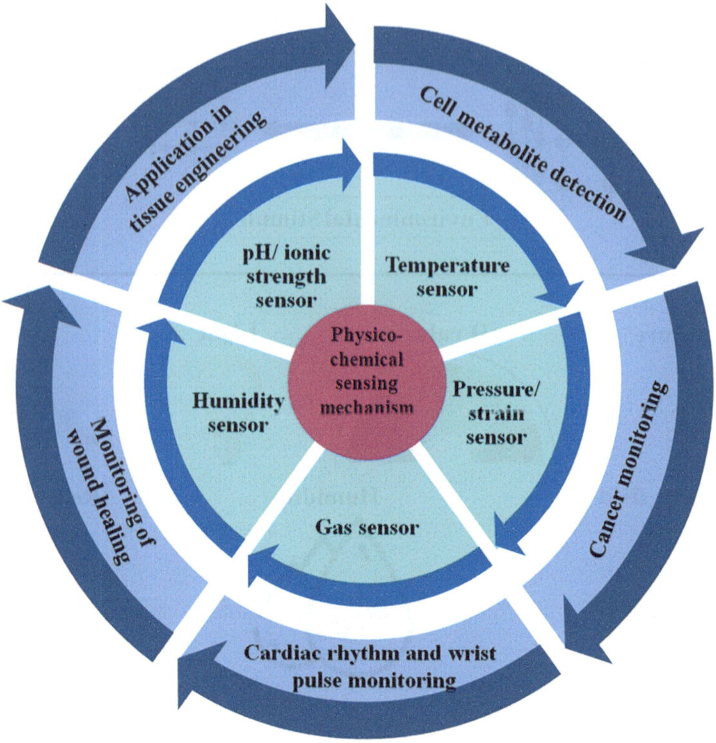

Fig. 2 Illustration of different physicochemical sensing mechanisms and application of hydrogel-based sensors in the biomedical field

for detecting the movement of different body organs such as the kidneys, heart, lungs, etc. [13]. Intelligent polymerized crystalline colloidal arrays and hydrogel photonic crystals are some of the most important materials used in temperature-sensitive systems. In photonic crystals, the dielectric constant may vary periodically and indicates that there is a formation of a photonic band structure. This change in photonic band structure under the effect of external stimuli allows photonic crystals to work as a thermo-responsive hydrogel [14]. Similarly, intelligent polymerized crystalline colloidal arrays of polymer spheres are useful for detecting the changes occurring in any biological system [15]. The assembly of colloidal spheres can work similarly to photonic crystals. Wu et al. [16] designed a thin-film sandwich hydrogel temperature sensor using κ-carrageenan, polyacrylamide, and *N*,*N*′-methylenebisacrylamide. The % transmittance of the sample was found to be more than 90%. The fast response time (0.19 s), a wide-ranging detection range (28–95.3 °C), high resolution (0.8 °C), small recovery time (0.08 s), and excellent stability of the fabricated device can comfortably attach to the human skin and analyze the skin temperature, along with the real-time monitoring of various human motions.

Pand et al. [17] designed a flexible, bearable, temperature-responsive sensor using polyvinylpyrrolidone (PVP), poly(*N*-isopropylacrylamide-*co*-acrylamide), *N,N*-methylene diacrylamide (MBAA), tannic acid (TA), and Fe^{3+} ions. The incorporation of TA, PVP, and Fe^{+} endowed the gel matrix with exceptional conductivity and stretchability. Moreover, the hydrogel can be used to control the environmental temperature by incorporating poly(*N*-isopropylacrylamide-*co*-acrylamide) with a lower critical solution temperature. This can further be applied as a tool to detect tissue hyperthermia or fever as a wearable temperature sensor in the human body. In another study, Lee et al. [17] designed a hydrogel network system using polyvinyl alcohol, glycerol, ethylene glycol, and $CaCl_2$ aqueous solution. Glycerol was used as an antifreeze humectant. Silver nanofibers were incorporated inside the gel matrix to add the moisturizing and temperature-sensitive properties inside the hydrogel network. The conductivity of synthesized hydrogel was found to show a linear change when there is a change in temperature (5–55 °C). There are many other examples of hydrogels which can be used as temperature sensors in healthcare applications.

2.2 Pressure/Strain Sensors

In preclinical practices, the monitoring of artery pulses such as monitoring of cardiac rhythm, monitoring of blood pressure, and diagnosis of possible cardiovascular disease are extremely important parameters [18]. Nowadays, such type of monitoring is feasible by the use of hydrogel-based flexible strain sensors [19, 20]. They can closely fit themselves on the surface/curvature of the skin and can monitor the pulse rate linked with cardiac rhythm, wrist pulse, and blood pressure. In the literature, there are many examples of hydrogel-based flexible strain sensors that can monitor joint motions. For example, Li et al. [21] fabricated a double network hydrogel-based flexible strain sensor using agar, acrylic acid, and Fe^{3+} ions through the integration of ionic and covalent cross-linking. The synthesized double network demonstrated excellent stretchability (i.e., up to 3174.3% break strain potential, gauge factor = 0.83 under 1000% strain). Moreover, the self-healing efficiency and 3D printability of the hydrogel network demonstrated its potential for fabricating soft robots, strain sensors, and electronic skin. Double network hydrogels usually contain one brittle network which is tightly cross-linked and one loosely cross-linked ductile matrix. Chen et al. [22] synthesized polyacrylamide and agar-based double network hydrogel containing one brittle network containing hydrogen bonds and one ductile network containing covalent bonds. In most of the double network-based hydrogel matrices, the breakage of covalent cross-linking in case of a brittle network is irreversible and can lead to the permanent destruction of the gel matrix. Therefore, the double network hydrogels can be synthesized via the introduction of reversible bonds to add some extent of physical cross-linking inside the brittle matrix which can further add self-healing capability [23]. Xia et al. [24] 2018 synthesized physically cross-linked double network hydrogel by using sodium alginate, acrylamide, and sodium dodecylsulfate. The designed network showed 2422% fracture strain, 935 kPa fracture stress, 54 kPa elastic modulus, and 10,075 kJ/m^3 toughness. In addition to this, the addition of Ca^{2+} to the matrix introduced good electrical conductivity and expanded its excellent potential to be used as a soft wearable strain sensor in multiple applications. Table 1 summarizes some recent self-healable, soft sensors showing their sensing mechanism and other target parameters.

The most important step while designing an appropriate strain sensor is the choice of the starting material, selection of the fabrication method, and properties of the designed sample. Many conducting materials including nanoparticles [41], nanofibers [42], nanowires [43], nanotubes [44], graphenes [45], conducting polymers [46], hybrid nanostructures [47], MXenes [48], and carbon blacks [49] have been used for developing such types of flexible strain sensors. Sometimes, the nanomaterials are directly incorporated inside the stretchable cross-linked hydrogel matrix. The suitable choice of conductive nanomaterial and its possible interactions with the gel matrix significantly affect its sensing properties. Moreover, reliable, facile, low-cost, and scalable manufacturing strategies are required for industrial-scale production of wearable strain sensors. The fabrication methods for the production of polymer-based strain sensors may include chemical synthesis, liquid-phase blending, transfer patterning, coating and sputtering, electrospinning, filtration, printing, and laser micromachining. The strain sensors can be categorized as resistive/capacitive/optical strain sensors. The sensing mechanism operating in case of capacitive and optical sensors may be their dimensional change while stretching. Similarly, the sensing mechanism operating in case of resistive types of stretchable strain sensors may include the surface interactions with the supporting material, and they may be designed for disconnection of nanomaterials, creation of controlled microcracks, and tunneling effects arising in the sensing film [50].

Metal nanoparticles possess unique optical, magnetic, and conductive properties. The incorporations of such types of nanoparticles (NPs) inside the gel matrix form NPs/hydrogel nanocomposite which further brings the conformational change inside the surface bounded stimuli sensitive polymers. Finally, the change in optical property of the material improves its mechanical properties and makes them a suitable choice to be used in different sensing platforms. For example, silver nanoparticles are most widely incorporated inside the cross-linked hydrogels containing –CN, –SH, and $-NH_2$ functional groups that can form dynamic covalent bonds responsible for different stimulus activities in strain sensors. The silver nanoparticles can show self-assembly in an amine rich environment (e.g., amine rich chitosan hydrogel) through horseradish peroxidase immobilization [51]. In addition to silver nanoparticles, magnetic Fe_3O_4 also plays an important role for the development of pH sensor that can display a reversible and repeatable response to any change in pH. For example, Song et al. [52] developed a pH sensor using Fe_3O_4 magnetic nanoparticles functionalized with poly(methacrylic acid-*co*-acrylamide) hydrogel. The sensor showed sufficient sensitivity to analyze 0.1 unit alteration in pH. Moreover, the fabricated sensor can be configured for wireless observations and implantations.

Table 1 Examples of recent self-healable strain sensors with their targeted parameters and sensing mechanism

Soft material	Signal	Self-healing mechanism	Sensing mechanism	References
Graphene oxide loaded self-healable hydrogel	Pressure	Hydrogen bonding	Piezoelectric material containing PVDF nanofibers For the generation of current	[25]
Copper and zinc loaded hydrogel	Strain	Hydrogen bonding	Battery based sensor	[26]
Self-healing conducting polymer blend doped with polypyrrole	Strain	Hydrogen bonding	Piezoresistance	[27]
Self-healing hydro-/organogel	Temperature	Hydrogen bonding and ionic interaction	Change in resistance with temperature	[28–30]
MXene and polyvinyl alcohol-based hydrogel	Strain	Hydrogen bonding	Change in capacitance with strain	[31]
Self-healing polymer blend	Chemical sensor	Hydrogen bonding	Change of organic field transistor characteristics	[32]
Self-healing polymer incorporated with carbon nanomaterial	Temperature	Hydrogen bonding and van der Waals interactions	Change in resistance with temperature	[33, 34]
Self-healable ionic hydrogel conductor	Touch	Hydrogen bonding	Electrostatic and triboelectric induction effects	[35, 36]
Self-healing hydrogel derived from oxidized dextran and chitosan and embedded with hollow nanospheres of CeO_2/MnO_2 and glucose oxidase	Glucose	Schiff base bonding	Cyclic voltammetry and amperometry based sensor	[37]
PEDOT:PSS-ZnO heterojunction coated on an agarose and PVA based self-healing hydrogel	UV light	Hydrogen bonding	Production of electric carriers under the exposure of UV light	[38]
Self-healing bio-based polymer incorporated with PANI-modified conductive carbon fibers	pH	Hydrogen bonding	Electrochemical sensor	[39]
Self-healing gelatin hydrogels loaded with AgNWs	Stain/pressure	Hydrogen bonding	Battery based sensor	[40]

2.3 Gas Sensor

Gas sensing devices are extremely important in environmental, industrial, military, and public safety applications [53]. Hydrogel-based soft materials can be used mainly for three purposes. They can be employed as a passive protective coating in case of sensor/electrode which can be related to their protein-repellent and anti-adhesive characteristics [54]. Alternatively, they can be used for the sensing of carbon dioxide which is related to their stimuli-responsive swelling characteristics [3, 55]. Finally, the hydrogel-based soft materials can be modified with the help of some gas-sensitive molecules including fluorescent dyes [56]. This is possible in case of pH-sensitive hydrogel, where the CO_2 diffusion shows a change in the pH of the electrolyte solution. For example, Herber et al. [57] designed pH-sensitive hydrogel disk that can swell/deswell depending upon the change in pH of the swelling medium and was utilized for the detection of CO_2. Another example of hydrogel-based gas sensor is the development of polyaniline-coated CNT films which are sensitive to ammonia and can be employed for public safety/explosive detections. On a self-healing polydimethylsiloxane and 1,3,5-triformylbenzene platform (PDMS-TFB), the resulting hydrogel sensor was able to show both self-healing and sensor properties [58]. The response of the synthesized sensor to ammonia was $23 \pm 3\%$ at 5 ppm concentration when measured up to five consecutive cycles of testing. In another study, a high capacitance self-healing PEI/PHPMA blend was fabricated, and it was used as a gate-insulating material for designing other organic field transistor-based gas sensors [32]. Such sensors showed attractive properties such as very low detection limit, flexibility, printability, low-voltage operation, and self-healing ability.

Hydrogel-based gas sensors are usually flexible, biocompatible, and portable. They also allow the gas and vapors to permeate through their pores easily. In addition, the deformations allow the sensor to fit directly on the curved human skin and allow the movement of mussels without any mechanical damage. The hydrogel-based sensors can show sensitivity against O_2, CO_2, NH_3, and NO_2. The hydrogel-based gas sensors are important in various practical applications such as respiratory analysis, the development of face masks, electronic skin, environmental monitoring, wireless monitoring, and medical health [59]. The operational functions of wearable vapor sensors may require complex environments, such as harsh and dry conditions and subzero temperatures, depending upon the living environment of the wearer. However, in the case of traditional hydrogel-based sensors, the low operational temperature may result in the freezing of water molecules into ice, brittleness, stiffness, and loss of working ability. Therefore, such parameters should be considered while designing a hydrogel-based gas sensor. The evaporation of water molecules from the hydrogel surface can be prevented by incorporating some elastic materials. For example, Yuk et al. [60] designed a hybrid hydrogel elastomer inspired by lymphatic and blood vessels in mammalian skins. The fabricated device increases the water retention ability along with stretchability, which further improves its working ability in a dry atmosphere. Further, the proposed method opened a new pathway for synthesizing flexible bioelectronic devices.

2.4 Humidity Sensors

After decades of research, the antifreeze agents and humectants have made excellent progress. The research on the antifreeze and water retention capacity of hydrogels has attracted the attention of scientists in this particular area. Polyacrylamide-based hydrogels have shown remarkable water retention properties and are employed for the development of humidity sensors. They are extremely demanding because the presence of humidity in our environment can reduce the performance of electronic devices due to its interference with most of the environmental/electric signals [61, 62]. The humidity in the environment can also reduce the shelf life of food products. Therefore, it is very important to design humidity sensors to reduce the degradation of food products. Hydrogel-based soft sensors play a very crucial role in designing such sensors. Guo et al. [63] synthesized the first self-healing humidity sensor by connecting single-walled carbon nanotubes with a polymeric network of poly(2-hydroxymethyl methacrylate). The self-healing ability of the composite hydrogel was retained even in a very dry environment (20% relative humidity). Due to the hydrophobicity of the hydroxyl groups, the composite hydrogel exhibited high humidity adsorption capacity under normal conditions. Ionically conductive self-healing polymers can be used to fabricate transparent humidity sensors. In this approach, the adsorption of water can reduce the concentration of ions inside the polymers, which leads to an increase in the conductor's resistance. A transparent, ultra-stretchable, self-healing humidity sensor was synthesized using polyacrylamide/k-carrageenan double network organohydrogels [64]. To stimulate anti-drying and humidity-sensitive properties, glycerol and hygroscopic ethylene glycol were introduced inside the hydrogel matrix. The hydrogel displayed a wide relative humidity detection range (4–90%), excellent stability, fast response time (0.27 s), and excellent recovery time (0.3 s), along with remarkable stretchability (1225%). An increase in relative humidity from 4 to 90% was found to increase the hydrogel conductance by 453 times. The hydrogel-based self-healable humidity sensor adds additional advantages, i.e., it can work under different mechanical deformations such as twisting, bending, and stretching without any loss in sensor efficiency, which is a crucial factor for wearable applications.

Many hydrogel-based novel materials are being discovered that can show stimuli-responsive properties, which are important for the generation of humidity sensors in different biomedical applications. Regarding sensing mechanism, the humidity sensor uses the swelling properties of hydrogel that can induce change in reflectance, optical path length, and tendency of ion transportation. For example, Liu et al. [65] showed that the absorption of moisture on the hydrogel coating changes the refractive index that further causes a shift in wavelength, and therefore, the relative humidity can be measured with ±2.3% precision.

2.5 pH/Ionic Strength Sensors

Stimuli-responsive hydrogels play an important role in sensing any change in pH or changes in the concentration of ions [66]. Richter et al. [67] published a comprehensive work on the physical and hypothetical characteristics of pH-responsive hydrogels and their design rules. Zhao et al. [68] *2018 synthesized hydrogel-based optical fiber surface plasmon resonance (SPR) sensor,* which was recognized with a multimode–single mode–multimode fiber splicing structure (MMF–SMF–MMF). On increasing the pH, the refractive index of hydrogel decreases and results in the resonance wavelength experiencing a blue shift. Further, the resulting sensor was able to detect a wide range of pH values (1–12) along with remarkable stability, great reproducibility, and excellent sensitivity (13 nm/pH). The working of hydrogel-based ion sensors is similar to that of pH-sensitive sensors. Sannino et al. [69] designed cellulose-based hydrogels on quartz crystal microbalance plates, which produce a thin superabsorbent layer. Polyelectrolyte carboxymethylcellulose induces the sorption sensitivity to the modifications in pH and ionic strength of the external solution. The synthesized thin layer of hydrogel film reduces the water diffusion through the material and enhances fast swelling response. Tokuyama et al. [70] synthesized zirconia nanoparticle-loaded hydrogel coated with quartz crystal microbalance to detect arsenic in an aqueous solution. Riboflavin functionalized hydrogel based on sodium alginate and dextrin was successfully synthesized for effectively removing and sensing copper and nickel ions in real water samples. The sensitivity of the designed sensor was found to be 0.257 μM and 0.445 μM in the case of nickel and copper ions, respectively [5]. Similarly, Sharma et al. [71] fabricated a hydrogel film of polyvinyl alcohol, quinine, and dextrin that can effectively sense the presence of lead and copper ions in real water samples. The detection limit was found to be 0.0521 μM and 0.0421 μM in case of lead and copper ions, respectively. The primary advantages of hydrogel-based sensors for detecting heavy metal ions are associated with their specific characteristic features such as stimuli-responsive properties, ease of preparation, efficiency, ease of degradation, and reusability of the synthesized material.

Nanocomposite hydrogels are usually designed for the developing colorimetric sensors. In different sensing platforms, the addition of magnetic Fe_3O_4 nanoparticles enhances the sensing ability of the hydrogel. For example, Jia et al. [72] synthesized photonic hydrogels using polyacrylamide, 2-hydroxyethyl methacrylate, and Fe_3O_4 nanocrystals through a one-step polymerization reaction. The hydrolyzed Fe_3O_4 based hydrogel system showed full color reversible switching properties by adjusting the ionic strength and pH of the aqueous solution. Moreover, the active –COOH functional groups responsible for sensing were controlled by regulating the number of poly 2-hydroxyethyl methacrylate units. The pH-responsive hydrogels are important in drug delivery applications. Shabnam et al. [73] designed multi-stimuli-responsive hydrogel beads using chitosan and carboxymethyl cellulose cross-linked with epichlorohydrin. The fabricated beads were suitable for the controlled release of doxorubicin and metronidazole. The incorporation of Fe_3O_4

nanoparticles enhances the release efficiency of hydrogel by 65.5% within 12 h in neutral medium, whereas switching the pH from acidic to basic medium, the efficiency for metronidazole drug release was found to enhance by 71%. The cytotoxicity of the beads was tested against Balb/c 3T3 and Hela cells through MMT assay. They were found to be least toxic compared to the free drugs doxorubicin and metronidazole.

3 Biomedical Applications of Hydrogel-Based Sensors

The unique characteristics of hydrogels, such as porosity, stiffness, stress relaxation, ease of fabrication, tissue compatibility, and mechanical strength, enable them to be used in biomedical applications. The hydrogel-based sensors are recently used for detecting cell metabolite, cancer cell monitoring, cardiac rhythm and wrist pulse monitoring, monitoring of wound healing, and tissue engineering applications. Such systems are generated through the process of cross-linking, which is also termed gelation. Chemical, physical, or hybrid bonding is possible during the synthesis of hydrogels through different routes such as solution casting, solution mixing, free radical mechanism, bulk polymerization, radiation method, and interpenetrating network formation. The gelation process may be carried out through physical cross-linking or chemical cross-linking methods. During physical cross-linking, different types of weak as well as strong intermolecular bonding may occur between the cross-linking chains, whereas covalent bonding is more predominant in the case of chemical cross-linking methods. Covalent bonding involves the reaction of different polymers and cross-linkers in the presence of free radicals. The common examples of cross-linkers involved in the synthesis of hydrogel include genipin, *N*,*N*′-methylene bisacrylamide, formaldehyde, diethyl squarate, glycidyl ether, glutaraldehyde, etc. Hydrogels find significant application perspective in tissue engineering, soft robotics, and flexible electronics. Hydrogel-based nanocomposites can show promising applications in biomedical fields because they can improve toughness, mechanical strength, and electric and magnetic properties. This chapter provides an overview of hydrogel-based devices for different sensing applications.

3.1 Cell Metabolite Detection

There are many disadvantages of classical methods used to detect cell metabolite. Media culture collection and cultivation of intermittent cells are among them. Usually, a large quantity of reagents/cells and costly procedures are required. Moreover, it is impossible to extract complete information related to the time fruition and dynamics of the analyzed system. In this direction, hydrogel-based smart materials are being discovered to monitor the changes observed in cell sites in our biological system [74]. For example, Zhang et al. [75] synthesized an enzyme-entrapped hydrogel-based biosensor for the colorimetric detection of metabolites. The detection limits in the case of glucose and hydrogen peroxide were 1.6 mM and 8.9 μM, respectively. The analysis was simple and the detection was achieved very rapidly. Another example of such technology was provided in the work carried out by Yan et al. [76]. They developed an enzyme-based optical biosensor for detecting hydrogen peroxide produced by stimulated macrophages. The designed system consists of a polyethylene hydrogel microstructure. The Amplex Red and horseradish peroxidase were entrapped inside the cross-linked gel matrix to create a sensing platform of the glass surface. Similarly, Yan et al. [77] designed a biosensor platform based on conductive polymer hydrogel reinforced with platinum nanoparticles on a glassy carbon electrode. The designed system exhibited excellent performance with rapid response and for detecting triglyceride, cholesterol, and uric acid. The major limitation of hydrogel-based biosensors is their long-term stability which creates a barrier for their uses over some time.

A new concept of designing stimuli-responsive hydrogels uses the tissue generation of reactive oxygen species (ROS). The reactive oxygen species may include hydroxyl radicals, hydrogen peroxide, and anion radicals. ROS plays an important role during cell signaling pathways and is abundantly found at the sites of tissue healing and inflammation [78]. For example, ROS are produced during chemotherapy and radiotherapy, which favors apoptosis and inhibits the growth of cancer cells. Monitoring the metabolites is of extreme importance in embryo screening, drug screening, and cancer research. Hydrogel-based sensors have provided a revolutionary advancement in understanding the basics of cell biology and important applications in medicine.

3.2 *Cancer Monitoring*

The biggest challenge in the present era is to develop a suitable sensing material for detecting and monitoring cancer. Sensor development is usually associated with sensitivity, selectivity, and biocompatibility with living tissues. In this regard, hydrogel-based sensors have shown remarkable efficiencies in analyzing and detecting the presence of cancer cells. Wang et al. [79] developed an antifouling peptide hydrogel-derived electrochemical biosensor that can be used for the analysis of breast cancer biomarker HER2 present in human serum. They functionalized the Phe–Glu–Lys–Phe short peptide with the fluorene methoxycarbonyl group. The functionalization of peptide chains enhances its activity and stability and develops a conductive substrate. The detection limit of the synthesized biosensor was 45 pg mL^{-1} along with a wide response range, i.e., 0.1 ng mL^{-1} to 1.0 μg mL^{-1}. The synthesized biosensor was adaptable to the complex biological samples and provided acceptable clinical accuracy for detecting HER2 in human serum. In another study, Kim et al. [80] designed a pyrophosphate-responsive self-reporting hydrogel for the naked-eye monitoring of HeLa and KG-1 cancer cells. The sensor was developed by the incorporation of copper-immobilized polymer dots into a polyacrylic acid hydrogel (PD-Cu^{2+}@PAA). The sensitivity of the hydrogel was dependent upon its shape, stretchability, viscoelasticity, changes in fluorescence, and electrical conductivity. After treatment with cancer cells, the detachment of copper by pyrophosphate produced soft, fluidlike, and injectable hydrogel when treated with cancer cells. The detachment of copper decreases the conductivity and enhances the fluorescence, which promotes the self-reporting strategy. In another study, Wu et al. [81] synthesized a stimulus-responsive fluorescent hydrogel by cross-linking 5,6-dicarboxylic fluorescein with partially ammoniated polyacrylamide and used it for the naked-eye sensing of glutathione. In fact, glutathione contains a thiol group that can impart a potential role in many physiological functions, such as antioxidant defense and nutrient metabolism. Many diseases such as Parkinson's, Alzheimer's, heart attack, and cancer are suspected if the level of glutathione is imbalanced inside the body [82, 83]. Glutathione is an endogenous signaling molecule in tumors. There are many other examples of hydrogel that show their importance for detecting and diagnosing cancer sites.

In many cancer types, the potential for cancer deformity can significantly influence the life quality of patients. For example, in the case of breast cancer patients, mastectomy creates unexpected effects on the social life of patients. Moreover, the utilization of currently available prostheses is extremely unsafe because of the extreme health risks of the patients. In mammoplasty, both injectable and prosthesis tissue replacements are applicable. One such soft tissue filter is polyacrylamide recently used for breast augmentation. Instead of so many advantages, such filters can still pose challenges because it is sometimes difficult to distinguish between the radiologic images of infected rectoceles and injected polyacrylamide. Despite so many challenges, hydrogel-based sensing platforms are still used for the monitoring of cancer cells. For example, Fang et al. [84] have prepared temperature-responsive hydrogels using a layer of cell-adherent arginine–glycine–aspartate (RGD) peptides (Fig. 3). The cancer cells were first allowed to bind onto the surface of gel matrix which then underwent a chemical transformation due to temperature change. Finally, the cancer cells get ruptured due to the physical forces generated by the expended hydrogel. In another study, Robby et al. [85] designed a strain pressure hydrogel-based sensor using polyacrylamide-based hydrogel incorporated with disulfide-cross-linked polymer dots, which was initially loaded with polydopamine nanoparticles. The existence of glutathione in cancer cells was the driving force to stimulate the self-healing ability and electric response caused due to selective strain pressure after the dissociation of disulfide bond in disulfide-cross-linked polymer dot. In comparison with the treatment of hydrogel with the normal cells, the self-healing reversibility and sensitivity response were found to be prominent in the case of cancer cells treated with the designed hydrogel composite. Further, the hydrogel can be connected to a wireless electronic device to monitor the pressure sensing via a smartphone. Thus, the fabricated system was suitable for rapid cancer cell detection. The hydrogel-based sensing platforms for the detection of cancer cells are more reliable and cheaper compared with the traditional sensing methods such as MRI, PET-CT, biopsy, and fluorescence imaging. Moreover, these methods possess many

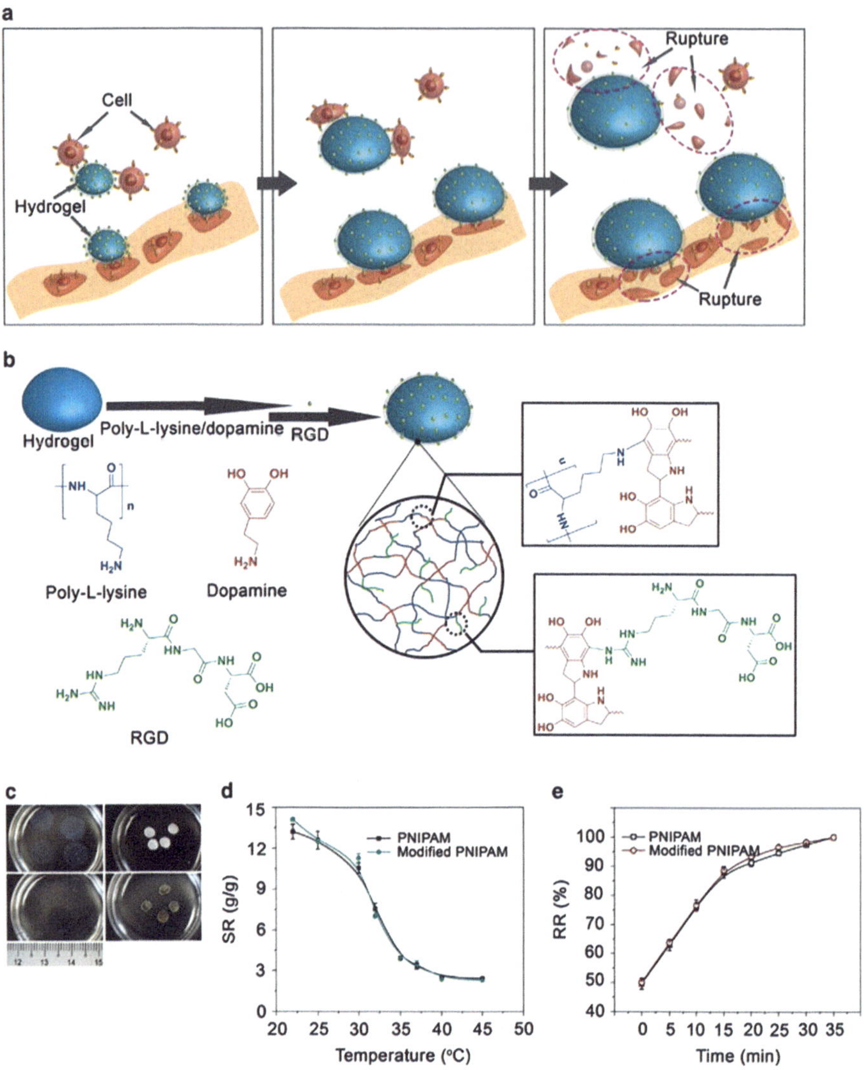

Fig. 3 Cancer cell rupture through stimuli-responsive hydrogels. (**a**) Illustration of the interaction of cancer cells onto the surface of stimuli-responsive hydrogels; the expansion of hydrogel under the impact of temperature and the rupture of cancer cells. (**b**) Modification of hydrogel surface with polydopamine, polylysine, and cell-adherent arginine–glycine–aspartate (RGD) peptides. (**c**) Digital images of PNIPAM hydrogel at different temperatures. (**d**) Comparative analysis of swelling ratio and (**e**) recovery rate percentage of PNIPAM hydrogel (Reprinted with permission from [84])

limitations, such as skilled professionals, the need for specialized tools, radiation exposure, low photostability, and low sensitivity. Hydrogels have become promising biomaterials that can show remarkable features such as biocompatibility, flexibility, and biodegradability. Therefore, such materials are widely used for the creation of artificial skin, tissue engineering, and the development of biosensors, particularly for recognizing cancer biomarkers such as extracellular pH, alkaline phosphatase, glutathione, and cancer microenvironment [85].

3.3 *Cardiac Rhythm and Wrist Pulse Monitoring*

In preclinical practice, the monitoring of artery pulse is an important indicator of many parameters of the cardiovascular system including cardiac rhythm and arterial blood pressure. The monitoring of artery pulse can help extract important information for noninvasive diagnosis of existing cardiovascular diseases. Therefore, a soft, flexible, and smart sensor is required for this application. Stimuli-responsive hydrogel-based soft sensors are the appropriate choice for closely fitting the curvature of the human skin and precisely detecting the sensitivity of pulse waves such as cardiac rhythm and blood pressure [86]. A large number of hydrogel-based soft sensors for the monitoring of wrist pulse and cardiac rhythm already exist in the literature. For example, Wang et al. [87] synthesized a conductive hydrogel based on polyaniline and poly(acrylamide-*co*-hydroxyethyl methyl acrylate). The fabricated hydrogel system imparted excellent strength, good sensitivity (gauge factor = 11), and toughness to cyclic loading. The sensitivity of the system showed a linear relationship with the strain. The fabricated sensor was useful for the detection of subtle vibrations and large strain including the movement of pulse, voiceprints, and human joints. Hydrogel-based prototype 2D sensor array was applicable as a promising device for biosensors, touchpads, biomedical implants, human–machine interfaces, and wearable electronic devices. In another study, Han et al. [88] utilized a two-way strategy to synthesize an adhesive and tough hydrogel from clay, polydopamine, and polyacrylamide. Dopamine was limitedly oxidized between the clay nanosheets containing free catechol groups. The hydrogel was able to stick directly on the skin surface without producing any inflammatory response, and it can be easily detached from the skin without causing any adverse effects. The designed system demonstrated excellent proliferation, cell attachment, and high cell affinity of polydopamine and can be used for cardiac rhythm monitoring. To further improve the sensing performance of hydrogels, Zhang et al. [89] introduced MXene-based hydrogel composites that can enhance unique features among similar systems. They incorporated MXene inside the polyvinyl alcohol-based hydrogel matrix that introduced remarkable stretchability (>3400%), adhesiveness, good self-healing ability, and comfortability over different surfaces including the human skin. Moreover, the excellent strain sensitivity of the fabricated hydrogel demonstrated its promising application in high-performance strain sensors that can be useful in wrist pulse monitoring. In another study, Zhang et al. [90] fabricated a digital light-processed hydrogel based on polyacrylamide and poly(ethylene glycol) diacrylate. The hydrogel displayed electrical conductivity on adding LiCl. The addition of 2,4,6-trimethylbenzoyl diphenylphosphine oxide made the hydrogelnetwork UV

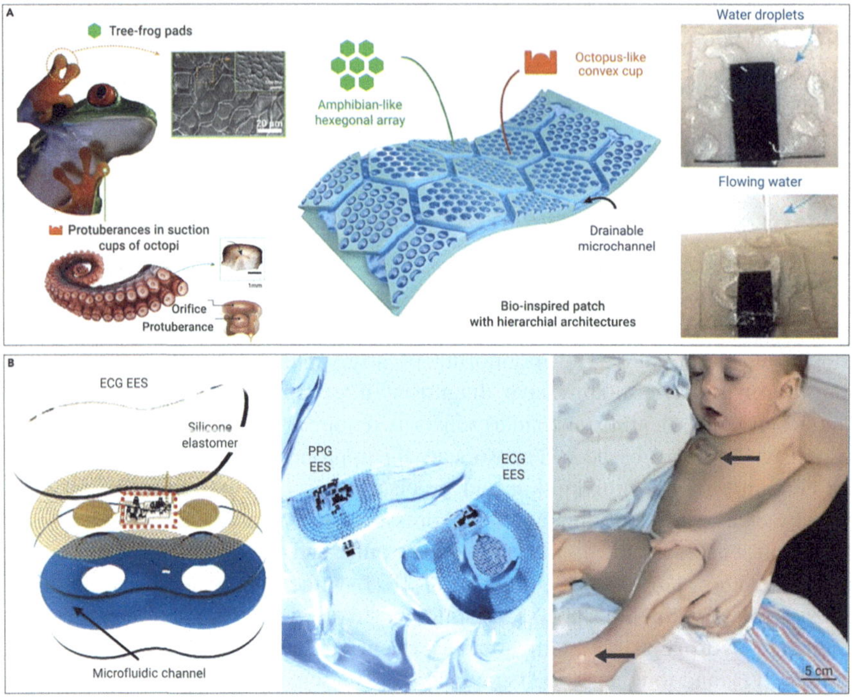

Fig. 4 (**a**) Reusable, water-drainable, and air-permeable patch synthesized from octopi and tree frogs; (**b**) neonatal intensive care using ultrasoft, ultrathin, and skin type electrodes. (Reprinted with permission from [91])

curable and compatible with the 3D printing technology by enhancing its resolution and fidelity (≈7 μm). The fabricated hydrogel system was stretchable up to 1300% and had biocompatibility. The excellent optical clarity of the designed hydrogels displayed their applicability for designing 3D printable contact lenses. The 3D-printed conductive hydrogels can be fabricated by selecting different material compositions depending on the specific requirements of their applications.

Further, the advancement in semiconductor performance and electronics miniaturization has significantly decreased the overall size and area footprint of wearable sensors, further creating new market opportunities for medical health monitoring and consumers. Recent advancements in flexible epidermal electronic technologies have addressed many issues related to heart rate monitoring devices and apparel-based cardiac devices. The Nobel epidermal devices are usually comprised of soft and wearable sensors embedded in ultrathin layers that can easily couple with the skin tissues. Zhang et al. [91] demonstrated the importance of flexible electronics in cardiovascular healthcare monitoring using bio-based polymers (Fig. 4). Such types of materials play a crucial role in fabricating virtual medicines, wireless design, multifunction, miniaturization, customized healthcare, and artificial intelligence assisted diagnosis. In addition to better personalized treatment, they are the best indicators for identifying different cardiovascular diseases.

3.4 Monitoring of Wound Healing

The monitoring of wound healing is another important application of hydrogels in the biomedical field. During wound healing, the destroyed tissues of the skin/wound are replaced by the living tissues. Wound healing is an essential process in the human body, and sometimes, biomedical tools are also used to activate/stimulate human cells. The tunable features of hydrogels and their stimuli-responsive behavior make them a suitable choice to activate immune cells and speed up the process of wound healing [92]. The monitoring of bacterial infection plays a crucial role during wound care treatment. The hydrogel-based wound care devices usually monitor physical parameters such as moisture level, pH level, and temperature. Further, the existence of any unusual compound may indicate the presence of bacterial infection.

Rahimi et al. [93] synthesized a hydrogel-based flexible and cost-effective sensor array for monitoring the wound healing process. The fabricated sensor was made of polymer-coated palette paper and contained two screen-printed electrodes. Ag/AgCl acted as a reference electrode, and the second electrode (carbon electrode) was coated with the polyaniline polymeric membrane. The pH sensitivity of the synthesized sensor was −50 mV/pH, and a linear relationship was observed between the output voltage and the pH ($R^2 = 0.9734$). The rise and fall time of the sensor was 12–36 s for a pH swing of 8–6–8. Moreover, the sensor was biocompatible with human keratinocyte cells, and hence, it was applicable for the assessment of wound healing. Jankowska et al. [94] designed a smart wound pad using a biocompatible hydrogel matrix that can detect the changes observed in glucose concentration and pH in wound serum. The device can provide both the colorimetric and fluorescence response. Horseradish peroxidase- and glucose oxidase-based metabolite sensing system was used to monitor the glucose concentration, whereas fluorescent dye carboxynaphthofluorescein was employed as a pH indicator. In another study, Zheng et al. [95] synthesized a carbon dot-doped colorimetric hydrogel sensor array using polydimethylsiloxane and agarose hydrogel matrix. The sensor was able to monitor the essential wound condition indicators (i.e., pH, total protein, urea, uric acid, and glucose). In addition to flexibility, lightweight, and long-term stability, the recovery rate of the sensor was very high (91.5–113.1%), indicating its clinical relevance for detecting all five wound biomarkers. The fabricated sensor can easily differentiate visually between the healing and nonhealing wounds. Table 2 summarizes some recent biomedical applications of hydrogel-based sensors.

Recently, hydrogels have become a promising material for designing intelligent wound dressing materials because they can show mechanical characteristics, injectability, self-healing behavior, and superior biocompatibility. In addition to the progressive development of flexible and wearable electronic devices, a new class of intelligent wound dressing materials has emerged that can integrate with various

Table 2 Self-healing hydrogels for different biomedical applications

Biomedical application	Measured property	System particulates	References
Cell metabolite detection	Molecule detection	Nanostructured conductive hydrogel electrode modified with platinum nanoparticles for the detection of human metabolite	[77]
	Molecule detection	Detection of H_2O_2 using optical horseradish peroxidase-based poly(ethylene glycol) hydrogel biosensor	[76]
	Molecule detection	Aptamer hydrogel functionalized metamaterial for the molecular specific detection of terahertz	[96]
Cancer monitoring	Molecule detection	Carbon dots imbibed chitosan hydrogel network	[97]
	Ion detection	Detection of superoxide ions released by cancer cells using poly(ethylene glycol) diacrylate hydrogel	[98]
	Molecule detection	Florescent imaging and cancer therapeutics with ruthenium–albumin hydrogel	[99]
Cardiac rhythm and wrist pulse monitoring	Strain	Polyacrylamide polydopamine and tannic acid-based hydrogel	[100]
	Strain	Agar/NaCl/polyacrylamide hydrogel-based high-performance ionic strain sensor	[101]
	Strain	Sodium alginate-based ionic hydrogel fibers	[102]
	Strain	3D nanocomposite of polydopamine/titanium dioxide/silver organohydrogel	[103]
Monitoring of wound healing	Visual monitoring	Guar gum, polyvinyl alcohol, borax, and tannic acid-based hydrogel dressing	[104]
	Molecule detection	Polysaccharide hydrogel matrix functionalized with glucose oxidase/horseradish peroxide and carboxynaphto-fluorscin	[94]
	Visual monitoring	Cerium-based MOF nanozyme hydrogel for the healing of diabetic wounds	[105]
Application in tissue engineering	Molecule detection	Gelatin and hyaluronic acid-based composite hydrogel for scaffold monitoring	[106]
	Molecule detection	Cross-linked hydrogel based on oxidized dextrin and peptide for MMP detection	[107]
	Molecule detection	Optical fiber containing sol-gel coated hydrogel	[108]

types of electronic sensors and is suitable for in situ wound sensing applications and related on-demand treatments [109]. Such types of sensors can provide the real-type analysis of wound status using different biomarkers such as uric acid, temperature, pH, etc. Moreover, mechano-responsive drug delivery is also possible through UV-responsive or NIR-triggered hydrogels. An optimal drug release system should show a release pattern of therapeutic agents depending upon the spatiotemporal and physiological requirements during the different wound healing stages. The release profile of hydrogel-based wound healing systems may occur either through the active or passive approaches of the drug release system. A passive approach generally favors the continuous release of drugs through the hydrogel matrix without using any external stimulus. On the other hand, an active approach generally uses an external stimulus, such as temperature, pressure, electric field, light, etc., to trigger the drug release. Smart hydrogel-based wound healing systems consider the need for specific drug release by getting the signal through different biomarkers present in the wound environment and stimulating the drug release whenever required [110]. Passive drug release systems are based on transmitting healing factors through a drug transporter to reach the wound bed. The drug transporter used for wound healing may be organic (including hydrogels or liposomes), or it may be inorganic (e.g., metallic nanoparticles, carbon nanotubes, ceramics, etc.). The primary objective of passive drug release devices is to control the kinetics of drug release depending upon degradability, hydrophilicity, pore size, and relative electrostatic charge in comparison with the drug. In most of the drug delivery systems, a sustained release is preferred over a burst release. The sustained release usually reduces the local availability of the therapeutic throughout the required time period. Polyesters, liposomes, and nanoparticles are the traditional sensing platforms for monitoring wound healing. On the other hand, hydrogel-based sensors have gained significant interest in monitoring wound healing because of their intelligent and stimuli-responsive properties.

3.5 Application in Tissue Engineering

This is a new emerging research field that involves the uses of cells, smart materials, and engineering to replace/improve the uses of existing tissue scaffolds to form new tissues. Hydrogel-based sensors have emerged as an innovative class of materials that can detect and regulate the concentration of some specific molecules, their interaction, and kinetics. They help monitor the growth of some specific target tissues [111, 112]. For example, Wang et al. [108] designed an optical sensing device for the in situ measurement of extracellular adenosine triphosphate (ATP) in biological systems. The sensor contains two layers of sol-gel coating at the end of the optical fiber probe and can sense a wide range of ATP concentrations (10^{-3} mM to 1.5 mM). In addition to ATP, matrix metalloproteinase (MMP) plays an important role as a potential biomarker. Biela et al. [107] fabricated a disposable sensor that was capable of selectively detecting the presence of MMP-

9 in the presence of MMP-2 within a very short period (5 min). MMP-9 is a peripheral biomarker of neuroinflammation in sclerosis. The sensor designed for the sensing of MMP-9 was prepared with oxidized dextran and cross-linked with the peptide sequence (Leu–Gly–Arg–Met–Gly–Leu–Pro–Gly–Lys). Nowadays, implantable hydrogel-based bioelectronics are being used to diagnose diseases and monitor human health. For example, Chen et al. [113] designed a gelatin-based metamaterial hydrogel film with a very high conformality for monitoring ultrasoft tissues. Gaharwar et al. [114] showed the synthesis of bio-nanocomposite hydrogels obtained from chitosan, silicate, and poly(ethylene oxide) for bone repair applications. Silicate and chitosan were cross-linked with poly(ethylene oxide) to make it adhesive to the cells and protein. The fabricated nanocomposite hydrogel showed enhanced mechanical properties and structural stability. It also sustained the differentiation of osteoblast cells. On increasing the chitosan content, the transparency of nanocomposite hydrogel decreased, which may be attributed to the synthesis of silicate aggregates. The structural stability of the nanocomposite hydrogel was due to the presence of ionic interactions between silicate and chitosan. A silicate/chitosan concentration ratio significantly influenced the entrapped drug's release kinetics. An increase in chitosan concentration was found to slow the drug release, which was due to the increase in the cross-linking of chitosan. The cell experimental results indicated that the fabricated nanocomposite hydrogel promoted the proliferation and adhesion of osteoblasts, further stimulating osteogenesis. In another study, Paul et al. [115] synthesized biocompatible nanocomposite hydrogels using gelatin methacryloyl (a photocross-linkable matrix) and photocross-linking the same with nanosilicates. The presence of nanosilicates inside the hydrogel matrix significantly stimulates the osteogenic differentiation of human mesenchymal stem cells, which was indicated by the enhancement in alkaline phosphate activity.

The 3D printing technology is extremely important in cardiac tissue engineering and plays an important role in designing the multifarious structure of myocardial tissues. For example, Lee et al. [116] fabricated a bio ink using methacryloyl gelatin (GelMA) and methacryloyl-substituted recombinant human collagen (MeTro) (Fig. 5). The resilience of composite hydrogels was found to enhance while incorporating MeTro. It can be modified to gain similar support cell adhesion, soft tissue mechanics, and cell proliferation. During the 3D printing process, human umbilical vein endothelial cells and cardiac fibroblasts were incorporated in GelMA bioink. The designed 3D-printed hydrogel system could perform the functions of heart tissues.

Hydrogel-derived scaffolds usually offer desired and controllable parameters such as biocompatibility, inherent adaptability, biodegradability, swellability, and tunable mechanical stiffness and strength. All such parameters are extremely crucial for designing engineered tissue scaffolds. Different cross-linking strategies are usually employed to overcome the disadvantages of conventional hydrogels, such as low mechanical durability. Due to the dynamic microenvironment of hydrogels compatible with the structural compatibility of the native tissue, they can easily modify their features to suit the targeted applications in tissue engineering [118]. Stimuli-

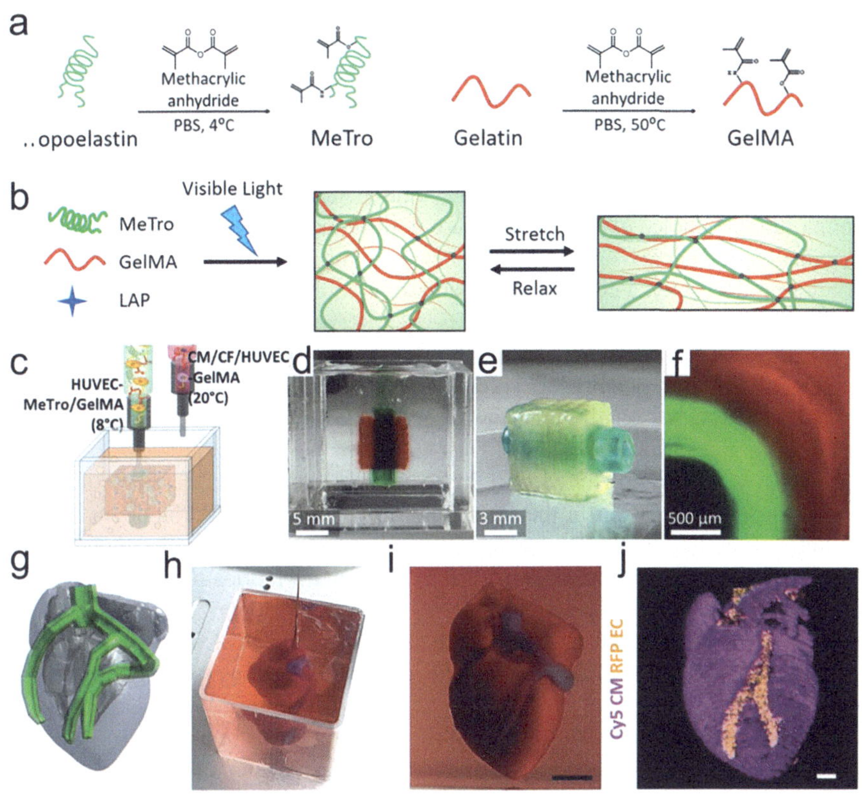

Fig. 5 (**a**, **b**) Synthesis pathway for designing methacryloyl gelatin (GelMA) and methacryloyl-substituted recombinant human collagen (MeTro) hydrogel composite; (**c**) illustration of 3D bio-printing of vascularized cardiac builds with CM/CF/HUVEC-laden GelMA and MeTro/GelMA bioink; (**d–f**) fabrication method of vascularized cardiac tissue; (**g**) CAD model of the human heart; (**h**, **i**) demonstration of the 3D printable heart within a supporting bath and confocal image depicting printed heart. (Reprinted with permission from [117])

responsive smart hydrogel networks possess many features for developing more suitable tissue-engineered contracts with more efficient and promising applications. Further, the tailoring of hydrogel properties can benefit from some extra features such as degradability, bidirectional adhesion, self-healing ability, and stretchability. Such types of smart hydrogel scaffolds are self-adherent and adjustable to sweating, varying temperature difference, bleeding, and smooth to rough conditions.

4 Conclusion and Future Perspectives

Over the past few decades, hydrogel-based sensing devices have witnessed exclusive development in different fields. This may be possible due to different intrinsic properties, chemical diversity, and tunability of the different physicochemical parameters of the hydrogel matrix. The unique features of the hydrogels made them a suitable choice of materials for use in different fields such as environmental remediation, the energy sector, soft robotics, humidity sensors, medicine, and health monitoring. Recently, a considerable increase in publications related to developing hydrogels in sensing applications has been observed. However, instead of great advancements and promises in the field of hydrogel-based sensors, there are many challenges with huge potential and plenty of scope to carry out future research work. This chapter focuses on the present status and recent advancement of hydrogel-based sensors in various fields.

Hydrogels can be converted into approximate skin tissues; therefore, they are drawing a lot of attention in the preparation of biosensors. The biodegradability, flexibility, accuracy, durability, and stability of the hydrogels depict the applicability of the biosensor. Nowadays, a new research direction for fabricating hydrogel-based flexible and biocompatible strain sensors is most demanding. The drawbacks of conventional hydrogel sensors, such as their weak mechanical strength and short-term stability, can be enhanced by synthesizing self-healing hydrogel networks with integrated sensing networks. From a future perspective, the research should be carried out to synthesize stimuli-responsive multifunctional hydrogel-based sensors that can perform complex actions, including actuation, measurement of optical/electric response, and sensing. To further enhance the sensor performance recovery and deformation of the matrix, the synthetic strategies should be combined with computational modeling and simulations at macroscopic levels.

Acknowledgments One of the authors is extremely thankful to Chandigarh University, Mohali, India, for providing infrastructure facilities. The authors are also thankful to NIT Jalandhar and IIT Mandi.

CRediT Authorship Contribution Statement **Amit Kumar Sharma**: conceptualization, writing-original draft preparation, figure preparation, writing-reviewing and editing; **Garima Agrawal**: writing-reviewing and editing; **Rohit**: writing-reviewing and editing; **Balbir Singh Kaith**: writing-reviewing and editing.

Funding Declaration There are no sources of funding and the authors are aware about the submission.

References

1. Dincer, C., Bruch, R., Costa-Rama, E., Fernández-Abedul, M.T., Merkoçi, A., Manz, A., Urban, G.A., Güder, F.: Disposable sensors in diagnostics, food, and environmental monitoring. Adv. Mater. **31**, 1806739 (2019)

2. Pinelli, F., Magagnin, L., Rossi, F.: Progress in hydrogels for sensing applications: a review. Mater Today Chem. **17**, 100317 (2020)
3. Buenger, D., Topuz, F., Groll, J.: Hydrogels in sensing applications. Prog. Polym. Sci. **37**, 1678–1719 (2012)
4. Bhalla, P., Sharma, A.K., Kaith, B.S., Chandel, K., Sethi, V., Waraich, I., Singh, A.: Bifunctional gelatin/dextrin hybrid backbone based fluorescent chemo-sensor for the detection of tannic acid and removal of eosin yellow dye. Mater. Chem. Phys. **254**, 123304 (2020)
5. Sharma, A.K., Priya, Kaith, B.S., Isha, Singh, A., Chandel, K., Vipula: Riboflavin functionalized dextrin–sodium alginate based fluorescent sensor: detoxification of Cu^{2+} and Ni^{2+} ions. ACS Appl Polym Mater. **1**, 3084–3094 (2019)
6. Sharma, A.K., Kaith, B.S., Arora, S.: Synthesis of gelatin and green tea based stretchable self-healing material of biomedical importance. React. Funct. Polym. **172**, 105188 (2022)
7. Kaith, B.S., Singh, A., Sharma, A.K., Sud, D.: Hydrogels: synthesis, classification, properties and potential applications—a brief review. J. Polym. Environ. **29**, 3827–3841 (2021)
8. Correa, S., Grosskopf, A.K., Lopez Hernandez, H., Chan, D., Yu, A.C., Stapleton, L.M., Appel, E.A.: Translational applications of hydrogels. Chem. Rev. **121**, 11385–11457 (2021)
9. Lee, S., Inoue, Y., Kim, D., Reuveny, A., Kuribara, K., Yokota, T., Reeder, J., Sekino, M., Sekitani, T., Abe, Y.: A strain-absorbing design for tissue–machine interfaces using a tunable adhesive gel. Nat. Commun. **5**, 5898 (2014)
10. Nakajima, S., Kawano, R., Onoe, H.: Stimuli-responsive hydrogel microfibers with controlled anisotropic shrinkage and cross-sectional geometries. Soft Matter. **13**, 3710–3719 (2017)
11. Sharma, A.K., Kaith, B.S., Gupta, B., Shanker, U., Lochab, S.P.: Microwave assisted in situ synthesis of gum Salai guggal based silver nanocomposites-investigation of anti-bacterial properties. Cellulose. **26**, 991–1011 (2019)
12. Döring, A., Birnbaum, W., Kuckling, D.: Responsive hydrogels-structurally and dimensionally optimized smart frameworks for applications in catalysis, micro-system technology and material science. Chem. Soc. Rev. **42**, 7391–7420 (2013)
13. Ashammakhi, N., Hernandez, A.L., Unluturk, B.D., Quintero, S.A., de Barros, N.R., Hoque Apu, E., Bin Shams, A., Ostrovidov, S., Li, J., Contag, C.: Biodegradable implantable sensors: materials design, fabrication, and applications. Adv. Funct. Mater. **31**, 2104149 (2021)
14. Hou, J., Li, M., Song, Y.: Recent advances in colloidal photonic crystal sensors: materials, structures and analysis methods. Nano Today. **22**, 132–144 (2018)
15. Li, B., Zhou, D., Han, Y.: Assembly and phase transitions of colloidal crystals. Nat Rev Mater. **1**, 1–13 (2016)
16. Wu, Z., Ding, H., Tao, K., Wei, Y., Gui, X., Shi, W., Xie, X., Wu, J.: Ultrasensitive, stretchable, and fast-response temperature sensors based on hydrogel films for wearable applications. ACS Appl. Mater. Interfaces. **13**, 21854–21864 (2021)
17. Pang, Q., Hu, H., Zhang, H., Qiao, B., Ma, L.: Temperature-responsive ionic conductive hydrogel for strain and temperature sensors. ACS Appl. Mater. Interfaces. **14**, 26536–26547 (2022)
18. Casabianca, A.B., Becker, D.E.: Cardiovascular monitoring: physiological and technical considerations. Anesth. Prog. **56**, 53–60 (2009)
19. Lin, J., Fu, R., Zhong, X., Yu, P., Tan, G., Li, W., Zhang, H., Li, Y., Zhou, L., Ning, C.: Wearable sensors and devices for real-time cardiovascular disease monitoring. Cell Rep Phys Sci. **2**, 100541 (2021)
20. Bayoumy, K., Gaber, M., Elshafeey, A., Mhaimeed, O., Dineen, E.H., Marvel, F.A., Martin, S.S., Muse, E.D., Turakhia, M.P., Tarakji, K.G.: Smart wearable devices in cardiovascular care: where we are and how to move forward. Nat. Rev. Cardiol. **18**, 581–599 (2021)
21. Li, H., Zheng, H., Tan, Y.J., Tor, S.B., Zhou, K.: Development of an ultrastretchable double-network hydrogel for flexible strain sensors. ACS Appl. Mater. Interfaces. **13**, 12814–12823 (2021)
22. Chen, Q., Zhu, L., Zhao, C., Wang, Q., Zheng, J.: A robust, one-pot synthesis of highly mechanical and recoverable double network hydrogels using thermoreversible sol-gel polysaccharide. Adv. Mater. **25**, 4171–4176 (2013)

23. Gong, J.P.: Why are double network hydrogels so tough? Soft Matter. **6**, 2583–2590 (2010)
24. Xia, S., Song, S., Gao, G.: Robust and flexible strain sensors based on dual physically cross-linked double network hydrogels for monitoring human-motion. Chem. Eng. J. **354**, 817–824 (2018)
25. Haraguchi, K., Uyama, K., Tanimoto, H.: Self-healing in nanocomposite hydrogels. Macromol. Rapid Commun. **32**, 1253–1258 (2011)
26. Hou, J., Liu, M., Zhang, H., Song, Y., Jiang, X., Yu, A., Jiang, L., Su, B.: Healable green hydrogen bonded networks for circuit repair, wearable sensor and flexible electronic devices. J. Mater. Chem. A. **5**, 13138–13144 (2017)
27. Chen, J., Liu, J., Thundat, T., Zeng, H.: Polypyrrole-doped conductive supramolecular elastomer with stretchability, rapid self-healing, and adhesive property for flexible electronic sensors. ACS Appl. Mater. Interfaces. **11**, 18720–18729 (2019)
28. Wu, J., Wu, Z., Wei, Y., Ding, H., Huang, W., Gui, X., Shi, W., Shen, Y., Tao, K., Xie, X.: Ultrasensitive and stretchable temperature sensors based on thermally stable and self-healing organohydrogels. ACS Appl. Mater. Interfaces. **12**, 19069–19079 (2020)
29. Lee, J., Tan, M.W.M., Parida, K., Thangavel, G., Park, S.A., Park, T., Lee, P.S.: Water-processable, stretchable, self-healable, thermally stable, and transparent ionic conductors for actuators and sensors. Adv. Mater. **32**, 1906679 (2020)
30. Wu, J., Han, S., Yang, T., Li, Z., Wu, Z., Gui, X., Tao, K., Miao, J., Norford, L.K., Liu, C.: Highly stretchable and transparent thermistor based on self-healing double network hydrogel. ACS Appl. Mater. Interfaces. **10**, 19097–19105 (2018)
31. Zhang, J., Wan, L., Gao, Y., Fang, X., Lu, T., Pan, L., Xuan, F.: Highly stretchable and self-healable MXene/polyvinyl alcohol hydrogel electrode for wearable capacitive electronic skin. Adv Electron Mater. **5**, 1900285 (2019)
32. Huang, W., Besar, K., Zhang, Y., Yang, S., Wiedman, G., Liu, Y., Guo, W., Song, J., Hemker, K., Hristova, K.: A high-capacitance salt-free dielectric for self-healable, printable, and flexible organic field effect transistors and chemical sensor. Adv. Funct. Mater. **25**, 3745–3755 (2015)
33. Kim, J.W., Park, H., Lee, G., Jeong, Y.R., Hong, S.Y., Keum, K., Yoon, J., Kim, M.S., Ha, J.S.: Paper-like, thin, foldable, and self-healable electronics based on PVA/CNC nanocomposite film. Adv. Funct. Mater. **29**, 1905968 (2019)
34. Yang, H., Qi, D., Liu, Z., Chandran, B.K., Wang, T., Yu, J., Chen, X.: Soft thermal sensor with mechanical adaptability. Adv. Mater. **28**, 9175–9181 (2016)
35. Lai, Y.C., Wu, H.M., Lin, H.C., Chang, C.L., Chou, H.H., Hsiao, Y.C., Wu, Y.C.: Entirely, intrinsically, and autonomously self-healable, highly transparent, and superstretchable triboelectric nanogenerator for personal power sources and self-powered electronic skins. Adv. Funct. Mater. **29**, 1904626 (2019)
36. Parida, K., Kumar, V., Jiangxin, W., Bhavanasi, V., Bendi, R., Lee, P.S.: Highly transparent, stretchable, and self-healing ionic-skin triboelectric nanogenerators for energy harvesting and touch applications. Adv. Mater. **29**, 1702181 (2017)
37. Liang, Z., Zhang, J., Wu, C., Hu, X., Lu, Y., Wang, G., Yu, F., Zhang, X., Wang, Y.: Flexible and self-healing electrochemical hydrogel sensor with high efficiency toward glucose monitoring. Biosens. Bioelectron. **155**, 112105 (2020)
38. Tsai, M.-S., Shen, T.-L., Wu, H.-M., Liao, Y.-M., Liao, Y.-K., Lee, W.-Y., Kuo, H.-C., Lai, Y.-C., Chen, Y.-F.: Self-powered, self-healed, and shape-adaptive ultraviolet photodetectors. ACS Appl. Mater. Interfaces. **12**, 9755–9765 (2020)
39. Yoon, J.H., Kim, S.-M., Park, H.J., Kim, Y.K., Oh, D.X., Cho, H.-W., Lee, K.G., Hwang, S.Y., Park, J., Choi, B.G.: Highly self-healable and flexible cable-type pH sensors for real-time monitoring of human fluids. Biosens. Bioelectron. **150**, 111946 (2020)
40. Wang, J., Tang, F., Wang, Y., Lu, Q., Liu, S., Li, L.: Self-healing and highly stretchable gelatin hydrogel for self-powered strain sensor. ACS Appl. Mater. Interfaces. **12**, 1558–1566 (2019)

41. Ketelsen, B., Schlicke, H., Schulze, V.R., Bittinger, S.C., Wu, S.D., Hsu, S.H., Vossmeyer, T.: Nanoparticle-based strain gauges: anisotropic response characteristics, multidirectional strain sensing, and novel approaches to healthcare applications. Adv Funct Mater. **33**, 2210065 (2023)
42. Ding, Y., Yang, J., Tolle, C.R., Zhu, Z.: A highly stretchable strain sensor based on electrospun carbon nanofibers for human motion monitoring. RSC Adv. **6**, 79114–79120 (2016)
43. Amjadi, M., Pichitpajongkit, A., Lee, S., Ryu, S., Park, I.: Highly stretchable and sensitive strain sensor based on silver nanowire–elastomer nanocomposite. ACS Nano. **8**, 5154–5163 (2014)
44. Santos, A., Amorim, L., Nunes, J., Silva, A., Viana, J.: Carbon nanotubes based multi-directional strain sensor. Mater. Sci. Eng. B. **285**, 115937 (2022)
45. Chen, H., Zhuo, F., Zhou, J., Liu, Y., Zhang, J., Dong, S., Liu, X., Elmarakbi, A., Duan, H., Fu, Y.: Advances in graphene-based flexible and wearable strain sensors. Chem. Eng. J. **464**, 142576 (2023)
46. Wang, L., Wang, H., Wan, Q., Gao, J.: Recent development of conductive polymer composite-based strain sensors. J. Polym. Sci. **61**, 3167–3185 (2023)
47. Kim, S.J., Mondal, S., Min, B.K., Choi, C.-G.: Highly sensitive and flexible strain–pressure sensors with cracked paddy-shaped MoS_2/graphene foam/ecoflex hybrid nanostructures. ACS Appl. Mater. Interfaces. **10**, 36377–36384 (2018)
48. Xiao, J., He, J., Wang, W., Chen, M., Guo, R.: A multifunctional flexible strain sensor based on an excellent sensing performance PDMS-MXene@ CNT/TPU nanofiber membrane with hydrophobic and photothermal conversion performance. New J. Chem. **47**, 19265–19275 (2023)
49. Chang, X., Sun, S., Sun, S., Liu, T., Xiong, X., Lei, Y., Dong, L., Yin, Y.: ZnO nanorods/carbon black-based flexible strain sensor for detecting human motions. J. Alloys Compd. **738**, 111–117 (2018)
50. Amjadi, M., Kyung, K.U., Park, I., Sitti, M.: Stretchable, skin-mountable, and wearable strain sensors and their potential applications: a review. Adv. Funct. Mater. **26**, 1678–1698 (2016)
51. Luo, X.-L., Xu, J.-J., Zhang, Q., Yang, G.-J., Chen, H.-Y.: Electrochemically deposited chitosan hydrogel for horseradish peroxidase immobilization through gold nanoparticles self-assembly. Biosens. Bioelectron. **21**, 190–196 (2005)
52. Song, S., Park, J., Chitnis, G., Siegel, R., Ziaie, B.: A wireless chemical sensor featuring iron oxide nanoparticle-embedded hydrogels. Sens Actuators B Chem. **193**, 925–930 (2014)
53. Khatib, M., Zohar, O., Saliba, W., Srebnik, S., Haick, H.: Highly efficient and water-insensitive self-healing elastomer for wet and underwater electronics. Adv. Funct. Mater. **30**, 1910196 (2020)
54. Hart, J., Abass, A.: A disposable amperometric gas sensor for sulphur-containing compounds based on a chemically modified screen printed carbon electrode coated with a hydrogel. Anal. Chim. Acta. **342**, 199–206 (1997)
55. Severinghaus, J.W.: First electrodes for blood PO_2 and PCO_2 determination. J. Appl. Physiol. **97**, 1599–1600 (2004)
56. Kojima, J., Nakayama, Y., Takenaka, M., Hashimoto, T.: Apparatus for measuring time-resolved light scattering profiles from supercritical polymer solutions undergoing phase separation under high pressure. Rev. Sci. Instrum. **66**, 4066–4072 (1995)
57. Herber, S., Olthuis, W., Bergveld, P., van den Berg, A.: Exploitation of a pH-sensitive hydrogel disk for CO_2 detection. Sens Actuators B Chem. **103**, 284–289 (2004)
58. Zhang, B., Zhang, P., Zhang, H., Yan, C., Zheng, Z., Wu, B., Yu, Y.: A transparent, highly stretchable, autonomous self-healing poly(dimethyl siloxane) elastomer. Macromol. Rapid Commun. **38**, 1700110 (2017)
59. Luo, Y., Li, J., Ding, Q., Wang, H., Liu, C., Wu, J.: Functionalized hydrogel-based wearable gas and humidity sensors. Nano Micro Lett. **15**, 136 (2023)
60. Yuk, H., Zhang, T., Parada, G.A., Liu, X., Zhao, X.: Skin-inspired hydrogel–elastomer hybrids with robust interfaces and functional microstructures. Nat. Commun. **7**, 12028 (2016)

61. Moustafa, H., Morsy, M., Ateia, M.A., Abdel-Haleem, F.M.: Ultrafast response humidity sensors based on polyvinyl chloride/graphene oxide nanocomposites for intelligent food packaging. Sens Actuators A Phys. **331**, 112918 (2021)
62. Jalal, A.H., Alam, F., Roychoudhury, S., Umasankar, Y., Pala, N., Bhansali, S.: Prospects and challenges of volatile organic compound sensors in human healthcare. ACS Sens. **3**, 1246–1263 (2018)
63. Guo, K., Zhang, D.L., Zhang, X.M., Zhang, J., Ding, L.S., Li, B.J., Zhang, S.: Conductive elastomers with autonomic self-healing properties. Angew. Chem. Int. Ed. **54**, 12127–12133 (2015)
64. Wu, J., Wu, Z., Xu, H., Wu, Q., Liu, C., Yang, B.-R., Gui, X., Xie, X., Tao, K., Shen, Y.: An intrinsically stretchable humidity sensor based on anti-drying, self-healing and transparent organohydrogels. Mater Horiz. **6**, 595–603 (2019)
65. Liu, Y., Wang, L., Zhang, M., Tu, D., Mao, X., Liao, Y.: Long-period grating relative humidity sensor with hydrogel coating. IEEE Photon. Technol. Lett. **19**, 880–882 (2007)
66. Mishra, S.K., Zou, B., Chiang, K.S.: Wide-range pH sensor based on a smart-hydrogel-coated long-period fiber grating. IEEE J Sel Topics Quantum Electron. **23**, 284–288 (2016)
67. Shin, J., Braun, P.V., Lee, W.: Fast response photonic crystal pH sensor based on templated photo-polymerized hydrogel inverse opal. Sens Actuators B Chem. **150**, 183–190 (2010)
68. Zhao, Y., Lei, M., Liu, S.-X., Zhao, Q.: Smart hydrogel-based optical fiber SPR sensor for pH measurements. Sens Actuators B Chem. **261**, 226–232 (2018)
69. Sannino, A., Pappada, S., Giotta, L., Valli, L., Maffezzoli, A.: Spin coating cellulose derivatives on quartz crystal microbalance plates to obtain hydrogel-based fast sensors and actuators. J. Appl. Polym. Sci. **106**, 3040–3050 (2007)
70. Tokuyama, H., Kitamura, E., Seida, Y.: Development of zirconia nanoparticle-loaded hydrogel for arsenic adsorption and sensing. React. Funct. Polym. **146**, 104427 (2020)
71. Sharma, A.K., Kaith, B.S., Singh, A., Chandel, K.: Enzymatic construction of quinine derivative of dextrin/PVA based hybrid gel film for the simultaneous detection and removal of copper and lead ions in real water samples. Chem. Eng. J. **382**, 122891 (2020)
72. Jia, X., Wang, K., Wang, J., Hu, Y., Shen, L., Zhu, J.: Full-color photonic hydrogels for pH and ionic strength sensing. Eur. Polym. J. **83**, 60–66 (2016)
73. Nezami, S., Nematidil, N., Farzan, F., Mirzaie, F., Sadeghi, H., Sadeghi, M.: pH-sensitive drug delivery systems based on CMC-ECH-CTS and CMC-ECH-CTS/Fe_3O_4 beads. Polym. Test. **97**, 107144 (2021)
74. Herrmann, A., Haag, R., Schedler, U.: Hydrogels and their role in biosensing applications. Adv. Healthc. Mater. **10**, 2100062 (2021)
75. Zhang, Y., Xu, Q., Wang, F., Gao, T., Wei, T.: Enzyme powered self-assembly of hydrogel biosensor for colorimetric detection of metabolites. Sens Actuators B Chem. **375**, 132942 (2023)
76. Yan, J., Sun, Y., Zhu, H., Marcu, L., Revzin, A.: Enzyme-containing hydrogel micropatterns serving a dual purpose of cell sequestration and metabolite detection. Biosens. Bioelectron. **24**, 2604–2610 (2009)
77. Li, L., Wang, Y., Pan, L., Shi, Y., Cheng, W., Shi, Y., Yu, G.: A nanostructured conductive hydrogels-based biosensor platform for human metabolite detection. Nano Lett. **15**, 1146–1151 (2015)
78. Ooi, H.W., Hafeez, S., Van Blitterswijk, C., Moroni, L., Baker, M.B.: Hydrogels that listen to cells: a review of cell-responsive strategies in biomaterial design for tissue regeneration. Mater Horiz. **4**, 1020–1040 (2017)
79. Wang, W., Han, R., Chen, M., Luo, X.: Antifouling peptide hydrogel based electrochemical biosensors for highly sensitive detection of cancer biomarker HER2 in human serum. Anal. Chem. **93**, 7355–7361 (2021)
80. Kim, S.G., Robby, A.I., Kim, E.H., Jin, E.-J., Park, S.Y.: Pyrophosphate-responsive viscoelasticity and conductive signaling of self-reporting hydrogel sensor for detection of cancer cells. Chem. Eng. J. **472**, 145069 (2023)

81. Wu, R., Ge, H., Liu, C., Zhang, S., Hao, L., Zhang, Q., Song, J., Tian, G., Lv, J.: A novel thermometer-type hydrogel senor for glutathione detection. Talanta. **196**, 191–196 (2019)
82. Hobani, Y.H.: The role of oxidative stress in koenimbine-induced DNA damage and heat shock protein modulation in HepG2 cells. Integr. Cancer Ther. **16**, 563–571 (2017)
83. Cecchi, C., Latorraca, S., Sorbi, S., Iantomasi, T., Favilli, F., Vincenzini, M.T., Liguri, G.: Gluthatione level is altered in lymphoblasts from patients with familial Alzheimer's disease. Neurosci. Lett. **275**, 152–154 (1999)
84. Fang, Y., Tan, J., Lim, S., Soh, S.: Rupturing cancer cells by the expansion of functionalized stimuli-responsive hydrogels. NPG Asia Mater. **10**, e465 (2018)
85. Robby, A.I., Lee, G., Lee, K.D., Jang, Y.C., Park, S.Y.: GSH-responsive self-healable conductive hydrogel of highly sensitive strain-pressure sensor for cancer cell detection. Nano Today. **39**, 101178 (2021)
86. Chen, S., Qi, J., Fan, S., Qiao, Z., Yeo, J.C., Lim, C.T.: Flexible wearable sensors for cardiovascular health monitoring. Adv. Healthc. Mater. **10**, 2100116 (2021)
87. Wang, Z., Chen, J., Cong, Y., Zhang, H., Xu, T., Nie, L., Fu, J.: Ultrastretchable strain sensors and arrays with high sensitivity and linearity based on super tough conductive hydrogels. Chem. Mater. **30**, 8062–8069 (2018)
88. Han, L., Lu, X., Liu, K., Wang, K., Fang, L., Weng, L.-T., Zhang, H., Tang, Y., Ren, F., Zhao, C.: Mussel-inspired adhesive and tough hydrogel based on nanoclay confined dopamine polymerization. ACS Nano. **11**, 2561–2574 (2017)
89. Zhang, Y.-Z., Lee, K.H., Anjum, D.H., Sougrat, R., Jiang, Q., Kim, H., Alshareef, H.N.: MXenes stretch hydrogel sensor performance to new limits. Sci. Adv. **4**, eaat0098 (2018)
90. Zhang, B., Li, S., Hingorani, H., Serjouei, A., Larush, L., Pawar, A.A., Goh, W.H., Sakhaei, A.H., Hashimoto, M., Kowsari, K.: Highly stretchable hydrogels for UV curing based high-resolution multimaterial 3D printing. J. Mater. Chem. B. **6**, 3246–3253 (2018)
91. Zhang, T., Liu, N., Xu, J., Liu, Z., Zhou, Y., Yang, Y., Li, S., Huang, Y., Jiang, S.: Flexible electronics for cardiovascular healthcare monitoring. Innovation. **4**, 100485 (2023)
92. Xiang, J., Shen, L., Hong, Y.: Status and future scope of hydrogels in wound healing: synthesis, materials and evaluation. Eur. Polym. J. **130**, 109609 (2020)
93. Rahimi, R., Ochoa, M., Parupudi, T., Zhao, X., Yazdi, I.K., Dokmeci, M.R., Tamayol, A., Khademhosseini, A., Ziaie, B.: A low-cost flexible pH sensor array for wound assessment. Sens Actuators B Chem. **229**, 609–617 (2016)
94. Jankowska, D., Bannwarth, M., Schulenburg, C., Faccio, G., Maniura-Weber, K., Rossi, R., Scherer, L., Richter, M., Boesel, L.: Simultaneous detection of pH value and glucose concentrations for wound monitoring applications. Biosens. Bioelectron. **87**, 312–319 (2017)
95. Zheng, X.T., Zhong, Y., Chu, H.E., Yu, Y., Zhang, Y., Chin, J.S., Becker, D.L., Su, X., Loh, X.J.: Carbon dot-doped hydrogel sensor array for multiplexed colorimetric detection of wound healing. ACS Appl. Mater. Interfaces. **15**, 17675–17687 (2023)
96. Zhou, J., Zhao, X., Huang, G., Yang, X., Zhang, Y., Zhan, X., Tian, H., Xiong, Y., Wang, Y., Fu, W.: Molecule-specific terahertz biosensors based on an aptamer hydrogel-functionalized metamaterial for sensitive assays in aqueous environments. ACS Sens. **6**, 1884–1890 (2021)
97. Mohammadi, S., Mohammadi, S., Salimi, A.: A 3D hydrogel based on chitosan and carbon dots for sensitive fluorescence detection of microRNA-21 in breast cancer cells. Talanta. **224**, 121895 (2021)
98. Crulhas, B.P., Recco, L.C., Delella, F.K., Pedrosa, V.A.: A novel superoxide anion biosensor for monitoring reactive species of oxygen released by cancer cells. Electroanalysis. **29**, 1252–1257 (2017)
99. Zhao, Z., Hu, R., Shi, H., Wang, Y., Ji, L., Zhang, P., Zhang, Q.: Design of ruthenium–albumin hydrogel for cancer therapeutics and luminescent imaging. J. Inorg. Biochem. **194**, 19–25 (2019)
100. Li, Y., Zhao, Z., Veronica, A., Yu Yeung, S., Hsing, I.M.: Skin-adherent elastomer-hydrogel patch for continuous 12-Lead cardiac ambulatory monitoring during physical activities. Adv Mater Technol. **8**, 2300326 (2023)

101. Hou, W., Sheng, N., Zhang, X., Luan, Z., Qi, P., Lin, M., Tan, Y., Xia, Y., Li, Y., Sui, K.: Design of injectable agar/NaCl/polyacrylamide ionic hydrogels for high performance strain sensors. Carbohydr. Polym. **211**, 322–328 (2019)
102. Tong, R., Ma, Z., Gu, P., Yao, R., Li, T., Zeng, M., Guo, F., Liu, L., Xu, J.: Stretchable and sensitive sodium alginate ionic hydrogel fibers for flexible strain sensors. Int. J. Biol. Macromol. **246**, 125683 (2023)
103. Alkabes, H.A., Elksass, S., El-Kelany, K.E., El-Kemary, M.: A wirelessly multi stimuli-responsive ultra-sensitive and self-healable wearable strain sensor based on silver quantum dots of 3D organo-hydrogel nanocomposites. J. Mater. Chem. C. **9**, 17291–17306 (2021)
104. Zhang, H., Li, W.X., Tang, S., Chen, Y., Lan, L.M., Li, S., Xiong, M., Hu, X., Liu, Y.H., Sun, J.: A boron-based probe driven theranostic hydrogel dressing for visual monitoring and matching chronic wound healing. Adv. Funct. Mater. **33**, 2305580 (2023)
105. Chen, Z., Song, S., Zeng, H., Ge, Z., Liu, B., Fan, Z.: 3D printing MOF nanozyme hydrogel with dual enzymatic activities and visualized glucose monitoring for diabetic wound healing. Chem. Eng. J. **471**, 144649 (2023)
106. Liang, Y., Bar-Shir, A., Song, X., Gilad, A.A., Walczak, P., Bulte, J.W.: Label-free imaging of gelatin-containing hydrogel scaffolds. Biomaterials. **42**, 144–150 (2015)
107. Biela, A., Watkinson, M., Meier, U.C., Baker, D., Giovannoni, G., Becer, C.R., Krause, S.: Disposable MMP-9 sensor based on the degradation of peptide cross-linked hydrogel films using electrochemical impedance spectroscopy. Biosens. Bioelectron. **68**, 660–667 (2015)
108. Wang, C., Huang, C.-Y., Lin, W.: Optical ATP biosensor for extracellular ATP measurement. Biosens. Bioelectron. **43**, 355–361 (2013)
109. Wang, L., Zhou, M., Xu, T., Zhang, X.: Multifunctional hydrogel as wound dressing for intelligent wound monitoring. Chem. Eng. J. **433**, 134625 (2022)
110. Farahani, M., Shafiee, A.: Wound healing: from passive to smart dressings. Adv. Healthc. Mater. **10**, 2100477 (2021)
111. Pandit, A.H., Mazumdar, N., Ahmad, S.: Periodate oxidized hyaluronic acid-based hydrogel scaffolds for tissue engineering applications. Int. J. Biol. Macromol. **137**, 853–869 (2019)
112. Thomas, M.S., Pillai, P.K., Thomas, S., Pothen, L.A.: Biopolymer-based nanofibers-synthesis, characterization, and application in tissue engineering and regenerative medicine. Funct Biomater Des Dev Biotechnol Pharmacol Biomed. **1**, 269–285 (2023)
113. Chen, Y., Zhou, Y., Hu, Z., Lu, W., Li, Z., Gao, N., Liu, N., Li, Y., He, J., Gao, Q.: Gelatin-based metamaterial hydrogel films with high conformality for ultra-soft tissue monitoring. Nano-Micro Lett. **16**, 34 (2024)
114. Gaharwar, A.K., Schexnailder, P.J., Jin, Q., Wu, C.-J., Schmidt, G.: Addition of chitosan to silicate cross-linked PEO for tuning osteoblast cell adhesion and mineralization. ACS Appl. Mater. Interfaces. **2**, 3119–3127 (2010)
115. Paul, A., Manoharan, V., Krafft, D., Assmann, A., Uquillas, J.A., Shin, S.R., Hasan, A., Hussain, M.A., Memic, A., Gaharwar, A.K.: Nanoengineered biomimetic hydrogels for guiding human stem cell osteogenesis in three dimensional microenvironments. J. Mater. Chem. B. **4**, 3544–3554 (2016)
116. Lee, S., Sani, E.S., Spencer, A.R., Guan, Y., Weiss, A.S., Annabi, N.: Human-recombinant-elastin-based bioinks for 3D bioprinting of vascularized soft tissues. Adv. Mater. **32**, 2003915 (2020)
117. Wu, T., Liu, W.: Functional hydrogels for the treatment of myocardial infarction. NPG Asia Mater. **14**, 9 (2022)
118. Khan, M.U.A., Stojanović, G.M., Abdullah, M.F.B., Dolatshahi-Pirouz, A., Marei, H.E., Ashammakhi, N., Hasan, A.: Fundamental properties of smart hydrogels for tissue engineering applications: a review. Int. J. Biol. Macromol. **254**, 127882 (2023)

Soft Materials for Biomedical Engineering

Mitesh Chakraborty and Joydip Dutta

Abstract This chapter discusses the different types and synthesis methods of soft materials. We have also reported biomaterials synthesized from natural proteins. In this context, we have discussed hydrogels, their structural formation, and their application in soft robotics for biomedical purposes. Biomedical applications of advanced biomaterials, shape memory alloys, soft robotics, the design of soft materials for drug delivery, biomimetics, and hydrogel-based biomedical sensors are explored in the present discussion. We have also analyzed hydrogels' applicability and performance parameters based on volumetric changes and elasticity factors. Though we have discussed the applicability of hydrogels in sensing, actuation devices, robotics, and advanced surgical procedures, there is still scope of new paradigms for uses of hydrogels in multidisciplinary fields of material and medical sciences.

Keywords Hydrogels · Polymers · Sensors · Actuators · Biomaterials

1 Introduction

Soft materials refer to a wide range of materials, viz., polymers, gels, and foams. They are distinguished from other materials due to their flexible nature. They can be easily deformed or molded without any external high energy or mechanical pressure [1]. Soft matter materials are usually formed of bigger building blocks, which make their physical and chemical properties unique and distinct from metals and ceramics. Soft materials can be used for various purposes, from making food items to preparing medicines [2]. Nowadays, intensive research is conducted in different

M. Chakraborty
Department of Physics, St. Xavier's College, Ranchi, Jharkhand, India

J. Dutta (✉)
Central Research Facility, Indian Institute of Technology (Indian School of Mines) Dhanbad, Dhanbad, Jharkhand, India

V. Kumar, Y. K. Mishra (eds.), *Soft Materials for Functional Applications*,
Advanced Structured Materials 225, https://doi.org/10.1007/978-981-97-9468-3_4

laboratories across the globe to study the complex behavior of these materials using spectroscopic techniques and microscopy to develop new materials with tailored properties [3]. In the fields of biology, translational medicine, bioelectronics, and personal healthcare, it has been observed that soft materials present unique opportunities as possible contenders for human–machine interactions [4]. Advanced diagnosis and treatments have been made possible on a large scale using soft materials. Nevertheless, there are lots of scientific and technological hindrances to overcome with intensive research and investigation in this field. The lack of technological progress and new theoretical models obstructs the development of sophisticated soft materials. As a result, scientists and academicians are presently ignoring many promising bio-applications.

This chapter aims to support and compile the most recent developments in soft materials. We have reported various methods of design, synthesis, characterization, production, theoretical analysis, and flexible biological applications. Present challenges and further scope of technological progress are also discussed in this chapter. Modern multifunctional soft materials are the main topic of this section since they offer solutions to pressing problems in bioengineering and medicine.

2 Methods for the Synthesis of Soft Materials

Scientists are doing intensive research about food and biomaterials, often termed polymer materials, and are interchanging them as gels or hydrogels. According to their mechanical strength of steady-state flow, gels are distinguished as strong or weak. Gels are defined as cross-linked polymer structures [5, 6]. As per the reported study, hydrocolloids or gelling polysaccharides are the primary source of edible gels that are used in the food sector [7]. As studied by Rosiak and Yoshii [8], the word "hydrogel" is said to be three-dimensional network structures made of a class of synthetic and/or natural polymers that have a substantial capacity to absorb and hold onto water. The hydrophilic groups or domains found in a polymeric network during hydration in water give rise to the hydrogel structure [8]. The process of chemically combining macromolecular chains is known as gelation. Depending on the structure and conformation of the starting material, this process first produces increasingly larger branched yet soluble polymers. "Sol" refers to the mixture of such polydisperse soluble branching polymers. The branched polymer gets bigger and less soluble the longer the linking process is carried out. Finite branched polymers are woven throughout this "infinite polymer," which is also referred to as the "gel" or "network." The "sol-gel transition" (also known as "gelation") is the change from a system with finite branching polymer to infinite molecules, and the "gel point" is the crucial point at which gel first occurs [9].

In general, hydrogels are classified into two groups:

1. Permanent/chemical gel: networks that are covalently cross-linked (i.e., where a stronger and more persistent covalent bond replaces a hydrogen bond) are

referred to as "permanent" or "chemical" gels [10]. They reach an equilibrium swelling condition, determined by the cross-link density and the polymer–water interaction parameter [11].

2. Reversible/physical gel: when molecular entanglements and/or secondary forces, such as ionic interaction, hydrogen bonding, or hydrophobic interactions, hold the networks together, they are referred to as "reversible" or "physical" gels.

   Physical interactions between distinct polymer chains in physically cross-linked gels inhibit dissolution [12]. These interactions are reversible and susceptible to interruptions from stressors or external environment modifications [10, 12].

3 Characteristics of a Hydrogel

The two key properties of a hydrogel are its permeability and water-holding capacity [13]. When polar hydrophilic groups come into touch with water, they become hydrated first, which causes primary-bound water to develop. The network swells, exposing the hydrophobic groups that can interact with the water molecules. As a result, "secondary bound water," also known as hydrophobically bound water, is created. "Total bound water" is the term used to refer to combined primary and secondary bound water. The network will absorb more water because the network's chains are driven toward infinite dilution by osmotic forces.

The hydrogel's third most essential necessary characteristic attribute is biocompatibility. The term "biocompatibility" refers to the hydrogel's immune system compatibility and the nontoxic breakdown products that result from it. Ideally, they can be eliminated by the renal filtering process or converted into innocuous compounds. Since their hydrophilic surface has a low interfacial free energy when in contact with bodily fluids, hydrogels generally have excellent biocompatibility since proteins and cells have a lower inclination to cling to these surfaces. Additionally, hydrogels' rubbery, soft texture reduces tissue irritation [14, 15]. The cross-links between the various polymer chains produce a gel-like substance and sometimes pure elastic or viscoelastic behavior.

3.1 Alginate

Brown seaweed is a cheap source of alginate, a water-soluble sodium salt of alginic acid, an anionic natural polymer. Because of its biocompatibility, low toxicity, and quick gelation in the presence of divalent cations, such as calcium cations like Ca^{2+}, it has been observed from a literature survey that alginate has been used in medical applications. The mannuronic acid (M) and guluronic acid (G) blocks that make up the chemical structure of alginate form a linear copolymer comprising various lengthy sequences of these acids [16]. According to reported works, variations in

M and/or G concentrations and variations in them are dependent upon the original sources [17]. G sequences in the alginate copolymer are also thought to be in charge of gel formation.

3.2 Chitosan

Regarding the many potential medicinal uses, chitosan is thought to be the most desirable biopolymer among polysaccharides due to its great biocompatibility, nontoxicity, and biodegradability, among other qualities. Chitosan has been shown to exhibit a variety of bioactive potentials that can restrict the growth of bacteria, fungi, and even anticancer agent properties, as reported by Nada et al. [18]. Chitosan has also been explored in, viz., medical field, food sector, cosmetics, textile industry, etc. [19].

Chitosan-Based Hydrogel Formation

Chitosan has been used in hydrogels to create liquid gels, microspheres, nanospheres, saturation films, capsules, and beads, among other forms. A physical or chemical cross-linking procedure can create networks in any chitosan hydrogel. The chemical structure of chitosan is reversible, and the physical cross-linking hydrogel can be formed under several factors that regulate the gel's behavior [20, 21]. The cationic amino groups of chitosan can form an ionic complex with negatively charged molecules or anions, such as sodium sulfate ions [22]. Metal ions like platinum, penta sodium tripolyphosphate [23], and many other anions form non-covalent cross-linked chitosan networks. The resulting hydrogel from those ionic interactions has a small lifetime; it is reversible and unstable when subjected to temperature and pH.

3.3 Polymer Gels

Even though the gel is widely used and readily available, it is challenging to categorize polymer gels as a single kind of material. A state of materials halfway between a solid and a liquid is called gel. According to Jen et al. [24], gels are defined as diluted cross-linked matrix systems that contain fluids and exhibit no flow when it is in their steady-state form. When an external medium, such as water or hydrogel, oil, or air, fills the pores of the network, these systems can be cross-linked through physical or chemical connections [25]. Theoretically, the gel network expands to reach the equilibrium state, which is determined by the interaction of osmotic forces. According to Singh et al. [26], this kind of polymer extension results in a high swelling capacity and medium absorption that can reach up to 10–20 times its molecular weight. It is important to note that gel expands but does not dissolve; the hydrophilic functional groups affixed to the polymeric chains allow the gel to expand, while the cross-links between the polymer chains give the gel its resistance to dissolution. This encourages gel to be viewed as the perfect material for absorbing substances used in many applications, like plant soil and baby diapers [26].

3.4 Cellulose

The most prevalent natural polymer, like cellulose, is the primary constituent of other plants and cotton and linen. The glucose units that make up cellulose's chemical structure are linked to one another by 1,4-beta-glucosidic bonds. Certain bacteria, such as *Acetobacter xylinum*, can also biosynthesize cellulose, although their physical and macromolecular characteristics differ slightly. Both plant and bacterial sources can yield cellulose, exhibiting a high degree of crystallinity, typically 40–60% for plant cellulose and 61–72% for bacterial cellulose. The prolonged hydrogen bonding along the cellulose chains, which results in highly tight and well-packed chains, also prevents cellulose dissolution in common solvents and is the cause of the high crystallinity [27].

3.5 Hyaluronic Acid

Hyaluronic acid (HA) is a liner copolymer of alternative repeating disaccharide units, namely, beta-1′,4′-D-glucuronic acid and beta-1,3-*N*-acetyl-D-glucosamine. In many areas of the human body, extracellular tissues naturally contain HA. Because of its superior ability to retain water, HA is involved in numerous biological processes that involve the preservation of moisture in tissues such as the skin, joints, and eyes. Furthermore, physiological enzymes easily break down HA, producing safe by-products. Consequently, among many other polysaccharides, HA is considered a promising biopolymer since it exhibits improved biocompatibility for tissue-related applications [28].

3.6 Starch

Potatoes, rice, and other locally accessible resources are used to make starch, a renewable polysaccharide. Two distinct forms of glucose with repeating units make up starch, branching amylopectin, which is linear and helical amylose. According to Reis et al. [29], starch has certain drawbacks, including poor processability, and is insoluble in cold water.

4 Biomaterials Synthesized from Natural Proteins

4.1 Silk Polymers

Due to its use in the textile industry, silk is one of the oldest insect-derived sophisticated biomaterials. Initially, numerous insects produced silk for various uses, including cocoons for silkworm survival and orb webs. Similarly some of the hymenoptera like bees, wasps, and ants build their nests for spiders to catch prey [30]. The dragline silk of spiders (*Nephila clavipes* and *Araneus diadematus*) and the cocoon silk of the mulberry silkworm *Bombyx mori* are currently the best-characterized silks. Researchers have intensively investigated their remarkable properties.

4.2 Other Types of Biomaterials

The main proteins in the extracellular matrix (ECM) are collagens, distinguished by their triple-helical molecular structure, made up of repeating amino acid sequences. They contain a lot of proline and necessitate the hydroxyproline (Hyp) that has been posttranslationally modified to facilitate stability. According to Chattopadhyay and Raines [31] and other studies, the most prevalent collagens are extensively utilized to create axially periodic fibrils in the tendon, bone, cartilage, and other tissues, and they play a significant role in cell signaling and development [32].

5 Advanced Biomaterials for Biomedical Engineering Applications

Natural protein-based biopolymers of good quality and synthetic polymers include collagen, elastin, and silk. These materials are better alternatives for biomedical and pharmaceutical applications and support cell attachment and tissue development. The biomechanical properties of natural protein-based biopolymers are distinctive and diverse, and their controlled biodegradability and strong biocompatibility make them attractive candidates for use as advanced biomaterials in wound healing, drug delivery, and tissue engineering [33, 34].

To increase their uses as sophisticated biomaterials, the produced protein-based biopolymers can be further processed into a variety of forms, including films, fibers, scaffolds, and hydrogels. Leal-Egana and Scheibel [35], in their research, have observed that among these natural polymers, silk-based biomaterials derived from silkworm cocoon silk have been utilized for sutures, and if sericin is removed, the core fibers of these biomaterials are biocompatible with the majority of regularly used biomaterials. Additionally, natural spider silks are shown to be noncytotoxic, lowly immunogenic, and non-inflammatory. Because of the wax coatings applied to silk fibers, silk is categorized as an advanced biomaterial that is not biodegradable.

Nevertheless, recent investigations into the enzymatic breakdown of silk have demonstrated that it is prone to proteolytic degradation, gradually disintegrating into smaller polypeptides and free amino acids when exposed to protease, which breaks down the antiparallel β-sheet structures of silks into nanofibrils and then into nanofilaments and α-chymotrypsin, which cleaves the less crystalline regions of the silk protein into peptides [36, 37].

5.1 Sustained Drug Delivery

To deliver the drug at the defined therapeutic region, sustained drug delivery is one of the potential alternatives for chronic use medications with effective clinical benefits such as reduced side effects, low toxicity, and improvement in patient convenience and compliance. This approach to drug delivery is cost-effective and highly efficient [38].

One of the most important requirements for both drug delivery and regenerative medicine applications is the regulated release of pharmaceuticals from the drug delivery system following accumulation at a designated site. Environmental factors such as pH and temperature can initiate the process. Innovative delivery methods for various bioactive substances, such as genes, small molecules, and biological drugs, have been studied for natural polymers made from proteins, such as collagen, elastin, and silk [39]. These biomaterials are biocompatible and biodegradable, have antibody properties, are low toxic, and have effective drug loading and release transfer at the target site. Additionally, they can form, combine in addition to their abilities to form, emulsify, freeze, and bind water. These materials are also used as drug carriers in cancer treatment, cartilage restoration, and vascular grafting [40].

Future enhanced drug delivery will need to take into account a variety of factors, including material production, processing conditions, drug compounds employed, and drug release kinetics and processes, to meet the demands of controlled drug delivery effectively. The design of specific sizes for drug-loaded nanoparticles is one of the most crucial requirements in protein biopolymer-based nanoparticle delivery systems to cross epithelial barriers, circulate in blood vessels before reaching the target site, and prevent inflammatory or immunological reactions [41]. To accomplish site-specific delivery and bioactive drug release at the necessary rate and amount, biopolymer-based nanoparticles' sizes, morphologies, solubility, biodegradability, and surface features must be considered for cellular internalization via endocytosis [41].

Because of their special mechanical qualities, controlled biodegradation into non-inflammatory by-products, aqueous-based ambient purification and processing options, biocompatibility with sterilization methods, and usefulness in drug stabilization, silk protein-based materials have been taken into consideration for advanced drug delivery systems [42]. Many silk-fiber broin-based formats, including hydrogels, films, tubes, nano-/microspheres, and transdermal microneedles, have been investigated from the aqueous silk-fiber broin solution and processed into materials for advanced drug delivery [43].

Different drug release behavior has been seen in silks utilized with varying hydrophobicity and charge due to investigations on drug molecules' distribution and loading efficiency using silk micro- and nanospheres with adjustable diameters [44]. The application of silk film coating for delivering small molecules was also investigated. Controlled drying and silk film treatment were used to govern the following drug release, with drug retention times ranging from a few hours to more than 10 days.

5.2 Gene Delivery and Its Related Therapy

Gene therapy has been used to treat a variety of illnesses, including cancer, AIDS, and cardiovascular disorders. It is thought that human diseases can be healed by introducing genetic materials into particular patient cells to replace damaged genes [41]. Using either bare DNA or a viral vector, the delivery of therapeutic genes into the patient's target cells is a potential method for treating various disorders. To accomplish this, the sophisticated gene delivery system incorporating gene encapsulation must be sufficiently small to enter cells and travel to the nucleus, avoid being processed by endosomes and lysosomes after endocytosis, and ultimately shield the gene until it reaches its intended location. Because they can form a hydrogel when injected into the body while still liquid at ambient temperature, the stimuli-responsive elastin-like polypeptides (ELPs) and protein-coding genes (SELPs) have the potential to be used as a sophisticated polymeric matrix for gene delivery. This is an appealing property for any injectable system application. For instance, regulated gene delivery via adenoviral vector delivery and recombinant SELP-47K hydrogels have been investigated. These reports demonstrate inverse DNA release and diffusivity linked to the molecular weight of the plasmids utilized [45]. Further investigations into the binding mechanism of DNA to SELPs were conducted by examining the effects of several parameters, including molecular weight, ionic strength, DNA concentration, and SELP concentration. As the ionic strength of the buffer is increased or the concentration of the polymers is lowered, the results indicate a greater release of DNA and adenovirus bound inside polymeric matrices [46].

The most successful path for adenoviral gene therapy in the treatment of cancer was shown by SELP hydrogels loaded with adenovirus in a mouse model, which showed a larger reduction in tumor volume than control injections of adenovirus in saline solution [41, 46].

5.3 Tissue Regeneration and Tissue Engineering

The scarcity of donors and rising morbidity severely restrict the therapeutic options for organ failure, and it remains difficult to replicate the in vivo physiological milieu in a way that will elicit the necessary cellular responses in functional tissue regeneration. Effective communication between the implanted device and the host is essential when replacing a biological element of the body with a material. Tissue engineering aims to regenerate tissue within an appropriate scaffold, so the created tissue can be implanted at the intended location modeled as cutting-edge biomaterials. As an architectural template, a 3D tissue engineering scaffold might offer an appropriate microenvironment [47]. Since structural proteins, such as collagen, elastin, and silk, are found in natural tissues, using them as porous 3D tissue scaffolds in tissue engineering makes sense.

Functional parameters such as surface topography, mitogenesis, cell differentiation, and cell attachment phenomena are important for the regeneration and implantation of tissues. Researchers have explored the formation of functional tissues and organs using biomaterials that use extracellular matrix (ECM) for different requirements of tissues [48]. One of the biomaterial illustrations is silk protein fibroin. This is engineered due to its strength, flexibility, and high compatible percentage with mammalian cells. This is used to aid in advancing biological devices [49]. Tissue engineering scaffolds can be made via solid free-form fabrication, freeze drying, porogenic leaching, and silk porous 3D sponges. All of these provide excellent control over porosity and pore sizes. The resultant sponges have pore diameters ranging from 60 to 250 μm, depending on the type of organic solvents, pH, and freezing temperature. Through surface chemical modification and genetic engineering, biomaterials can be customized to create sophisticated hybrid and composite systems that can meet the specific needs of each tissue [50]. For instance, dentin matrix protein can mineralize calcium phosphate (CaP), and the mineralizing domain R5 can perform bone-like properties in spider silks [51]. Silk protein properties can also be improved by binding and delivering cell signaling factors like RGD-functionalized SLP, which improves cell adhesion [52].

With its non-immunogenic, biocompatible, and biodegradable qualities, elastin is a major component of the ECM and a promising source of advanced materials for tissue engineering for elastin-derived biomaterials. Easy purification by using their ITT from bacteria that produce large amounts of ELP proteins makes ELPs an excellent option for liver, vascular, ocular, and cartilaginous tissue regeneration. To meet the needs of the application, ELPs can be further processed to create a variety of advanced material types, including hydrogels, films, and fibers. For instance, ELP hydrogels were employed to simulate ECM-like 3D environments for cell encapsulation in tissue engineering. In one of the investigations by McHale et al. the lysine-enzymatic cross-linking of ELPs causes chondrocytes to be encapsulated and a collagen II-rich hyaline cartilage-like substrate to develop [53]. The ELP hydrogels can be further designed with gelation rate and gel stiffness controlled by H_2O_2 and protein concentration by using cysteine-based disulfide bridge crosslinks. This allows for cell encapsulation and the in situ formation of soft gels useful for tissue engineering [53]. ELP cross-linked hydrogels can also be molded into elastic films, which can further be cross-linked by utilizing chemicals to join lysine residues and fused with functional modules and cell-adhesive motifs. For instance, in the basic surface treatments of biochemical effects on cells, ELPs fused with RGD and CS5 domain have been produced, generally enhancing cell-interactive characteristics of ELP scaffold surfaces [54] as reported by Rodriguez et al. Studies by Benitez et al. have reported that a large surface area of ELP fibers derived from concentrated protein solutions can be seen in the cell signaling modules more clearly.

The exceptional mechanical properties of resilin and RLPs can be explored as prospective advanced biomaterials for tissue engineering. As per investigation on the chimeric material, RZ10 coupled with the RGD sequence, which is obtained from fibronectin, has revealed that the faster cells spread to human mesenchymal stem cells (hMSCs) with well-organized actin structures. Resilin-based biomaterials (RZ10), for example, have demonstrated an undefined compressive modulus similar to that of human cartilage. These results suggest that, as described by Renner et al. the RGD sequence can be selectively identified and used to facilitate cell adhesion and spread within resilin-based materials [55].

Furthermore, at high frequencies corresponding to the human voice, Resilin RLP12 hydrogels demonstrated extensibility and mechanical strength comparable to native vocal fold tissues [56]. Moreover, resilin RLP24 supports the viability and spreading of encapsulated human aortic adventitial fibroblasts, which are useful for cardiovascular applications [56].

6 Shape Memory Polymers and Their Biomedical Applications

Smart and adaptive materials like shape memory polymers (SMPs) can regain their shape in response to an external input. This feature has attracted much interest in biological applications as it shows high biocompatible properties of polymers [57]. This technological progress will pave the way for the potential integration of SMPs into device design, which could result in significant technological advancements within the biomedical industry [58].

Numerous instances of stimulus-responsive systems can be found in the natural world. One of the plants with the biological name *Mimosa pudica* has a thigmonastic response, which means leaves will fold when physically touched. Another example of this plant physiology is the Venus flytrap with the biological name *Dionaea muscipula*, which possesses a mechanism for shutting its leaves when insects rub their hairs on the leaf surface repeatedly [59, 60]. These illustrations inspire the creation of artificially intelligent materials, whose stimuli-responsiveness may be explored in different situations. Due to its adaptability, improved biocompatibility, and potential biodegradability compared to other materials, soft stimuli-responsive polymeric materials have received a lot of attention among the wide variety of stimuli-responsive materials [61, 62].

SMPs are intelligent, flexible materials that may regain their original property in response to external effects like temperature, light, magnetic fields, electrical currents, etc. or may show new characteristics [63]. SMPs are not limited to pure polymeric systems; they can also be found in polymer networks like hydrogels and polymer blends and composites (SMPCs) [64]. Unlike form memory alloys, shape memory soft materials are less expensive and lighter and can withstand higher recoverable strains. Due to these characteristics, SMPs can be used in household and commercial activities, aerospace technology, textiles, actuators, sensors, microfluidic systems, and biomedical applications, viz., cardiovascular stents and vascular grafts [65, 66].

As early as the 1960s, heat-shrinkable polyethylene tubes were utilized as wire wraps for electrical insulation, utilizing the concept of the shape memory effect in polymers in commercial products. Reports of SMPs that are activated by various stimuli and display various properties have been published more frequently during the last three decades. Thus, a categorical review that highlights the most recent developments in this field is required.

6.1 Design Consideration of SMPs

Biocompatibility, or the capacity of a material to be used in biomedical applications, is among the most crucial design factors of any content that must be performed in vivo while eliciting a suitable host reaction. Therefore, it is envisaged that SMPs intended for biological applications will not be cytotoxic as a fundamental requirement. As a first assessment of biocompatibility, in vitro, cytotoxicity investigations using cell viability assays and cell membrane integrity assays are frequently employed. Commonly used tests for determining cell viability include MTT and MTS assays [67], based on reducing a tetrazolium molecule into an insoluble, colored formazan product. Moreover, techniques for evaluating cell membrane integrity using trypan blue tests 47, 48 and lactate dehydrogenase (LDH) assays [68] are well-established.

SMPs have a wide range of mechanical and physical characteristics that make it possible to better tailor adaptive materials to the needs of human tissue in biomedical systems [69]. Specifically, SMPs have demonstrated significant promise in emulating the physical characteristics of soft biological tissue, thereby extending the biomedical applicability of smart materials beyond the previously investigated domains of hard biological tissue, a field in which shape memory alloys have the upper hand.

The term "biodegradation" has been used in the biomedical area to refer to degradation processes, including hydrolytic and oxidation reactions, despite its different definitions that vary depending on the domain in issue [70]. When designing a shape memory polymer, biodegradability is a crucial factor that heavily relies on the desired permanence of the finished product. To withstand the body's breakdown and long functional use, materials utilize permanent biomedical items that are not biodegradable. However, biodegradability will provide a transitory existence for goods that are only needed to fulfill a single purpose, negating the need for surgical removal once utilized [71].

Preclinical sterilizing is a necessary precaution for shape memory polymers developed for biomedical applications as required for all medical devices [72]. The SMP has to be sterilized without losing any of its quality. In this context, parameters like chemical and thermal stability must be considered in addition to the shape memory effect activation mechanism. The US Food and Drug Administration's (FDA) handbook lists steam exposure, ethylene oxide, and radiation as sterilization techniques [73]. As each method is limited, the challenge remains to accommodate all types of shape memory polymers.

6.2 Biomedical Applications of SMPs

Generally speaking, thermoplastic physically cross-linked systems and chemically cross-linked thermoset networks are the two primary types of SMPs. A chemically cross-linked electroactive SMP made of six-armed polylactic acid (PLA) and aniline trimer (AT) was recently reported by Xie et al. [74]. It was demonstrated that the biodegradable PLA/AT thermoset was electroactive and had superior mechanical and shape memory qualities. Furthermore, immunofluorescence, enzyme activity, and relative gene expression tests supported the copolymer's better osteogenic differentiation capabilities, proving that it is a promising material for bone tissue engineering applications.

Physically cross-linked SMP systems can be remolded after synthesis by overcoming the physical contacts through processing and reprocessing using solvents or heat [75]. Because of this, physically cross-linked SMP systems can be created in various shapes to fit a range of needs. Among the most prominent types are SMP foams and micro- and nanofibers, which have a lot of potential for use in biological applications. The increased interest in nanostructured materials for biomedical purposes has led to increased use of certain processing techniques, including electrospinning. In their investigation, Kai et al. have reported various possible uses of electrospun nanofibers, such as biosensors, medication delivery, and tissue engineering [76, 77].

Chen et al. had previously reported using triethoxysilane end-capped copolymer that is subsequently cross-linked after electrospinning to create an electrospun SMP film with a reversible fibrous structure. Comparing the SMP film to bulk polymer samples, the scientists discovered that the SMP film had a faster shape recovery reaction due to its stable fibrous structure [78]. This discovery provides a foundation for further improvement in the efficiency of the surface and shape memory qualities of SMPs; hence, it can be further explored as potential applications in biological systems.

The possibility of electrospinning D-lactide copolymerized with trimethylene carbonate to create a biodegradable shape memory copolymer was investigated by Zhang et al. [79] as presented in Fig. 1. The result was a biomimetic scaffold with a shape memory effect supported by the complementary qualities of SMPs and nanofibers. The resulting scaffold demonstrated good osteoblast proliferation, shape fixity, and recovery (Fig. 3), suggesting its potential use as a scaffold for less invasive medical treatments involving bone regeneration.

In addition to nanofibers, biological applications may also benefit from porous SMP structures like SMP foams. According to research by Marino et al. [80], PCL-diacrylate SMP foams coated with polydopamine increase the production of osteogenic proteins in human mesenchymal stem cells (MSCs).

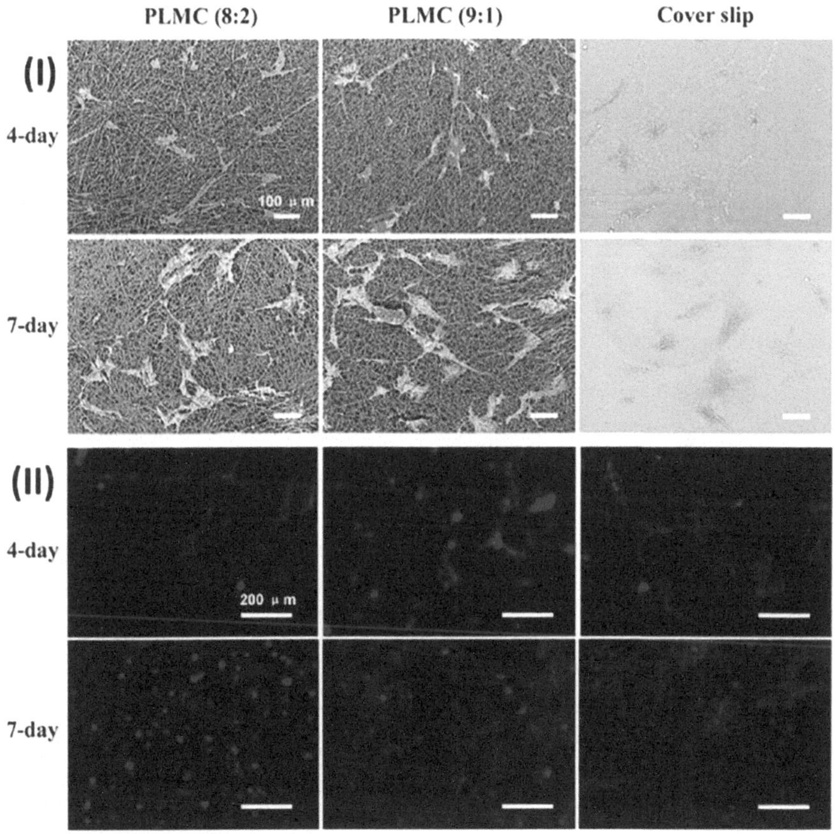

Fig. 1 (**I**) SEM images and (**II**) stained images of electrospun poly(D,L-lactide-*co*-trimethylene carbonate) nanofibers characterized by fluorescence spectroscopy. The images show high-quality proliferation and osteoblast attachment [Reproduced from ACS Appl. Mater. Interfaces 2014; 6(4), 2611–2621 with permission from the American Chemical Society Publishing]

7 Biomedical Applications of Soft Robotics

The progress of sensitive device-related machines has been made possible in various shapes and sizes due to technological advancement in mechanical engineering. Delicate robots due to their flexible mechanical properties have strongest appeal for therapeutic applications [81] as shown in Fig. 2. The sensitive materials used in the design of robots for biomedical applications must be biocompatible and biomimetic, depending upon human requirements. It has come to notice that the Da Vinci surgical robot [82] can help surgeons improve their precision, predictions, and reproducibility. (Intuitive Surgical) Yet, aside from wrist articulation, this process still calls for stiff and inflexible instruments. Flexible tools, on the other hand, can be utilized in conjunction with MIS techniques and have a high intrinsic flexibility. However, because they are frequently long and flexible with a swiveling tip, they

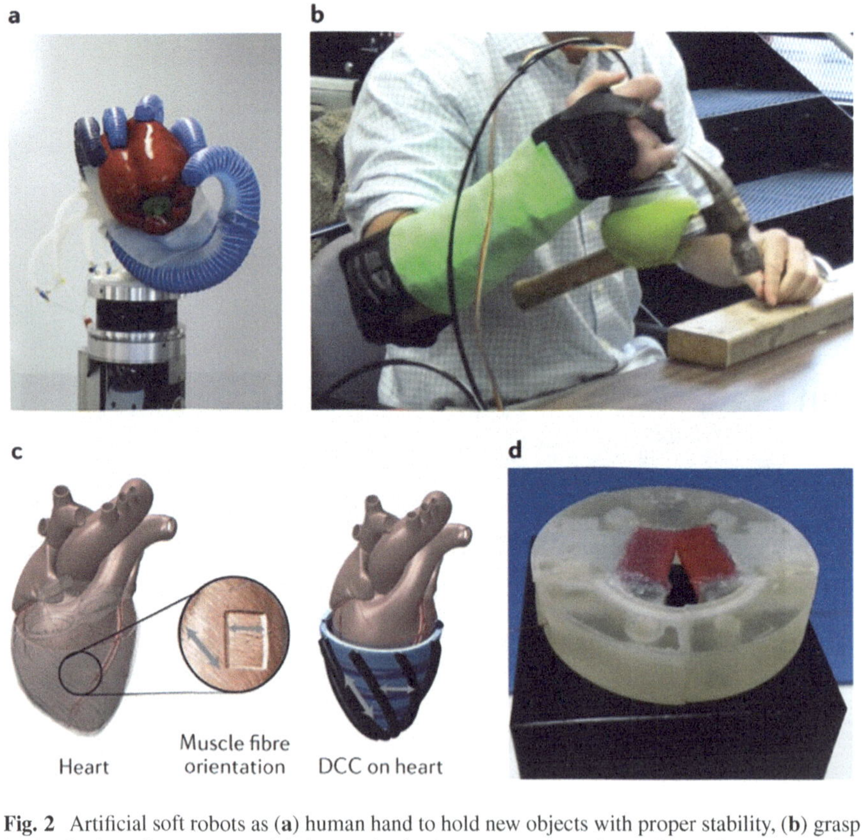

Fig. 2 Artificial soft robots as (**a**) human hand to hold new objects with proper stability, (**b**) grasp holder for jamming purposes, (**c**) thermoplastic sleeve to continue the working of a nonoperating human heart, and (**d**) the simulating device of human vocal folds. (Reproduced from Nat Rev. Mater, 2018, **3**, 143–153 with permission from the Nature Reviews Materials Publishing)

will likely have limited dexterity once they are at the surgical site. Additionally, they are unstable and have a limited range of motion. From a surgical and medical perspective, rigid instruments are the primary requirements for achieving precision during surgery [83]. Numerous solutions for active soft actuation have been investigated to incorporate dexterity, safety, accuracy, and flexibility in soft surgical instruments, with the most innovative materials as flexible fluidic actuators (FFAs) and novel routing techniques for cable-driven flexible mechanisms are examples of promising technologies [84]. Different types of actuation technologies in soft robotics are given in Table 1. The advantages of cable-driven approaches are easy miniaturization if the motors are placed externally and are lightweight. The limitations are that they are typically costly, sterilization issues, and mechanical limitations like fatigue, nonlinear friction, backlash hysteresis, and inadequately transmitted forces [85, 86].

Because they solve some of these problems, smart materials have been investigated for catheter-like operations that call for miniaturization in addition to more conventional endoscopic or surgical procedures. For instance, shape memory alloys (SMAs) have been the subject of much research due to their exceptional qualities, including great corrosion resistance, biocompatibility, and nonmagnetic behavior [83]. They can also be produced in a variety of forms and sizes, have a high energy density, and apply localized stresses. On a modest scale, SMAs provide direct transmission for onboard actuation. Furthermore, because of the way they function, they operate more effectively and quickly when arranged in small wire configurations, which maximizes heat exchange and reduces the amount of current needed for the Joule effect [84, 86].

Table 1 Actuation technologies of soft robotics

Technology	Sections	Mechanism
Flexible fluid actuators	Air muscle, artificial muscles	Modeling of materials into the required
	Fluidic elastomer actuators	Shapes for biological applications
	Pneunets, Pneuflex	A flexible structure that works on fluid
Shape memory materials	Shape memory polymers	Deformation of desired crystal structure
	Shape memory alloys	Through heat energy. This is dependent
	Shape memory composites	Upon specific mechanical properties
Electroactive polymers	Electric and ionic materials	Shape change in polymers due to
		Applied electric field
Tendon driven actuators	Tendons	Small flexural rigidity used in soft robots
Material jamming	Granular and layers	Elastomeric membrane and a filter material
Electric and magnetic	Electric and magnetic	Materials that incorporate magnetic and
Rheological materials	Dependent elastomers	Electrical particles

Another parameter of discussion in SMAs is the change in mechanical properties with temperature. SMAs have limited strain, low speed, and low bandwidth. These features of SMAs are mainly dependent upon thermal regulations and the requirement of high currents to attain the activation temperature. As SMAs have biological applications, they are also known as "muscle actuators" and are applied in super-elastic design [85]. Due to its elastic property, SMAs, when induced by external load, are withdrawn; they regain their original state at constant room temperature. However, when the external load effect is increased and the alloy moves to the plastic region, some deformation in the shape of these materials is observed. Hence, these materials are said to be super-elastic. These types of materials are nowadays used to design robots. The design of these medical devices allows for improving the dimension of small-diameter catheters [82–86].

Soft and shrinkable FFAs [87] are primarily used in colonoscopy [88]. It has been observed from reported studies that FFAs made of pneumatic bellows provide self-propelled motion, and those FFAs that are made of balloons shape increase the friction between the device and the intestine walls. These properties have been explored in inchworm-like active colonoscopy systems. Present-day FFAs are advantageous in surgical procedures as they are workable in the external effects of radioactive exposure and magnetic fields. This is possible because of their mechano-elastic nature. Further, as they are workable in the absence of direct electric current, their functional optimization is enhanced with magnetic resonance imaging. Also, as there is very little stiffness in the motion of FFAs, chamber leaks diminish in the presence of human tissues [89, 90]. It has been observed that the movement of FFAs is complex [91]. Lastly, FFAs are biocompatible materials that make their contact safe with adjacent organs and tissues [92]. In addition, FFAs that use pressure gas are easy to activate, as they are lightweight and flexible. Actuation technologies can further be explored to control the rigidity of these alloys.

7.1 Soft Materials for Drug Delivery

Nowadays, hydrogels and biocompatible and biodegradable materials are used as soft materials in the engineering of soft robotics for medicine transportation in the human body. These soft materials are devised because endoscopic and other surgical tools cannot be effective at remote locations in the human physiological system, viz., the inner portion of the brain, liver, and pancreas, or if the treatment is related to the chronic release of a medicinal drug using the surgical device. It has also been observed that wired surgical instruments cannot deliver medicines precisely to reach minute body tissues or organs. In soft materials, it is possible to administer the drugs at the target location systematically as shown in Fig. 3.

Drug delivery methods can be derived from thin films, nanoparticles, microcarriers, and micelle forms [93]. An active investigation by researchers leads to the introduction of thin-film techniques to overcome the issues of swallowing pills and capsules [94]. Thin films have varying thicknesses ranging from a few nanometers

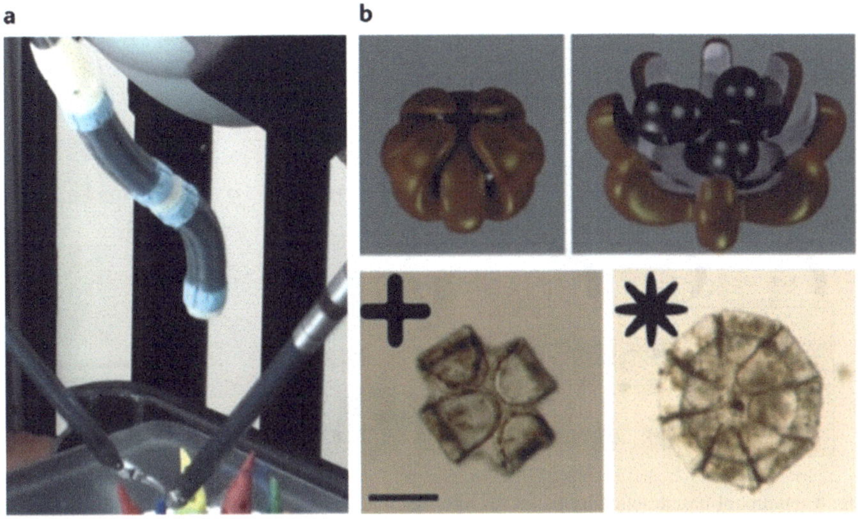

Fig. 3 Artificially prepared soft robots for surgical and drug transportation at specific human organs, tissues, and cells. (**a**) Flexible robot for minimal invasion surgery; (**b**) drug delivery system as displayed with concept at the top image illustration at the bottom on a scale of 100 μm. (Reproduced from Nat Rev. Mater, 2018, **3**, 143–153 with permission from the Nature Reviews Materials Publishing)

to some micrometers. Thin films are highly flexible, have adhesive properties, have high molecular permeability, have a high surface-to-volume ratio, and have good drug-carrying capacity. A polymer material can also be introduced in thin films to transform the host into a highly responsive matrix whose property can be controlled by adding suitable nanofillers for magnetic and optical studies. This further enhances the materials' ability to implement material responses and gives the material new functions.

Because of their straightforward operation, drug delivery systems are the primary application for electro-driven actuators. These systems, also known as electro-simulated drug release devices (EDRDs), release the drug either by diffusion or by shrinking the gels in response to an electric field [95, 96] as reported in Fig. 4. In one experimental study by Liu et al. hydrogel application was explored for deep brain stimulation, low-voltage actuation, and the development of drug release devices using electrical stimulation. In the experimental setup, a polyelectrolyte hydrogel was placed in between two electrodes. Under direct current (DC) voltage, the hydrogen becomes de-swelled and starts to release water [97, 98]. However, there are certain drawbacks to this approach. Firstly, the gels used for electro-response experience stress, which deforms the hydrogel, and this results in time lag and decreases the work efficiency. In order to overcome this problem, researchers are using sol-gels or smart glasses, and decreasing the size of the hydrogels leads

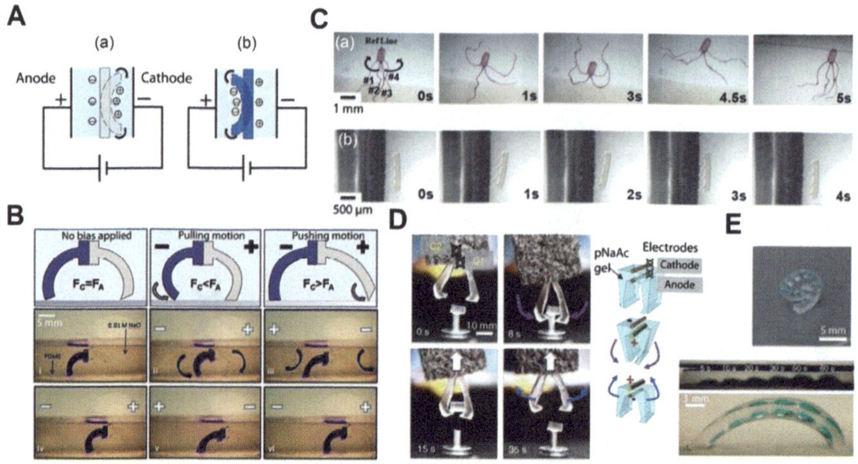

Fig. 4 Hydrogel-based actuators based on the principle of electrode system. (**A**) Illustration of electrodes working with (a) anions and (b) cations in the form of solutions prepared by gels. Both anions and cations can drift in opposite directions by applying an external electric field. (**B**) Image representation of the cationic and anionic motion of the prepared hydrogel. An external field of 5 V/cm was applied, as shown above. (**C**) Image showing the motion of an anion gel (a) octopus and (b) walker under external applied electric fields in solution. In the present case, alternating voltage was applied (+7 V/−15 V) and ± 15 V, respectively. (**D, E**) Representation of one printing using hydrogels. (**D**) Grip holder to suck and release gels. (**E**) Image showing effect on bending due to specified imprint [Reproduced from Biomimetics (Basel). 2018 Jul 10;3(3): 15 with permission from the Biomimetics Publishing]

to improvement in the electro-performance of this system. Secondly, optimal drug delivery depends on the mechanochemical stress factor of hydrogel units.

It is well known that materials that generate motion in response to external magnetic fields are known as magneto-driven actuators. The demand for these types of actuators is increasing day by day in the biomedical field for drug delivery and cell design. The working principle of these actuators is that they move freely in the liquid medium when direct current source magnetic fields are applied and grasp in the presence of alternating current magnetic forces. The actuator system bends, expands, and can even deform in various types of magnetic particles placed in the actuators in the presence of an external magnetic field [99, 100]. From reported studies, it can be said that magnetic actuators are highly efficient in controlling drug release. For drug regulation purposes, in one of the investigations by Barbucci et al. a hybrid magnetic carboxymethylcellulose (CMC) hydrogel was used with $CoFe_2O_4$ nanoparticles, which functions as cross-linkers and improved the performance of magnetic actuators [101, 102]. In recent studies, multi-walled nanotubes (MWNTs) have been incorporated in hydrogel nanocomposites like magneto-responsive hydrogels which is applied in tissue engineering [103] and drug release in controlled way [104, 105]. In Fig. 5, we have shown the basic principles of design and development of magneto-responsive hydrogel-based actuating systems.

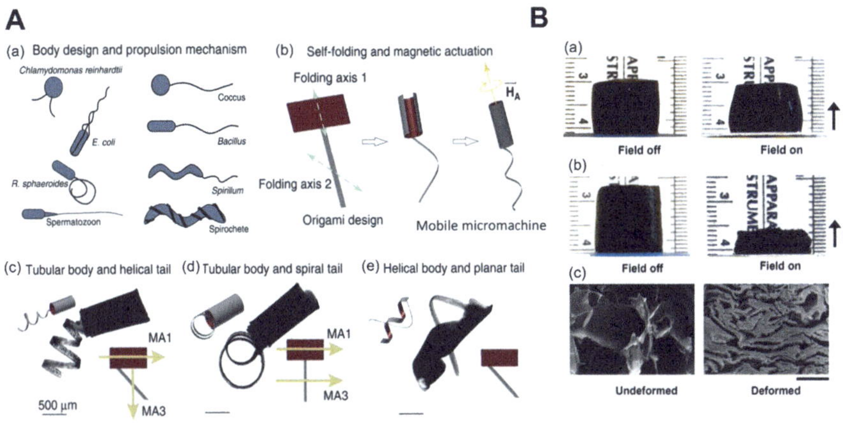

Fig. 5 Image illustration of the magnet dependent hydrogel-based actuators. (**A**) Re-orientable planning of body prepared by soft hydrogel-based small machines. (a) Many types of body styles and propeller systems based on hydrogels. (b) The three-dimensional functioning micromachines are developed through isotropic swelling behavior regulated by magnetic nanoparticle alignment along predetermined three-dimensional paths and selective patterning of supporting layers. The direction in which each compartment should fold is indicated by the folding axes 1 and 2. When a magnetic field is applied, the micromachine's motility is determined by its several distinct magnetic axes. In the presence of a uniform rotating magnetic field, the micromachine which comprises magnetic nanoparticles performs fixed oscillations in all the three dimensions. (**c**–**e**) Optical images of thrashed soft micromachines with complex body plans. The magnetic axes MA1 and MA3 represent the magnetic axes along the head and tail direction, respectively, graduated in a scale of 500 μm [Reproduced from Biomimetics (Basel). 2018 Jul 10; 3(3): 15 with permission from the Biomimetics Publishing]

7.2 *Biomimetics*

Biomimetics is a term often used to solve complex human problems. With technological advancements, biomimetics has provided biological solutions at the macro- and nanoscales, like when the bone marrow was artificially designed using synthetic polyethylene glycol (PEG) hydrogels [106, 107]. Simulation software SOFA can be employed to design artificial models for soft robots and simulated to study the response time effects. In Fig. 6, we have shown some designs of biomimetics for soft robotic applications. Hydrogels are ideal materials to be used in biomimetics as they are biocompatible and have good ability to respond to external effects. Their mechanical parameters like rigidity, strain factors, modulus of elasticity, etc. can be further improved by synthesis of composites at the macro and nano level. For example, as per reported studies, chitosan hydrogels which were designed for tissues and cells can also be used against strong mechanical stress [108–111]. It is of common knowledge that researchers have developed a reversible, temperature-dependent double layer hydrogel similar to the characteristics of *Mimosa pudica*. This plant can move in subtle ways such as nastic, tropisms, and other strategic

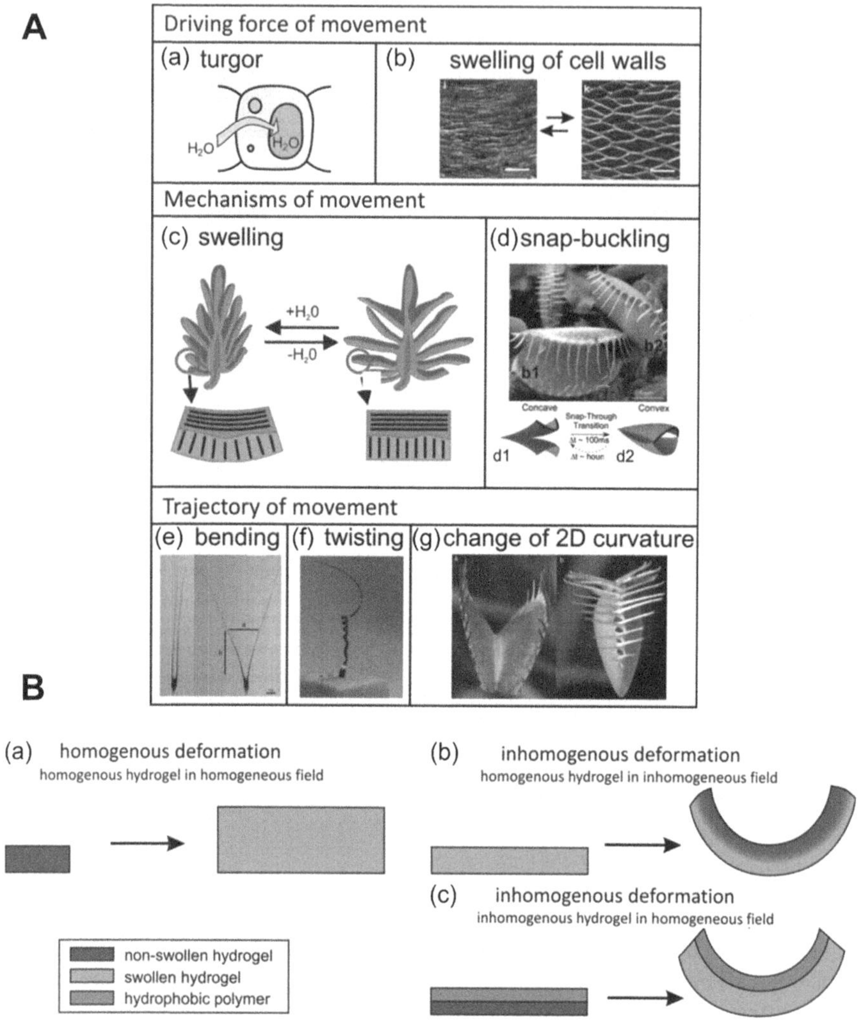

Fig. 6 (**A**) Illustration of reversible movement in plants concerning the applied external force in a (a) turgor and (b) swelling of cell walls; mechanism: (c) swelling and (d) snap-buckling; and trajectory (e) bending, (f) twisting, and (g) orientational change in two-dimensional curvature. (**B**) Different scenarios of swelling of hydrogels (a) uniform deformation in hydrogel (b) non-uniform deformation of homogeneous hydrogel in non-homogenous magnetic field (c) inhomogeneous deformation of inhomogeneous hydrogel in a homogenous magnetic field [Reproduced from Biomimetics (Basel). 2018 Jul 10; 3(3): 15 with permission from the Biomimetics Publishing]

motions. These natural movements of plants can be explored to design biomimetic actuators. Employing hydrogels in biomimetics also has certain defects, viz., intolerance to load-bearing stress after a certain limit, heavy water content, sterilization in hydrogels is a cumbersome process and sterility is the fundamental requirement for use of hydrogels in different sections.

7.3 *Hydrogel-Based Biomedical Sensors*

Biomedical sensors, also known as transducers, are devices that change outside pressure or stimuli like chemical, optical, magnetic, mechanical, acoustic, or biological into bio electromechanical signals using different input systems [112, 113]. For example, photonic crystal hydrogels have the ability to change color when exposed to water; hence, they are used as humidity sensors [114]. Similarly, hydrogels are also used in gas sensors as a structural layer to detect certain specific gases [115, 116]. These material developments are helpful in the diagnosis procedures and medical treatments related to external and internal body movements. A Flowchart description of a hydrogel-based sensor is reported in Fig. 7.

Hydrogels that respond to specific external factors or parameters are often used as sensing material. This is because the individual gel units in the hydrogel are very sensitive to mechanical stress and shape deformations. These gel units, hydrogel shapes, and sizes are also responsible for response time and delay factors. Apart from sensor application, hydrogels can be explored in sensors as stabilizers, structural protective cover, or coats to shield sensor components from unwanted chemicals or other cells. In biosensors, hydrogels are used to protect sensor parts from external effects. They are also used as immobilizing sensors [117, 118]. There are various factors, viz., parametric measurements, biomedical processes, material composition, research progress, etc., for advancement in medical treatments [119–122]. From earlier reported studies, it is known that boronic acid is a glucose-sensitive monomer. Therefore, hydrogel sensors can monitor glucose and treat hyper- or hypoglycemia. A capacitive transducer developed as a microelectromechanical system (MEMS) can measure the dielectric behavior in

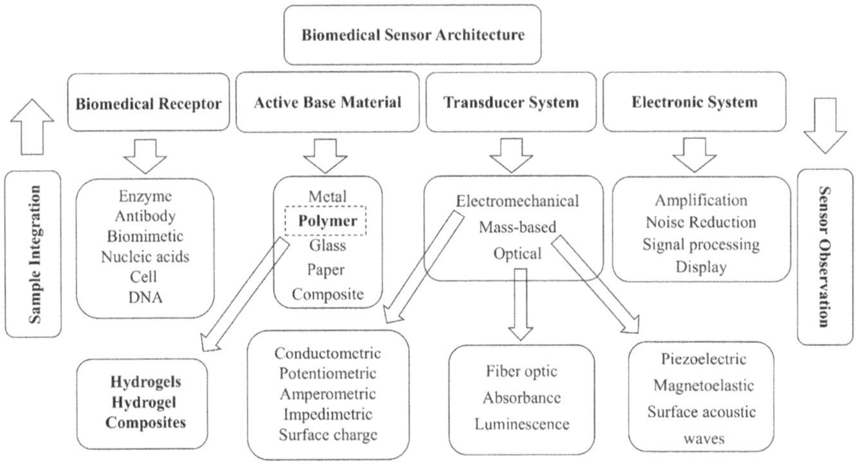

Fig. 7 Flowchart description of a hydrogel-based sensor architecture. [Reproduced from Biomimetics (Basel). 2018 Jul 10; 3(3): 15 with permission from the Biomimetics Publishing]

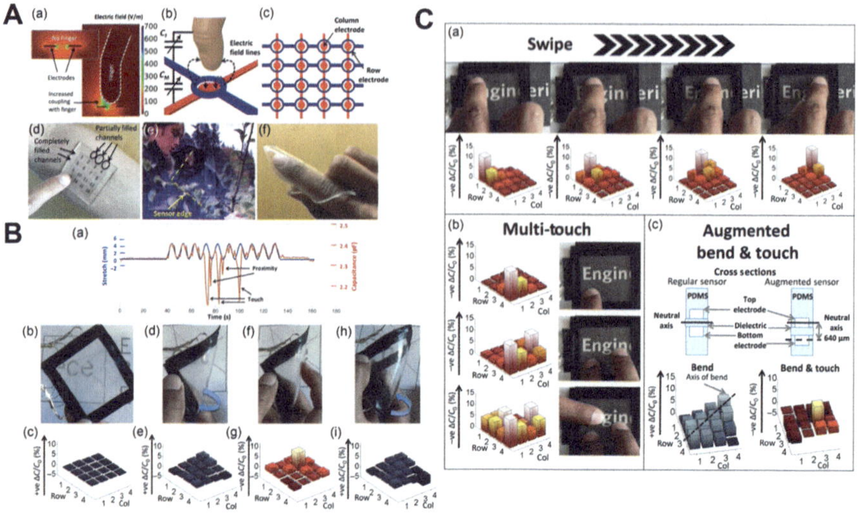

Fig. 8 Exemplification of the development of a touch sensor using elastic and conducting hydrogel electrodes with a projection of an electric field above the sensor to couple with and sense a finger. (**a**) Working mechanism and basic features of the touch sensor. (**b**) Illustration of touch feeling in the sensor. (**c**) Multiple touch features with swipe and bending detection. [Reproduced from Biomimetics (Basel). 2018 Jul 10; 3(3): 15 with permission from the Biomimetics Publishing]

hydrogel due to molecules of glucose that diffuse into the sensor and form chemical bonding with boronic acids through electron affinity [123]. This can be used for the evaluation of glucose concentration. It has also come to notice that using robotic surgery, including minimally invasive surgery (MIS), and during incision procedures, surgeons lose their touch-sensing capacity, and these may lead to deeper cuts and damage to nearby body organs. Hence, biological pressure sensors are also important [124]. By their regular features, all hydrogels alter their surface-to-volume ratio. Therefore, different gel units and other materials are used with electroactive hydrogels to create touch-detecting devices/pressure sensors [125]. Pressure sensors having piezo property were used to monitor mechanical pressures of hydrogels in the presence of external parameters [126]. One such illustration of a touch sensor is reported in Fig. 8.

Potentiometer sensors are one form of pH sensor consisting of two electrodes. One of the electrodes is used for reference purposes. Hydrogels can be designed with different materials for pH measurements. pH-sensitive materials are used in hydrogels for high-accuracy measurements as the performance of hydrogels is dependent upon the surface-to-volume ratio, rate of swelling/shrinking when placed in a solvent, and elastic constants of the material. The swelling/shrinking in hydrogels transforms the volume, which can be transduced to the desired electric force. An investigation by researchers has revealed that silicon-based piezoresistive materials can be used for the conversion of these signals.

In one of the investigations, a hydrogel was combined with a silicon-based piezoresistive sensor, and when the hydrogel's deflection was transferred to the sensor, electric signals were produced. Then, it was observed how the output voltage changed with pH [127]. In a further study, the deflections were impersonated using an analysis system and a hydrogel-based piezoresistive pH sensor [128]. Measurement of pH using a single comparable sensor is simple to operate. Making polyelectrolyte hydrogels using weakly acidic ionizing chemical reagents is also possible. Close to their acid dissociation constant (pK_a), the hydrogels go through a phase transition within a limited range. This gel's ionizable group is the main source of pH sensitivity [129].

Reese et al. [130] illustrated the application of pH-sensitive hydrogels based on an intelligent polymerized crystalline colloidal array (IPCCA). Moreover, pH-responsive hydrogels are employed in both medication delivery and actuation. In one instance, these hydrogels were utilized for medication delivery and targeting in a magnetically activated soft microrobot.

Both biological processes and the operation of biological machinery are significantly influenced by temperature. Using a hospital as an example, measuring the temperature of blood, respiratory air, incubators, cooling systems, and many other things is necessary. Temperature sensors are, therefore, expected to be constructed and developed with the highest level of precision and accuracy possible. Because thermoresponsive hydrogels' shape and volume change in response to temperature, it is possible to compute the temperature precisely, and their application in temperature sensors is widespread. pNIPAM and other cross-linkable polymers like *N*,*N*-dimethylacrylamide (DMAAm) and 2-(dimethyl maleimido)-*N*-ethyl-acrylamide (DMIAAm) are the most widely utilized thermoresponsive hydrogels. In biomedicine, a great diversity of swelling behaviors about temperature changes, demands, and temperature measurement ranges can be discovered. Furthermore, certain techniques have been proposed to enhance the thermoresponsive hydrogels' characteristics, including pNIPAM in chemical sensors [131]. The other techniques for measuring temperature involve fluorescence and luminescence, which are produced due to these hydrogels' spectrum characteristics at various wavelengths. As an illustration, a temperature sensor that detects changes in quantum dots' spectrum shifts was created [132]. Dipyrazolyltriazine tris(β-diketonate) europium(III) complexes were employed in a different investigation as luminescence temperature sensors, and they were coupled with a polyurethane hydrogel matrix to ensure the probes' adequate stability [133]. Intriguing studies have also been conducted on the dual luminescence of hydrogel nanoparticles, which are sensitive to pH and temperature and emit strong, independent emissions. The blue emission indicated a linear response between pH values of 6.5 and 7.6, whereas the red emission related to a temperature shift from 10 to 80 °C [134]. IPCCA temperature-dependent volume fluctuations resulted in a color transformation in one investigation because of diffraction [112]. Numerous investigations on temperature sensors have been conducted. For instance, materials known as hydrogel PCs swell in response to temperature changes; as a result, the lattice spacing rises, and the photonic stop band varies proportionally to the temperature change. Additionally, these hydrogels serve as stabilizers in these sensors [135].

Hydrogels can also be explored by chemical interaction with molecules, antigens, and antibodies apart from development as sensing material. They can be explored as biochemical sensors. They can also be applied in developing living sensors to be used as biosensors using composites at the bulk and/or nano level. This type of bioengineering can harness effective results in deformed cells. In a reported study, it was shown that hydrogel-elastomer matrices were combined with programmable living cells, which resulted in stretchable materials One such illustration is pictorially shown in Fig. 9. These can be designed as living sensors, genetic circuits, and wearable devices [136].

Even in these living and cell sensors, a hydrogel coating can be used to prevent deposits from entering the sensor. Additionally, there has been a development in lanthanide luminescence-based sensing of enzymes in the body with the help of hydrogels, as this is advantageous over fluorescence-based sensing because of interference. Recently, there was a breakthrough at the Massachusetts Institute of Technology (MIT), where wearable living sensors with programmed cells in the form of gloves and bandages were designed to detect certain chemicals by glowing.

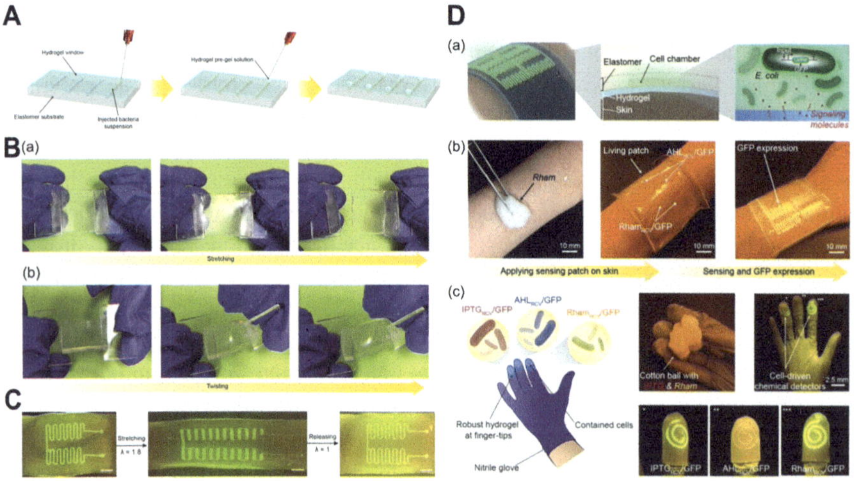

Fig. 9 Design of a device for detection of living materials based on the basic characteristics of stretching strong and biocompatible elastomer hydrogels that can hold genetically developed bacteria cells. (**a**) Picture representation of the suspension of a cell with sealing of injection points. (**b**) Shape irregularity of an agar-based living devices. (**c**) Living device in working condition under the application of a one-directional stretch. (**d**) Biofriendly devices for strongly matching human condition [Reproduced from Biomimetics (Basel). 2018 Jul 10; 3(3): 15 with permission from the Biomimetics Publishing]

8 Conclusion and Outlook

In this book chapter, we have briefly covered the structures of hydrogels and their uses as actuators and sensors in soft robotics for biomedical applications, mostly from a structural mechanics perspective. Hydrogels will become an increasingly important material in soft robotics shortly, even though their use and chemical processing are currently limited. This conclusion can be drawn from the studies discussed regarding various sources used for advanced surgical procedures, novel materials used for sensing, and robot design. Hydrogels are increasingly preferred for use in actuating and sensing mechanisms for bio-applications because of their hydrophilic nature, swelling/bending capabilities, high biocompatibility, self-healing abilities, and stimuli-responsiveness to various environmental factors. These properties set them apart from conventional materials like silicone elastomers. Another advantage of using them is the potential and resources to create hybrid composites that combine different polymers and are mechanically stable. Hydrogels' spatial gel inhomogeneous density distribution, which lowers optical clarity and strength, is a major drawback, even though many of the promises shown here have prospective applications in soft robotics and the related biomedical field. As a result, hydrogels are currently exclusively employed as coatings, and sterilizing hydrogels is challenging. The other challenges in developing hydrogel robots are their high cost, nonadherence, and weak mechanical strength due to their high water content.

Depending on the intended application and associated biocompatibility and biomimicry requirements, choosing or developing materials for biomedical soft robotics poses several challenges. Medical equipment that is used in surgery, endoscopy, and medication delivery must safely interact with the human body either acutely or semi-chronically. Wearable soft robots and assistive robots must be able to make seamless and efficient contact and coupling with human limbs and body components. Finally, the mechanical characteristics, mobility, and functionality of artificial organs and body part simulators must be modeled after those of human tissues. The key characteristics of materials for biomedical soft robots are controllability and tunability of the mechanical properties, a controllable response to external stimuli, and morphological adaptability to physiological environments. These features enable biomedical soft robots to meet biomedical needs that rigid link robots cannot fully address. Therefore, the development of biomedical soft robots in the future is heavily dependent on advancements in the design of smart soft materials.

In conclusion, although we have discussed avenues for hydrogel potential and challenges from a mechanical stability standpoint, there are still many inter- and multidisciplinary intensive research to be made for this technology to further advance in addressing the twenty-first-century biomedical challenges.

References

1. Li, J., Wong, W., Tao, X.: Recent advances in soft functional materials: preparation, functions and applications. Nanoscale. **12**, 1281–1306 (2020)
2. Tanford, C.: The hydrophobic effect: formation of micelles and biological membranes, 2nd edn. Wiley, New York (1980)
3. Chan, C.L.C., Taylor, J.M., Davidson, E.C.: Design of soft matter for additive processing. Nat Synth. **1**, 592–600 (2022)
4. Peppas, N.A., Hiltz, J.Z., Khademhosseini, A., Langer, R.: Hydrogels in biology and medicine: from molecular principles to nanobiotechnology. Adv. Mater. **18**, 1345–1360 (2006)
5. Ferry, J.D.: Viscoelastic properties of polymers, pp. 486–544. Wiley, New York (1980)
6. Chaterji, S., Kwon, K., Park, K.: Smart polymeric gels: redefining the limits of biomedical devices. Prog. Polym. Sci. **32**, 1083–1112 (2007)
7. Phillips, G.O., Williams, P.A.: Handbook of hydrocolloids. In: Murphy, P. (ed.) Starch. Woodhead Publishing, Cambridge (2000)
8. Rosiak, J.M., Yoshii, F.: Hydrogels and their medical applications. Nucl. Inst. Methods Phys. Res. B. **151**, 56–64 (1999)
9. Rubinstein, M., Colby, R.H.: Polymer physics. Oxford University Press, Oxford (2003)
10. Hennink, W.E., Nostrum, C.F.V.: Novel crosslinking methods to design hydrogels. Adv. Drug Deliv. Rev. **54**, 13–36 (2002)
11. Rosiak, J.M.: Gel/sol analysis of irradiated polymers. Radiat. Phys. Chem. **51**, 13–17 (1998)
12. Hoare, T.R., Kohane, D.S.: Hydrogels in drug delivery: progress and challenges. Polymer. **49**, 1993–2007 (2008)
13. Thakur, V.K., Thakur, M.K.: Recent advances in green hydrogels from lignin: a review. Int. J. Biol. Macromol. **72**, 834–847 (2015)
14. Anderson, J.M., Langone, J.J.: Issues and perspectives on the biocompatibility and immune toxicity evaluation of implanted controlled release systems. J. Control. Release. **57**, 107–113 (1999)
15. Smetana, K.: Cell biology of hydrogels. Biomaterials. **14**, 1046–1050 (1993)
16. Augst, A.D., Kong, H.J., Mooney, D.J.: Alginate hydrogels as biomaterials. Macromol. Biosci. **6**, 623–633 (2006)
17. Pawar, S.N., Edgar, K.J.: Alginate derivatization: a review of chemistry, properties and applications. Biomaterials. **33**, 3279–3305 (2012)
18. Nada, A.A., James, R., Shelke, N.B., Harmon, M.D., Awad, H.M., Nagarale, R.K., Kumbar, S.G.: A smart methodology to fabricate electrospun chitosan nanofiber matrices for regenerative engineering applications. Polym. Adv. Technol. **25**, 507–515 (2014)
19. Ibrahim, N.A., Eid, B.M., Elmaaty, T.M.A., El-Aziz, E.A.: A smart approach to add antibacterial functionality to cellulosic pigment prints. Carbohydr. Polym. **94**, 612–618 (2013)
20. Alves, N.M., Mano, J.F.: Chitosan derivatives obtained by chemical modifications for biomedical and environmental applications. Int. J. Biol. Macromol. **43**, 401–414 (2008)
21. Sudheesh, K.P.T., Lakshmanan, V.K., Anilkumar, T.V., Ramya, C., Reshmi, P., Unnikrishnan, A.G., Nair, S.V., Jayakumar, R.: Flexible and microporous chitosan hydrogel/nano ZnO composite bandages for wound dressing: in vitro and in vivo evaluation. ACS Appl. Mater. Interfaces. **4**, 2618–2629 (2012)
22. Bacaita, E.S., Ciobanu, B.C., Popa, M., Agop, M., Desbrieres, J.: Phases in the temporal multiscale evolution of the drug release mechanism in IPN-type chitosan based hydrogels. Phys. Chem. Chem. Phys. **16**, 25896–25905 (2014)
23. Csaba, N., Köping-Höggård, M., Alonso, M.J.: Ionically crosslinked chitosan/tripolyphosphate nanoparticles for oligonucleotide and plasmid DNA delivery. Int. J. Pharm. **382**, 205–214 (2009)

24. Jen, A.C., Wake, M.C., Mikos, A.G.: Review: Hydrogels for cell immobilization. Biotechnol. Bioeng. **50**, 357–364 (1996)
25. Kabiri, K., Omidian, H., Zohuriaan-Mehr, M.J., Doroudiani, S.: Superabsorbent hydrogel composites and nanocomposites: a review. Polym. Compos. **32**, 277–289 (2011)
26. Singh, A., Sharma, P.K., Garg, V.K., Garg, G.: Hydrogels: a review. Int J Pharm Sci Rev Res. **4**, 97–105 (2010)
27. Sannino, A., Madaghiele, M., Lionetto, M.G., Schettino, T., Maffezzoli, A.: A cellulose-based hydrogel as a potential bulking agent for hypocaloric diets: an in vitro biocompatibility study on rat intestine. J. Appl. Polym. Sci. **102**, 1524–1530 (2006)
28. Bhattacharyya, S., Guillot, S., Dabboue, H., Tranchant, J.-F., Salvetat, J.: Carbon nanotubes as structural nanofibers for hyaluronic acid hydrogel scaffolds. Biomacromolecules. **9**, 505–509 (2008)
29. Reis, A.V., Guilherme, M.R., Moia, T.A., Mattoso, L.H.C., Muniz, E.C., Tambourgi, E.B.: Synthesis and characterization of a starch-modified hydrogel as potential carrier for drug delivery system. J Polym Sci Part A Polym Chem. **46**, 2567–2574 (2008)
30. Woltje, M., Bobel, M.: 12-Natural biodegradable medical polymers: silk. In: Science and principles of biodegradable and bioresorbable medical polymers, pp. 351–376. Elsevier (2017)
31. Chattopadhyay, S., Raines, R.T.: Review collagen-based biomaterials for wound healing. Biopolymers. **101**(8), 821–833 (2014)
32. Faraj, K.A., Kuppevelt, T.H., Daamen, W.F.: Construction of collagen scaffolds that mimic the three-dimensional architecture of specific tissues. Tissue Eng. **13**(10), 2387–2394 (2007)
33. Qin, G., Kai, X.: Advanced materials for biomedical engineering applications. In: Design, fabrication, properties and applications of smart and advanced materials, pp. 376–412. Taylor & Francis Group, LLC (2016)
34. Han, S., Cruz, S., Park, S., Shin, S.: Nano-biomaterials and advanced fabrication techniques for engineering skeletal muscle tissue constructs in regenerative medicine. Nano Converg. **10**(48), 1–19 (2023)
35. Leal-Egana, A., Scheibel, T.: Silk-based materials for biomedical applications. Biotechnol. Appl. Biochem. **55**, 155–167 (2010)
36. Horan, R.L., Antle, K., Collette, A.L., Wang, Y., Huang, J., Moreau, J.E., Volloch, V., Kaplan, D.L., Altman, G.H.: In vitro degradation of silk fibroin. Biomaterials. **26**, 3385–3393 (2005)
37. Numata, K., Kaplan, D.L.: Differences in cytotoxicity of beta-sheet peptides originated from silk and amyloid beta. Macromol. Biosci. **11**, 60–64 (2011)
38. Pritchard, E.M., Hu, X., Finley, V., Kuo, C.K., Kaplan, D.L.: Effect of silk protein processing on drug delivery from silk films. Macromol. Biosci. **13**, 311–320 (2013)
39. Yucel, T., Cebe, P., Kaplan, D.L.: Vortex-induced injectable silk fibroin hydrogels. Biophys. J. **97**, 2044–2050 (2009)
40. Liu, L., Zhang, X., Liu, X., Liu, J., Lu, G., Kaplan, D.L., Zhu, H., Lu, Q.: Biomineralization of stable and monodisperse vaterite microspheres using silk nanoparticles. ACS Appl. Mater. Interfaces. **7**, 1735–1745 (2015)
41. Nitta, S.K., Numata, K.: Biopolymer-based nanoparticles for drug/gene delivery and tissue engineering. Int. J. Mol. Sci. **14**, 1629–1654 (2013)
42. Yucel, T., Lovett, M.L., Kaplan, D.L.: Silk-based biomaterials for sustained drug delivery. J. Control. Release. **190**, 381–397 (2014b)
43. Valluzzi, R., Winkler, S., Wilson, D., Kaplan, D.L.: Silk: molecular organization and control of assembly. Philos. Trans. R Soc. Lond. B Biol. Sci. **357**, 165–167 (2002)
44. Wang, X., Wenk, E., Matsumoto, A., Meinel, L., Li, C., Kaplan, D.L.: Silk microspheres for encapsulation and controlled release. J. Control. Release. **117**, 360–370 (2007)
45. Megeed, Z., Winters, R.M., Yarmush, M.L.: Modulation of single-chain antibody affinity with temperature-responsive elastin-like polypeptide linkers. Biomacromolecules. **7**, 999–1004 (2006)

46. Swierczewska, M., Hajicharalambous, C.S., Janorkar, A.V., Megeed, Z., Yarmush, M.L., Rajagopalan, P.: Cellular response to nanoscale elastin-like polypeptide polyelectrolyte multilayers. Acta Biomater. **4**, 827–837 (2008)
47. Kundu, B., Rajkhowa, R., Kundu, S.C., Wang, X.: Silk fibroin biomaterials for tissue regenerations. Adv. Drug Deliv. Rev. **65**, 457–470 (2013)
48. Annabi, N., Mithieux, S.M., Camci-Unal, G., Dokmeci, M.R., Weiss, A.S., Khademhosseini, A.: Elastomeric recombinant protein-based biomaterials. Biochem. Eng. J. **77**, 110–118 (2013)
49. Desai, M.S., Lee, S.W.: Protein-based functional nanomaterial design for bioengineering applications. Wiley Interdiscip. Rev. Nanomed. Nanobiotechnol. **7**, 69–97 (2015)
50. Vepari, C., Kaplan, D.L.: Silk as a biomaterial. Prog. Polym. Sci. **32**, 991–1007 (2007)
51. Wong Po Foo, C., Patwardhan, S.V., Belton, D.J., Kitchel, B., Anastasiades, D., Huang, J., Naik, R.R., Perry, C.C., Kaplan, D.L.: Novel nanocomposites from spider silk-silica fusion (chimeric) proteins. Proc. Natl. Acad. Sci. USA. **103**, 9428–9433 (2006)
52. Bini, E., Foo, C.W., Huang, J., Karageorgiou, V., Kitchel, B., Kaplan, D.L.: RGD functionalized bioengineered spider dragline silk biomaterial. Biomacromolecules. **7**, 3139–3145 (2006)
53. McHale, M.K., Setton, L.A., Chilkoti, A.: Synthesis and in vitro evaluation of enzymatically cross-linked elastin-like polypeptide gels for cartilaginous tissue repair. Tissue Eng. **11**, 1768–1779 (2005)
54. Rodriguez-Cabello, J.C.: Smart elastin-like polymers. Adv. Exp. Med. Biol. **553**, 45–57 (2004)
55. Renner, J.N., Cherry, K.M., Su, R.S., Liu, J.C.: Characterization of resilin-based materials for tissue engineering applications. Biomacromolecules. **13**, 3678–3685 (2012)
56. Li, L., Tong, Z., Jia, X., Kiick, K.L.: Resilin-like polypeptide hydrogels engineered for versatile biological functions. Soft Matter. **9**, 665–673 (2013)
57. Chan, B.Q.Y., Low, Z.W.K., Heng, S.J.W., Chan, S.Y., Owh, C., Loh, X.J.: Recent advances in shape memory soft materials for biomedical applications. ACS Appl. Mater. Interfaces. **8**, 10070–10087 (2016)
58. Meng, H., Mohamadian, H., Stubblefield, M., Jerro, D., Ibekwe, S., Pang, S.-S., Li, G.: Various shape memory effects of stimuli-responsive shape memory polymers. Smart Mater. Struct. **22**(9), 093001 (2013)
59. Lu, H., Huang, W.M., Yao, Y.: Review of chemo-responsive shape change/memory polymers. Pigm. Resin Technol. **42**(4), 237–246 (2013)
60. Zhao, Q., Behl, M., Lendlein, A.: Shape-memory polymers with multiple transitions: complex actively moving polymers. Soft Matter. **9**(6), 1744–1755 (2013)
61. Lendlein, A., Langer, R.: Biodegradable, elastic shape-memory polymers for potential biomedical applications. Science. **296**(5573), 1673–1676 (2002)
62. Kagami, Y., Gong, J.P., Osada, Y.: Shape memory behaviors of crosslinked copolymers containing stearyl acrylate. Macromol. Rapid Commun. **17**(8), 539–543 (1996)
63. Sahoo, N.G., Jung, Y.C., Cho, J.W.: Electroactive shape memory effect of polyurethane composites filled with carbon nanotubes and conducting polymer. Mater. Manuf. Process. **22**(4), 419–423 (2007)
64. Felton, S.M., Tolley, M.T., Shin, B., Onal, C.D., Demaine, E.D., Rus, D., Wood, R.J.: Self-folding with shape memory composites. Soft Matter. **9**(32), 7688–7694 (2013)
65. Luo, H., Li, Z., Yi, G., Wang, Y., Zu, X., Wang, H., Huang, H., Liang, Z.: Temperature sensing of conductive shape memory polymer composites. Mater. Lett. **140**, 71–74 (2015)
66. Balasubramanian, A., Morhard, R., Bettinger, C.J.: Shape-memory microfluidics. Adv. Funct. Mater. **23**(38), 4832–4839 (2013)
67. Mosmann, T.: Rapid colorimetric assay for cellular growth and survival: application to proliferation and cytotoxicity assays. J. Immunol. Methods. **65**(1), 55–63 (1983)
68. Mitchell, D.B., Santone, K.S., Acosta, D.: Evaluation of cytotoxicity in cultured cells by enzyme leakage. J. Tissue Cult. Methods. **6**(3–4), 113–116 (1980)

69. El Feninat, F., Laroche, G., Fiset, M., Mantovani, D.: Shape memory materials for biomedical applications. Adv. Eng. Mater. **4**(3), 91 (2002)
70. Calmon-Decriaud, A., Bellon-Maurel, V., Silvestre, F.: Standard methods for testing the aerobic biodegradation of polymeric materials. Review and perspectives. In: Blockcopolymers-polyelectrolytes-biodegradation. Advances in polymer science, vol. 135, pp. 207–226. Springer, Berlin (1998). https://doi.org/10.1007/3-540-69191-X_3
71. Sung, H.-J., Meredith, C., Johnson, C., Galis, Z.S.: The effect of scaffold degradation rate on three-dimensional cell growth and angiogenesis. Biomaterials. **25**(26), 5735–5742 (2004)
72. Wong, Y., Kong, J., Widjaja, L.K., Venkatraman, S.S.: Biomedical applications of shape-memory polymers: how practically useful are they? Sci. China Chem. **57**(4), 476–489 (2014)
73. Kanjickal, D., Lopina, S., Evancho-Chapman, M.M., Schmidt, S., Donovan, D.: Effects of sterilization on poly(ethylene glycol) hydrogels. J. Biomed. Mater. Res., Part A. **87**(3), 608–617 (2008)
74. Xie, M., Wang, L., Ge, J., Guo, B., Ma, P.X.: Strong electroactive biodegradable shape memory polymer networks based on star-shaped polylactide and aniline trimer for bone tissue engineering. ACS Appl. Mater. Interfaces. **7**(12), 6772–6781 (2015)
75. Chan, B.Q.Y., Liow, S.S., Loh, X.J.: Organic–inorganic shape memory thermoplastic polyurethane based on polycaprolactone and polydimethylsiloxane. RSC Adv. **6**, 34946–34954 (2016)
76. Kai, D., Prabhakaran, M.P., Chan, B.Q.Y., Liow, S.S., Ramakrishna, S., Xu, F., Loh, X.J.: Elastic poly(ε-caprolactone)–polydimethylsiloxane copolymer fibers with shape memory effect for bone tissue engineering. Biomed. Mater. **11**(1), 015007 (2016)
77. Kai, D., Liow, S.S., Loh, X.J.: Biodegradable polymers for electrospinning: towards biomedical applications. Mater. Sci. Eng. C. **45**, 659–670 (2014)
78. Chen, H., Cao, X., Zhang, J., Zhang, J., Ma, Y., Shi, G., Ke, Y., Tong, D., Jiang, L.: Electrospun shape memory film with reversible fibrous structure. J. Mater. Chem. **22**(42), 22387–22391 (2012)
79. Bao, M., Lou, X., Zhou, Q., Dong, W., Yuan, H., Zhang, Y.: Electrospun biomimetic fibrous scaffold from shape memory polymer of PDLLA-*co*-TMC for bone tissue engineering. ACS Appl. Mater. Interfaces. **6**(4), 2611–2621 (2014)
80. Erndt-Marino, J.D., Munoz-Pinto, D.J., Samavedi, S., Jimenez-Vergara, A.C., Diaz-Rodriguez, P., Woodard, L., Zhang, D., Grunlan, M.A., Hahn, M.S.: Evaluation of the osteoinductive capacity of polydopamine-coated poly(ε-caprolactone) diacrylate shape memory foams. ACS Biomater Sci. Eng. **1**(12), 1220–1230 (2015)
81. Vyas, L., Aquino, D., Kuo, C.-H., Dai, J.S., Dasgupta, P.: Flexible robotics. BJU Int. **107**, 187–189 (2011)
82. Loeve, A., Breedveld, P., Dankelman, J.: Scopes too flexible… and too stiff. IEEE Pulse. **1**, 26–41 (2010)
83. Le, H.M., Do, T.N., Phee, S.J.: A survey on actuators-driven surgical robots. Sens. Actuators A. Phys. **247**, 323–354 (2016)
84. Ikuta, K., Tsukamoto, M., Hirose, S.: In: Proceedings 1988 IEEE international conference on robotics and automation. pp. 427–430. Philadelphia, PA, USA, (1988)
85. Robinson, G., Davies, J.B.C.: Proceedings 1999 IEEE international conference on robotics and automation, pp. 2849–2854 (1999)
86. Burgner-Kahrs, J., Caleb, R.D., Choset, H.: Continuum robots for medical applications: a survey. IEEE Trans. Rob. **31**, 1261–1280 (2015)
87. Polygerinos, P., et al.: Soft robotics: review of fluid driven intrinsically soft devices; manufacturing, sensing, control, and applications in human–robot interaction. Adv. Eng. Mater. **19**, 1700016 (2017)
88. Phee, L., et al.: Analysis and development of locomotion devices for the gastrointestinal tract. I.E.E.E. Trans. Biomed. Eng. **49**, 613–616 (2002)
89. Bertetto, A.M., et al.: A novel fluidic bellows manipulator. J. Robot. Mechatron. **16**, 604–612 (2004)

90. Volder, M.D., et al.: Production and characterization of a hydraulic microactuator. J. Micromech. Microeng. **15**, S15–S21 (2005)
91. Howell, L.L., Magleby, S.P., Olsen, B.M.: Handbook of compliant mechanisms. Wiley (2013)
92. De Greef, A., Lambert, P., Delchambre, A.: Towards flexible medical instruments: review of flexible fluidic actuators. Precis. Eng. **33**, 311–321 (2009)
93. Karki, S., et al.: Thin films as an emerging platform for drug delivery. Asian J. Pharm. Sci. **11**, 559–574 (2016)
94. Borodkin, S., Tucker, F.E.: Drug release from hydroxypropyl cellulose-polyvinyl acetate films. J. Pharm. Sci. **63**, 1359–1364 (1974)
95. Di Domenico, E.D., Hobot, C.M., Stokes, K.B., Coury, A.J., Doan, P.D., Sandstrom, R.D.: Medical electrical lead with polymeric monolithic controlled release device and method of manufacture. U.S. Patent Application No. 4,972,848. (1990)
96. Cortese, R., Theeuwes F.: Osmotic device with hydrogel driving member. U.S. Patent Application No. 4,327,725. (1982)
97. Liu, Y., Servant, A., Guy, O., Al-Jamal, K.T., Williams, P., Hawkins, K., Kostarelos, K.: An electric-field responsive microsystem for controllable miniaturised drug delivery applications. Sens. Actuators B Chem. **175**, 100–105 (2012)
98. O'Grady, M.L.: Conductive polymers for biological applications. Harvard University, Cambridge (2010)
99. Pessot, G., Schümann, M., Gundermann, T., Odenbach, S., Löwen, H., Menzel, A.M.: Tunable dynamic moduli of magnetic elastomers: from characterization by X-ray micro-computed tomography to mesoscopic modeling. J. Phys. Condens. Matter. **30**, 125101 (2018)
100. Harne, R.L., Deng, Z., Dapino, M.J.: Adaptive magnetoelastic metamaterials: a new class of magnetorheological elastomers. J. Intell. Mater. Syst. Struct. **29**, 265 (2017). https://doi.org/10.1177/1045389X17721037
101. Barbucci, R., Giani, G., Fedi, S., Bottari, S., Casolaro, M.: Biohydrogels with magnetic nanoparticles as crosslinker: characteristics and potential use for controlled antitumor drug-delivery. Acta Biomater. **8**, 4244–4252 (2012)
102. Barbucci, R., Pasqui, D., Giani, G., De Cagna, M., Fini, M., Giardino, R., Atrei, A.: A novel strategy for engineering hydrogels with ferromagnetic nanoparticles as crosslinkers of the polymer chains. Potential applications as a targeted drug delivery system. Soft Matter. **7**, 5558–5565 (2011)
103. Leijten, J., Seo, J., Yue, K., Trujillo-de Santiago, G., Tamayol, A., Ruiz-Esparza, G.U., Shin, S.R., Sharifi, R., Noshadi, I., Álvarez, M.M., et al.: Spatially and temporally controlled hydrogels for tissue engineering. Mater. Sci. Eng. R. Rep. **119**, 1–35 (2017)
104. Satarkar, N.S., Johnson, D., Marrs, B., Andrews, R., Poh, C., Gharaibeh, B., Saito, K., Anderson, K.W., Hilt, J.Z.: Hydrogel-MWCNT nanocomposites: synthesis, characterization, and heating with radiofrequency fields. J. Appl. Polym. Sci. **117**, 1813–1819 (2010)
105. Zhang, Y., Bai, Y., Yan, B.: Functionalized carbon nanotubes for potential medicinal applications. Drug Discov. Today. **15**, 428–435 (2010)
106. Huang, H.W., Sakar, M.S., Petruska, A.J., Pané, S., Nelson, B.J.: Soft micromachines with programmable motility and morphology. Nat. Commun. **7**, 12263 (2016)
107. Zhao, X., Kim, J., Cezar, C.A., Huebsch, N., Lee, K., Bouhadir, K., Mooney, D.J.: Active scaffolds for on-demand drug and cell delivery. Proc. Natl. Acad. Sci. USA. **108**, 67–72 (2011)
108. Ji, C., Khademhosseini, A., Dehghani, F.: Enhancing cell penetration and proliferation in chitosan hydrogels for tissue engineering applications. Biomaterials. **32**, 9719–9729 (2011)
109. Rafat, M., Li, F., Fagerholm, P., Lagali, N.S., Watsky, M.A., Munger, R., Matsuura, T., Griffith, M.: PEG-stabilized carbodiimide crosslinked collagen–chitosan hydrogels for corneal tissue engineering. Biomaterials. **29**, 3960–3972 (2008)
110. Tan, H., Chu, C.R., Payne, K.A., Marra, K.G.: Injectable in situ forming biodegradable chitosan–hyaluronic acid based hydrogels for cartilage tissue engineering. Biomaterials. **30**, 2499–2506 (2009)

111. Suh, J.K.F., Matthew, H.W.: Application of chitosan-based polysaccharide biomaterials in cartilage tissue engineering: a review. Biomaterials. **21**, 2589–2598 (2000)
112. Buenger, D., Topuz, F., Groll, J.: Hydrogels in sensing applications. Prog. Polym. Sci. **37**, 1678–1719 (2012)
113. Topuz, F., Buenger, D., Tanaka, D., Groll, J.: Hydrogels in biosensing applications. In: Comprehensive biomaterials. Elsevier, New York (2011)
114. Tian, E., Wang, J., Zheng, Y., Song, Y., Jiang, L., Zhu, D.: Colorful humidity sensitive photonic crystal hydrogel. J. Mater. Chem. **18**, 1116–1122 (2008)
115. Arregui, F.J., Ciaurriz, Z., Oneca, M., Matıas, I.R.: An experimental study about hydrogels for the fabrication of optical fiber humidity sensors. Sens. Actuators B Chem. **96**, 165–172 (2003)
116. Herber, S., Bomer, J., Olthuis, W., Bergveld, P., Van Den Berg, A.: A miniaturized carbon dioxide gas sensor based on sensing of pH-sensitive hydrogel swelling with a pressure sensor. Biomed. Microdevices. **7**, 197–204 (2005)
117. Qu, F., Zhang, Y., Rasooly, A., Yang, M.: Electrochemical biosensing platform using hydrogel prepared from ferrocene modified amino acid as highly efficient immobilization matrix. Anal. Chem. **86**, 973–976 (2014)
118. Edmondson, R., Broglie, J.J., Adcock, A.F., Yang, L.: Three-dimensional cell culture systems and their applications in drug discovery and cell-based biosensors. Assay Drug Dev. Technol. **12**, 207–218 (2014)
119. Russell, R.J., Pishko, M.V., Gefrides, C.C., McShane, M.J., Cote, G.L.: A fluorescence-based glucose biosensor using concanavalin A and dextran encapsulated in a poly(ethylene glycol) hydrogel. Anal. Chem. **71**, 3126–3132 (1999)
120. Zhan, W., Seong, G.H., Crooks, R.M.: Hydrogel-based microreactors as a functional component of microfluidic systems. Anal. Chem. **74**, 4647–4652 (2002)
121. Koh, W.G., Pishko, M.V.: Fabrication of cell-containing hydrogel microstructures inside microfluidic devices that can be used as cell-based biosensors. Anal. Bioanal. Chem. **385**, 1389–1397 (2006)
122. Quinn, J.G., O'Neill, S., Doyle, A., Mc Atamney, C., Diamond, D., Mac Craith, B.D., O'Kennedy, R.: Development and application of surface plasmon resonance-based biosensors for the detection of cell–ligand interactions. Anal. Biochem. **281**, 135–143 (2000)
123. Shang, J., Yan, J., Zhang, Z., Huang, X., Maturavongsadit, P., Song, B., Jia, Y., Ma, T., Li, D., Xu, K., et al.: A hydrogel-based glucose affinity microsensor. Sens. Actuators B Chem. **237**, 992–998 (2016)
124. Liu, C.: Bioinspired touch sensors for medical applications. In: Proceedings of the 2013 IEEE SENSORS. pp. 1–4. Baltimore. (2013)
125. Sarwar, M.S., Dobashi, Y., Preston, C., Wyss, J.K., Mirabbasi, S., Madden, J.D.W.: Bend, stretch, and touch: locating a finger on an actively deformed transparent sensor array. Sci. Adv. **3**, e1602200 (2017)
126. Orthner, M., Buetefisch, S., Magda, J., Rieth, L., Solzbacher, F.: Development, fabrication, and characterization of hydrogel based piezo resistive pressure sensors with perforated diaphragms. Sens. Actuators A. Phys. **161**, 29–38 (2010)
127. Gerlach, G., Guenther, M., Sorber, J., Suchaneck, G., Arndt, K.F., Richter, A.: Chemical and pH sensors based on the swelling behavior of hydrogels. Sens. Actuators B Chem. **111**, 555–561 (2005)
128. Trinh, Q.T., Gerlach, G., Sorber, J., Arndt, K.F.: Hydrogel-based piezoresistive pH sensors: design, simulation and output characteristics. Sens. Actuators B Chem. **117**, 17–26 (2006)
129. Richter, A., Paschew, G., Klatt, S., Lienig, J., Arndt, K.F., Adler, H.J.P.: Review on hydrogel-based pH sensors and microsensors. Sensors. **8**, 561–581 (2008)
130. Reese, C.E., Baltusavich, M.E., Keim, J.P., Asher, S.A.: Development of an intelligent polymerized crystalline colloidal array colorimetric reagent. Anal. Chem. **73**, 5038–5042 (2001)

131. Guenther, M., Gerlach, G., Corten, C., Kuckling, D., Muller, M., Shi, Z., Sorber, J., Arndt, K.F.: Application of polyelectrolytic temperature-responsive hydrogels in chemical sensors. In: Macromolecular symposia, vol. 254, pp. 314–321. Wiley, Hoboken (2007)
132. Maruyama, H., Masuda, T., Honda, A., Arai, F.: Local temperature measurement using spectra shift of quantum-dot hydrogel sensor. In: Proceedings of the 2011 11th IEEE conference on nanotechnology (IEEE-NANO), Portland, OR, USA. pp. 319–322. (2011)
133. Borisov, S.M., Wolfbeis, O.S.: Temperature-sensitive europium(III) probes and their use for simultaneous luminescent sensing of temperature and oxygen. Anal. Chem. **78**, 5094–5101 (2006)
134. Zhao, Y., Shi, C., Yang, X., Shen, B., Sun, Y., Chen, Y., Xu, X., Sun, H., Yu, K., Yang, B., et al.: pH-and temperature-sensitive hydrogel nanoparticles with dual photoluminescence for bioprobes. ACS Nano. **10**, 5856–5863 (2016)
135. Wang, H., Zhang, K.Q.: Photonic crystal structures with tunable structure colour as colorimetric sensors. Sensors. **13**, 4192–4213 (2013)
136. Liu, X., Tang, T.C., Tham, E., Yuk, H., Lin, S., Lu, T.K., Zhao, X.: Stretchable living materials and devices with hydrogel–elastomer hybrids hosting programmed cells. Proc. Natl. Acad. Sci. USA. **114**, 2200–2205 (2017)

Soft Materials for Wastewater Treatment

Amit Kumar Sharma, Balbir Singh Kaith, Garima Agrawal, Sunil, Khushi Choudhary, and Harshdeep Tiwari

Abstract In the present era, treatment of industrially polluted wastewater is a challenging task as it requires the use of toxic chemicals and complex procedures during industrial-scale treatment. The contamination of water with toxic metal ions and dyes not only destroys the aquatic species but also produces many adverse effects on many living and nonliving species. Soft materials, including hydrogels, microgels, nanogels, and related nanocomposites, have provided a better alternative to resolve the issues related to water contamination. Recently, the soft materials obtained from natural backbones have become popular due to their unique features such as biodegradability and cost-effectiveness. The desired properties among such materials can be introduced by controlling the cross-linking strategies, fabrication methods, and choice of monomers. This chapter highlights the recent applications and perspectives of soft material-based devices for wastewater treatments. Also, an overview of the challenges faced during wastewater treatments and alternatives to such problems is provided.

A. K. Sharma (✉)
Department of Chemistry, University Institute of Sciences, Chandigarh University, Mohali, Punjab, India

University Centre for Research and Development, Chandigarh University, Mohali, Punjab, India

B. S. Kaith
Department of Chemistry, Dr B R Ambedkar National Institute of Technology Jalandhar, Jalandhar, Punjab, India

G. Agrawal
School of Chemical Sciences, Indian Institute of Technology Mandi, Mandi, Himachal Pradesh, India

Indian Knowledge System and Mental Health Applications Centre, IIT Mandi, Mandi, Himachal Pradesh, India

Technology Innovation Hub in Human-Computer Interaction (iHub), Mandi, Himachal Pradesh, India

Sunli · K. Choudhary · H. Tiwari
Department of Chemistry, University Institute of Sciences, Chandigarh University, Mohali, Punjab, India

V. Kumar, Y. K. Mishra (eds.), *Soft Materials for Functional Applications*,
Advanced Structured Materials 225, https://doi.org/10.1007/978-981-97-9468-3_5

Keywords Soft materials · Wastewater · Hydrogel · Dye removal · Nanocomposite · Cross-linking

1 Introduction

In the present era, the most important aspect of sustainable development is environmental protection, which may be contaminated by discharging toxic effluents from wastewater. Rapid growth in industrialization and urbanization has become the source of many toxic contaminants in the form of organic and inorganic waste [1]. The major sources of water pollution include textile dyes, aromatic hydrocarbons, and heavy metal ions. The long-term discharge of these pollutants inside the water resources reduces the BOD and COD values and retards the growth of aquatic species. Moreover, the indirect exposure of these chemicals to humans and other living species also creates serious health issues [2]. It has been reported in many studies that there are around 1,00,000 toxic textile dyes that are used in different industries, including paper, printing, cosmetics, leather, paint and varnishing, food, and pharmaceuticals (Fig. 1). Acute exposure to these dyes can create serious issues for the survival of fish and other aquatic species. Furthermore, the toxicity, nonbiodegradability, and stability against different oxidizing agents also make it difficult to separate or degrade these dye molecules on an industrial scale [3].

In addition to dyes, many heavy metal ions equally contribute to the contamination of our natural water resources. The foremost contributor to the release of heavy metal ions includes metallurgic/mining industries [5]. It is also important to note that in comparison with other harmful chemicals, the heavy metal ions degrade very slowly and persevere for a longer duration of time in the environment. The traces of heavy metal ions can cause adulteration of water and adversely affect living species. Therefore, the US Environmental Protection Agency and the World

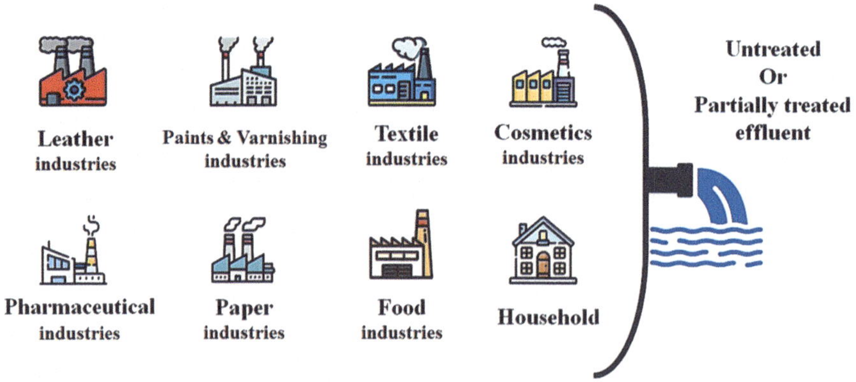

Fig. 1 Discharge of dyes into the aquatic environment through different sources. (Reprinted with permission from [4])

Health Organization have enforced strict regulations to monitor the permissible concentration of heavy metal ions in drinking water resources [6]. To date, the existing approaches for detecting the concentration of heavy metal ions comprise colorimetric analysis, surface-enhanced Raman spectroscopy, atomic absorption spectroscopy, photo-electrochemical sensing, inductively coupled plasma-atomic emission spectrometry, microelectromechanical sensing, fluorescence spectroscopy, potentiometric electrochemical methods, label-free sensors, and inductively coupled plasma mass spectrometry [7]. Similarly, different removal/degradation methods employed for the sequestration of heavy metal ions and dyes include adsorption, photocatalysis, nanofiltration, chemical degradation, and coagulation [8]. Out of these approaches, adsorption is the most frequent and suitable technique to separate the contaminated toxic waste from water resources. The important advantages of adsorption include cost-effectiveness, adsorbent reusability, high efficiency, and versatility. Despite so many advantages, this technique also has certain limitations (such as the indirect production of a huge amount of noxious waste that needs to be disposed of). Furthermore, advanced analytical methods are required to destroy the residual chemicals/side products formed during practical applications on an industrial scale. Traditional inorganic adsorbents such as activated carbon, activated alumina, iron oxide, and other nanomaterials showed some specific limitations during adsorption, which may be due to their high preparation cost, complicated synthesis procedures, inconvenience in uses, and low biodegradability [9, 10]. In this regard, hydrogel-based soft materials are highly acceptable for detoxifying pollutants from wastewater. The important characteristics of such materials, including sustainability, biodegradability, eco-friendliness, and low cost, make them potential adsorbents for water purification.

Hydrogel-based soft materials constitute a cross-linked network comprised of hydrophilic monomers that can accumulate water molecules or provide the sites for the binding of metal ions/dyes inside their pores [11]. The hydrogels derived from bio-based origin such as chitin, cellulose, dextrin, gelatin, starch, etc. are more economical and environmentally friendly for the sequestration of toxic pollutants [12]. Some of such materials also show regeneration and reusability potential and can be used multiple times to effectively absorb some specific pollutants. Therefore, hydrogel-based soft adsorbents are more efficient and environmentally friendly. Based on the source of origin, the hydrogel-based adsorbents can be categorized as natural, synthetic, or hybrid. In recent years, composite hydrogels containing inorganic nanoparticles inside the porous cross-linked hydrogel matrix have been more favorable due to their increased efficiency for adsorption. This chapter provides an overview of hydrogel-based soft materials for the adsorption of toxic pollutants.

2 An Overview of Hydrogels

In 1894, Lee, Known, and Park published an article, and at that time, the term "hydrogel" first came into existence, which was mentioned as a colloidal gel derived from inorganic salt. Later, in 1960, Wichterle and Lim, for the first time, synthesized soft contact lenses made up of poly(2-hydroxyethylmethacrylate). Slowly, the research on hydrogels became more popular in the biomedical field, and they were the first materials used inside patients in the form of implants [13]. Hydrogels have substantial potential for accumulating water molecules inside the network of cross-linked polymers. Nowadays, these materials find extensive applications in different areas such as in the cosmetic industry for developing hygiene products, agriculture fields for the sustained release of pesticides and fertilizers, pharmaceutical industries for the development of different drugs, the packaging industry, and the environmental sector for monitoring different types of toxic pollutants [14–16]. The excellent properties of hydrogels are due to their stimuli-responsive behavior, which means that they can swell/shrink when external stimuli such as temperature, pressure, or pH of the swelling medium are changed [17]. Some hydrogels can also change their color while exposed to external stimuli. Therefore, such exciting features of polymer-based soft materials also make them a suitable choice for developing sensors. The choice of monomers used for synthesizing a particular hydrogel depends upon the type of application, where it will be used. For example, during the synthesis of a hydrogel matrix for in vitro drug-delivery application, the choice of monomer should be natural to favor biocompatibility and biodegradability. However, synthetic monomers or hybrid hydrogels are sometimes preferred for developing a hydrogel-based dressing material to provide durability to the matrix.

In recent years, research emphasis on wastewater treatment has been focused on developing new hydrogel-based soft matrices that can provide an efficient and economical treatment method. The detailed literature survey explored increasing publications on hydrogels engineered with magnetic nanoparticles for their application in wastewater treatment [18]. In addition, soft hydrogels engineered with other nanoparticles have also found progressive applications in water treatment plants [19]. The development and efficiency of engineered nanomaterials for water treatment still need to be explored on an industrial scale for municipal wastewater treatment systems.

Hydrogels have a high capacity to absorb or hold water and have drawn much attention in research during the last few decades. Hydrogels are important in many biomedical and engineering applications, such as fabrications of tissue-engineering scaffolds, actuators, sensors, soft contact lenses, drug-delivery systems, antifouling coatings, wastewater treatment methods, and soft robotics, because of their controllable properties and diverse functionalities as well as their easy preparation methods [20]. Such materials can be formed through the cross-linking of polymer chains called "gelation." It can be achieved either through physical or chemical cross-linking of polymer chains. In the case of physically cross-linked hydrogels, the gelation process is followed through strong and weak interactions among the polymer chains, such as hydrogen bonding, hydrophobic interactions, electrostatic interactions, and crystallization. Similarly, the chemical cross-linking in hydrogels is achieved by forming covalent bonds formed by complementary functional groups or free-radical polymerization of vinyl monomers with the cross-linking agents to connect the polymer chains.

Different properties of the hydrogels need to be thoroughly characterized to meet the needs and standards of diverse biological or technical applications. Physical properties (such as degradability, porosity, swelling ratio, thermal stability, etc.), chemical properties including chemical composition, mechanical properties (e.g., tensile strength/compressive strength, toughness, Young's modulus, etc.), rheological properties (such as viscosity, storage modulus, etc.), and biological properties such as biocompatibility are the five general aspects from which the properties of the hydrogel can be described [21]. It is crucial to understand the frequently used techniques for characterizing hydrogels. Although laboratory research has made considerable progress in the synthesis and application of hydrogels, they are still rarely used in wastewater treatment. While related studies have been conducted, we still lack a complete understanding of how the chemical composition of wastewater affects the performance of hydrogel absorbents.

Hydrogels' anti-interference properties can be enhanced in two ways. One advantage of hydrogels is their ability to self-clean. Hydrogel microspheres can successfully resist adhesion by forming a hydrated layer on their surface.

Heavy metal ions are adsorbed on the binding sites available on the adsorbent microspheres. Grafting polymer brushes onto the surface of superabsorbents can improve resistance. The polymer brushes can prevent organic macromolecule flocculants and nanoparticles from occupying the adsorption sites suitable for targeted metal ion harvesting [22]. The length of the fabricated brush is another important parameter for calculating the anti-interference. For example, long brushes can efficiently prevent the movement of floccules onto the superabsorbent surface, but this is not true for large particles. By taking advantage of the diversity of grafting, the hydrogels' performances can be considerably enhanced.

The term self-healing describes the capacity of any material to repair damage independently and reclaim its original characteristics [23]. The development of sophisticated self-healing hydrogels has generated a lot of interest in research since the ability to mend itself could not only increase longevity but also increase the hydrogels' dependability and toughness in a range of engineering and healthcare applications. The self-healing hydrogels can be classified as chemically or physically cross-linked, based on their self-healing methods, through reversible non-covalent interactions (such as host–guest interactions, metal–ligand coordination, hydrophobic interactions, hydrogen bonds, a mixture of numerous intermolecular interactions, etc.), physically self-healed hydrogels reconstruct the networks. Through dynamic covalent chemistry (e.g., Diels–Alder reactions, reversible radical reactions, phenylboronic ester complexation, disulfide bonds, acylhydrazone bonds, Schiff base, etc.), chemically cross-linked self-healed hydrogels rebuild the matrix structure [24].

The major sources of heavy metal ions are refineries, fertilizers, mining, smoking, pesticides, chemical industries, fungicides, paint, nuclear fission facilities, welding, electroplating, batteries, and vehicles [25]. Heavy metal ion contamination may impact the human organs and aquatic systems, including the brain, kidneys, nose, skin, gastrointestinal tract, central nervous system, and lungs. Numerous efforts have been made to eliminate the presence of heavy metal ions from contaminated water. Chemical precipitation is a technique used to extract heavy metal ions from polluted water, causing them to form sediments. Precipitates are created by forming complexes between heavy metal ions and coagulants or precipitation agents [26]. These can be detached by using either centrifugation or filtration. The chemical precipitation process requires the adjustment of pH. This strategy has both benefits and drawbacks. This approach is useful due to its cost-effectiveness. Eliminating the substantial quantity of sludge generated during the purification process necessitates further expenses.

The removal of heavy metal ions and monolayers of ions accumulated on the adsorbent surface is commonly accomplished via adsorption [27]. This is a fairly basic technique that works well to remove metal ions. This method is flexible and is not affected by harmful contaminants; therefore, the researchers have concentrated on designing appropriate adsorbents for the removal of toxic metal ions. Many polymer-based adsorbents like manganese dioxide-loaded biochar, polypyrrole, polypyrrole-polyaniline, aminated polyacrylonitrile, metal-organic frameworks, zeolite, *Eucalyptus camaldulensis*, graphene oxide embedded hydrogel particles, natural Jordanian, sodium alginate modified materials, and spherical chitosan-gelatin hydrogel particles are being used for designing hydrogel-based adsorbents [28].

Hydrogels can be fabricated through different techniques, such as bulk polymerization, which is a simple method that requires the use of only initiators and monomers. The reaction occurs at a high temperature and results in a highly viscous superabsorbent due to a high degree of monomer concentration brought on by the polymerization reaction.

3 Soft Materials for Dye Removal

3.1 Conceptual Idea Behind the Sequestration of Textile Dyes

Synthetic textile dyes represent a class of hydrophilic organic dispersible colorants employed in various industries such as food, textiles, cosmetics, and leather [29]. The excessive increase of these colorants in distinct fields is mainly due to increased urbanization and population growth. Dyes are toxic substances that, when present even with low concentration, can cause carcinogenic as well as mutagenic impacts on the water bodies as well as on human beings [30]. The impact on plant species present in water bodies is mainly related to the retardation of photosynthesis activity and penetration of sunlight. The impact on living species present in aquatic systems is mainly disturbed by the increase in concentration of chlorides, metals, and aromatic hydrocarbons produced as a result of the slower degradation of these colorants [31]. In addition, the organic colorants may destroy several microorganisms in water bodies.

Dyes belong to a significant category of contaminants, and when they find their way into the water, it becomes unsuitable and challenging to purify them. Dyes are man-made contaminants and possess intricate molecular configurations, rendering them more durable and harder to decompose. The occurrence of dyes in our natural water resources may lead to the reduction of dissolved oxygen content, thereby obstructing the diffusion of sunlight into these water bodies [32]. The excess concentration of dye molecules in the water resource may seriously threaten human life and the environment. Industrial wastewater often contains toxic substances, including cyanides, oil, fat, degreasing solvents, alkaline cleaning agents, and metals. The primary methods employed for wastewater treatment encompass adsorption, chemical analysis, coagulation, sedimentation, biological processes, and advanced oxidation procedures. Many sectors, including textiles, paper, and plastics, employ dyes to color their goods while simultaneously consuming large amounts of water. As a result, they produce a large number of colorful effluents. It is recognized that the discharge dyes have an incredible impact on the open discernment of water quality. Color is the primary contaminant to be analyzed in wastewater. Extremely colored species may be found in effluent released by the dye factory. These waste materials are not only anesthetic, but they also block light from entering the aquatic bodies, which may disrupt biological processes. Textile dyes have a very low ability to break down naturally; therefore, traditional biological treatment methods are not very successful in cleaning the dye wastewater [32]. Typically, wastewater treatment methods are handled through physical or chemical processes. However, these methods are expensive and are not suitable for the treatment of diverse types of dye wastewater.

Dyes can be categorized based on solubility, i.e., water-soluble dyes and water-insoluble dyes. The water-soluble dyes cover acidic, metal complex, direct, basic, mordant, and reactive dyes. Similarly, water-insoluble dyes can be vat, sulfur, azoic, and disperse dyes. It is important to mention that azo dyes are the most commonly utilized dyes, making up to 65–70% of all dyes manufactured [32].

Synthetic dyes represent a category of emerging pollutants characterized by their complex chemical structures and mechanisms of action. Despite their low concentrations, these dyes may pose a significant challenge in terms of their removal from the wastewater due to their intricate nature. Nowadays, environmental scientists are facing issues due to the leakage of synthetic dyes into the environment. Every year, more than 2% of the manufactured dyes are released directly into aqueous effluent. Globally, the entire dye usage in the textile sector is more than 10,000 tons per year, and roughly 100 tons per year are discharged directly into water streams [33].

It may be noted that the adsorbents cannot be recycled in most cases, and the process of chemiagglomeration leads to environmental contamination by creating colloidal substances in water waste. Traditional chemical oxidation processes, e.g., water treatment with chlorine, are slow and require reactive substances that are hazardous for transportation and storage. Additionally, these approaches are not very effective due to the occurrence of numerous chemical substances in water waste. Moreover, advanced oxidation techniques, such as ozonation, photofenton process, and photocatalysis, are expensive and not financially viable [34]. Thus, there is a demand for more efficient and affordable methods for treating industrially polluted wastewater that limits the use of toxic chemicals. A detailed analysis of the literature showed that dyes can be effectively removed using a variety of inexpensive, nonconventional adsorbents. For example, adsorption stands out as the preferred method as it is effective in eliminating various kinds of dyes. When the adsorption system is properly constructed, it leads to the generation of top-quality water treatment products. Getting rid of dye substances is challenging, and different methods are employed to lower the dye level to the standards set by the WHO. These methods are highly efficient in eliminating organic contaminants, like dyes, but they also create additional waste materials.

Therefore, it is very crucial to treat dye-containing effluent using sustainable means. Most of the colorants are synthetically designed and contain a complex molecular structure, so they are least degradable in water through conventional treatment methods. The major sources of dye discharge in water bodies include pharmaceutical industries, textile effluents, dye intermediate industries, kraft bleaching industries, tannery, and paper/pulp industries [35]. It has been reported that the textile industry is the leading source of water contamination, producing around 10,000 distinct colorants and 7.105 metric tons of annual production worldwide [36–38]. Some of the existing methods that are being widely used for the sequestration of toxic textile dyes are summarized in Fig. 2. Due to the presence of different complex hydrocarbons, including naphthalene, benzidine, and aromatic compounds, dyes can impart mutagenic effects and their degradation products are also highly toxic [39]. Therefore, modern techniques are needed for the sequestration of such pollutants in order to reduce their toxic impacts.

3.2 Hydrogel Derived Soft Materials: Technical Features

The attractive features of hydrogels for mimicking stimuli-responsive behavior are mainly governed by the choice of monomer, cross-linker, and grafted backbone. Such types of materials can be synthetic, natural, or hybrid. Most importantly, the hybrid hydrogels comprising the specific features of both synthetic and natural hydrogels are widely accepted for their applications in various fields. The existing methods for synthesizing hydrogels include solution polymerization, sol-gel synthesis, bulk polymerization, suspension polymerization, and free-radical polymerization [40]. The functional characteristics of hydrogels are governed by their adsorption capacity, porosity, durability, stability, degradability, photostability, and cost-effectiveness [41]. Therefore, different process parameters should be

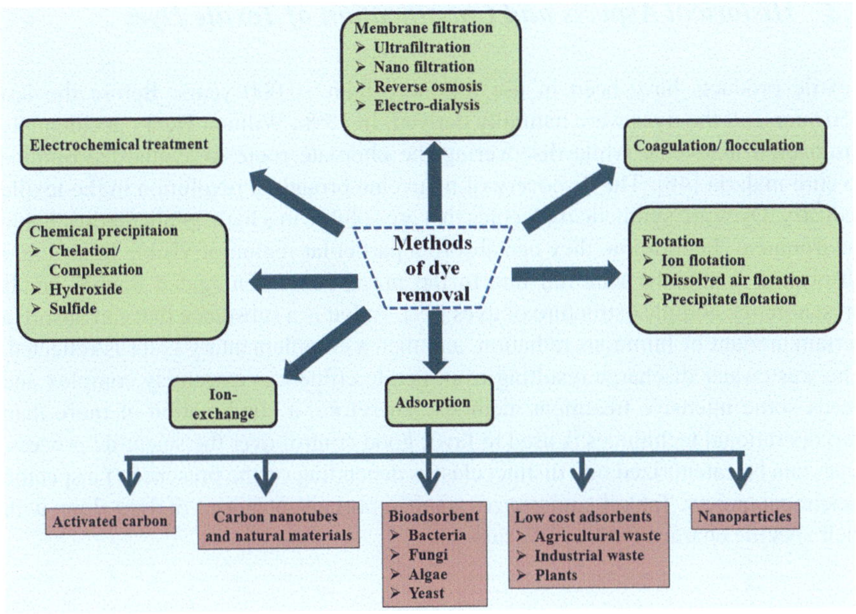

Fig. 2 Some important methods of dye removal

optimized in such a way as to achieve the desired properties of the hydrogel being synthesized.

Two important techniques for developing polymeric hydrogels can be achieved by the direct polymerization of aqueous monomers or by the functional modification of the existing polymers which may be artificial or natural. Free-radical polymerization is the most versatile and widely used method to derive hydrogels for distinct applications [42]. In 1998, a new category of hydrogels was developed and named superporous hydrogels (SPHs) [43]. Such types of hydrogels were widely applicable in agriculture as well as the cosmetic industry, where porous materials are required to absorb a huge amount of water. SPHs can absorb a large amount of water volume ranging from 10 to 1000 times of their initial dry weight in a very short period. SPHs can absorb water within a minute to hours, depending upon their granule size and choice of monomers used to synthesize the final product [44]. Acrylic acid, acrylamide, and salt of acrylic acid including 2-hydroxyethyl methacrylate are the commonly used monomers for synthesizing SPHs. Hydrophilic functional moieties on the surface of distinct polymers can induce hydrogen bonding, dipole–dipole interactions, complex formation, and other similar interactions with the dye molecules of metal ions present in industrial wastewater [45].

3.3 Historical Aspects and Classification of Textile Dyes

Textile products have been in use for more than 10,000 years. Before the last 150 years, all the dyes were naturally derived. In 1896, William Henry accidentally discovered mauveine while discovering the alternate route to synthesize quinine to cure malaria [46]. The discovery of mauveine brought a revolution in the textile industry. Dyes are synthetic molecules that are soluble in a hydrophilic/hydrophobic environment. In addition, they can absorb a particular region of visible light. Light absorption is mainly achieved due to the presence of conjugated double bonds present in the complex structure of dyes [47]. A dye is a substance that can absorb a certain amount of luminous radiation, and then a complementary color is reflected. The wastewater discharge resulting from textile effluent is extremely complex and needs some intensive treatment methods. Therefore, a combination of more than two operational techniques is used to favor good control over the scientific process. Dyes can be categorized into distinct classes depending on the presence of a specific functional moiety. Table 1 summarizes some important categories of dyes along with their specific characteristics and examples.

3.4 Application of Hydrogels for Dye Sequestration

Hydrogel-based dye sequestration devices are crucial for designing low-cost, biodegradable, efficient, and eco-friendly adsorbents [48]. In general, natural gums, in combination with some synthetic monomers, impart excellent dye removal properties. The synthetic strategies can be controlled to develop either IPNs or semi-IPNs [49]. Hydrogel-based adsorbents can show remarkable dye adsorption potential for the sequestration of selective cationic/anionic dyes [50]. For example, Hu et al. [51] designed nanocomposite hydrogels based on polyacrylic acid with 2000 mg g^{-1} maximum adsorption capacity for the sequestration of methylene blue dye in a neutral dye solution, which was favored due to the high water penetration (about 500 times) inside the nanocomposite matrix. A study carried out by Choudhary et al. [52] depicted the synthesis of gum gellan-based hydrogel GG-cl-poly(AA), depicting 531.9 mg g^{-1} and 552.48 mg g^{-1} adsorption capacities for the adsorption of methylene blue and malachite green dyes, respectively. The adsorption process was further benefited through adsorption-desorption cycles and reusability of the synthesized adsorbent. Moreover, the synthesis method favors the use of nontoxic material, which is easy to handle, cost-effective, and eco-friendly. Incorporating some nanostructured materials also favored the sequestration of dyes through the hydrogel. Taking advantage of these things, Sharma et al. [50] designed Fe_3O_4 incorporated hybrid hydrogel based on sodium alginate and κ-carrageenan biopolymers (CAG-NaAlg-cl-polyAA/Fe_3O_4). The integration of magnetic nanoparticles inside the cross-linked soft materials increased the tendency for dye removal from 95 to 99% in the case of cationic dyes. The

Table 1 Classification of textile dyes

Category	Characteristics	Examples
Azo dyes	They contain –N=N– functional group joining two symmetrical/asymmetrical identical or non-azo aryl/alkyl free radicals; they contain around 60–70% of all dyestuff included in the textile production	• Acid red 2 • Acid Orange 20 • Disperse yellow 7 • Direct black 22 • Methyl red • Direct blue 71 • Trypan blue
Sulfur dyes	These are the high molecular weight dye molecules that can be synthesized during the sulfurization of some organic compounds; out of total US dye production, 9.1% of dyes are sulfur dyes; total world production of sulfur dyes is around 110,000–120,000 tons per year	• 8-Anilino-5-(*p*-hydroxyanilino)-1-naphthalenesulfonic acid • Sulfur brilliant green, CI 53570 • Sulfur blue dye, CI 53235 • *N*-[*p*-(*p*-hydroxyanilino) phenyl] sulfanilic acid • Leuco sulfur black 1, C.I 53185 • Sulfur black, CI 53185
Acid dyes	The dye molecules containing a group of acid functional moieties such as COOH and SO_3H are designated as acid dyes; due to their acidic nature, they possess strong affinity to bind with the basic fabrics; the consumption of acid dyes is around 30–40% as compared to other dyes, and therefore, they are mostly used to color wool, nylon, and cotton	• C.I. Acid blue 7 • Acid blue 349 • Acid Orange IV • C.I. Acid red 249 dye • C.I. Acid blue 25 (AB25) • Acid red 337 • Acid blue 83 • Acid blue 147 • Acid yellow 36 • C.I. Acid blue 25 (anthraquinone) • Acid violet 17 • Acid Orange 7 • Acid Black-234
Basic dyes	Basic dyes usually contain a positive charge on the ammonium group or a delocalized charge in xanthenes, triarylmethane, and acridine dyes. The ionic solution of these dyes can easily bind with the surface of textile fiber containing an overall negative charge due to the strong electrostatic interactions	• Basic blue 11 • Basic red 9 monohydrochloride • Basic blue 11 • Basic blue 3 • Basic blue 6 • Basic blue 24 • Basic yellow 2 • Basic green 1 • Basic red 1

(continued)

Table 1 (continued)

Category	Characteristics	Examples
Vat dyes	A class of water-insoluble dyes specifically used to color cellulosic fiber are vat dyes; they are known for excellent properties such as color fastness, wet fastness, and excellent light properties	• Indigo carmine • Vat black 25 • Vat yellow 1 • Vat black 25 • Vat blue 1 • C.I. Vat green 1 • Vat acid blue 74 • Vat yellow 1
Reactive dyes	Reactive dyes showed excellent ability of the functional groups and their ability to react and bind with the textile fiber; this class of dyes shows ranges of hues, brilliance, and wet fastness	• C.I. Reactive red 239 • C.I. Reactive red 19 • C.I. Reactive red 120 • C.I. Reactive red 3 • Reactive red 6
Direct dyes	These dyes are also termed substantive dyes, and they can be directly attached to the textile fiber due to the presence of strong affinity of the dye molecules; their primary application includes the coloring of paper products	• Direct red 75 • Direct yellow 12 • Direct red 250 • Direct Orange 18 • Direct red 28 • Direct red 23 • Direct Orange 26 • Direct green 33 • Direct blue 78 • Direct black 150 • Direct yellow • Direct yellow 22
Disperse dyes	This class of dye contains a polar structure comprised of anthraquinone or azo groups; mainly applied on polyester or similar types of hydrophobic textile surfaces; they possess extremely low solubility in water but are highly soluble in nonpolar organic solvents; they are also the persistent class of dye due to their nonbiodegradation and recalcitrant nature	• Disperse blue 1 • Disperse blue 106 • C.I. Disperse blue 79 • C.I. Disperse Orange 3 • Disperse blue 14 dye • C.I. Disperse red 278 • Disperse red 17 • C.I. Disperse red 8 • Disperse blue 124 • C.I. Disperse red 7 • C.I. Disperse Orange 3

removal of dyes was favorable, with Langmuir adsorption favoring pseudo-second-order kinetics. In addition, the fabricated soft material was recyclable for up to five consecutive cycles. In another study, Njuguna et al. [53] synthesized xanthan gum-based hydrogels showing 99% dye removal efficiency at pH 8–9 with 502 mg g^{-1} dye sequestration potential in the case of Gentian Violet dye. The adsorption and desorption kinetics in the case of dye removal were favored via pseudo-first-order and second-order kinetics, respectively. In another study, Yan et al. [54] fabricated a mussel-inspired hydrogel based on cellulose-derived *co*-polydopamine@Pd nanoparticles and graphene oxide nanosheets, i.e., *p*(DACO-PDA@Pd NPs-*co*-AA-SA)@P-GOM). An interpenetrating polymer network was

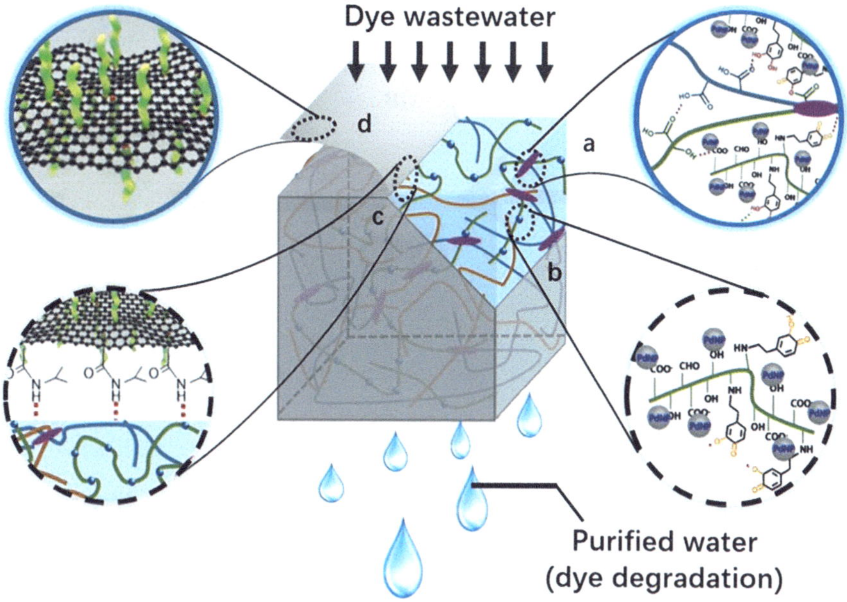

Fig. 3 Schematic illustration of dye degradation through *p*(DACO-PDA@Pd NPs-*co*-AA-SA)@P-GOM hydrogel. (**a**) Cross-linking sites between PAA with DACO-PDA@Pd NPs. (**b**) Immobilization of Pd nanoparticles on the active sites of DACO-PDA. (**c**) Cross-linking between *p*(DACO-PDA@Pd NPs-*co*-AA-SA) hydrogel surface and the PNIPAm-*co*-GO membrane. (**d**) PNIPAm-*co*-GO membrane (P-GOM). (Reprinted with permission from [54])

formed by the cross-linking of covalent and noncovalent bonds (Fig. 3). Moreover, excellent durability, reversibility, ductility, and self-healing ability was induced by the dynamic redox catechol system arising between hydroquinone, semiquinone, and quinone radicals initiated by Pd nanoparticles in the IPN network.

The fabricated hydrogel could catalytically degrade cationic (methylene blue) and anionic dyes (Congo red) with a more than 95% degradation rate. Further, the hydrogel showed recyclability without any deterioration in its structure and performance. After recyclability, the reduction time was found to increase from 20 (first cycle) to 23 min (fifth cycle), whereas, in the case of methylene blue, it increases from 3.5 (first cycle) to 4 min (fifth cycle).

The adsorption and removal of methylene blue were also carried out using pH-sensitive starch and polyacrylic acid-based hydrogels [55]. The polyacrylic acid chains were grafted over starch and cross-linked with *N*,*N*′-methylenebisacrylamide. During fabrication, the molar ratios of acrylic acid and *N*,*N*′-methylenebisacrylamide were 95:7, 94:6, and 93:5. The highest swelling of the hydrogel was 201.9% with 94:6 molar ratio at 7.4 pH. The water absorption in the hydrogel was regulated with the Fickian diffusion process, and the kinetic of swelling was Schott's pseudo-second-order. The maximum methylene blue dye adsorption (66.7%) was achieved with the hydrogel synthesized with a 93:5 molar

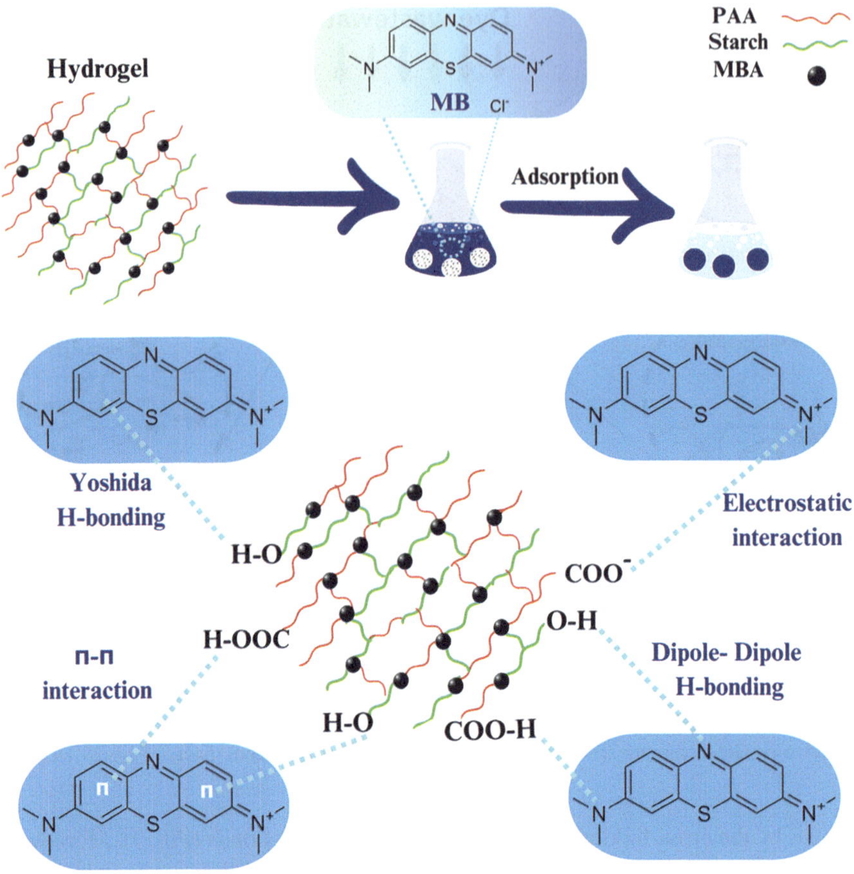

Fig. 4 Mechanism for the adsorption of MB on the synthesized starch/poly(acrylic acid) hydrogel. (Reprinted with permission from [55])

ratio of acrylic acid and *N,N′*-methylenebisacrylamide at initial dye concentration of 20 mg L^{-1} and pH 10. The adsorption of methylene blue onto the heterogeneous surface of the adsorbent was favorable through an exothermic process, as indicated by the Temkin adsorption isotherm model. Mechanistically, the adsorption occurs through dipole–dipole hydrogen bonding, electrostatic interactions, Yoshida hydrogen bonding, and π–π interactions (Fig. 4).

A detailed literature survey revealed that hydrogel-based soft materials in semi-IPNs, IPNs, and nanocomposites possess tremendous potential to separate large categories of ionic dyes from industrially polluted wastewater. Table 2 describes some important hydrogel-based dye adsorbents that can be beneficial to the textile industry as well as environmental pollution control agencies to overcome the specific challenges caused by water pollution through textile dyes.

Table 2 Hydrogel-based soft materials for the sequestration of textile dyes

Hydrogel-based adsorbent	Dye removed	Adsorption equilibrium time	Maximum adsorption capacity (mg g^{-1})	References
Starch/poly(alginic acid-cl-acrylamide) nanohydrogel	Coomassie brilliant blue	150 min	31.24	[56]
GT-CAG-cl-polyAA hydrogel	Methylene blue	12 h	297.3	[57]
	Coomassie brilliant blue	12 h	81.8	
Functionalized hydrogel based on cellulose nanofibers and wood pulp	Methylene blue	3000 min	340	[58]
Glutaraldehyde cross-linked hydrogel with activated carbon	Food blue 2	24 h	133.9	[59]
	Food red 17	24 h	155.1	
Superabsorbent hydrogels derived from gum arabic, polyacrylamide, and polyacrylate	Methylene blue	50 min	48	[60]
Bioconjugated graphene oxide hydrogel	Methylene blue	20 min	122.5	[61]
	Rhodamine B	40 min	62	
GO-CS composite hydrogels	Methylene blue	58 h	350	[62]
Polyacrylamide/cellulose nanocrystals	Methylene blue	250 min	326	[63]
β-CD/PAA/GO nanocomposite	Methylene blue	150 min	247.99	[64]
GO/PEI hydrogel	Methylene blue	240 min	300	[65]
P25-graphene hydrogel	Methylene blue	55 min	87.63	[66]
MRGO nanocomposite hydrogel	Rhodamine B	120 min	13.15	[67]
Nanocrystalline cellulose based hydrogel	Methylene blue	200 min	8.1	[68]
Chitosan/κ-carrageenan bioadsorbents	Eriochrome black-T	120 min	280	[69]

(continued)

Table 2 (continued)

Hydrogel-based adsorbent	Dye removed	Adsorption equilibrium time	Maximum adsorption capacity (mg g^{-1})	References
Date pits powder/sodium alginate hydrogel beads	Crystal violet	24 h	83.56	[70]
Polyethyleneimine functionalized apple pomace derived cellulose	Methyl orange	250 min	416.72	[71]
CuZn-HDS/alginate hydrogel	Methyl orange	60 h	763.3	[72]
Gum fenugreek galactomannan conjugated MnO_2 mesoporous hydrogel	Malachite green	180 min	526.31	[73]
Activated carbon/clay/sodium alginate hydrogel composite	Methyl red	30–360 min	248.13	[74]
Hydrogel derived from carboxymethyl Huangshui polysaccharide, sodium alginate, and Fe_3O_4	Methylene blue	180 min	186.57	[75]

4 Soft Materials for the Sequestration of Toxic Heavy Metal Ions

4.1 Toxicity of Heavy Metal Ions

Heavy metal ions, such as Cu, Fe, Zn, Mn, Ni, Pb, Co, Cd, and Hg, are present in trace amounts in our environment. Still, these are highly toxic to the environment, and they are also the major components of wastewater discharge. The components of these metals can be stored in soft tissues inside the body because they are not metabolized easily inside the body and can cause serious health hazards [76]. Approximately 80% of the total global wastewater is discharged into our ecosystem without any primary treatments [77]. The discharged wastewater usually contains a large fraction of heavy metal ions, which can produce serious impacts on humans, animals, as well as other nonliving species. For example, acute exposure to Hg can cause circulatory system disorder, nervous system disorder, and arthritis in humans [78], whereas exposure to Cu can damage liver tissues, produce sleep disorders, and adversely affect the growth of plants by inhibiting enzymatic activities in soil [79]. Similarly, exposure to Cu can cause anemia, whereas kidney failure, muscle impairment, and brain damage in infants are the common consequences of Pb poisoning [80]. In addition, the specific properties of heavy metal ions, including nonbiodegradability, stability, and high solubility in water, further stimulate their easy discharge in water causing detrimental impressions on plants retarding enzyme activity, seed germination, chlorophyll synthesis, and photosynthesis [81].

As in dyes, the separation of heavy metal ions is also more favorable through adsorptions rather than other methods such as coagulation, ion exchange, chemical precipitation, solvent extraction, membrane filtration, and electrochemical separation. This is due to the several limitations of other methods such as costly disposal procedures, sensitive operating conditions, high energy requirements, inadequate removal, and poor efficiency.

4.2 Role of Hydrogels for the Sequestration of Heavy Metal Ions

In the present era, hydrogels have emerged as promising adsorbents for sequestering many toxic metal ions in industrial wastewater. The main sources of heavy metal ions are fertilizers, pesticides, refineries, fungicides, nuclear fission plants, automobiles, buildings, chemical industries, welding, batteries, smoking, and mining. The excessive release of heavy metal ions into our freshwater resources can contaminate our aquatic ecosystem by showing an adequate impact on the skin, lungs, kidneys, gastrointestinal tract, nose, and brain of the aquatic species. Removal of hazardous heavy metal ions is done through different processes such as adsorption, biological treatment, microbial treatment, membrane separation, nanofiltration, ion exchange,

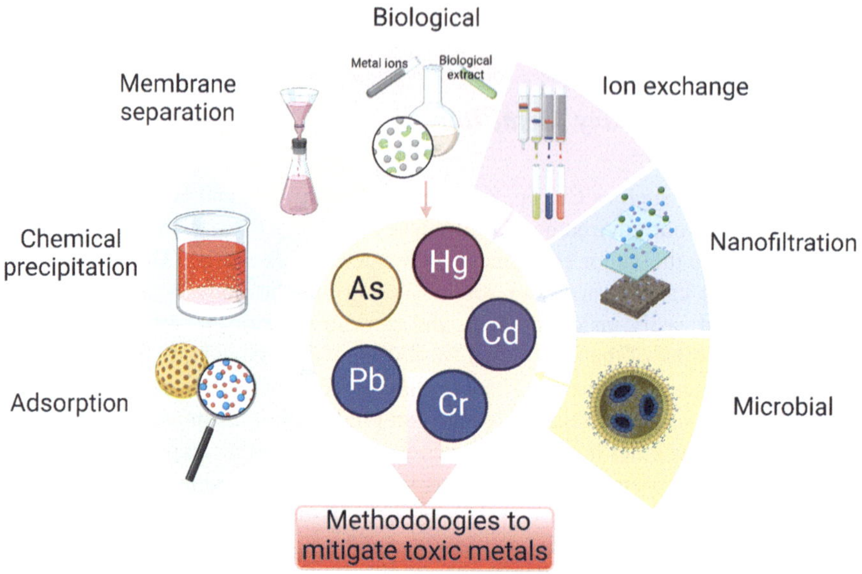

Fig. 5 Removal of different advanced treatment methods for the sequestration of heavy metal ions. (Reprinted with permission from [84])

and chemical precipitation (Fig. 5). The binding of these metal ions to the surface of hydrogels is favored due to the presence of possible electrostatic interactions as well as the availability of vacant sites for their interaction with the surface. The fabricated hydrogels can have different metal ion selectivity and adsorption capacity depending upon the type of matrix synthesized. For example, Yond et al. [82] synthesized composite beads based on alginate and UiO-66(SO_3H). The selectivity factor of the bead against Pb^{2+} against Zn^{2+}, Cd^{2+}, Cu^{2+}, and Ca^{2+} was 6.1, 5.7, 5.4, and 9.1. The adsorption performance of the beads depicted 400.5 mg g^{-1} adsorption capacity in the case of lead, and the beads can be reused for up to five cycles without any deterioration of the matrix. However, the efficiency of the bead reduces to 94.2% of the initial value after the fifth cycle. In another study, Perumal et al. [83] designed carbon dot doped hydrogel particles, which were further evaluated for the sequestration of different metal ions such as Hg^{2+}, Cd^{2+}, Pb^{2+}, and Cr^{2+}. Comparative results suggested that the hydrogel particles showed around 72% metal uptake in a single metal ion system for Pb^{2+} ions, whereas the uptake percentage increases to 99% in the case of multiple ion systems.

Naturally occurring biomass-based hydrogels have shown more promising features for uptaking metal ions from industrially polluted wastewater. For example, Zhang et al. [85] synthesized biomass-based hydrogel, which was derived from soybean dregs, and evaluated the potential of synthesized soft material for the removal of Zn^{2+}, Cu^{2+}, Fe^{2+}, and Cr^{2+} metal ions. pH-responsive uptake was shown by the fabricated hydrogels with maximum removal efficiency equal to

121.2 mg g^{-1}, 75.4 mg g^{-1}, 78.5 mg g^{-1}, and 41.7 mg g^{-1}, in the case of Zn^{2+}, Cu^{2+}, Fe^{2+}, and Cr^{2+}, respectively. Similarly, Sharma et al. [86] fabricated dextrin- and sodium alginate-based gel films that can detect and detoxify the metal ions present in wastewater contaminated with Cu^{2+} and Ni^{2+} ions. In 5 μM concentrated solution, the detoxification efficiency of Cu^{2+} and Ni^{2+} ions was found to be 95.54% and 91.30%, respectively. In another study, Sharma et al. [87] enzymatically designed dextrin and polyvinyl alcohol (PVA) based porous hydrogel film, exhibiting 92.89% and 95.77% sorption of toxic lead and copper ions in industrially polluted wastewater. In addition to this, the fabricated gel-based films can also detect the existence of lead and copper ions in the wastewater. The detection limit was 0.521 μM and 0.421 μM in the case of lead and copper ions, respectively. Polyvinyl alcohol-based hydrogels were further fabricated by Tassanapukdee et al. [88], combining them with polyvinylpyrrolidone (PVP) and chitosan (CS). The optimized ratio of PVP/PVA/CS was 0.3:0.6:0.3 using epichlorohydrin as a cross-linking agent and irradiating the reaction mixture for 3 min under 600 W of microwave radiation. The gel fraction and swelling degree of the synthesized hydrogel were 42.6% and 1627.4%, respectively. The hydrogel was reported for the removal of Ni(II), Pb(II), Cd(II), and Cu(II). The maximum removal capacities of Ni(II), Pb(II), Cd(II), and Cu(II) were 15.13 mg g^{-1}, 80.02 mg g^{-1}, 8.36 mg g^{-1}, and 33.7 mg g^{-1}, respectively. Moreover, the variation in pH from 4.0 to 8.0 did not affect the adsorption capacity. After adsorption and desorption cycles, using 0.1 M EDTA, more than 60% of the removal efficiency of the fabricated hydrogel was preserved.

Semi-interpenetrating polymer network hydrogels based on chitosan, α-ketoglutaric acid, polyacrylamide, and *N,N*-methylene bis (acrylamide) were designed for the sequestration of Pb(II), Cu(II), and Zn(II) ions [89]. The semi-IPN hydrogel was comprised of a porous structure having a high specific surface area for the exposure of more active sites for the selective adsorption of heavy metal ions. The adsorption of hydrogels was affected by adsorbent dosage, ionic strength, pH of solution, and concentration of metal ions. At pH 5.0 and temperature 30 °C, the theoretical maximum adsorption capacity of the adsorbent against Pb(II), Cu(II), and Zn(II) ions was 61.41 mg g^{-1}, 72.39 mg g^{-1}, and 51.89 mg g^{-1}, respectively. The adsorption data followed pseudo-second-order kinetics, and it fitted well with the Langmuir adsorption isotherm model. The recyclability analysis of the adsorption process showed that even after five consecutive adsorption-desorption cycles, the synthesized hydrogel exhibited 90% metal ion removal efficiency. Further, the adsorption was mainly due to the coordination and ion exchange process indicated by XPS analysis. In addition to semi-IPNs, the IPN hydrogels were also reported for the adsorption and sequestration of heavy metal ions. For example, Tanan et al. [90] synthesized an IPN hydrogel using poly(2-hydroxyethyl methacrylate-*co*-acrylamide) (P(HEMA-*co*-AM)) and poly(vinyl alcohol) (PVA) through a one-pot microwave-assisted synthesis. The fabricated system showed enhanced metal ion removal potential against Cu(II) and Pb(II) ions. The maximum water absorbency of HEMA IPN hydrogel was 1310%. Cu(II) and Pb(II) adsorption on the IPN hydrogel was favorable with pseudo-first-order adsorption kinetics following Freundlich adsorption isotherm with $R^2 = 0.999$. The maximum metal uptake capacity of IPN hydrogel against Pb(II) and Cu(II) ions was 292.5 mg g^{-1} and 200.4 mg g^{-1}.

In short, hydrogel-based adsorbents have shown tremendous potential for separating the toxic metal ions present in wastewater. They can be used on an industrial scale to fabricate novel technologies to purify industrially adulterated wastewater.

Table 3 summarizes recent hydrogel-based adsorbents for sequestering different toxic heavy metal ions.

4.3 Nanogel- and Microgel-Based Soft Materials for the Separation of Metal Ions

In addition to hydrogels, functional microgels and nanogels also play a crucial role in the separation of metal ions present in wastewater. Nanogel- and microgel-based superabsorbents are most widely applicable in agriculture farming for the controlled release of fertilizers and pesticides to promote plant growth. Simultaneously, they can be used to uptake certain types of metal ions, which can retard plant growth. In this way, functional microgels and nanogels provide a dual strategy to promote plant growth through the release of essential micronutrients and the uptake of metal ions. For example, Dhiman et al. [107] synthesized microgels based on alginate which were cross-linked with 3,3′-dithiopropionohydrazide showing a dual mechanism for release and uptake. The microgel showed 85% encapsulation efficiency for diuron herbicide and 100% release efficiency in 2 mM glutathione. Further, after degradation, the microgel captured a maximum of 34% and 66% of Hg^{2+} and Cu^{2+} ions, respectively. In another study, functionalized chitosan- and stearic acid-based microgels were prepared using 2,3-dihydroxybenzoic acid. The synthesized microgel was used to release imidacloprid insecticide, offering 75% release potential after 133 h along with 85% encapsulation efficiency. The degradation of the microgel network produces 2,3-dihydroxybenzoic acid, which can effectively bind with the Fe^{3+} ions and Hg^{2+} ions with removal efficiency equal to 90% and 97%, respectively [108].

Microgels are mostly preferred among different adsorbents due to their easy method of preparation, tunable features, high adsorption capacity, long-term stability, and reusability. The presence of different functional moieties such as carboxyl, amine, hydroxyl, and amide further explores their utility because they can be easily modified into a microgel network through various types of copolymerization reactions. Zheng et al. [109] designed a polydopamine-coated imprinted microgel adsorbent that was selective for the sequestration of Pb(II) ions and showed NIR activity. With the increase in temperature from 20 to 40 °C, the adsorption proficiency of imprinted microgel for removing lead increased from 128 to 135.2 mg g^{-1} following pseudo-second-order kinetics. Similarly, carboxy alkylated lignin-derived

Table 3 Hydrogel-based adsorbents for the sequestration of different toxic heavy metal ions

Name of the adsorbent	Metal ion removed	Adsorption equilibrium time	Maximum adsorption capacity (mg g^{-1})	References
Polydopamine, chitosan, and poly(*N*-isopropyl acrylamide) based hydrogel	Pb^{2+}	12 h	42.63	[91]
	Cu^{2+}	12 h	38.53	
Sulfomethylated lignin-grafted polyacrylic acid hydrogel	Co^{2+}	250 min	145.14	[92]
	Cu^{2+}	250 min	117.64	
	Cd^{2+}	250 min	277.89	
	Ni^{2+}	250 min	166.72	
	Pb^{2+}	250 min	344.85	
Chitosan/sodium alginate/calcium ion based hydrogel	Cu^{2+}	200 min	174.82	[93]
	Cd^{2+}	400 min	157.72	
Nanocomposite hydrogel based on poly(*N*-vinyl imidazole) and nitrogen-doped graphene quantum dot	Cd^{2+}	180 min	500	[94]
	Ni^{2+}	180 min	555	
	Cr^{6+}	180 min	357	
Malic acid cross-linked chitosan hydrogel	Cr^{4+}	18 min	427	[95]
Tragacanth cross-linked 2-hydroxyethyl methacrylate-*co*-acrylamide	Hg^{2+}	400 min	154.8	[96]
Xylan/graphene oxide composite	Cu^{2+}	24 h	228	[97]

(continued)

Table 3 (continued)

Name of the adsorbent	Metal ion removed	Adsorption equilibrium time	Maximum adsorption capacity (mg g^{-1})	References
Polyvinyl alcohol/alginate/chitosan hydrogel	Pb^{2+}	700 min	139.4	[98]
	Cr^{6+}	700 min	86.1	
Fluorescent hydrogel based on sulfhydryl-derived carbon dots	Hg^{2+}	1 h	662.25	[99]
Nano calcium carbonate doped chitin hydrogel	Cu^{2+}	48 h	194.61	[100]
	Cd^{2+}	48 h	191.58	
Fluorescent chitosan based hydrogel	Hg^{2+}/Hg^{+}	40 min	121	[101]
EDTA cross-linked chitooligosaccharide/sodium alginate/Ca^{2+} hydrogel	Zn^{2+}	12 h	160.19	[102]
	Ni^{2+}	12 h	72.67	
	Mn^{2+}	12 h	52.43	
Acrylic fiber-based hydrogel	Pb^{2+}	120 min	371.36	[103]
3D-printed montmorillonite nanosheet based hydrogel	Pb^{2+}	240 min	134	[104]
Titanate nanofiber fluorescent nanocellulose-based hydrogel	Cr^{4+}	180 min	648.4	[105]
Mesoporous alginate/β-cyclodextrin hydrogel beads	Pb^{2+}	6 h	21.09	[106]
	Cu^{2+}	6 h	15.54	
	Cd^{2+}	6 h	2.47	
	Ni^{2+}	6 h	2.68	

microgels with variable cross-linking degrees were synthesized, and the adsorption sequence for removing Zn^{2+}, Cd^{2+}, and Ni^{2+} was $Ni^{2+} > Zn^{2+} > Cd^{2+}$ [110]. In another study, Zhang et al. [111] used carbon dots to detect Hg^{2+}, Fe^{3+}, and Cu^{2+} ions. The carbon dots were embedded inside the microgel matrix and named as polyacrylamide-*co*-CD-*co*-2-acrylamido-2-methylpropanesulfonic acid. The sensitivity of the microgels was 2.3 nM, 3.2 μM, and 1.3 μM in the case of Hg^{2+}, Fe^{3+}, and Cu^{2+} ions, respectively. Further, the adsorption capacity of the synthesized microgels was 85.00 mg g^{-1} and 02.13 mg g^{-1} toward Cu^{2+} and Fe^{3+}, respectively. Reusability experiments exhibited better stability toward the absorption or detection of heavy metal ions.

In addition to microgels, the nanogels have shown an extreme potential for the sequestration of heavy metal ions. The uses of nanogels were further highlighted by Atta et al. [112] using sodium montmorillonite (Na-MMT). This was further functionalized with amphiphiles based on cross-linked nanogel polymers of acrylamide (AAm), sodium 2-acrylamido-2-methylpropane sulfonate (Na-AMPS), *N*-isopropylacrylamide (NIPAm), and (acrylamidopropyl)trimethylammonium chloride solution (APTAC) through surfactant-free techniques. The interactions between Na-MMT and nanogels were confirmed by FTIR analysis. The Na-MMT nanogels reduced the water surface tension of Co and Ni ions within 60 min. It was also reusable for four adsorption-desorption cycles. Another study carried out by Shoueir et al. [113] demonstrated that Cu^{2+} ions could be extracted from industrially polluted wastewater using poly(vinyl alcohol)/poly(acrylamide-*co*-*N*-isopropylacrylamide) core-shell nanogels which were synthesized through surfactant-free emulsion polymerization. The maximum Cu^{2+} ion removal was found to be 129.87 mg g^{-1} using 0.9 mol% PAA nanogel. The kinetic model demonstrated that pseudo-second-order kinetics was the favorable process for the removal of Cu^{2+} ions. Freundlich adsorption isotherm was the favorable adsorption isotherm. The nanogel was reusable for up to three cycles of adsorption and desorption. A novel method was developed for the trace-level detection of Pb^{2+} ions using a capsule membrane encapsulated smart hydrogel. It was comprised of a semipermeable membrane of calcium alginate and poly(*N*-isopropylacrylamide-*co*-acryloylamidobenzo-18-crown-6) (PNB) nanogels [114]. The membrane detected 10^{-9} mol L^{-1} of Pb^{2+} ions in the water sample. The provided strategy was facile and efficient for monitoring water contaminants. The existing studies have shown that hydrogel-based microgels and nanogels possess an extreme potential for the sequestration of heavy metal ions.

5 Conclusion and Future Perspectives

Hydrogel-based soft materials have been proven to be an excellent choice of materials for wastewater treatment, as they cause the least toxic effects during decontamination. Compared to other adsorbents, such as nanomaterials, they can be used multiple times for the sequestration of textile dyes. The bio-based soft materials add novel features such as cost-effectiveness, biodegradability, and ease of handling. Many biopolymers such as alginate, chitosan, dextrin, gelatin, and cellulose have been reported in the literature for wastewater treatment, particularly for dye removal and heavy metal ion sequestration. Hydrogel-based soft materials are suitable for eliminating different textile dyes, including acidic dyes, basic dyes, vet dyes, sulfur dyes, direct dyes, azo dyes, reactive dyes, and dispersed dyes. In comparison with the advantages of hydrogel-based sequestration devices, there are still some challenges associated with them, such as the selectivity of the adsorbent, efficiency of the degradation method, and disposal of the degraded product after multiple adsorption cycles.

Acknowledgments One of the authors is extremely thankful to Chandigarh University, Mohali, India, for providing infrastructure facilities. The authors are also thankful to NIT Jalandhar and IIT Mandi.

CRediT Authorship Contribution Statement **Amit Kumar Sharma**: conceptualization, writing-original draft preparation, figure preparation, writing-reviewing and editing; **Balbir Singh Kaith**: writing-reviewing and editing; **Garima Agrawal**: writing-reviewing and editing; **Sunil**: writing-original draft preparation; **Khushi Choudhary**: writing-original draft preparation; **Harshdeep Tiwari**: writing-original draft preparation.

Funding Declaration There are no sources of funding, and the authors are aware of the submission.

References

1. Vardhan, K.H., Kumar, P.S., Panda, R.C.: A review on heavy metal pollution, toxicity and remedial measures: current trends and future perspectives. J. Mol. Liq. **290**, 111197 (2019)
2. Garg, S., Chowdhury, Z.Z., Faisal, A.N.M., Rumjit, N.P., Thomas, P.: Impact of industrial wastewater on environment and human health. In: Advanced industrial wastewater treatment and reclamation of water: comparative study of water pollution index during pre-industrial. Industrial period and prospect of wastewater treatment for water resource conservation, pp. 197–209. Springer (2022)
3. Maheshwari, K., Agrawal, M., Gupta, A.: Dye pollution in water and wastewater. In: Novel materials for dye-containing wastewater treatment, pp. 1–25. Springer (2021)
4. Kheradmand, A., Negarestani, M., Mollahosseini, A., Shayesteh, H., Farimaniraad, H.: Low-cost treated lignocellulosic biomass waste supported with $FeCl_3/Zn(NO_3)_2$ for water decolorization. Sci. Rep. **12**, 16442 (2022)
5. Liu, Q., Li, Y., Chen, H., Lu, J., Yu, G., Möslang, M., Zhou, Y.: Superior adsorption capacity of functionalised straw adsorbent for dyes and heavy-metal ions. J. Hazard. Mater. **382**, 121040 (2020)
6. Frisbie, S.H., Mitchell, E.J.: Arsenic in drinking water: an analysis of global drinking water regulations and recommendations for updates to protect public health. PLoS One. **17**, e0263505 (2022)
7. Deepa, K., Kapoor, A., Sivaraman, P.: Applying noncarbon-based nanomaterials to detect environmental contaminants. In: Nanotechnology for environmental pollution decontamination: tools. Methods, and approaches for detection and remediation, p. 135. Taylor and Francis (2022)

8. Ajiboye, T.O., Oyewo, O.A., Onwudiwe, D.C.: Simultaneous removal of organics and heavy metals from industrial wastewater: a review. Chemosphere. **262**, 128379 (2021)
9. Chen, H., Sharma, S.K., Sharma, P.R., Yeh, H., Johnson, K., Hsiao, B.S.: Arsenic(III) removal by nanostructured dialdehyde cellulose–cysteine microscale and nanoscale fibers. ACS Omega. **4**, 22008–22020 (2019)
10. Rocky, M.M.H., Rahman, I.M., Biswas, F.B., Rahman, S., Endo, M., Wong, K.H., Mashio, A.S., Hasegawa, H.: Cellulose-based materials for scavenging toxic and precious metals from water and wastewater: a review. Chem. Eng. J. **472**, 144677 (2023)
11. Lim, J.Y., Goh, S.S., Liow, S.S., Xue, K., Loh, X.J.: Molecular gel sorbent materials for environmental remediation and wastewater treatment. J. Mater. Chem. A. **7**, 18759–18791 (2019)
12. Nawaz, S., Tabassum, A., Muslim, S., Nasreen, T., Baradoke, A., Kim, T., Boczkaj, G., Jesionowski, T., Bilal, M.: Effective assessment of biopolymer-based multifunctional sorbents for the remediation of environmentally hazardous contaminants from aqueous solutions. Chemosphere. **329**, 138552 (2023)
13. Holzapfel, B.M., Reichert, J.C., Schantz, J.-T., Gbureck, U., Rackwitz, L., Nöth, U., Jakob, F., Rudert, M., Groll, J., Hutmacher, D.W.: How smart do biomaterials need to be? A translational science and clinical point of view. In: Advanced drug delivery reviews, vol. 65, pp. 581–603. Elsevier (2013)
14. Ullah, F., Othman, M.B.H., Javed, F., Ahmad, Z., Akil, H.M.: Classification, processing and application of hydrogels: a review. Mater. Sci. Eng. C. **57**, 414–433 (2015)
15. Machado, T.O., Grabow, J., Sayer, C., de Araújo, P.H., Ehrenhard, M.L., Wurm, F.R.: Biopolymer-based nanocarriers for sustained release of agrochemicals: a review on materials and social science perspectives for a sustainable future of agri-and horticulture. Adv. Colloid Interface Sci. **303**, 102645 (2022)
16. Nascimento, D.M., Nunes, Y.L., Figueirêdo, M.C., de Azeredo, H.M., Aouada, F.A., Feitosa, J.P., Rosa, M.F., Dufresne, A.: Nanocellulose nanocomposite hydrogels: technological and environmental issues. Green Chem. **20**, 2428–2448 (2018)
17. Zhang, Y., Dong, L., Liu, L., Wu, Z., Pan, D., Liu, L.: Recent advances of stimuli-responsive polysaccharide hydrogels in delivery systems: a review. J. Agric. Food Chem. **70**, 6300–6316 (2022)
18. Singh, N., Riyajuddin, S., Ghosh, K., Mehta, S.K., Dan, A.: Chitosan-graphene oxide hydrogels with embedded magnetic iron oxide nanoparticles for dye removal. ACS Appl Nano Mater. **2**, 7379–7392 (2019)
19. Sharma, A.K., Kaith, B.S., Tanwar, V., Bhatia, J.K., Sharma, N., Bajaj, S., Panchal, S.: RSM-CCD optimized sodium alginate/gelatin based ZnS-nanocomposite hydrogel for the effective removal of biebrich scarlet and crystal violet dyes. Int. J. Biol. Macromol. **129**, 214–226 (2019)
20. Kaith, B.S., Singh, A., Sharma, A.K., Sud, D.: Hydrogels: synthesis, classification, properties and potential applications—a brief review. J. Polym. Environ. **29**, 3827–3841 (2021)
21. Wang, W., Narain, R., Zeng, H.: Hydrogels, polymer science and nanotechnology, pp. 203–244. Elsevier (2020)
22. Azzaroni, O.: Polymer brushes here, there, and everywhere: recent advances in their practical applications and emerging opportunities in multiple research fields. J. Polym. Sci. A Polym. Chem. **50**, 3225–3258 (2012)
23. Yang, Y., Ding, X., Urban, M.W.: Chemical and physical aspects of self-healing materials. Prog. Polym. Sci. **49**, 34–59 (2015)
24. Shi, L., Ding, P., Wang, Y., Zhang, Y., Ossipov, D., Hilborn, J.: Self-healing polymeric hydrogel formed by metal–ligand coordination assembly: design, fabrication, and biomedical applications. Macromol. Rapid Commun. **40**, 1800837 (2019)
25. Mishra, S., Bharagava, R.N., More, N., Yadav, A., Zainith, S., Mani, S., Chowdhary, P.: Heavy metal contamination: an alarming threat to environment and human health. In: Environmental biotechnology: for sustainable future, pp. 103–125. Springer (2019)

26. Benalia, M.C., Youcef, L., Bouaziz, M.G., Achour, S., Menasra, H.: Removal of heavy metals from industrial wastewater by chemical precipitation: mechanisms and sludge characterization. Arab. J. Sci. Eng. **47**, 5587–5599 (2022)
27. Chakraborty, R., Asthana, A., Singh, A.K., Jain, B., Susan, A.B.H.: Adsorption of heavy metal ions by various low-cost adsorbents: a review. Int. J. Environ. Anal. Chem. **102**, 342–379 (2022)
28. Perumal, S., Atchudan, R., Edison, T.N.J.I., Babu, R.S., Karpagavinayagam, P., Vedhi, C.: A short review on recent advances of hydrogel-based adsorbents for heavy metal ions. Metals. **11**, 864 (2021)
29. Ngulube, T., Gumbo, J.R., Masindi, V., Maity, A.: An update on synthetic dyes adsorption onto clay based minerals: a state-of-art review. J. Environ. Manag. **191**, 35–57 (2017)
30. Tkaczyk, A., Mitrowska, K., Posyniak, A.: Synthetic organic dyes as contaminants of the aquatic environment and their implications for ecosystems: a review. Sci. Total Environ. **717**, 137222 (2020)
31. Sharma, P., Bano, A., Singh, S.P., Sharma, S., Xia, C., Nadda, A.K., Lam, S.S., Tong, Y.W.: Engineered microbes as effective tools for the remediation of polyaromatic aromatic hydrocarbons and heavy metals. Chemosphere. **306**, 135538 (2022)
32. Al-Tohamy, R., Ali, S.S., Li, F., Okasha, K.M., Mahmoud, Y.A.-G., Elsamahy, T., Jiao, H., Fu, Y., Sun, J.: A critical review on the treatment of dye-containing wastewater: ecotoxicological and health concerns of textile dyes and possible remediation approaches for environmental safety. Ecotoxicol. Environ. Saf. **231**, 113160 (2022)
33. Singha, K., Pandit, P., Maity, S., Sharma, S.R.: Harmful environmental effects for textile chemical dyeing practice. In: Green chemistry for sustainable textiles, pp. 153–164. Elsevier (2021)
34. Giwa, A., Yusuf, A., Balogun, H.A., Sambudi, N.S., Bilad, M.R., Adeyemi, I., Chakraborty, S., Curcio, S.: Recent advances in advanced oxidation processes for removal of contaminants from water: a comprehensive review. Process Saf. Environ. Prot. **146**, 220–256 (2021)
35. Subhiksha, V., Kokilavani, S., Khan, S.S.: Recent advances in degradation of organic pollutant in aqueous solutions using bismuth based photocatalysts: a review. Chemosphere. **290**, 133228 (2022)
36. G. Saratale, S. Kalme, S. Govindwar, Decolorisation of textile dyes by *Aspergillus ochraceus* (NCIM-1146), (2006)
37. Daneshvar, N., Aleboyeh, A., Khataee, A.: The evaluation of electrical energy per order (EEo) for photooxidative decolorization of four textile dye solutions by the kinetic model. Chemosphere. **59**, 761–767 (2005)
38. Soloman, P.A., Basha, C.A., Velan, M., Ramamurthi, V., Koteeswaran, K., Balasubramanian, N.: Electrochemical degradation of Remazol black B dye effluent. CLEAN Soil Air Water. **37**, 889–900 (2009)
39. Selvaraj, V., Karthika, T.S., Mansiya, C., Alagar, M.: An over review on recently developed techniques, mechanisms and intermediate involved in the advanced azo dye degradation for industrial applications. J. Mol. Struct. **1224**, 129195 (2021)
40. Xu, S., Lu, H., Zheng, X., Chen, L.: Stimuli-responsive molecularly imprinted polymers: versatile functional materials. J. Mater. Chem. C. **1**, 4406–4422 (2013)
41. Qu, B., Luo, Y.: Chitosan-based hydrogel beads: preparations, modifications and applications in food and agriculture sectors—a review. Int. J. Biol. Macromol. **152**, 437–448 (2020)
42. Zhu, S., Wang, X., Cong, Y., Liu, L., Li, L.: Free radical polymerization of gold nanoclusters and hydrogels for cell capture and light-controlled release. ACS Appl. Mater. Interfaces. **13**, 19360–19368 (2021)
43. Omidian, H., Rocca, J.G., Park, K.: Advances in superporous hydrogels. J. Control. Release. **102**, 3–12 (2005)
44. Ladola, M.K.: Superporous hydrogel (SPH): a novel and advanced technique of oral controlled release drug delivery system. World J Pharm. Res. **3**, 297–319 (2014)
45. Morin-Crini, N., Winterton, P., Fourmentin, S., Wilson, L.D., Fenyvesi, E., Crini, G.: Water-insoluble β-cyclodextrin–epichlorohydrin polymers for removal of pollutants from aqueous

solutions by sorption processes using batch studies: a review of inclusion mechanisms. Prog. Polym. Sci. **78**, 1–23 (2018)
46. Abel, A.: The history of dyes and pigments: from natural dyes to high performance pigments. In: Colour design, pp. 557–587. Elsevier (2012)
47. Shabir, M., Yasin, M., Hussain, M., Shafiq, I., Akhter, P., Nizami, A.-S., Jeon, B.-H., Park, Y.-K.: A review on recent advances in the treatment of dye-polluted wastewater. J. Ind. Eng. Chem. **112**, 1–19 (2022)
48. Pereira, A.G., Rodrigues, F.H., Paulino, A.T., Martins, A.F., Fajardo, A.R.: Recent advances on composite hydrogels designed for the remediation of dye-contaminated water and wastewater: a review. J. Clean. Prod. **284**, 124703 (2021)
49. Li, J., Jia, X., Yin, L.: Hydrogel: diversity of structures and applications in food science. Food Rev. Int. **37**, 313–372 (2021)
50. Sharma, A.K., Gupta, A., Dhiman, A., Garg, M., Mishra, R., Agrawal, G.: Fe_3O_4 embedded κ-carrageenan/sodium alginate hydrogels for the removal of basic dyes. Colloids Surf. A Physicochem. Eng. Asp. **654**, 130155 (2022)
51. Hu, X.-S., Liang, R., Sun, G.: Super-adsorbent hydrogel for removal of methylene blue dye from aqueous solution. J. Mater. Chem. A. **6**, 17612–17624 (2018)
52. Choudhary, S., Sharma, K., Sharma, V., Kumar, V.: Performance evaluation of gum Gellan-based hydrogel as a novel adsorbent for the removal of cationic dyes: linear regression models. ACS Appl. Mater. Interfaces. **15**, 5942–5953 (2023)
53. Njuguna, D.G., Schönherr, H.: Xanthan gum hydrogels as high-capacity adsorbents for dye removal. ACS Appl Polym Mater. **3**, 3142–3152 (2021)
54. Yan, G., Feng, Y., Wang, H., Sun, Y., Tang, X., Hong, W., Zeng, X., Lin, L.: Interfacial assembly of self-healing and mechanically stable hydrogels for degradation of organic dyes in water. Commun Mater. **1**, 41 (2020)
55. Mohammadzadeh, F., Golshan, M., Haddadi-Asl, V., Salami-Kalajahi, M.: Adsorption kinetics of methylene blue from wastewater using pH-sensitive starch-based hydrogels. Sci. Rep. **13**, 11900 (2023)
56. Sharma, G., Naushad, M., Kumar, A., Rana, S., Sharma, S., Bhatnagar, A., Stadler, F.J., Ghfar, A.A., Khan, M.R.: Efficient removal of Coomassie brilliant blue R-250 dye using starch/poly(alginic acid-cl-acrylamide) nanohydrogel. Process Saf. Environ. Prot. **109**, 301–310 (2017)
57. Sharma, A.K., Dhiman, A., Nayak, A.K., Mishra, R., Agrawal, G.: Environmentally benign approach for the efficient sequestration of methylene blue and coomassie brilliant blue using graphene oxide emended gelatin/κ-carrageenan hydrogels. Int. J. Biol. Macromol. **219**, 353 (2022)
58. Harris, J.T., McNeil, A.J.: Localized hydrogels based on cellulose nanofibers and wood pulp for rapid removal of methylene blue. J. Polym. Sci. **58**, 3042–3049 (2020)
59. Gonçalves, J., Santos, J., Rios, E., Crispim, M., Dotto, G., Pinto, L.: Development of chitosan based hybrid hydrogels for dyes removal from aqueous binary system. J. Mol. Liq. **225**, 265–270 (2017)
60. Paulino, A.T., Guilherme, M.R., Reis, A.V., Campese, G.M., Muniz, E.C., Nozaki, J.: Removal of methylene blue dye from an aqueous media using superabsorbent hydrogel supported on modified polysaccharide. J. Colloid Interface Sci. **301**, 55–62 (2006)
61. Soleimani, K., Tehrani, A.D., Adeli, M.: Bioconjugated graphene oxide hydrogel as an effective adsorbent for cationic dyes removal. Ecotoxicol. Environ. Saf. **147**, 34–42 (2018)
62. Chen, Y., Chen, L., Bai, H., Li, L.: Graphene oxide–chitosan composite hydrogels as broad-spectrum adsorbents for water purification. J. Mater. Chem. A. **1**, 1992–2001 (2013)
63. Zhou, C., Wu, Q., Lei, T., Negulescu, I.I.: Adsorption kinetic and equilibrium studies for methylene blue dye by partially hydrolyzed polyacrylamide/cellulose nanocrystal nanocomposite hydrogels. Chem. Eng. J. **251**, 17–24 (2014)
64. Liu, J., Liu, G., Liu, W.: Preparation of water-soluble β-cyclodextrin/poly(acrylic acid)/graphene oxide nanocomposites as new adsorbents to remove cationic dyes from aqueous solutions. Chem. Eng. J. **257**, 299–308 (2014)

65. Guo, H., Jiao, T., Zhang, Q., Guo, W., Peng, Q., Yan, X.: Preparation of graphene oxide-based hydrogels as efficient dye adsorbents for wastewater treatment. Nanoscale Res. Lett. **10**, 1–10 (2015)
66. Hou, C., Zhang, Q., Li, Y., Wang, H.: P25-Graphene hydrogels: room-temperature synthesis and application for removal of methylene blue from aqueous solution. J. Hazard. Mater. **205**, 229–235 (2012)
67. Sun, H., Cao, L., Lu, L.: Magnetite/reduced graphene oxide nanocomposites: one step solvothermal synthesis and use as a novel platform for removal of dye pollutants. Nano Res. **4**, 550–562 (2011)
68. Samiey, B., Tehrani, A.D.: Study of adsorption of Janus green B and methylene blue on nanocrystalline cellulose. J. Chin. Chem. Soc. **62**, 149–162 (2015)
69. Karimi, M.H., Mahdavinia, G.R., Massoumi, B., Baghban, A., Saraei, M.: Ionically crosslinked magnetic chitosan/κ-carrageenan bioadsorbents for removal of anionic eriochrome black-T. Int. J. Biol. Macromol. **113**, 361–375 (2018)
70. Mokhtar, A., Abdelkrim, S., Hachemaoui, M., Boukoussa, B., Chaibi, W., Sardi, A., Djelad, A., Sassi, M., Issam, I., Iqbal, J.: Removal of crystal violet dye using a three-dimensional network of date pits powder/sodium alginate hydrogel beads: experimental optimization and DFT calculation. Int. J. Biol. Macromol. **251**, 126270 (2023)
71. Zhang, J., Li, Y., Dai, Y., Lin, T., Shao, H., Liu, Y., Liu, X.: Three-dimensional porous hydrogel based on hyperbranched polyethyleneimine functionalized apple pomace derived cellulose for efficient removal of methyl orange. Chem. Eng. J. **456**, 140995 (2023)
72. Wang, Y., Feng, W., Li, J., You, Z.: Facile one-step synthesis of CuZn-HDS/alginate hydrogel beads for highly efficient and selective removal of methyl orange. Colloids Surf. A Physicochem. Eng. Asp. **670**, 131561 (2023)
73. Sharma, P., Laddha, H., Agarwal, M., Gupta, R.: Selective and effective adsorption of malachite green and methylene blue on a non-toxic, biodegradable, and reusable fenugreek galactomannan gum coupled MnO_2 mesoporous hydrogel. Microporous Mesoporous Mater. **338**, 111982 (2022)
74. Ullah, N., Ali, Z., Khan, A.S., Adalat, B., Nasrullah, A., Khan, S.B.: Preparation and dye adsorption properties of activated carbon/clay/sodium alginate composite hydrogel membranes. RSC Adv. **14**, 211–221 (2024)
75. Wu, Z., Wu, J., Huang, M., Liang, H., Sun, B.: Preparation of reusable hydrogel spheres based on sodium alginate/Fe_3O_4 modified with carboxymethyl Huangshui polysaccharide and the efficient adsorption performance for methylene blue. Food Chem. **438**, 138064 (2024)
76. Fu, Z., Xi, S.: The effects of heavy metals on human metabolism. Toxicol. Mech. Methods. **30**, 167–176 (2020)
77. Dutta, D., Arya, S., Kumar, S.: Industrial wastewater treatment: current trends, bottlenecks, and best practices. Chemosphere. **285**, 131245 (2021)
78. Khan, F., Momtaz, S., Abdollahi, M.: The relationship between mercury exposure and epigenetic alterations regarding human health, risk assessment and diagnostic strategies. J. Trace Elem. Med. Biol. **52**, 37–47 (2019)
79. Renu, K., Mukherjee, A.G., Gopalakrishnan, A.V., Wanjari, U.R., Kannampuzha, S., Murali, R., Veeraraghavan, V.P., Vinayagam, S., Paz-Montelongo, S., George, A.: Protective effects of macromolecular polyphenols, metal (zinc, selenium, and copper)-polyphenol complexes, and pectin in different organs with an emphasis on arsenic poisoning: a review. Int. J. Biol. Macromol. **2023**, 126715 (2023)
80. Collin, M.S., Venkatraman, S.K., Vijayakumar, N., Kanimozhi, V., Arbaaz, S.M., Stacey, R.S., Anusha, J., Choudhary, R., Lvov, V., Tovar, G.I.: Bioaccumulation of lead (Pb) and its effects on human: a review. J Hazard Mater Adv. **7**, 100094 (2022)
81. Goyal, D., Yadav, A., Prasad, M., Singh, T.B., Shrivastav, P., Ali, A., Dantu, P.K., Mishra, S.: Effect of heavy metals on plant growth: an overview. In: Contaminants in agriculture: sources, impacts and management, pp. 79–101. Springer (2020)

82. Yang, C., Xu, M., Wang, Y., Li, S., Lv, X., Wang, H., Li, Z.: Recyclable hydrogel-MOFs composite beads for selective removal of Pb(II) from water. Chem. Eng. Res. Des. **193**, 540–554 (2023)
83. Perumal, S., Atchudan, R., Thirukumaran, P., Yoon, D.H., Lee, Y.R., Cheong, I.W.: Simultaneous removal of heavy metal ions using carbon dots-doped hydrogel particles. Chemosphere. **286**, 131760 (2022)
84. Farhan, A., Zulfiqar, M., Samiah, E.U., Rashid, S., Nawaz, H.M., Iqbal, T., Jesionowski, M., Bilal, J.Z.: Removal of toxic metals from water by nanocomposites through advanced remediation processes and photocatalytic oxidation. Curr. Pollut. Rep. **9**, 338–358 (2023)
85. Zhang, M., Song, L., Jiang, H., Li, S., Shao, Y., Yang, J., Li, J.: Biomass based hydrogel as an adsorbent for the fast removal of heavy metal ions from aqueous solutions. J. Mater. Chem. A. **5**, 3434–3446 (2017)
86. Sharma, A.K., Priya, Kaith, B.S., Isha, Singh, A., Chandel, K., Vipula: Riboflavin functionalized dextrin-sodium alginate based fluorescent sensor: detoxification of Cu^{2+} and Ni^{2+} ions. ACS Appl Polym Mater. **1**, 3084–3094 (2019)
87. Sharma, A.K., Kaith, B.S., Singh, A., Chandel, K.: Enzymatic construction of quinine derivative of dextrin/PVA based hybrid gel film for the simultaneous detection and removal of copper and lead ions in real water samples. Chem. Eng. J. **382**, 122891 (2020)
88. Tassanapukdee, Y., Prayongpan, P., Songsrirote, K.: Removal of heavy metal ions from an aqueous solution by CS/PVA/PVP composite hydrogel synthesized using microwaved-assisted irradiation. Environ Technol Innov. **24**, 101898 (2021)
89. Zhao, Z., Huang, Y., Wu, Y., Li, S., Yin, H., Wang, J.: α-Ketoglutaric acid modified chitosan/polyacrylamide semi-interpenetrating polymer network hydrogel for removal of heavy metal ions. Colloids Surf. A Physicochem. Eng. Asp. **628**, 127262 (2021)
90. Tanan, W., Saengsuwan, S.: A one-pot microwave-assisted synthesis of IPN hydrogels based on HEMA/AM/PVA blend for enhancing Cu(II) and Pb(II) ions removal. J. Environ. Chem. Eng. **8**, 103469 (2020)
91. Zhao, Z., Zhang, Z., Zhu, Z., Zou, X., Zhao, Y., Shi, J., Wang, J.: Photothermal responsive hydrogel for adsorbing heavy metal ions in aqueous solution. Colloids Surf. A Physicochem. Eng. Asp. **651**, 129425 (2022)
92. Jiao, G.-J., Ma, J., Li, Y., Jin, D., Zhou, J., Sun, R.: Removed heavy metal ions from wastewater reuse for chemiluminescence: successive application of lignin-based composite hydrogels. J. Hazard. Mater. **421**, 126722 (2022)
93. Sun, H., Zhan, J., Chen, L., Zhao, Y.: Preparation of CTS/PAMAM/SA/Ca^{2+} hydrogel and its adsorption performance for heavy metal ions. Appl. Surf. Sci. **607**, 155135 (2023)
94. Massoudi, S., Bagheri, M., Hosseini, M.: Poly(*N*-vinyl imidazole)/nitrogen-doped graphene quantum dot nanocomposite hydrogel as an efficient metal ion adsorbent of aqueous systems. Iran. Polym. J. **31**, 533–551 (2022)
95. Sethi, S., Thakur, S., Sharma, D., Singh, G., Sharma, N., Kaith, B.S., Khullar, S.: Malic acid cross-linked chitosan based hydrogel for highly effective removal of chromium(VI) ions from aqueous environment. React. Funct. Polym. **177**, 105318 (2022)
96. Thakur, S., Sharma, B., Thakur, A., Gupta, V.K., Alsanie, W.F., Makatsoris, C., Thakur, V.K.: Synthesis and characterisation of zinc oxide modified biorenewable polysaccharides based sustainable hydrogel nanocomposite for Hg^{2+} ion removal: towards a circular bioeconomy. Bioresour. Technol. **348**, 126708 (2022)
97. Sun, X.-F., Xie, Y., Shan, S., Li, W., Sun, L.: Chemically-crosslinked xylan/graphene oxide composite hydrogel for copper ions removal. J. Polym. Environ. **30**, 3999–4013 (2022)
98. Zhang, P., Zou, K., Yuan, L., Liu, J., Liu, B., Qing, T.-P., Feng, B.: A biomass resource strategy for alginate–polyvinyl alcohol double network hydrogels and their adsorption to heavy metals. Sep. Purif. Technol. **301**, 122050 (2022)
99. Li, M., Yang, M., Liu, B., Guo, H., Wang, H., Li, X., Wang, L., James, T.D.: Self-assembling fluorescent hydrogel for highly efficient water purification and photothermal conversion. Chem. Eng. J. **431**, 134245 (2022)

100. Dou, D., Wei, D., Guan, X., Liang, Z., Lan, L., Lan, X., Liu, P., Mo, H., Lan, P.: Adsorption of copper(II) and cadmium(II) ions by in situ doped nano-calcium carbonate high-intensity chitin hydrogels. J. Hazard. Mater. **423**, 127137 (2022)
101. Li, C., Duan, L., Cheng, X.: Facile method to synthesize fluorescent chitosan hydrogels for selective detection and adsorption of Hg^{2+}/Hg^{+}. Carbohydr. Polym. **288**, 119417 (2022)
102. Yue, H., Shang, Z., Xu, P., Feng, D., Li, X.: Preparation of EDTA modified chitooligosaccharide/sodium alginate/Ca^{2+} physical double network hydrogel by using of high-salinity oilfield produced water for adsorption of Zn^{2+}, Ni^{2+} and Mn^{2+}. Sep. Purif. Technol. **280**, 119767 (2022)
103. Aniagor, C.O., Taha, G.M., Badawy, S.M., El-Naggar, M.E., Hashem, A.: Preparation of a novel acrylic fiber-based hydrogel and its utilization for the removal of aqueous lead ion. J. Mater. Res. Technol. **18**, 1450–1459 (2022)
104. Miao, Y., Peng, W., Wang, W., Cao, Y., Li, H., Chang, L., Huang, Y., Fan, G., Yi, H., Zhao, Y.: 3D-printed montmorillonite nanosheets based hydrogel with biocompatible polymers as excellent adsorbent for Pb(II) removal. Sep. Purif. Technol. **283**, 120176 (2022)
105. Peng, J., Yuan, H., Ren, T., Liu, Z., Qiao, J., Ma, Q., Guo, X., Ma, G., Wu, Y.: Fluorescent nanocellulose-based hydrogel incorporating titanate nanofibers for sorption and detection of Cr(VI). Int. J. Biol. Macromol. **215**, 625–634 (2022)
106. Hassan, M., Naidu, R., Du, J., Qi, F., Ahsan, M.A., Liu, Y.: Magnetic responsive mesoporous alginate/β-cyclodextrin polymer beads enhance selectivity and adsorption of heavy metal ions. Int. J. Biol. Macromol. **207**, 826–840 (2022)
107. Dhiman, A., Sharma, A.K., Bhardwaj, D., Agrawal, G.: Biodegradable dual stimuli responsive alginate based microgels for controlled agrochemicals release and soil remediation. Int. J. Biol. Macromol. **228**, 323–332 (2023)
108. Dhiman, A., Bhardwaj, D., Goswami, K., Agrawal, G.: Biodegradable redox sensitive chitosan based microgels for potential agriculture application. Carbohydr. Polym. **313**, 120893 (2023)
109. Zheng, Z., Liu, J., Zhang, M., Li, H., Pan, J.: A crescent-shaped imprinted microgel adsorbent with near-infrared light-responsive performance for selective adsorption of lead(II). Sep. Purif. Technol. **328**, 124963 (2024)
110. Hopa, D., Kazzaz, A.E., Fatehi, P.: Fabrication of carboxyalkylated lignin derived microgels for adsorbing heavy metals. Ind. Crop. Prod. **187**, 115482 (2022)
111. Zhang, D., Li, H., Li, J., Xu, Z., Liu, H., Zhao, Y., Feng, X., Chen, L.: Hydrophilic P(Am-CD-AMPS) microgel for visual detection and removal metal ions in aqueous solution. Appl. Surf. Sci. **512**, 145668 (2020)
112. Atta, A.M., Al-Lohedan, H.A., ALOthman, Z., Abdel-Khalek, A.A., Tawfeek, A.M.: Characterization of reactive amphiphilic montmorillonite nanogels and its application for removal of toxic cationic dye and heavy metals water pollutants. J. Ind. Eng. Chem. **31**, 374–384 (2015)
113. Shoueir, K.R., Sarhan, A., Atta, A.M., Akl, M.A.: Adsorption studies of Cu^{2+} onto poly(vinyl alcohol)/poly(acrylamide-*co*-*N*-isopropylacrylamide) core–shell nanogels synthesized through surfactant-free emulsion polymerization. Sep. Sci. Technol. **51**, 1605–1617 (2016)
114. Liu, W.-Y., Ju, X.-J., Faraj, Y., He, F., Peng, H.-Y., Liu, Y.-Q., Liu, Z., Wang, W., Xie, R., Chu, L.-Y.: Capsule membranes encapsulated with smart nanogels for facile detection of trace lead(II) ions in water. J. Membr. Sci. **613**, 118523 (2020)

Soft Materials for Energy Applications

Mukhesh K. Ganesha, Rahuldeb Roy, Athira Chandran M, Pritha Dutta, and Ashutosh K. Singh

Abstract The advent of soft materials has revolutionized energy technologies, offering novel solutions for both energy harvesting and storage. Soft materials, such as polymers, gels, and organic compounds, exhibit exceptional mechanical properties like flexibility, deformability, and adaptability, making them ideal for diverse energy applications. In energy harvesting, soft materials enhance the efficiency of solar cells, piezoelectric, and triboelectric nanogenerators, converting mechanical vibrations and solar energy into usable power. These materials also play a pivotal role in catalytic processes, improving solar energy conversion. On the energy storage front, soft materials contribute to the development of advanced batteries and supercapacitors, enabling higher energy densities, better electrode performance, and increased safety. Their role in hydrogen storage further supports clean energy initiatives. The versatility of soft materials opens up possibilities in medical devices, offering compact and biocompatible energy storage solutions for implants. This chapter explores the transformative role of soft materials in sustainable energy technologies, highlighting their contributions to a greener future through enhanced performance and novel functionalities. The integration of soft materials into energy systems holds significant potential for driving future advancements in both energy harvesting and storage, shaping a more sustainable world.

Keywords Soft materials · Energy harvesting · Energy storage · Solar cells · Supercapacitors

Mukhesh K. Ganesha, Rahuldeb Roy, M. Athira Chandran and Pritha Dutta contributed equally with all other contributors.

M. K. Ganesha · R. Roy · A. Chandran M · P. Dutta · A. K. Singh (✉)
Centre for Nano and Soft Matter Sciences, Bangalore, India

Manipal Academy of Higher Education, Manipal, Karnataka, India
e-mail: aksingh@cens.res.in

V. Kumar, Y. K. Mishra (eds.), *Soft Materials for Functional Applications*, Advanced Structured Materials 225, https://doi.org/10.1007/978-981-97-9468-3_6

1 Introduction

In recent years, the energy field has undergone a remarkable transformation, driven by the growing need for sustainable and renewable power sources [1]. One of the key enablers of this transformation is the application of soft materials, which have emerged as promising candidates for various energy-related technologies. Soft materials encompass various materials with unique mechanical properties, such as polymers, gels, and organic compounds, that exhibit flexibility, deformability, and adaptability [2, 3]. The application of soft materials in energy focuses on two vital aspects: energy harvesting and energy storage. Energy harvesting involves the conversion of ambient energy sources into usable electrical energy, while energy storage aims to store and deliver energy efficiently. Soft materials offer significant advantages in these areas due to their unique properties and functionalities.

In energy harvesting, soft materials have revolutionized the development of efficient and sustainable energy conversion technologies. Solar cells, the most widely recognized form of energy harvesting, have been significantly improved by incorporating soft materials [4, 5]. These materials enhance the flexibility, lightweight nature, and cost-effectiveness of solar cells, making them more adaptable for a range of applications. Additionally, the use of soft materials in piezo/tribo nanogenerators has shown immense potential for harvesting mechanical energy from ambient vibrations and movements. Furthermore, photocatalytic and electrocatalytic processes, facilitated by soft materials, have opened up new avenues for harnessing solar energy and enabling efficient energy conversion. Similarly in energy storage, where soft materials play a critical role in the development of advanced energy storage devices. Batteries, the backbone of portable energy storage, have greatly benefited from the incorporation of soft materials. These materials contribute to enhanced electrode performance, increased energy density, and improved safety. Additionally, soft materials have revolutionized the field of supercapacitors, offering high power densities, fast charge/discharge rates, and excellent cycling stability. In the context of energy storage for medical devices and implants, soft materials have opened up new possibilities for compact, portable, and safe battery configurations. Their implementation in biomedical applications, from pacemakers to neurostimulators, has demonstrated their significance in healing and improving human lives. Moreover, soft materials have also made significant contributions to the field of hydrogen storage, enabling efficient and safe hydrogen storage as a clean energy carrier.

In summary, this chapter explores the wide-ranging applications of soft materials in energy, focusing on energy harvesting and energy storage. The utilization of soft materials in these areas has led to significant advancements in sustainable energy technologies, paving the way for a greener and more sustainable future. By understanding the unique properties and capabilities of soft materials, researchers and engineers can unlock new possibilities and innovations in the energy field, ultimately driving us toward a more sustainable and energy-efficient world.

2 Application of Soft Materials in Energy

The use of soft materials in the field of energy is classified broadly into two categories: (a) energy harvesting and (b) energy storage.

2.1 *Soft Materials in Energy Harvesting*

Soft Materials in Photovoltaics

Photovoltaic (PV) cells can convert solar energy into electrical energy. Many types of PVs are being studied, but only silicon-based solar cells have taken over the market. The perovskite PVs are unstable in atmospheric conditions, so they cannot come to market through higher photo-conversion efficiency (PCE) [6]. The other type is polymer PVs, which are the only hope that can replace the energy-intensive silicon-based PVs due to their simple and large-scale production capabilities [4, 7]. But due to the high excitonic binding energy of photoexcited electron–hole pairs, organic solar cells have not achieved the high PCE predicted by the theoretical calculations. By heterojunction studies, it has been found that effective transfer of charges can take place by selecting donors and acceptor films which are having a suitable offset of band positions and much attention has been paid to developing materials which can effectively separate the electrons and holes [8]. The initial organic photovoltaic (OPV) devices exhibited very low PCE; by the development of solution-based bulk heterojunction (BHJ) devices, PCE was increased up to ~10%. In BHJ, typically, a blend consisting of conjugated polymer donor and soluble molecular acceptor is sandwiched between an anode (ITO) and cathode (metal electrode), as shown in Fig. 1a. Since the exciton diffusion length is in the order of tens of nm, in the blend, the exited electron should see the acceptor layer within a few tens of nm, and this acceptor is connected to charge-collecting electrodes citation. Exciton binding energy plays an important role in OPVs—since it is more than the typical semiconductors (~0.5 eV), the offset between the lowest unoccupied molecular orbit (LUMO) of donor and acceptor should be of the same order for effective electron–hole separation at the interface, but while selecting the acceptor molecule, one should be careful and have to make sure acceptor has higher LUMO energy compared to donor highest occupied molecular orbit (HOMO) energy to yield large voltage. In fullerenes-based OPVs, the absorption coefficient in the visible range is very limited, and it is difficult to extend into the red and infrared region [4, 10].

Fullerenes were initially used as acceptors in BHJ OPVs. The widely used acceptors phenyl-C61-butyric acid methyl ester (PC61BM) and [6,6]-phenyl-C71-butyric acid methyl ester (PC71BM) were first synthesized by Hummelen et al. These low-lying LUMO molecules can separate the photoinduced electron and holes very effectively, thereby increasing the J_{sc}, but V_{oc} suffers due to this [11, 12]. To

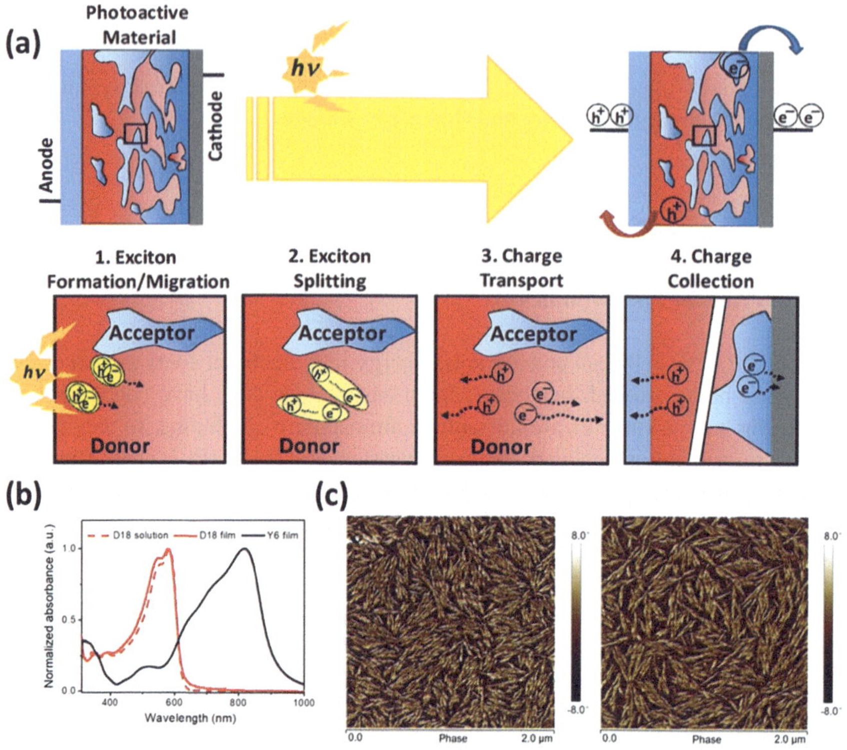

Fig. 1 (**a**) Schematic of BHJ organic solar cell showing energy production during illumination and processes involved in converting light into electrical energy (Reprinted with permission Ref. [5]). (**b**) Normalized absorbance of D18 and Y6 molecules (Reprinted with permission Ref. [9]). (**c**) Atomic force microscope phase images for D18:Y6 blend films with and without solvent vapor annealing during film treatment (Reprinted with permission Ref. [9])

increase the V_{OC}, Li et al. have developed fullerene bisadduct acceptors Indene-C60 Bisadduct (IC60BA) and Indene-C70 Bisadduct (IC70BA), which have increased the photo-conversion efficiency to 7.40% [13, 14]. Isomer-free fullerene bisadduct e-PPMF was developed by Ding et al. to eliminate the energetic disorder of isomers by regioselective synthesis and achieved the photo-conversion efficiencies of 8.11%, which is the highest reported value of PCE for organic solar cells based on fullerene bisadduct acceptors to date [15]. As discussed before, fullerene-based acceptors have very low absorption coefficients in visible light. Non-fullerene-based n-type semiconducting polymers can be taken as acceptor molecules due to their high absorption coefficients in the visible region [9]. The non-fullerene acceptors (NFAs) have provided effective ways to improve the performance of polymer solar cells due to their tunable energy levels by changing the electron donor groups and strong absorption in the near-infrared region (NIR). For OPV to work effectively, the acceptor molecule should have the highest HOMO and lowest LUMO energy levels that are correctly positioned with respect to the energy levels of the donor material. The initial NFAs developed are ITIC and IDIC. These are the fused-ring acceptor–donor–acceptor (A–D–A) small molecules, though these molecules had stronger visible-NIR light-harvesting capability with good electron mobility, J_{sc}, and PCE than fullerene acceptors, and work has been done to improve it further [16, 17]. Ding et al. have developed stronger light-harvesting A–D–A acceptors using highly electron-rich carbon–oxygen-bridged core units, mainly COi8DFIC acceptor molecule, which has 1.26 eV bandgap and strong absorption in 600–1000 nm [18]. The blend of Y6 and wide and gap donor can harvest most of the visible and NIR light (see Fig. 1b) and give excellent PCE up to 17% [9, 19]. High-performance donors matching these low bandgap NFAs are limited; the development of high-performance donors would keep increasing the PCEs of OPVs.

Organic solar cells are prepared using solution-based processes over other thin film fabrication techniques, which require expensive instruments and laborious processes. It gives the flexibility to control phase separation, molecular rearrangements, and morphology variation during solvent evaporation or while treating films. In polymer blends, morphology and phase separation are mainly controlled by the selection of solvents [20, 21]. In the MDMO-PPV:PCBM device, it was observed that there is a 2.5% increment in PCE when toluene is used instead of chlorobenzene, and the increment is attributed to the homogenous mixing of donor and acceptor molecules due to higher solubility in chlorobenzene [22]. Since the excitonic diffusion lengths are only a few tens of nanometers in organic solar cells, the continuously connected uniformly distributed donor and acceptor phases should be nearby. The J_{sc} and fill factor mainly depend on the ability of the organic solar cells to separate the charges and transport them to electrodes. In NFA-based organic solar cells, the effect of solvent vapor annealing with and without solvent annealing during film processing is studied, and it has been shown to increase the J_{sc} and fill factor. It is observed that in ITO/PEDOT:PSS/D18:Y6/PDIN/Ag solar cells, the SVA leads to the formation of bundle-like structures due to the promoted aggregation of nanofibers (see Fig. 1c). It led to the PCE of 18.22% and J_{sc} of 27.70%, with a 76.6% fill factor. EQE is >80% in the visible region, indicating effective photoelectron generation and isolation [9]. The increase in the performance may be due to increased charge carrier generation, reduced charge recombination, and good charge transport through nanofibers. Recently, Lei et al. reported that single junction organic solar cells based on PM6:D18:L8-BO showed over 19% [10]. Here, the efficient exciton production and dissociation, transport, and reduction of recombination losses were realized by morphological control of acceptor and sonar phases at nanoscales. The PM6 has a higher solubility than the D18, which is highly crystalline, due to which during film formation, D18 precipitates and initiates the polymeric crystallization. So, D18 enhances the crystallinity of the blend by forming a fibril network with increased fiber diameter and distribution. Fibrils are separated by 25 nm on average, which is a suitable length scale of phase separation to split excited and transport it. So densely packed donor–acceptor fibrils show excellent exciting separation and diffusion.

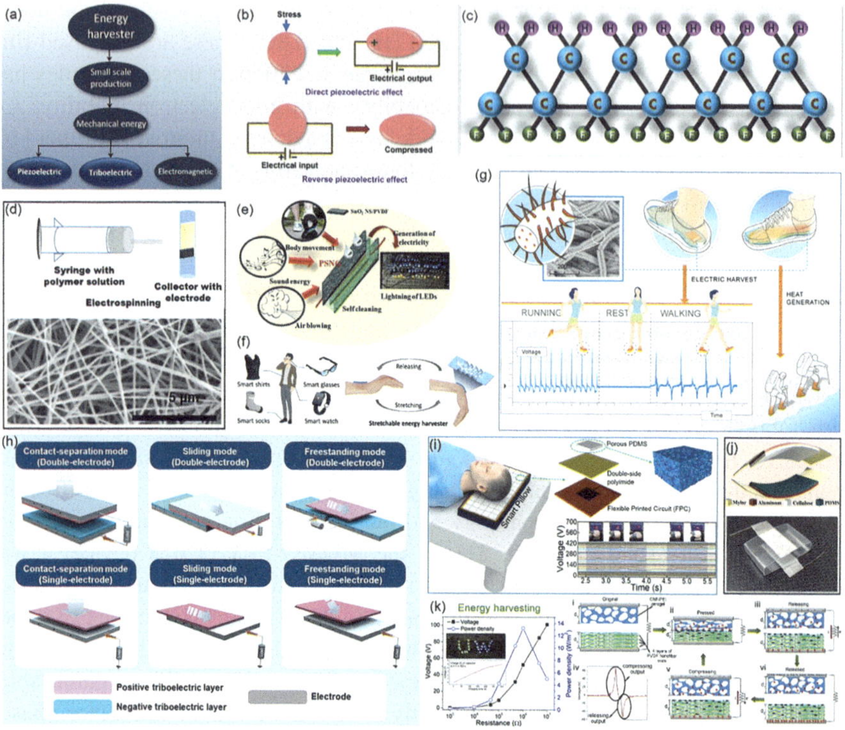

Fig. 2 (**a**) Classification of mechanical energy sources as an energy harvester. (**b**) Schematic diagram for direct and reverse piezoelectric effect. (**c**) β-Phase of polyvinylidene fluoride. (**d**) Electrospinning of polymer solution and FESEM image of P(VDF-TrFE) electrospun film +1% ZnO polymer nanocomposite (Reprinted with permission Ref. [23]). (**e**) 2D SnO_2 nanosheet/PVDF composite-based nanogenerator can power up to 85 LEDs and harvest energy based on human motion (Reprinted with permission Ref. [24]). (**f**) High-performance stretchable piezoelectric energy harvester in wearable-device applications (Reprinted with permission Ref. [25]). (**g**) A ZnO/PAN nanofabric-based wearable piezoelectric nanogenerator works in monitoring human activity and thermal management (Reprinted with permission Ref. [26]). (**h**) Three operational modes of triboelectric nanogenerator in two different types of electrode configurations (Reprinted with permission Ref. [27]). (**i**) A smart pillow made of a flexible and breathable triboelectric nanogenerator that acts as a self-powered pressure sensor (Reprinted with permission Ref. [28]). (**j**) The 3D schematic and actual image of electrospun cellulose acetate nanofibers and PDMS-based triboelectric nanogenerator can generate 400 V of output voltage with 0.9 W m^{-2} power (Reprinted with permission Ref. [29]). (**k**) High output voltage and power obtained from TENG, made of CNF/PEI based aerogel film paired with PVDF nanofiber mat, which can glow 26 red and green LEDs and schematic illustration of the step-by-step operation of the TENG (Reprinted with permission Ref. [30])

Soft Materials in Piezo/Tribo Electric Energy Harvesting

The basic and primary mechanical vibration energy harvesters are based on electromagnetic, electrostatic, and vibrational (Fig. 2a). In contrast, the performances of electrostatic and vibrational source-based energy generators, which are known as triboelectric and piezoelectric nanogenerators, are much higher than those of electromagnetic-based ones. When an external stress is applied to a piezoelectric material, the change in the centers of anions and cations generates electric dipoles, which consequently creates electric potential between two electrodes (Fig. 2b). The non-centrosymmetricity in structure and the higher values of piezoelectric coefficients [charge coefficients: longitudinal (d_{33}), transverse (d_{31}), and voltage coefficient—g_{33}] are the indications of achieving high power density from the piezoelectric material. The piezoelectric materials are of two types—lead zirconate titanate (PZT) [31, 32], zinc oxide (ZnO) [33, 34], and barium titanate (BT) [35–37] belong to piezoelectric ceramics; and the other type is of piezoelectric polymers. For all purposes and mainly to use on or in the human body, the flexible, stretchable, and wearable piezoelectric devices need piezoelectric soft materials instead of ceramics with high rigidity and brittleness (Fig. 2f) [25]. Among soft polymeric materials, polyvinylidene fluoride (PVDF) as a semicrystalline piezopolymer showed a new opportunity to explore piezoelectricity with high flexibility, low cost, biocompatibility (lead-free), and easy processing. Polymerization of $H_2C{=}CF_2$ monomer gives PVDF, which has four types of polymorphs—α, β, γ, and δ—among which β form can show the highest piezoelectric performance due to the existence of spontaneous polarization. Though possessing incomparable piezoelectric coefficient like ceramic materials, the enhancement of β-phase (shown in Fig. 2c) can directly improve the piezoelectric performance in PVDF. People are trying to achieve the same by utilizing different manufacturing strategies like the spin coating method, where proper control of parameters helps in crystal orientation toward β-phase [38] and making electrospun PVDF fibers, which shows the potential performance in energy harvesting applications [39]. Along with stretching and poling, different composites are conceptualized by adding piezoelectric fillers like metal oxides and carbon-based materials. PVDF and its copolymers are reported to show a high value of piezoelectric coefficient of 15–32 pC N^{-1}. To enhance the β-phase, different defect modifications are being employed in the polymer chains by introducing copolymers like PVDF–TrFE (Fig. 2d) [23, 40], P(VDF-HFP), and P(VDF-TrFE-CFE), where both piezoelectric coefficient and power density were observed to be higher than the single PVDF polymer. By aiming toward achieving high energy harvesting performance, multilayers are being employed, where a group of researchers designed 3D multilayer nanofiber mats (NFMs) using an electrospinning method and achieved 1.5 times voltage (10.3 V) and current (1.29 $\mu A\ cm^{-2}$) compared to ordinary multilayer PVDF films [41]. By utilizing different processes and different fillers, people have made PVDF-based nanogenerator devices of different sizes. Hoque et al. (2017) reported the PVDF-based piezoelectric nanogenerator (PENG) using Er^{3+}/Fe^{3+} as a catalytic agent for β-phase nucleation, achieving 160 mW cm^{-3} power density [42]. Shi et al. reported (2018) $2.5 \times 2.5\ cm^2$ PVDF-based PENG comprising graphene sheets and $BaTiO_3$, which can glow 15 LEDs with 112 V by finger pressing [43]. Kar et al. (2019) reported 2D SnO_2 nanosheet/PVDF composite-based PENG, which can glow 85 numbers of LEDs with 42 V voltage and 4900 W m^{-3} power density (Fig. 2e) [24]. Mondal et al. (2021) designed a

PENG by stacking PVDF on ZnO nanorods, which gives 21.5 V and 135.45 μW power. Recently, Pei et al. (2023) designed 3D-printed PVDF-based PENG with locked interfacial piezoelectric polarization, which showed effective performance as wearable electronics in healthcare monitoring applications [44].

Another polymer known as polyacrylonitrile (PAN) is also becoming a competitor of PVDF due to having higher thermal stability and stronger dipole moment (3.5 Debye) than PVDF (2.1 Debye) [45–47]. High voltage (~6 V) and current (5.1 μA) could be achieved from the PAN fibrous membrane, made by the electrospinning method [45]. Yuan et al. reported that the 15 wt% addition of piezoelectric $BaTiO_3$ with PAN nanofiber membrane can attain three times higher voltage than pristine PAN [48]. Due to the improvement in planar zigzag conformation in microstructures of PAN, greater deformation in PAN nanofibers, and intrinsic piezoelectric property, the integration of ZnO nanorods with PAN was reported to give twice the output power value than the similar ZnO-PVDF composite (Fig. 2g) [26]. Recently, Shi et al. reported a high piezoelectric voltage of 5.7 V by developing PAN composite fiber, which enhanced the planar zigzag conformation of PAN [49]. It is worth mentioning that cellulose is the most natural ample polymer to show piezoelectric characteristics. The highly crystalline structure created from the hydrogen bonding between the hydroxyl groups present in the repetitive glucose unit and the net dipole moment arising from the non-centrosymmetrically ordered hydroxyl groups are the main reasons for piezoelectricity in cellulose [50, 51]. The combination of the absence of a center of symmetry in monoclinic and triclinic structures with the electron-donating ability of hydroxyl groups also helps in enhancing piezoelectric activity. Due to the solubility and processability issues of natural cellulose, it shows a low value of longitudinal piezoelectric coefficient (d_{33}) (0.4 pC N^{-1}). Different piezoelectric fillers and different methods are reported to enhance the piezoelectric property of natural cellulose up to 20–23 pC N^{-1} [51–53]. Due to the shortcomings of natural cellulose, people are more interested in nanocellular structures like cellulose nanofibers (CNFs) and cellulose nanocrystals (CNCs), which can show almost 15 times higher d_{33} value than natural cellulose due to non-centrosymmetric crystallinity and dipole arrangement. Various smart and flexible energy harvesting piezoelectric nanogenerators (PENGs) are being made by utilizing the great piezoelectric features of cellulose. Zheng et al. fabricated a high-performance PENG device with a film made of 2,2,6,6-tetramethylpiperidine-1-oxyl (TEMPO)-oxidized cellulose nanofibrils (TOCNs) and PDMS. The 1 cm × 2 cm × 0.48-mm device showed high performance (60 V, 10.1 μA, and 6.3 mW cm^{-3}) against 10 MΩ load and 10 Hz, 0.05 MPa external force [54]. Alam et al. mixed PDMS and MWCNT with cellulose microcrystalline (CMC), and the 3.3 cm × 2.5 cm × 2-mm device showed an open circuit voltage of 30 V, short-circuit current of 500 nA, and 9 μW cm^{-3} power density under 30 MΩ load [55]. Despite having the strong potential of piezoelectric polymers such as PVDF, PAN, and cellulose to withstand competition with ceramic piezoelectric materials, they often suffer from low mechanical strain and poor elasticity, which hinders the application in terms of unstable flexibility and stretchability. In aiming toward the application of stretchable energy harvesting, Owusu et al. developed piezoelectric elastomer by blending polar polynorbornene nanoparticles into a chemically cross-linked PDMS (polydimethylsiloxane) matrix. After electric poling, the material showed piezoelectricity with a high d_{31} value of 37 pC N^{-1} [56]. Kachroudi et al. tried to control the shape and size and made cellular microstructured PDMS with a d_{33} value of 350 pC N^{-1} [57]. In another study, Hu et al. demonstrated a PDMS-based sandwich material, where PTFE (polytetrafluoroethylene) nanoparticles are embedded in a PDMS matrix and after poling the material showed a high d_{33} value of 1574 pC N^{-1} [58]. The miniaturization and blending of the elastomers with other materials will certainly make them stand potentially as energy harvesters in terms of enhancement of piezoelectricity. Biocompatible electronic devices can be implanted into the body and degrade after a certain period of time by dissolving in the body without any harmful effect or surgery. Technology should also provide biocompatible nanogenerators to supply power to biocompatible electronic devices. By aiming toward it, Kim et al. introduced the natural polymer, silk-fibroin-based nanogenerators, due to the excellent mechanical properties of the polymer. The nanogenerators made of silk-fibroin-based biocompatible nanocomposite exhibited 2.2 V output voltage and 0.12 $\mu A\ cm^{-2}$ current density as 2D thin films and 1.8 V output voltage and 0.1 $\mu A\ cm^{-2}$ current density as 1D wire [59].

Like PENG, one more popular energy harvester based on triboelectrification is known as triboelectric nanogenerators (TENGs), which can also convert numerous mechanical energies into electricity. The basic working mechanism is based on mainly contact charge and electrostatic induction. It can generate two opposite charges on the surface of two materials, when they are in close contact, while developing potential difference after separation of them. Different working modes exist for both single and double electrodes, like vertical contact-separation mode, freestanding mode, and in-plane contact sliding mode (Fig. 2h) [27]. In this era of flexible and wearable electronics, flexible TENGs are exhibiting tremendous impact by using oppositely charged soft materials as the main triboelectric materials. By using optimized structures, and suitable soft materials with low-cost and simple production, TENGs have proven their ability to be a potential energy source for flexible and wearable electronics. The selection of suitable materials with opposite charges will affect the output performance of a TENG and help it achieve high output power. Fan et al. have reported a low-cost paper-based rollable TENG, which uses PTFE (polytetrafluoroethylene) and Cu as two electrification layers. It gives 121 mW m^{-2} power density with an acoustic pressure of 117 dB, which can act as a self-powered microphone [60]. Zhu et al. reported a micro-grating TENG made of PTFE thin films, which works on sliding electrification. For a 10 m s^{-1} sliding velocity, it can deliver 3 W of output power, sufficient to operate a light bulb [61]. Zheng et al. reported the application of implanted TENG (iTENG) made of PDMS, and due to the flexibility and biocompatibility nature of PDMS, the iTENG can harvest energy from animal breathing and drive a pacemaker prototype by giving the power density of up to 8.44 mW m^{-2} [62]. Dong et al. reported a textile-structured TENG made of silicone hollow tube and gel electrodes. This fiber-based deformable TENG can supply up to 106 V of output voltage while solving the complex fabrication process associated with usual triboelectric fibers [63]. The simple electrospinning method was adopted by Lin et al. to fabricate poly(3-hexylthiophene) (P3HT)/poly(vinylidene fluoride-*co*-hexafluoropropylene) (PVDF-HFP) based triboelectric nanogenerators (TENGs) consisting of semiconductive nanofiber mat. These TENGS were proven to lighten 500 red LEDs by giving 0.55 mW power and 78 V output voltage. The ultrahigh power density of 13.3 W m^{-2} with 106 V was achieved in a TENG made of porous CNF/PEI (cellulose nanofibrils/polyethyleneimine) aerogels and PVDF nanofiber mats (Fig. 2k) [30]. Thus, various soft materials, with proper combination, were extensively utilized by fabricating through simple and low-cost processes to make flexible, stretchable, wearable, and self-powering TENGs (Fig. 2i, j), which can certainly power various small electronics and have the potential to replace the batteries for such a low requirement of power [28, 29].

Soft Materials as Photo-/Electrocatalyst for Energy Harvesting

Photocatalytic hydrogen production has emerged as one of the suitable candidates for the renewable energy use of natural sunlight or artificial indoor illumination [64, 65]. The term "photocatalysis" refers to the presence of a catalyst and photons required for the reaction. After the discovery of photocatalysis, researchers have put tremendous effort into semiconductors [66]. Generating hydrogen by photocatalytic water splitting surely can bring down the energy demand in the future. Reducing CO_2 in various products, such as CH_4, CH_3OH, etc., will solve environmental issues caused by the overload of CO_2 in the atmosphere. However, photocatalysis is a thermodynamically downhill process where $\Delta G < 0$. It mainly consists of four steps: (1) absorption of light, (2) separation of the charges, (3) transport of charges to the catalytic surface, and (4) surface catalytic reaction. The combination of the four steps decides the efficiency of a photocatalytic material. For inorganic semiconductor material, the absorption of light is decided by the bandgap, i.e., the difference between the conduction and valance bands. However, for organic semiconductor materials, the light absorption is decided by the difference between the highest occupied molecular orbital (HOMO) and the lowest unoccupied molecular orbital (LUMO) (Fig. 3) [67, 68]. After getting excited by light absorption, the generated electron–hole pair has to be prevented from recombination.

Excitons have to be separated from the active materials; thus, diffusion length and band bending play a major role. Compared to inorganic semiconductor materials, organic materials such as soft materials are relatively complex in determining the intrinsic parameter, which determines the efficiency of the photocatalysis activity. The exact position of the band in the electrolyte, diffusion length, bulk or grain boundary resistance, and band bending between material and electrolyte are difficult to identify accurately. Soft materials such as carbon nitrides, π-conjugated polymers, covalent organic frameworks (COFs), covalent triazine frameworks (CTFs), etc. have become potential candidates for photocatalysis due to their environmental friendliness. Among them, researchers paid much attention to carbon nitrides and π-conjugated polymers due to earth abundance, easy synthesis process, and less

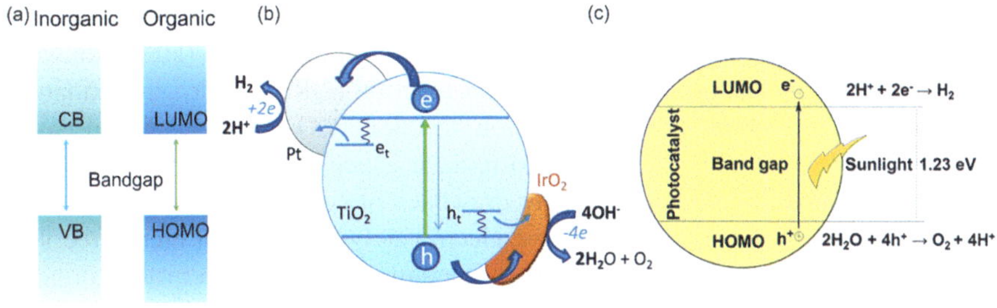

Fig. 3 (**a**) Inorganic and organic materials have a bandgap defined by the difference between the valance band and conduction band and the difference between HOMO and LUMO, respectively. Schematic of the photocatalysis activity of (**b**) inorganic (Reprinted with permission Ref. [67]) and (**c**) organic semiconductors (Reprinted with permission Ref. [68])

electronic bandgap. The charge separation efficiency is high for π-conjugated polymers due to low bandgap and utilizes the visible spectra, whereas inorganic materials are photocatalytic active at UV range only. π-Conjugated polymers were reported for photocatalysis for fuel cells under UV light by Yanagida et al. [69] Since then, multiple strategies have been implemented to increase the photocatalytic activity, most interestingly increasing the surface area of the photocatalytic material. COFs have a very high surface area by utilizing the porous structure, which enhances several properties compared to their solid counterparts. These materials have a very high surface-to-volume ratio, an easy synthesis process, and recyclability. Furthermore, the higher order of crystallinity supports the charge separation and transport, making them promising materials in photocatalytic activity. For the first time, Lotch et al. developed a crystalline hydrazone-based TFPT-COF for photocatalytic activity, which is competitive with the best nonmetallic photocatalysis with a surface area of 1603 $m^2\ g^{-1}$ [70]. In addition, the performance of COFs depends on the comonomer's length. Schmidt et al. show that the photocatalytic hydrogen evolution activity of TP-BDDA with acetylene (TP-EDDA) and phenyl (TP-DTP) containing COFs demonstrated the major impact of diacetylene moieties [71]. Another class of COF, covalent triazine framework (CTF), is getting much attention among researchers due to its significant performance in photocatalysis, CO_2 capture, and energy storage. Thomas et al. show that CTF-1 has very high photocatalytic activity with pretreatment of CTF with molten $ZnCl_2$ with a hydrogen evolution rate of 1072 $\mu mol\ h^{-1}\ g^{-1}$ under visible light of <420 nm [72]. A grand challenge for photocatalysis is understanding the defects, disorders, and functional groups to enable surface engineering to build the artificial photocatalytic system. Long cyclic stability is a key challenge for soft materials in particular COFs as photocatalytic activity.

Similarly, electrocatalysis is one of the leading areas in energy conversion technologies. It comprises oxidation and reduction reactions at the electrode–electrolyte interface by transfer of electrons. The designed electrocatalyst undergoes the catalytic reaction in a particular reaction pathway. Research is going on in the field of electrocatalysis for designing and engineering the best electrocatalyst for different catalytic reactions such as hydrogen evolution reaction (HER), oxygen evolution reaction (OER), oxygen reduction reaction (ORR), hydrogen oxidation reaction (HOR), carbon dioxide reduction reaction (CO_2RR), etc. Metal-based electrocatalysts are highly researched to replace expensive commercial catalysts. It undergoes corrosion, which is a major drawback of commercialization due to its short lifespan [73]. Alternative low-cost electrocatalysts based on metal-free soft matter are another option. Porous polymer-based materials such as covalent organic polymer, conductive polymer, supramolecular polymer, covalent organic frameworks (COFs), etc. possess high surface area with a large number of active sites, low resistance, and good stability which is suitable for a good electrocatalyst. Porphyrin-based metal-free covalent organic polymer TpPAM with triformyl phloroglucinol (Tp) and 5,10,15,20-tetra(4-aminophenyl)-21*H*,23*H*-porphyrin (PAM) showed the mechanism for HER in an acidic environment by the protonation of two pyrrolic N atoms at the porphyrin ring with a near zero ΔG_{H*} value of 0.08 eV as shown in Fig. 4a, b. It has an overpotential value of 250 mV at -10 mAcm^{-2} with a Tafel slope value of 106 mV dec^{-1} (Fig. 4c, d) [74]. Here, the high-temperature treatment required for metal porphyrin catalysts, which increases the production cost, can be resolved by developing metal-free porphyrin-based catalysts. Conductive polymers polyaniline, polypyrrole, poly(tyramine), poly(-amidoamine), and poly(8-hydroxyquinoline) not only provide high stability with low resistance and high surface area but also transfer electrons as well as protons which enable them to act as a redox mediator between electrode and electrochemical reactants. This behavior makes them a competitor as an electrocatalyst for HER and for fuel cells as a Nafion membrane replacement [77]. Liu et al. used poly(2-amino-5-mercapto-1,3,4-thiadiazole) (PAMT) alone as an electrocatalyst for HER where –N= present in

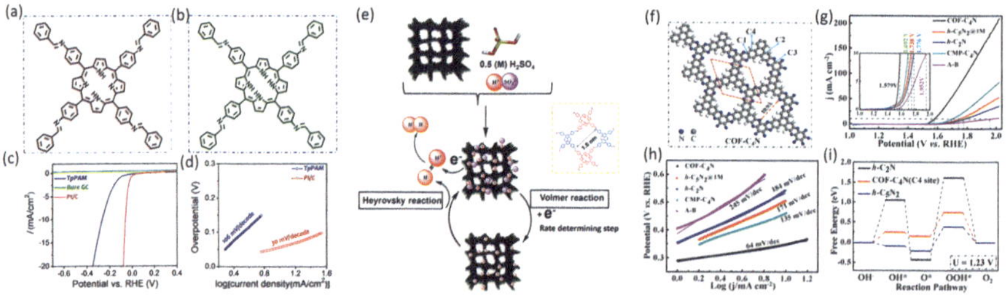

Fig. 4 (**a**, **b**) TpPAM polymer model with and without hydrogen, respectively, explains the mechanism for HER. (c) Linear sweep voltammetry showing the HER activity of TpPAM. (**d**) Tafel plot showing the Tafel slope value for TpPAM (Reprinted with permission Ref. [74]). (**e**) SB-PORPy-COF showing Volmer–Heyrovsky reaction mechanism for HER (Reprinted with permission Ref. [75]). (**f**) COF-C_4N monolayer supercell structure indicates the pore structure and size. (**g**) Linear sweep voltammetry for OER activity testing for COF-C_4N. (**h**) Tafel plot of COF-C_4N and its control samples are used to understand the reaction kinetics and mechanism. (**i**) Free energy diagram from DFT calculation for understanding the OER activity of COF-C_4N and its control sample (Reprinted with permission Ref. [76])

pairs, which helps in H–H bond formation in HER [78]. Organic polymer catalysts also exist as metalized polymers and metal-incorporated polymers. Porphyrin forms a metalized polymer having ($M–N_4$) square planar structure with strong collation, forming the active center [79]. Bimetallic Co–Ni 2-D sheet polymer is a reported electrocatalyst for HER with an overpotential of only 105 mV at -10 mA cm^{-2} [80]. Similarly, it also acts as an OER [81] ORR [82], and CO_2RR catalyst [83]. A metal-free COF as a catalyst for HER is first reported by Subhajit Bhunia et al. where they developed pyrene-porphyrin comprised of conjugated imine-based COF (SB-PORPy). The porous structure of COF with a uniform porosity of 1.7 nm has the imine nitrogen group (–N=) as an active center for protonation. It has good HER activity and a Tafel slope value of 116 mV dec^{-1} indicating the Volmer–Heyrovsky reaction pathway as shown in Fig. 4e [75]. Chunhua Yang et al. designed OER active COF, phenazine-linked 2DCOF (COF-C4N), by solvothermal reaction of triphenylene hexamine (TPHA) and hexaketocyclohexane (HKH). C_4N stoichiometry and hexagonal-shaped pores with 10.9 Å pore size also paved the way for higher OER activity with 349 mV at 10 mA cm^{-2} having a Tafel slope value of 64 mV dec^{-1} (Fig. 4g–i) [76].

The gel is another soft matter with a low density, porous structure with many active sites, high specific surface area, an abundant defect for mass transfer, and tunable composition, making it a suitable material as an electrocatalyst [84]. Substrate-free electrocatalysts are developed using hydrogel/aerogel, which are freestanding 3D porous structures. N-doped graphene aerogel [85] has an overpotential of 128 mV at -10 mA cm^{-2} and molybdenum sulfide/N-doped graphene (NG–Mo) hybrid hydrogel with 140.6 mV at -10 mA cm^{-2} for HER [86]. The hydrophilic nature of 3D NG makes a smooth contact at the electrode–electrolyte interface, creating fast kinetics and improved activity. The structural properties and compositional tunability have made gel-derived alloys, oxides, phosphides, etc. a hot topic in electrocatalysis. After carbonization, transition metal ion chelation with EDTA in agarose hydrogel produced Co_2P–N-doped carbon frameworks (NC). This was studied by Bo You et al. as an electrocatalyst for HER, having an attractive overpotential of about 139 mV, a Tafel slope of 45 mV dec^{-1}, and durability of 1000 cycles [87]. The wettability of the electrode for improving the electrolyte access to the active site is achieved by engineering the material to a super hydrophilic nature. Hu et al. have successfully executed this by developing superhydrophilic phytic acid (PA)-doped polypyrrole (PPy; PA–PPy) hydrogels as an electrocatalyst for superior OER performance. A lower value of the Tafel slope (54.9 mV dec^{-1}) is observed for the catalyst due to its faster charge transfer kinetics. This is due to the PA molecules that are anchored by porous conductive PPy hydrogel. This material is noted as a metal-free and binder-free hydrogel electrocatalyst having excellent OER activity with a small overpotential of 340 mV [88]. Heteroatom-doped graphene and carbon nanotubes, which require a complex synthesis process, acids, and oxidizing agents, are replaced by polymer gel-derived carbon framework with porous 3D network structure for oxygen reduction reaction (ORR). It helps in fast electron transfer and effective oxygen diffusion. Using 3D interconnected PANI hydrogel, a 3D N, P co-doped mesoporous nanocarbon

(NPMC) foams were synthesized using a low-cost and simple method. The high surface area introduced by the gel material and surplus amount of edge-like graphitic structures of NPMC enabled high ORR activity with a high value of kinetic current, an onset potential of 0.94 V versus RHE, and a halfwave potential of 0.85 V versus RHE [89].

2.2 *Soft Materials in Energy Storage Applications*

Soft Materials in Batteries and Supercapacitors

In the last few decades, batteries achieved colossal success in energy storage applications to power various portable electronics such as laptops, mobile phones, watches, and even electric vehicles [90–92]. Batteries generally comprise a cathode and an anode separated by a separator soaked in electrolyte. With the ability to be in wide applications, it is desired that batteries be made safer and more sustainable. Presently, lithium-ion batteries (LIB) are triumphant in battery application, but the high cost of the cathode materials and safety issues stop them from becoming superior in the field. Recently, much attention has been devoted to soft materials such as organic materials because of their abundance, environment friendliness, and structural flexibility [93, 94]. Smart devices have emerged by integrating flexible energy storage devices as power sources (Fig. 5a) [95]. This type of battery is devoted to multiple considerations, such as implantable and wearable devices used in the human body. Stretchable batteries also show multiple degrees of freedom for various applications. Furthermore, few biomedical devices are permanently planted inside the human body, which needs constant power to function. Hence, soft materials are suitable for use as different battery parts in this application because of their bio-friendliness characteristics. In this section, we discuss the advancement of batteries using soft materials such as cathode, anode, and electrolyte in terms of flexibility, sustainability, and high electrochemical performance. The majority of flexible energy storage equipment such as batteries rely on ultrathin flexible and stretchable plastic substrates such as PET, polyimide, PDMS, polyurethane, etc. [99, 100]. Generally, flexible and stretchable energy storage devices are accomplished by depositing the active materials onto the substrates using various techniques such as solution, chemical, physical, and lithography processes. Cathode materials are called active materials for batteries where ions are stored while charging, and it will release the ions at the time of discharging. Inorganic and organic semiconductor materials are mostly used as cathode materials for batteries. For the serge of environmental friendliness, flexible and wearable batteries and organic semiconductors are attracting much attention among researchers. Various organic semiconductors such as conducting polymers, colloids, etc. are explored as cathode material in Li, Na, and Zn-ion batteries [101–104]. Conducting polymers (CPs) consist of alternating single and double bonds, which create overlapping *p*-orbital delocalized pi-electrons, leading to interesting electrical and optical properties. The

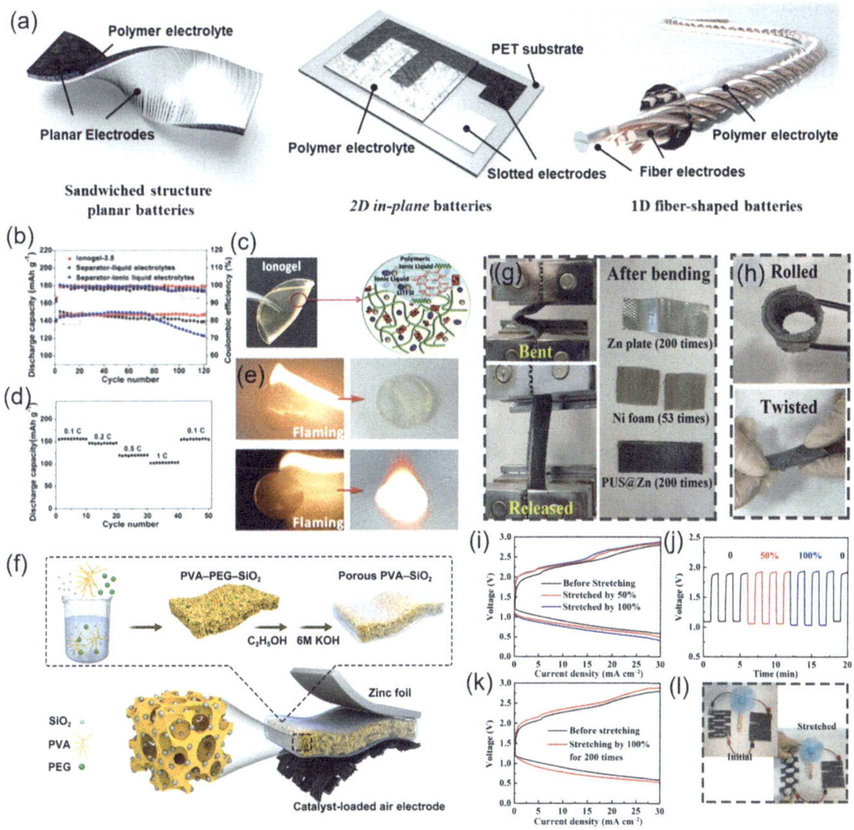

Fig. 5 (**a**) Device configuration of flexible batteries such as sandwiched structure, 2D in-plane, and 1D fiber-shaped structure (Reprinted with permission Ref. [95]). (**b**–**e**) Cycling performance, digital image, rate capability, and exposure to the flame of ionogel-based Li-ion battery (Reprinted with permission Ref. [96]). (**f**) Flexible Zn-ion battery of PVA-SiO_2 nanocomposite electrolyte (Reprinted with permission Ref. [97]). (**g**–**l**) Flexible and stretchable Zn-ion battery with galvanostatic charge-discharge performance in different strain conditions (Reprinted with permission Ref. [98])

conductivity of CP is in the range of 0.01–500 S cm^{-1}. Most impressively, CPs used for battery applications are polyaniline (PANI), polypyrrole (PPy), polythiophene (PTh), PEDOT, and their various derivatives [101, 105, 106]. The electrochemical activity and conductivity make them applicable in rechargeable Li, Na, and Zn-ion batteries. CPs can charge/discharge corresponding to doped/dedoped processes with mobility of anions or cations. Because of their good electrochemical stability in the p-doped state, CPs are used as cathode materials in Li, Na, K, and Zn-ion batteries. The specific capacity offered through CP mainly depends on the doping level. Typically, CPs possess a doping level of 0.3–0.5 (0.3–0.5 number of ions per monomer ring of CP). The higher doping level will lead to the irreversible reaction of the CPs. However, CPs are suffering from the absence of plateau regions in the charging and discharging curve. These drawbacks can be overcome by functionalizing the polymer matrix or integrating suitable inorganic materials into the polymer to increase the doping/dedoping process of ions and additional species enriching into capacity. Chen et al. prepared PEDOT with Sn_3P_4@CNT, achieving superior performance and cyclic stability in Li-ion batteries [107]. PPy consists of a repeating pyrrole unit and exhibits charge storage property with a flat charge/discharge plateau. However, it offers high power density but suffers from low energy density. Making composite with PPy improves energy density. α-MnO_2/PPy is an excellent zinc-ion storage for aqueous zinc ion batteries that exhibits 270 mAh g^{-1} at 0.1 A g^{-1} [108]. PPy-LFP cathodes show enhancement in capacity in Li-ion batteries compared to bare LFP cathodes [109]. However, using side chain functionalization, PTh recently attained attention in energy storage due to its stable n-doping state. Jiang et al. reported a series of conjugate polymers with different linkage patterns of anthraquinone and benzene for high-performance Na-ion batteries [110].

Electrolytes transport ions from the anode to the cathode and vice versa. The electrolyte properties, such as ionic conductivity and interfacial resistance between electrode and electrolyte, highly affect the battery's overall performance. Solid polymer electrolytes (SPEs) are required to make flexible and wearable batteries; hence, soft materials play a significant role in developing the electrolytes. SPE can be made using various processes such as physical cross-linking, chemical cross-linking, etc. Different approaches can be taken to improve the SPE performance, such as enhancement of the mobility of the polymer chains, an increment in the carrier density, and a decrement in the interfacial resistance between the active material and SPE. The combination of salt and polymer-based SPE can achieve ionic conductivity of about 1×10^{-3} S cm^{-1} at room temperature. However, the salt tends to crystallize at lower temperatures, which prevents its practical application. Furthermore, SPEs have very high tensile strength, outstanding elasticity, and excellent electrochemical performance, as shown by Guan et al. [111]. To make the flexible and wearable battery, nontoxicity and nonflammability measures are desired. Sun et al. show the highly conductive, self-healable, flexible, and nonflammable electrolytes for high-performing wearable Li-ion batteries as shown in Fig. 5b–e [96]. The healable electrolytes show a specific 150 mAh g^{-1} capacity with excellent rate capability. This polymer-based electrolyte can serve as an electrolyte and separating membrane simultaneously, which shows great potential to be used in the next generation of flexible and healable batteries. The additives in the polymer electrolyte enhance the mechanical and electrical properties of the electrolytes. Hu et al. show that the SiO_2 fillers in PVA gel polymer electrolyte enhance the mechanical properties and conductivity up to 57 mS cm^{-1} with good cyclic stability in Zn-ion battery, as shown in Fig. 5f [97]. With the various functionalities in the battery, stretchability is also an important factor, for which soft materials are promising candidates. A full-cell battery is to be flexible and stretchable; all the components of the battery have to be flexible and stretchable. Hu et al. show the flexible and stretchable Zn-ion battery with bending, twisting, and rolling capabilities without losing the battery's performance, as shown in Fig. 5g [98]. SPE-based flexible batteries have attracted significant attention, but unwanted side reactions at the anode side lead to dendrite formation. Recently, the concept of water in salt (WIS) in polymer has been proposed to overcome the dendrite formation issue at the anode by blocking undesired side reactions. Due to the high concentration of the salt and polymer matrix, the free water molecules will be absent, reducing the ionic conductivity of the electrolyte, which increases the electrochemical potential window. Kuila et al. fabricated a flexible and durable Zn-ion battery with a very high level of safety and also able to run a 1.8 V LED [112]. The utilization of soft materials in batteries is high in biomedical and medical studies. Portable and compact configurations in primary and secondary batteries are suitable for various developments in medical instruments. It starts from wheelchairs to sensors and pacemakers to neurostimulators everywhere. Schneider et al. show that PVP was used as an electrolyte in Li-ion batteries for pacemakers in 1984. The contribution of soft materials to achieve the successful development and implementation of batteries in biomedical devices is remarkable and helps to enable healing from human disease [113].

Supercapacitor

Similarly, the fabrication of flexible, wearable, lightweight, portable, and fireproof modern energy storage technologies like supercapacitors (SCs) relies heavily on soft materials. Due to their superior power and energy densities when compared to traditional capacitors, fuel cells, and batteries, SCs have established themselves as vital components in common electronic devices. To address the rising need for portable and adaptable electronics, there has been an increase in interest in flexible wearable energy-storing technology in recent years [114]. As a result, considerable work has been done to develop soft materials appropriate for different SC components, including electrodes and electrolytes. Developing electrode materials for wearable SCs, including carbon nanotubes (CNTs), graphene, transition metal oxides, conducting polymers, and their composites, is one of the main areas of the current study. With its single-layer hexagonal sp^2 hybridized carbon atom structure, enormous surface area (2630 $m^2\ g^{-1}$), and high thermal and electrical conductivity, graphene is a well-known material that makes an excellent electrode for wearable SCs [115]. Due to their low cost and simplicity in modification and assembly into strong films or other flexible substrates, freely generated flexible and graphene-based electrodes have been successfully developed for flexible SCs. The most popular electrode material for SCs is reduced graphene oxide (rGO), created via the chemical redox technique and which has many favorable features. Flexible and stretchable graphene-based electrodes for SCs have been made possible by creative manufacturing techniques using flexible and stretchable substrates, including elastic substrates and textiles, [116] for example, rGO films (reduced by laser-scribe) on a membrane of nitrocellulose and polyethylene terephthalate (PET) with 0.4 μm pore size [117]. When employed as SC electrodes, these films had a high specific surface area (1520 $m^2\ g^{-1}$) and conductivity (1738 $S\ m^{-1}$), which led to good electrochemical performance. In addition, an ultrathin supercapacitor using a PET substrate and an ultrathin graphene electrode were developed. The strong affinity between the graphene nanosheets, which restricts the use of graphene's enormous surface area, is one drawback to employing them. Researchers have looked into porous 3D graphene as a solution to this problem because it has outstanding electrical conductivity, strong chemical stability, and a sizable open surface area. A core-sheath fiber wrapped in 3D graphene sheet that has a high exposed surface area and outstanding electrical conductivity was successfully made by Meng et al. With a specific capacitance of 30–40 F g^{-1}, such designed fibers are flexible and compressible, making them appropriate for wearing SCs even during cyclic bending. Because electrode materials in supercapacitors may be bent or stretched, durability is a key consideration when choosing materials for energy storage devices in wearable electronics. A potential fix for this problem is the use of self-healing materials, which can automatically fix both internal and outward flaws. SWCNTs and a self-healing substrate were used by Wang et al. to create an electrical and mechanical self-healing wearable electrochemical capacitor [118]. These self-healing materials depend on supramolecular linkages to sustain TiO_2 nanostructures, which allow the electrode materials to fix damage on their own. By using a PVP-H_2SO_4 gel electrolyte, it is possible to boost this electrode material's capacity for self-healing, leading to an 85% increase in specific capacitance despite repeated cutting. SCs have widely used conductive polymers as electrode materials, including polypyrrole (PPy), PANi, polythiophene (PTh), and poly(3,4-ethylenedioxythiophene) (PEDOT) [119]. Compared to carbon- or metal oxide-based materials, they are better prospects for SCs due to their outstanding Young's modulus, deformability, and conductivity. The conductivity of conductive polymers can be enhanced up to 104 S cm^{-1} by adding different doping amounts or dopants. However, elements including polymer polarization state (doped or undoped state), doping agent substance, pH, temperature, ionic strength, and solvent significantly impact the conducting state of such polymers. Despite having beneficial features, conductive polymers have difficulties because of their poor stability, which causes their backbones to contract and deteriorate during the electrochemical process. As a result, SC devices' power and energy densities are reduced, restricting their new uses. The most popular techniques for conductive polymers to improve their electrochemical performance include manufacturing them on a structured surface or mixing them with other electroactive materials, including carbon or metal compounds. This method has been successfully used to SCs, allowing the development of stretchy and flexible electrodes. The fabrication of a flexible and elastic electrode employing vertical PPy nanotube arrays and carbon fabric is one noteworthy example [120]. With a capacitance density of 64 F g^{-1} and outstanding capacitance retention (99%) at a 50% strain after 500 cycles, these stretchy SCs are perfect for real-world applications because they provide a consistent energy supply under a variety of bending circumstances. Additionally, printing technologies have been used to create conductive polymer-based SC electrodes. For instance, a printable SC with a high energy density and capacitance was created using a low-cost, air-stable PANi ink created utilizing a straightforward assemble-disperse technique. An intense study has been done on hydrogel electrodes [121]. While many hydrogels that are with nonconductive polymer skeletons fall short of SC criteria, conductive polymers like PTh, PEDOT, PANi, and PPy can be used to create hydrogels with unique structural and electrical properties [122]. The high capacitance and exceptional conductivity of PANi hydrogels are achieved by adding the compound phytic acid as the gelator dopant [123]. In order to produce electrode hydrogels free of insulating materials, multivalent ions (Fe^{3+}, Mg^{2+}, and Cu^{2+}) were used as the gelator dopant. This improved the electrical characteristics of conductive polymer-based hydrogels. New approaches, including double networks and low-temperature procedures, were used to develop good hydrogels with hierarchical micro-/nanostructures in order to address the issue of decreasing mechanical characteristics in conductive polymer-based hydrogels. Although a slight reduction in conductivity was noticed after introducing phytic acid, these hydrogels show enhanced durability (29-fold enhancement) and electrochemical performance. Researchers are still working to develop conductive polymer-based hydrogels that have great stretchability and good conductivity.

Soft Materials for Hydrogen Storage

Hydrogen, which has a high gravimetric energy density of 142 MJ kg^{-1}, can replace fossil fuels [124]. One of the major drawbacks of attaining a hydrogen economy that aims toward decarbonization is the efficient storage of the produced H_2. An energy-efficient storage method in a limited space in near atmospheric conditions is the major challenge confronted by the research community [125]. The US Department of Energy (DoE) has set a target of 55 g of H_2 per kg for onboard hydrogen storage systems by 2025 [126]. A volumetric density of 70 g L^{-1} and gravimetric density of 7.5 wt% of hydrogen at an intermediate pressure of 12 bar and within a temperature range of −40 and 85 °C and in a minimum fueling time. Commercialization looks for low-cost materials that are durable and safe to handle for hydrogen storage. The methods of hydrogen storage are low-temperature liquid hydrogen at 20 K, high-pressure gaseous hydrogen at 700 bar, and storage of atomic or molecular hydrogen in solid materials as shown in Fig. 6a [127]. The third method of storing hydrogen by adsorption on materials can be either physisorption where molecular hydrogen adsorbs weakly with meV range of binding energy which favors low-energy desorption. It can also be chemisorption where atomic hydrogen binds strongly with 2–4 eV binding energy which requires a higher temperature for desorption or quasi-molecular bonding at which the H_2 bond weakens with an intermediate binding energy (0.1–0.8 eV) as shown in Fig. 6b–d [128]. Due to the complexity of the development and management of these systems, the physisorption of H_2 at low temperatures on microporous material opens a new path for hydrogen storage. Low-density and high surface area materials capable of reservable adsorption of H_2 at lower pressure can even be used as an inner lining for H_2 tank for improved volumetric density of

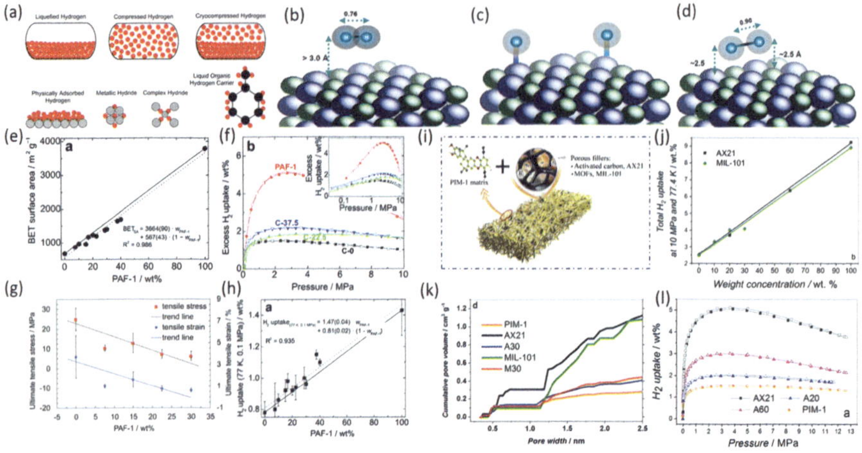

Fig. 6 (**a**) Different hydrogen storage methods (Reprinted with permission Ref. [127]). (**b–d**) Methods of storing molecular hydrogen on solid materials by physisorption, chemisorption, and quasi-molecular bonding, respectively (Reprinted with permission Ref. [128]). (**e**) BET surface area of PIM-1/PAF-1 composite as a function of different PAF-1 wt%. (**f**) Excess H_2 uptake at 10 MPa for PIM-1, PAF-1, and composites with different concentrations of PAF-1. (**g**) Ultimate tensile stress of PIM-1/PAF-1 composite at different concentrations of PAF-1 up to 30 wt%. (**h**) Hydrogen uptake at different PAF-1 concentrations; dotted and solid lines indicate experimental and theoretical values (Reprinted with permission Ref. [129]). (**i**) Schematic representation of PIM-1 with activated carbon fillers. (**j**) Total H_2 uptake with respect to the weight percent of fillers. (**k**) Cumulative pore volume calculated using NLDFT. (**l**) High-pressure H_2 isotherm, symbols representing experimental data points and solid lines indicating fitted curve (Reprinted with permission Ref. [130])

hydrogen storage [131]. The characteristics of the material used are the efficiency for hydrogen loading and unloading, low cost, being easily processable, and high stability. Metal-organic frameworks [132–135], covalent organic frameworks [136–138], activated carbon [139–141], zeolites [142, 143], and porous aromatic frameworks [144, 145] are some of the materials studied in this area. Polymers are having lower surface area due to high conformational and rotational freedom for efficient packing. This is a major limitation to use it as hydrogen storage material. At the same time, they are having the advantage in terms of their low density, capability to form self-standing film, and good mechanical properties. One of the promising materials under research is polymers of intrinsic microporosity (PIM) [130]. They are rigid, fused-ring compounds with high surface area having molecular-sized voids, chemical homogeneity, and thermal and chemical stability. They are reported to show low-temperature moderate-pressure hydrogen storage by physisorption. Any one of the monomers used for making PIM should contain a spiro-center or a rigid nonplanar unit which can be a site for contortion. PIM-1 is the result of the reaction between 5,5′,6,6′-tetrahydroxy-3,3,3′,3′-tetramethyl-1,1′-spirobisindane and tetrafluoroterephthalonitrile [146]. It is having H_2 Langmuir surface area of 540 $m^2\ g^{-1}$ and shows H_2 uptake of 1.45% mass at 10 bar pressure.

One of the interesting properties of this polymer is that according to the selection of the monomer, its micropore structure can be engineered. A monomer with intrinsic cavities can provide more active sites for H_2 adsorption. CTC-PIM having bowl-shaped cyclotricatechylene incorporation with PIM shows 1.7% of mass, and hexaazatrinaphthylene-(HATN-) based PIM shows 1.56% of mass hydrogen uptake at 10 bar. Triptycene-based polymer (Trip-PIM) shows enhanced hydrogen uptake of 2.71% at 10 bar and 1.6% at 1 bar. The increase in hydrogen uptake is achieved by engineering the polymer to the one with a higher surface area. Here HATN-PIM has a surface area of 590 $m^2\ g^{-1}$, CTC-PIM has 630 $m^2\ g^{-1}$, and Trip-PIM has 1100 $m^2\ g^{-1}$. Similarly, pore size is also a dependent factor. Volumetric capacity may reduce for hydrogen storage on macroporous materials because of higher empty voids [147]. Composite materials are designed to enhance the properties of H_2 physisorption. PIM is doped with high surface area adsorbents such as hyper-cross-linked nanoparticle fillers, porous organic cage molecules, zeolites, etc. Sebastien Rochat et al. synthesized porous aromatic framework (PAF) doped PIM-1 and processed it to form a freestanding film. BET surface area increases from 681 to 1639 $m^2\ g^{-1}$ with an increase in PAF content from 0 to 37.5 wt%, respectively, as shown in Fig. 6e. Similarly, total pore volume increases from 0.48 to 0.97 $cm^3\ g^{-1}$. All these factors influenced the increment in the H_2 uptake from 2.60 to 4.79 wt%. Here, they have shown the concept of tuning the porosity, surface area, H_2 uptake, and mechanical strength of PIM by doping with PAF. These fully reversible isotherms without any hysteresis have shown good charge/discharge stability with an optimum 20 wt% PAF contents as indicated Fig. 6f, h showing the increase in H_2 uptake with PAF concentration. But with the increase in PAF filler concentration, the tensile strength of the composite weakens and the film becomes brittle (Fig. 6g). Young's modulus reduces by 20% and the flexible nature of the membrane is lost [129].

Metal-organic framework (MOF), which is a well-known H_2 storage material, is used along with PIM as a filler material to obtain a composite of high surface area and great H_2 uptake, along with a mechanically stable self-standing film. Materials Institute Lavoisier chromium(III) terephthalate [MIL-101(Cr)] monoliths along with cross-linked polymer of intrinsic microporosity (PIM-1) formed PIM-1/MIL-101(Cr). From N_2 adsorption–desorption isotherms, the pore size was observed to be in the mesoporous range, and pore volume increased from 0.53 to 1.20 $cm^3\ g^{-1}$, with increase in MIL-101(Cr) content from 0 to 80 wt%. Similarly, BET surface area and H_2 uptake increased from 784 to 2720 $m^3\ g^{-1}$ and 1.02 to 1.76 wt%, respectively. By designing the powder form of MOF to a self-standing composite with PIM-1, an efficient material is formed which can be a part of hybrid hydrogen storage cylinders. Similarly, activated carbon, which is a well-known porous filler, is used to make a composite with PIM-1 (AX21/PIM-1), as shown in the schematic in Fig. 6i. It has shown BET surface area of 2034 $m^2\ g^{-1}$ and total pore volume of 1.22 $cm^3\ g^{-1}$ at 60 wt% AX21/PIM-1 which was increasing with increase in the percentage of AX21. It has shown a huge value of H_2 uptake up to 6.36 wt% at 10 MPa, which can meet the US DOE target of 5.5 wt% for a complete system (Fig. 6j–l). Since a larger surface area is the prerequisite for a better hydrogen storage material, Jonathan Germain et al. carried out experiments for surface area calculation (m^2 g^{-1}) and H_2 capacity (wt%) using different commercial resins. They found that all polymers are not showing a trend of higher H_2 capacity for the ones with higher BET surface area. One of the reasons for this inflated value of surface area comes from the swelling of the polymer during the adsorption cycle. The best performance occurs when a near-monolayer of gas gets adsorbed when the gas is at a temperature higher than its critical temperature. So, as a better option for accurate measurement, the surface area is calculated using the Langmuir equation from H_2 isotherm, which showed significant variation compared to BET surface area. The same group synthesized hyper-cross-linked polystyrene gel with pores of about 2 nm, which showed higher surface area than commercial resin and H_2 adsorption. It can have 2.2 wt% H_2 adsorption at 1.2 MPa [148]. Selection of the suitable monomer and the dissolution in suitable solvent play a very important role in the structure as surface area and H_2 uptake. A more promising material for commercialization can be prepared by synthesizing a PIM with a suitable monomer combination and an easy and large-scale synthesis method. This is one of the best materials that can be designed to meet the target of 55 g of H_2 per kg by 2025.

3 Summary and Future Outlook

The application of soft materials in energy harvesting and storage holds great promise for the future of sustainable energy technologies. This chapter has provided an overview of the various subtopics within these areas, highlighting the significant contributions of soft materials in each domain. Looking ahead, future advancements and innovations can be expected in energy harvesting. Solar cells, already benefiting from soft materials, will continue to improve in efficiency, flexibility, and cost-effectiveness. Integrating soft materials in piezo/tribo nanogenerators will likely lead to the development of more efficient devices for harvesting mechanical energy from ambient vibrations and movements. Additionally, the use of soft materials in photocatalytic and electrocatalytic processes will continue to drive advancements in solar energy conversion, facilitating more efficient and sustainable energy generation. In energy storage, soft materials will play a crucial role in advancing battery technology. Incorporating soft materials will enhance electrode performance, energy density, and safety, leading to the development of more powerful and reliable batteries. Similarly, soft materials will continue to revolutionize the field of supercapacitors, enabling high power densities, rapid charge/discharge rates, and improved cycling stability. Moreover, soft materials will continue to enable efficient and safe hydrogen storage, supporting the utilization of hydrogen as a clean energy carrier. As the energy field continues to evolve, researchers and engineers are expected to explore novel soft materials with enhanced properties and functionalities, leading to even more significant breakthroughs. Developing multifunctional soft materials capable of simultaneously harvesting and storing energy may revolutionize how we approach energy systems. Integrating soft materials with emerging technologies, such as flexible electronics and wearables, will create new energy harvesting and storage possibilities in various applications. Applying soft materials in energy holds tremendous potential for shaping a greener and more sustainable future. By harnessing the unique properties of soft materials, researchers and engineers can continue to drive advancements in energy harvesting and storage, ultimately leading us toward a more sustainable and energy-efficient world.

References

1. Gielen, D., et al.: The role of renewable energy in the global energy transformation. Energ. Strat. Rev. **24**, 38–50 (2019)
2. Dutta, P., et al.: Highly stable poly(*o*-methoxyaniline)/WO_3-nanoflower composite-based electrochromic supercapacitors with real-time charge indication. ACS Appl. Polym. Mater. **5**, 4088–4099 (2023)
3. Cherumannil Karumuthil, S., et al.: Fabrication of dual-functional electrochromic smart window based on low-cost hybrid transparent electrode coated with a solution-processable polymer. J. Mater. Chem. A. **10**, 23265–23273 (2022)
4. Li, G., et al.: Polymer solar cells. Nat. Photonics. **6**, 153–161 (2012)
5. Facchetti, A.: Polymer donor-polymer acceptor (all-polymer) solar cells. Mater. Today. **16**, 123–132 (2013)
6. Wang, R., et al.: A review of perovskites solar cell stability. Adv. Funct. Mater. **29**, 1808843 (2019)
7. Li, Y., et al.: Polymer photovoltaic materials with broad absorption band and high charge carrier mobility. Adv. Mater. **20**, 2952–2958 (2008)
8. Schlenker, C.W., et al.: The molecular nature of photovoltage losses in organic solar cells. Chem. Commun. **47**, 3702–3716 (2011)
9. Liu, Q., et al.: 18% Efficiency organic solar cells. Sci. Bull. **65**, 272–275 (2020)
10. Zhu, L., et al.: Single-junction organic solar cells with over 19% efficiency enabled by a refined double-fibril network morphology. Nat. Mater. **21**, 656–663 (2022)
11. Hummelen, J.C., et al.: Preparation and characterization of Fulleroid and Methanofullerene derivatives. J. Org. Chem. **60**, 532–538 (1995)
12. Yu, G., et al.: Polymer photovoltaic cells: enhanced efficiencies via a network of internal donor-acceptor heterojunctions. Science. **270**, 1789–1791 (1995)
13. He, Y., et al.: Indene—C60 bisadduct: a new acceptor for high-performance polymer solar cells. J. Am. Chem. Soc. **132**, 1377–1382 (2010)
14. Guo, X., et al.: High efficiency polymer solar cells based on poly(3-hexylthiophene)/indene-C70 bisadduct with solvent additive. Energy Environ. Sci. **5**, 7943–7949 (2012)
15. Xiao, Z., et al.: Development of isomer-free fullerene bisadducts for efficient polymer solar cells. Energy Environ. Sci. **9**, 2114–2121 (2016)
16. Yan, C., et al.: Non-fullerene acceptors for organic solar cells. Nat. Rev. Mater. **3**, 1–19 (2018)
17. Lin, Y., et al.: An electron acceptor challenging fullerenes for efficient polymer solar cells. Adv. Mater. **27**, 1170–1174 (2015)
18. Xiao, Z., et al.: Ternary organic solar cells offer 14% power conversion efficiency. Sci. Bull. **62**, 1562–1564 (2017)
19. Yuan, J., et al.: Single-junction organic solar cell with over 15% efficiency using fused-ring acceptor with electron-deficient core. Joule. **3**, 1140–1151 (2019)
20. Li, G., et al.: Manipulating regioregular poly(3-hexylthiophene):[6,6]-phenyl-C61-butyric acid methyl ester blends-route towards high efficiency polymer solar cells. J. Mater. Chem. **17**, 3126–3140 (2007)

21. Hoppe, H., et al.: Morphology of polymer/fullerene bulk heterojunction solar cells. J. Mater. Chem. **16**, 45–61 (2006)
22. Shaheen, S.E., et al.: 2.5% efficient organic plastic solar cells. Appl. Phys. Lett. **78**, 841–843 (2001)
23. Cherumannil Karumuthil, S., et al.: Electrospun poly(vinylidene fluoride-trifluoroethylene)-based polymer nanocomposite fibers for piezoelectric nanogenerators. ACS Appl. Mater. Interfaces. **11**, 40180–40188 (2019)
24. Kar, E., et al.: 2D SnO_2 nanosheet/PVDF composite based flexible, self-cleaning piezoelectric energy harvester. Energy Convers. Manag. **184**, 600–608 (2019)
25. Kim, Y.G., et al.: Improved performance of stretchable piezoelectric energy harvester based on stress rearrangement. Sci. Rep. **12**, 19149 (2022)
26. Sun, Y., et al.: Enhanced energy harvesting ability of ZnO/PAN hybrid piezoelectric nanogenerators. ACS Appl. Mater. Interfaces. **12**, 54936–54945 (2020)
27. Kim, W.G., et al.: Triboelectric nanogenerator: structure, mechanism, and applications. ACS Nano. **15**, 258–287 (2021)
28. Kou, H., et al.: Smart pillow based on flexible and breathable triboelectric nanogenerator arrays for head movement monitoring during sleep. ACS Appl. Mater. Interfaces. **14**, 23998–24007 (2022)
29. Varghese, H., et al.: A high-performance flexible triboelectric nanogenerator based on cellulose acetate nanofibers and micropatterned PDMS films as mechanical energy harvester and self-powered vibrational sensor. Nano Energy. **98**, 107339 (2022)
30. Mi, H.Y., et al.: High-performance flexible triboelectric nanogenerator based on porous aerogels and electrospun nanofibers for energy harvesting and sensitive self-powered sensing. Nano Energy. **48**, 327–336 (2018)
31. Dagdeviren, C., et al.: Conformable amplified lead zirconate titanate sensors with enhanced piezoelectric response for cutaneous pressure monitoring. Nat. Commun. **5**, 4496 (2014)
32. Jaffe, H.: Piezoelectric ceramics. J. Am. Ceram. Soc. **41**, 494–498 (1958)
33. Tao, K., et al.: Piezoelectric ZnO thin films for 2DOF MEMS vibrational energy harvesting. Surf. Coat. Technol. **359**, 289–295 (2019)
34. Khan, A., et al.: Analysis of direct and converse piezoelectric responses from zinc oxide nanowires grown on a conductive fabric. Phys. Status Solidi (A) Appl. Mater. Sci. **212**, 579–584 (2015)
35. Dai, Y.J., et al.: Piezoelectric and ferroelectric properties of Li-doped (bi 0.5Na0.5)TiO_3-(Bi0.5K 0.5)TiO_3-$BaTiO_3$ lead-free piezoelectric ceramics. J. Am. Ceram. Soc. **93**, 1108–1113 (2010)
36. Roberts, S.: Dielectric and piezoelectric properties of barium titanate. Phys. Rev. **71**, 890 (1947)
37. Zhao, C., et al.: Practical high piezoelectricity in barium titanate ceramics utilizing multiphase convergence with broad structural flexibility. J. Am. Chem. Soc. **140**, 15252–15260 (2018)
38. Shaik, H., et al.: Towards β-phase formation probability in spin coated PVDF thin films. J. Polym. Res. **24**, 1–6 (2017)
39. Sharma, M., et al.: Outstanding dielectric constant and piezoelectric coefficient in electrospun nanofiber mats of PVDF containing silver decorated multiwall carbon nanotubes: assessing through piezoresponse force microscopy. RSC Adv. **6**, 6251–6258 (2016)
40. Yuan, X., et al.: The large piezoelectricity and high power density of a 3D-printed multilayer copolymer in a rugby ball-structured mechanical energy harvester. Energy Environ. Sci. **13**, 152–161 (2020)
41. Li, Y., et al.: Multilayer assembly of electrospun/electrosprayed PVDF-based nanofibers and beads with enhanced piezoelectricity and high sensitivity. Chem. Eng. J. **388**, 124205 (2020)
42. Hoque, N.A., et al.: Er^{3+}/Fe^{3+} stimulated electroactive, visible light emitting, and high dielectric flexible PVDF film based piezoelectric nanogenerators: a simple and superior self-powered energy harvester with remarkable power density. ACS Appl. Mater. Interfaces. **9**, 23048–23059 (2017)

43. Shi, K., et al.: Synergistic effect of graphene nanosheet and $BaTiO_3$ nanoparticles on performance enhancement of electrospun PVDF nanofiber mat for flexible piezoelectric nanogenerators. Nano Energy. **52**, 153–162 (2018)
44. Pei, H., et al.: 3D printing of PVDF-based piezoelectric nanogenerator from programmable metamaterial design: promising strategy for flexible electronic skin. Nano Energy. **109**, 108303 (2023)
45. Wang, W., et al.: Unexpectedly high piezoelectricity of electrospun polyacrylonitrile nanofiber membranes. Nano Energy. **56**, 588–594 (2019)
46. Wu, Q.Y., et al.: Structure and performance of polyacrylonitrile membranes prepared via thermally induced phase separation. J. Membr. Sci. **409**, 355–364 (2012)
47. Martins, P., et al.: Electroactive phases of poly(vinylidene fluoride): determination, processing and applications. Prog. Polym. Sci. **39**, 683–706 (2014)
48. Yuan, L., et al.: Piezoelectric PAN/$BaTiO_3$ nanofiber membranes sensor for structural health monitoring of real-time damage detection in composite. Compos. Commun. **25**, 100680 (2021)
49. Shi, Q., et al.: Enhanced the dielectric and piezoelectric properties of polyacrylonitrile piezoelectric composite fibers filled with ionic liquids. J. Appl. Polym. Sci. **140**, e53824 (2023)
50. Zeng, J., et al.: Fiber-like TiO_2 nanomaterials with different crystallinity phases fabricated via a green pathway. ACS Appl. Mater. Interfaces. **3**, 2074–2079 (2011)
51. Chae, I., et al.: Review on electromechanical coupling properties of biomaterials. ACS Appl. Bio. Mater. **1**, 936–953 (2018)
52. Sriplai, N., et al.: Enhancing piezoelectric properties of bacterial cellulose films by incorporation of $MnFe_2O_4$ nanoparticles. Carbohydr. Polym. **231**, 115730 (2020)
53. Yoon, S., et al.: Effect of process orientation on the mechanical behavior and piezoelectricity of electroactive paper. Materials. **13**, 204 (2020)
54. Baytekin, B., Baytekin, H.T., Grzybowski, B.A.: Retrieving and converting energy from polymers: deployable technologies and emerging concepts. Energy Environ. Sci. **6**, 3467–3482 (2013)
55. Alam, M.M., et al.: Native cellulose microfiber-based hybrid piezoelectric generator for mechanical energy harvesting utility. ACS Appl. Mater. Interfaces. **8**, 1555–1558 (2016)
56. Owusu, F., et al.: Stretchable high response piezoelectric elastomers based on polable polynorbornene fillers in a polydimethylsiloxane matrix. Adv. Funct. Mater. **32**, 2207083 (2022)
57. Kachroudi, A., et al.: Piezoelectric cellular micro-structured PDMS material for micro-sensors and energy harvesting. J. Phys. Conf. Ser. **660**, 012040 (2015)
58. Hu, H., et al.: Enhanced piezoelectric properties of flexible PDMS sponges modified by PTFE solid films and nanoparticles. Mater. Lett. **326**, 132989 (2022)
59. Kim, K.N., et al.: Silk fibroin-based biodegradable piezoelectric composite nanogenerators using lead-free ferroelectric nanoparticles. Nano Energy. **14**, 87–94 (2015)
60. Fan, X., et al.: Ultrathin, rollable, paper-based triboelectric nanogenerator for acoustic energy harvesting and self-powered sound recording. ACS Nano. **9**, 4236–4243 (2015)
61. Zhu, G., et al.: A shape-adaptive thin-film-based approach for 50% high-efficiency energy generation through micro-grating sliding electrification. Adv. Mater. **26**, 3788–3796 (2014)
62. Zheng, Q., et al.: In vivo powering of pacemaker by breathing-driven implanted triboelectric nanogenerator. Adv. Mater. **26**, 5851–5856 (2014)
63. Dong, L., et al.: Deformable textile-structured triboelectric nanogenerator knitted with multifunctional sensing fibers for biomechanical energy harvesting. Adv. Fiber Mater. **4**, 1486–1499 (2022)
64. Chen, X., et al.: Semiconductor-based photocatalytic hydrogen generation. Chem. Rev. **110**, 6503–6570 (2010)
65. Willkomm, J., et al.: Dye-sensitised semiconductors modified with molecular catalysts for light-driven H_2 production. Chem. Soc. Rev. **45**, 9–23 (2016)

66. Fujishima, A., et al.: Electrochemical photolysis of water at a semiconductor electrode. Nature. **238**, 37–38 (1972)
67. Ibhadon, A.O., et al.: Heterogeneous photocatalysis: recent advances and applications. Catalogue. **3**, 189–218 (2013)
68. Xu, C., et al.: Porous organic polymers: an emerged platform for photocatalytic water splitting. Front. Chem. **6**, 592 (2018)
69. Yanagida, S., et al.: Poly(*p*-phenylene)-catalysed photoreduction of water to hydrogen. J. Chem. Soc. **8**, 474–475 (1985)
70. Stegbauer, L., et al.: Hydrazone-based covalent organic framework for photocatalytic hydrogen production. Chem. Sci. **5**, 2789–2793 (2014)
71. Pachfule, P., et al.: Diacetylene functionalized covalent organic framework (COF) for photocatalytic hydrogen generation. J. Am. Chem. Soc. **140**, 1423–1427 (2018)
72. Kuecken, S., et al.: Fast tuning of covalent triazine frameworks for photocatalytic hydrogen evolution. Chem. Commun. **53**, 5854–5857 (2017)
73. Huot, J.-Y., et al.: Hydrogen evolution and interface phenomena on a nickel cathode in 30 w/o KOH: I. Kinetics parameters and electrode impedance between 303 and 363 K. J. Electrochem. Soc. **136**, 1993 (1989)
74. Patra, B.C., et al.: A metal-free covalent organic polymer for electrocatalytic hydrogen evolution. ACS Catal. **7**, 6120–6127 (2017)
75. Bhunia, S., et al.: Electrochemical stimuli-driven facile metal-free hydrogen evolution from pyrene–porphyrin–based crystalline covalent organic framework. ACS Appl. Mater. Interfaces. **9**, 23843–23851 (2017)
76. Yang, C., et al.: Theory-driven design and targeting synthesis of a highly-conjugated basal-plane 2D covalent organic framework for metal-free electrocatalytic OER. ACS Energy Lett. **4**, 2251–2258 (2019)
77. Antolini, E., et al.: Carbon supports for low-temperature fuel cell catalysts. Appl. Catal. **365**, 1–2 (2009)
78. Liu, L., et al.: Nitrogen- and sulfur-rich conductive polymer for electrocatalytic evolution of hydrogen in acidic electrolytes. Int. J. Hydrog. Energy. **39**, 14712–14719 (2014)
79. Yang, D.H., et al.: Porous organic polymers for electrocatalysis. Chem. Soc. Rev. **51**, 761–791 (2022)
80. Chandran Ranjeesh, K., et al.: Metalloporphyrin two-dimensional polymers via metal-catalyst-free C–C bond formation for efficient catalytic hydrogen evolution. ACS Appl. Energy Mater. **1**, 6442–6450 (2018)
81. Meng, J., et al.: Porphyrin-based NiFe porous organic polymer catalysts for the oxygen evolution reaction. ChemCatChem. **13**, 1396–1402 (2021)
82. Li, B.Q., et al.: One-pot synthesis of framework porphyrin materials and their applications in bifunctional oxygen electrocatalysis. Adv. Funct. Mater. **29**, 1901301 (2019)
83. Lin, S., et al.: Covalent organic frameworks comprising cobalt porphyrins for catalytic CO_2 reduction in water. Science. **349**, 1208–1213 (2015)
84. Fang, Z., et al.: Gel electrocatalysts: an emerging material platform for electrochemical energy conversion. Adv. Mater. **32**, 2003191 (2020)
85. Tian, Y., et al.: Three-dimensional N-doped, plasma-etched graphene: highly active metal-free catalyst for hydrogen evolution reaction. Appl. Catal. A Gen. **529**, 127–133 (2017)
86. Chen, S., et al.: Molybdenum sulfide clusters-nitrogen-doped graphene hybrid hydrogel film as an efficient three-dimensional hydrogen evolution electrocatalyst. Nano Energy. **11**, 11–18 (2015)
87. You, B., et al.: Universal molecular-confined synthesis of interconnected porous metal oxides–N–C frameworks for electrocatalytic water splitting. Nano Energy. **48**, 600–606 (2018)
88. Hu, Q., et al.: Superhydrophilic phytic-acid-doped conductive hydrogels as metal-free and binder-free electrocatalysts for efficient water oxidation. Angew. Chem. Int. Ed. **58**, 4318–4322 (2019)

89. Zhang, J., et al.: A metal-free bifunctional electrocatalyst for oxygen reduction and oxygen evolution reactions. Nat. Nanotechnol. **10**, 444–452 (2015)
90. Lee, S.H., et al.: Self-powered flexible inorganic electronic system. Nano Energy. **14**, 111–125 (2015)
91. Shi, Q., et al.: More than energy harvesting—combining triboelectric nanogenerator and flexible electronics technology for enabling novel micro-/nano-systems. Nano Energy. **57**, 851–871 (2019)
92. Sun, Y., et al.: Inorganic semiconductors for flexible electronics. Adv. Mater. **19**, 1897–1916 (2007)
93. Windmiller, J.R., et al.: Wearable electrochemical sensors and biosensors: a review. Electroanalysis. **25**, 29–46 (2013)
94. Bandodkar, A.J., et al.: Soft, stretchable, high power density electronic skin-based biofuel cells for scavenging energy from human sweat. Energy Environ. Sci. **10**, 1581–1589 (2017)
95. Mo, F., et al.: Recent progress and challenges of flexible Zn-based batteries with polymer electrolyte. Batteries. **8**, 59 (2022)
96. Guo, P., et al.: Healable, highly conductive, flexible, and nonflammable supramolecular ionogel electrolytes for lithium-ion batteries. ACS Appl. Mater. Interfaces. **11**, 19413–19420 (2019)
97. Fan, X., et al.: Porous nanocomposite gel polymer electrolyte with high ionic conductivity and superior electrolyte retention capability for long-cycle-life flexible zinc–air batteries. Nano Energy. **56**, 454–462 (2019)
98. Qu, S., et al.: Kirigami-inspired flexible and stretchable zinc–air battery based on metal-coated sponge electrodes. ACS Appl. Mater. Interfaces. **12**, 54833–54841 (2020)
99. Wan, J., et al.: Ultrathin, flexible, solid polymer composite electrolyte enabled with aligned nanoporous host for lithium batteries. Nat. Nanotechnol. **14**, 705–711 (2019)
100. Zhai, Q., et al.: Recent advances in flexible/stretchable batteries and integrated devices. Energy Storage Mater. **33**, 116–138 (2020)
101. Zhou, D., et al.: Polymer electrolytes for lithium-based batteries: advances and prospects. Chem. **5**, 2326–2352 (2019)
102. Gebert, F., et al.: Polymer electrolytes for sodium-ion batteries. Energy Storage Mater. **36**, 10–30 (2021)
103. Yang, H.-L., et al.: Progress and challenges for all-solid-state sodium batteries. Adv. Energy Sustain. Res. **2**, 2000057 (2021)
104. Lu, K., et al.: Hydrogel electrolytes for quasi-solid zinc-based batteries. Front. Chem. **8**, 546728 (2020)
105. Hareendrakrishnakumar, H., et al.: Ion-selective PEDOT:PSS-decorated separator as a potential polysulfide immobilizer for lithium-sulfur batteries. Ionics. **27**, 1087–1099 (2021)
106. Chao, C., et al.: Stable and dendrite-free Zn anode enabled by a PEDOT:PSS layer for high-performance Zn-ion capacitors. Ind. Eng. Chem. Res. **62**, 1350–1357 (2023)
107. Zhao, Q., et al.: Sn_4P_3 encapsulated in carbon nanotubes/poly(3,4-ethylenedioxythiophene) as the anode for pseudocapacitive lithium-ion storage. ACS Appl. Energy Mater. **5**, 2412–2420 (2022)
108. Xu, J.W., et al.: High-performance reversible aqueous zinc-ion battery based on iron-doped alpha-manganese dioxide coated by polypyrrole. J. Colloid Interface Sci. **598**, 419–429 (2021)
109. Oriňáková, R., et al.: Effect of PPy/PEG conducting polymer film on electrochemical performance of $LiFePO_4$ cathode material for Li-ion batteries. Chem. Pap. **67**, 860–875 (2013)
110. Luo, L.W., et al.: Synthetic control of electronic property and porosity in anthraquinone-based conjugated polymer cathodes for high-rate and long-cycle-life Na-organic batteries. ACS Nano. **16**, 14590–14599 (2022)
111. Guan, J., et al.: Stretchable ionic-liquid-based gel polymer electrolytes for lithium-ion batteries. Ind. Eng. Chem. Res. **56**, 12456–12463 (2017)

112. Samanta, P., et al.: Molecular crowded "water-in-salt" polymer gel electrolyte for an ultra-stable Zn-ion battery. ACS Appl. Mater. Interfaces. **14**, 1138–1148 (2022)
113. Schneider, A. A., *et al.* Primary cells and iodine containing cathodes therefor. U.S. Patent No. 3,674,562. (1972)
114. An, T., et al.: Recent progress in stretchable supercapacitors. J. Mater.Chem. A. **6**, 15478–15494 (2018)
115. Wei, M., et al.: Engineering reduced graphene oxides with enhanced electrochemical properties through multiple-step reductions. Electrochim. Acta. **258**, 735–743 (2017)
116. Wu, Y., et al.: Three-dimensional micro/nano-interconnected scaffold graphene-based micro-supercapacitors with high electrochemical performance. Electrochim. Acta. **427**, 140864 (2022)
117. Chee, W.K., et al.: Flexible graphene-based supercapacitors: a review. J. Phys. Chem. C. **120**, 4153–4172 (2016)
118. Wang, H., et al.: A mechanically and electrically self-healing supercapacitor. Adv. Mater. **26**, 3638–3643 (2014)
119. Wang, Y., et al.: Conductive polymers for stretchable supercapacitors. Nano Res. **12**, 1978–1987 (2019)
120. Hong, S.H., et al.: Polypyrrole nanofoam/carbon nanotube multilayered electrode for flexible electrochemical capacitors. ACS Appl. Energy Mater. **5**, 4059–4069 (2022)
121. Zhao, Y., et al.: Self-healing and shape-editable wearable supercapacitors based on highly stretchable hydrogel electrolytes. Adv. Sci. **9**, 2201039 (2022)
122. Sardana, S., et al.: Conducting polymer hydrogel based electrode materials for supercapacitor applications. J. Energy Storage. **45**, 103510 (2022)
123. Zhao, Z., et al.: Ultrathin tape-supercapacitor with high capacitance and durable flexibility via repeated in-situ polymerizations of polyaniline. Chem. Eng. J. **471**, 144721 (2023)
124. Yao, D., et al.: A strategy for preparing high-efficiency and economical catalytic electrodes toward overall water splitting. Nanoscale. **13**, 10624–10648 (2021)
125. Schlapbach, L., et al.: Hydrogen-storage materials for mobile applications. Nature. **414**, 353–358 (2001)
126. Target Explanation Document: Onboard Hydrogen Storage for Light-Duty Fuel Cell Vehicles. www.uscar.org. (2017)
127. Usman, M.R.: Hydrogen storage methods: review and current status. Renew Sustain. Energ. Rev. **167**, 112743 (2022)
128. Jena, P.: Materials for hydrogen storage: past, present, and future. J. Phys. Chem. Lett. **2**, 206–211 (2011)
129. Rochat, S., et al.: Hydrogen storage in polymer-based processable microporous composites. J. Mater. Chem. A. **5**, 18752–18761 (2017)
130. Tian, M., et al.: Nanoporous polymer-based composites for enhanced hydrogen storage. Adsorption. **25**, 889–901 (2019)
131. Schlichtenmayer, M., et al.: Nanosponges for hydrogen storage. J. Mater. Chem. **22**, 10134–10143 (2012)
132. Suresh, K., et al.: Optimizing hydrogen storage in MOFs through engineering of crystal morphology and control of crystal size. J. Am. Chem. Soc. **143**, 10727–10734 (2021)
133. Varghese, A.M., et al.: An in-situ-grown Cu-BTC metal–organic framework/graphene oxide hybrid adsorbent for selective hydrogen storage at ambient temperature. Ind. Eng. Chem. Res. **61**, 6200–6213 (2022)
134. Purewal, J., et al.: Estimation of system-level hydrogen storage for metal-organic frameworks with high volumetric storage density. Int. J. Hydrog. Energy. **44**, 15135–15145 (2019)
135. Zhao, D., et al.: Porous metal-organic frameworks for hydrogen storage. Chem. Commun. **58**, 11059–11078 (2022)
136. Ghosh, S., et al.: Hydrogen adsorption in pyridine bridged porphyrin-covalent organic framework. Int. J. Hydrog. Energy. **44**, 1782–1796 (2019)

137. Braunecker, W.A., et al.: Thermal activation of a copper-loaded covalent organic framework for near-ambient temperature hydrogen storage and delivery. ACS Mater. Lett. **2**, 227–232 (2020)
138. Tong, M., et al.: An easy way to identify high performing covalent organic frameworks for hydrogen storage. Chem. Commun. **56**, 6376–6379 (2020)
139. Sethia, G., et al.: Activated carbon with optimum pore size distribution for hydrogen storage. Carbon. **99**, 289–294 (2016)
140. Sevilla, M., et al.: Energy storage applications of activated carbons: supercapacitors and hydrogen storage. Energy Environ. Sci. **7**, 1250–1280 (2014)
141. Doğan, M., et al.: Activated carbon synthesis from tangerine peel and its use in hydrogen storage. J. Energy Inst. **93**, 2176–2185 (2020)
142. Langmi, H.W., et al.: Hydrogen storage in ion-exchanged zeolites. J. Alloys Compd. **404**, 637–642 (2005)
143. Dong, J., et al.: Hydrogen storage in several microporous zeolites. Int. J. Hydrog. Energy. **32**, 4998–5004 (2007)
144. Lan, J., et al.: High-capacity hydrogen storage in porous aromatic frameworks with diamond-like structure. J. Phys. Chem. Lett. **1**, 978–981 (2010)
145. Wu, X., et al.: Ultrahigh hydrogen storage capacity of novel porous aromatic frameworks. J. Mater. Chem. A. **3**, 10724–10729 (2015)
146. McKeown, N.B., et al.: Towards polymer-based hydrogen storage materials: engineering ultramicroporous cavities within polymers of intrinsic microporosity. Angew. Chem. Int. Ed. Engl. **118**, 1836–1839 (2006)
147. McKeown, N.B., et al.: Microporous polymers as potential hydrogen storage materials. Macromol. Rapid Commun. **28**, 995–1002 (2007)
148. Germain, J., et al.: High surface area nanoporous polymers for reversible hydrogen storage. Chem. Mater. **18**, 4430–4435 (2006)

Polymer-Dispersed Liquid Crystal Films for Smart Windows to Gas Sensors: Recent Advances and Development

Manpreet Kaur, Ashwani Kumar Singh, Harsh Sharma, Poonma Malik, Khushboo, and Praveen Malik

Abstract Polymer-dispersed liquid crystal (PDLC) films have gained significant attention from both researchers and industry professionals due to their wide-ranging applications. Recent progress in PDLC technology has led to notable enhancements in performance, the exploration of innovative operating modes, and the expansion of application areas. This chapter begins with an introduction to the basics, historical evolution, and classification of liquid crystals (LCs), followed by an in-depth examination of PDLCs, covering their fundamental working principles, preparation techniques, and droplet configurations. Following this, the chapter systematically reviews the latest development in PDLC applications, including privacy windows and stimuli-responsive sensors such as gas and biosensors. By providing a thorough overview of the recent developments, the chapter highlights new fabrication methods, enhancements in physical properties, and the diverse applications of PDLCs, particularly in sensing and smart window technologies. It also discusses the potential for future advancements and the global commercialization of PDLC based technology.

Keywords PDLC · Droplets · Phase separation · Smart windows · Gas sensor · Biosensor

---

M. Kaur · H. Sharma · P. Malik (✉)
Liquid Crystals Lab, Department of Physics, Dr B R Ambedkar National Institute of Technology, Jalandhar, Punjab, India
e-mail: malikp@nitj.ac.in

A. K. Singh
School of Basic Sciences and Technology, Intellectual Institute of Management & Technology, Meerut, India

P. Malik
Advance Research Material and Solutions, Technology Business Incubator, Indian Institute of Science Education and Research Mohali, Mohali, Punjab, India

Khushboo
A-702, Dr B R Ambedkar National Institute of Technology, Jalandhar, Punjab, India

V. Kumar, Y. K. Mishra (eds.), *Soft Materials for Functional Applications*, Advanced Structured Materials 225, https://doi.org/10.1007/978-981-97-9468-3_7

1 Introduction

Liquid crystal (LC) is a mesophase (intermediate) between anisotropic crystal and isotropic liquid, as shown in schematic Fig. 1. LCs are smart and functional materials that combine the properties of both liquids (fluidics) and crystals (ordering) [1, 2]. The most exciting point is that this class of materials possesses self-assembling properties which can be used for advanced applications in optics, optoelectronics, and photonics [2].

1.1 Historical Background

In 1888, Austrian botanist Friedrich Reinitzer discovered LCs and observed that cholesteryl benzoate had two distinct melting points [3]. At 145.5 °C, the substance turned into a milky liquid, and at 178.5 °C, it became a clear isotropic phase. Later, German physicist Otto Lehmann used a polarizing microscope to study these materials, identifying various optical textures and coining the term "liquid crystals." Unlike regular crystals, LCs exhibit long-range orientational order and short-range positional order above their melting point [4, 5]. The pioneering work of Reinitzer and Lehmann laid the foundation for extensive research and a wide range of applications for LCs.

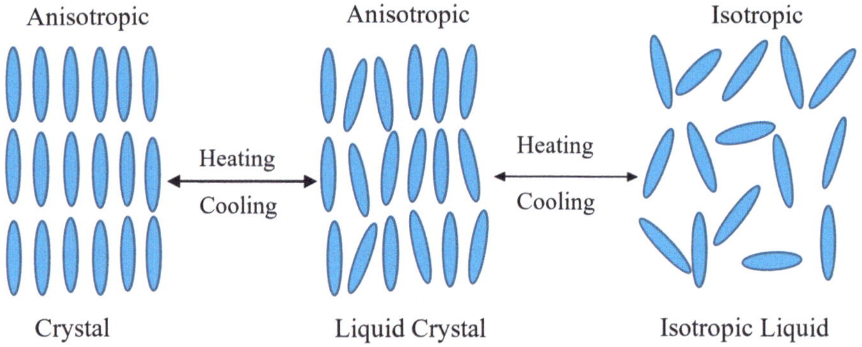

Fig. 1 Phase transition sequence with increasing and decreasing temperature

1.2 Classification of LCs

LCs are categorized into three types based on their chemical composition:

- Thermotropic LCs
- Lyotropic LCs
- Polymeric LCs

Thermotropic LCs

Thermotropic LCs are the most prevalent type within the LC family. These materials exhibit characteristic behaviors when heated or cooled, making them ideal for various applications [6]. Depending on the shape anisotropy of their molecules, thermotropic LCs are further divided into three major categories:

- Calamitic LCs (rodlike)
- Discotic LCs (disclike)
- Bent-core LCs (banana-like)

Calamitic LCs

In calamitic LCs, one molecular axis is rodlike and significantly longer than the other two. These molecules maintain an elongated shape and consist of four parts—the core unit, linking groups, terminal chains, and lateral substituents—as illustrated in Fig. 2a [7–9]. Each component is crucial in defining the LC subphases and their characteristics.

The core units, typically aromatic rings (such as 1,4-phenyl, biphenyl, terphenyl, or 2,6-naphthyl) or alicyclic structures (like *trans*-1,4-cyclohexyl, *trans*-2,5-dioxanyl), provide rigidity to the molecular structure and influence electrical and elastic properties [10, 11]. These core units are directly connected or linked by groups that maintain polarizability, anisotropy, and linearity [11].

Linkage groups (including esters, ethylene, methylenedioxy, azo, etc.) stabilize the LCs against chemicals, UV radiation, and moisture. Terminal groups (alkyl or alkoxy chains) are small polar or nonpolar substituents that affect the dielectric constant and anisotropy of LCs [11]. These chains, often branched, determine the elastic constant and transition temperature, providing flexibility essential for molecular alignment and lowering the melting point. The length of the terminal chain significantly influences the presence of smectic and nematic phases.

Various lateral substituents (F, Cl, CN, and NO_2) modify the LC's physical properties, including birefringence, dielectric anisotropy, and transition temperature. The size, polarity, and position of these substituents play a crucial role, with fluoro substituents being particularly important due to their small size and high electronegativity.

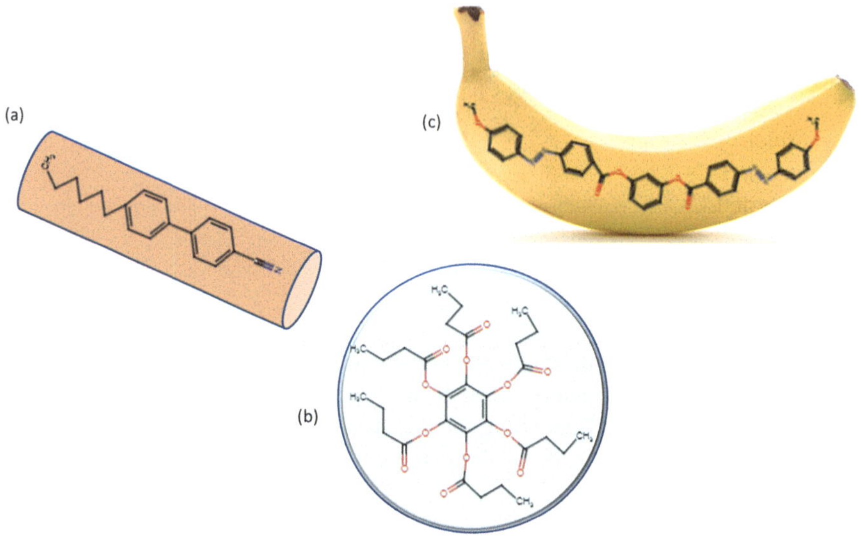

Fig. 2 Schematic picture of (**a**) calamitic LCs (rod-shaped), (**b**) discotic LCs (disc-shaped), and (**c**) bent-core LCs (banana-shaped)

Discotic LCs

Discotic LCs (DLCs) are composed of advanced disc-shaped molecules where one axis is shorter than the other two, as depicted in Fig. 2b. Chandrasekhar et al. [12] discovered the mesomorphic properties of DLCs. The two most well-known discotic mesophases are columnar and nematic. In the nematic discotic mesophase, the molecules align with their short axes parallel to each other. In the columnar phase, the molecules stack into columns. DLCs can form rectangular, hexagonal, tetragonal, and oblique phases, depending on their symmetry and packing. Their potential applications in optoelectronics and other fields have made DLCs a focal point of interest.

Bend-Core LCs

Bent-core LCs are a unique class of LCs characterized by self-organized superstructures of biaxial molecules with a bent, banana-like shape, hence the name "banana-shaped LCs" [12]. These molecules consist of two rodlike mesogenic units (arms) connected at a bending angle to a central unit, as shown in Fig. 2c. The presence of the liquid crystalline phase in bent-core materials depends on two main factors: sufficient charge distribution along the arms and an appropriate bending angle. The stability of the LC phase decreases as the bending angle increases, and the connecting groups also influence this angle. Various mesophases, designated as B1 through B8, have been identified in bent-core compounds based on optical textures and changes in X-ray diffraction patterns [13, 14].

Bent-core LCs have attracted significant research interest due to their unique properties. Although the molecules are not chiral, their shape induces chirality in smectic layers. Bent-core LCs exhibit polarity, spontaneous polarization, and fast response switching and stabilize the blue phase [14].

Lyotropic LCs

Unlike thermotropic LCs, which form at specific temperatures, lyotropic LCs emerge when a particular concentration of amphiphilic molecules is present in suitable solvents [13]. Lowering the temperature or increasing the concentration of these molecules leads to forming larger structures known as micelles. These micelles form when the concentration surpasses a critical threshold, typically between 0.01 and 0.1% by weight. Amphiphilic molecules, which have both polar and nonpolar regions, assemble into micelles [14]. Like thermotropic LCs, lyotropic LCs exhibit various phases, such as lamellar, micellar, micellar cubic, hexagonal, and bicontinuous cubic phases, depending on the degree of molecular ordering and arrangement within the solvent matrix.

The solvent concentration is crucial in determining the occurrence and type of lyotropic LC phase. Additionally, temperature can alter the lyotropic LC phase at a fixed concentration. Dierking and Al-Zangana [15] reported that mixing anisotropic nanomaterials into isotropic media can induce lyotropic LC phases. Although lyotropic LCs are less well-known than thermotropic LCs, they are commonly encountered in everyday life, such as in soaps and detergents in water, biological membranes, and medical treatments.

Polymeric LCs

To date, dozens of LCs have been thoroughly investigated for various applications. However, achieving optimal properties for specific applications with pure LCs alone is nearly impossible [15]. Using nano-objects as dopants in small concentrations effectively tailors LC properties to meet desired specifications [15–19]. Nanoparticles (NPs) primarily influence the physical properties of LCs due to their unique characteristics. Over the past decade, the design and study of advanced functional materials, particularly LC nanocomposites and LC/polymer composites, have become prominent research areas [20–25].

Since their discovery, LC/polymer composites have been recognized as electrically controllable materials [26]. These composites are homogeneous blends of low-molecular-weight LCs and high molecular weight polymers, with the polymer matrix providing mechanical and structural stability to the devices [27, 28]. Based on polymer content, LC/polymer composites are categorized into two main classes:

polymer-stabilized liquid crystals (PSLCs) and polymer-dispersed liquid crystals (PDLCs) [29].

PSLCs have less than 10% polymer content, featuring a continuous LC phase with a small amount of cross-linked polymer within the anisotropic fluid [30]. Despite the successful development of PSLCs into light shutters, phase modulators, infrared reflectors, and various optical devices, these materials face challenges related to the poor film-forming properties and limited durability of the polymer matrix due to the low polymer content (typically below 10 wt%) [31]. In contrast, PDLCs contain a higher polymer content, ranging from 20 to 80% [32].

PDLCs have become versatile and dynamic materials with a broad range of applications, including smart windows and sensors. These composites successfully merge the optical properties of LCs with the mechanical stability and process ability of polymers, leading to the creation of PDLCs [33, 34]. The pioneering PDLC film was created using an epoxy resin and LCs through a phase separation technique, marking a significant advancement in the field [35]. Figure 3 schematically represents the introduction, working principle, droplet configuration, preparation methods, doping, and application of PDLCs.

1.3 Working Principle of PDLC

In PDLCs, LCs in the form of small-size droplets are distributed within the polymer matrix, imparting greater durability than PSLCs due to the higher polymer content [36, 37]. PDLC requires special attention due to their extraordinary properties such as light-controlling effects under an applied electric field. The PDLC film shows an opaque state when no electric field is applied. This opaque state is due to mismatching of the refractive index between the polymer matrix and the LC [36]. In this state, the director of the LC droplets is randomly oriented, causing high light scattering due to refractive index mismatch, resulting in an opaque state [38, 39]. However, in applying an external electric field, the refractive index of the polymer aligns with the LC, leading to a transparent state as the LC droplets orient along the electric field [40–42]. The working of PDLC is illustrated in Fig. 4.

1.4 Preparation Methods

PDLCs are prepared using encapsulation and phase separation methods. In the encapsulation, numerous small and nonuniformly sized LC capsules are dispersed within the bulk polymer, while in the phase separation method, LC droplets are formed through polymerization induced by heat, light, or solvent [43].

Fig. 3 Schematic representation of introduction, working principle, droplet configuration, preparation methods, doping, and advantages of PDLC

Encapsulation

Initially, PDLC composites were created using a microencapsulation technique. In this method, an LC material is combined with a water-soluble polymer. Both the LC and polymer are dissolved in water. As the water evaporates, the LC droplets become enclosed by a layer of polymer, resulting in numerous tiny LC capsules distributed throughout the polymer matrix. A significant drawback of this method is the nonuniformity, small size, and interconnectedness of the LC droplets produced. These composites are nematic curvilinear aligned phases (NCAP) [26, 28].

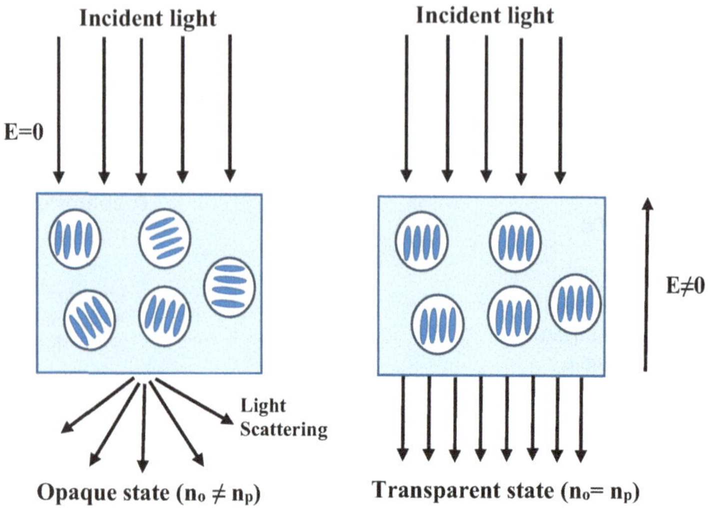

Fig. 4 Working principle of PDLC; in off-state, PDLC film appears opaque and in on-state, it gets transparent

Phase Separation

Various methods are employed to separate materials to produce PDLC composites, including temperature changes, solvent use, and monomer polymerization. The process begins by mixing the LC with a prepolymer, followed by techniques to separate the mixture into droplets before it solidifies [26, 44]. The phase separation method includes polymerization-induced phase separation (PIPS), thermal-induced phase separation (TIPS), and solvent-induced phase separation (SIPS) [45–47].

Thermal-Induced Phase Separation (TIPS)

In the TIPS process, a LC material is mixed with a thermoplastic polymer at elevated temperatures. As the mixture cools slowly, it separates into two distinct phases, with the polymer solidifying and creating LC domains. The cooling rate affects the size of the LC droplets, with faster cooling resulting in smaller droplets. TIPS is typically used when thermoplastics melt below their decomposition temperature. Examples of polymer materials that can create droplets with LC materials include polymethylmethacrylate (PMMA) and polyvinyl formal (PVF) [48, 49].

Solvent-Induced Phase Separation (SIPS)

In the SIPS technique, an LC material and a polymer are combined in the presence of a solvent. As the solvent evaporates, the LC material separates into droplets, forming a PDLC composite. This method is employed when thermoplastics melt above the decomposition temperature of the LC or when coating techniques are used [50]. The LC and polymer are first dissolved in a common solvent, and as the solvent evaporates, the two components separate, resulting in the polymer solidifying into a PDLC composite [51].

Polymerization-Induced Phase Separation (PIPS)

PIPS is a commonly used phase separation method, especially when a monomer is miscible with low-molecular-weight LCs. In this process, the monomer and LC are mixed together and then subjected to UV curing [52–54]. The polymerization that follows induces phase separation, resulting in the formation of a PDLC film. Among these techniques, PIPS is usually preferred for PDLC preparation due to its solvent-free nature and lower cost compared to TIPS and SIPS; the LC/polymer mixture is cured under UV light for polymerization [28, 37].

1.5 Droplet Configurations in PDLC

Morphological studies of PDLCs have revealed various configurations of LC droplets based on the director profile. Three main configurations of PDLC droplets have been identified, radial, axial, and bipolar, shown in Fig. 5 [33, 55]. LC molecules are directed along the indium tin oxide (ITO) glass substrate for planar anchoring in nematic and form a bipolar droplet configuration, as depicted in Fig. 5(a). These types of configurations are characterized by two-point defects (boojums) diametrically opposite to each other on the droplet surface [56–58]. Conversely, for homeotropic anchoring in nematics (LC molecules are oriented perpendicular to the ITO surface), a radial configuration with a single-point defect (hedgehog) at the center of the droplets is seen, Fig. 5(b). When one polarizer aligns with the optic axis, radial droplets exhibit a fanlike pattern under crossed polarizers, known as a maltese pattern, with four crossed brushes, shown in Fig. 5(c) [59].

1.6 Applications of PDLC

In recent years, PDLCs have emerged as a fascinating field within LCs due to their exceptional properties and potential applications across diverse areas. One of the most exciting applications of PDLCs is in smart windows, where the electrically

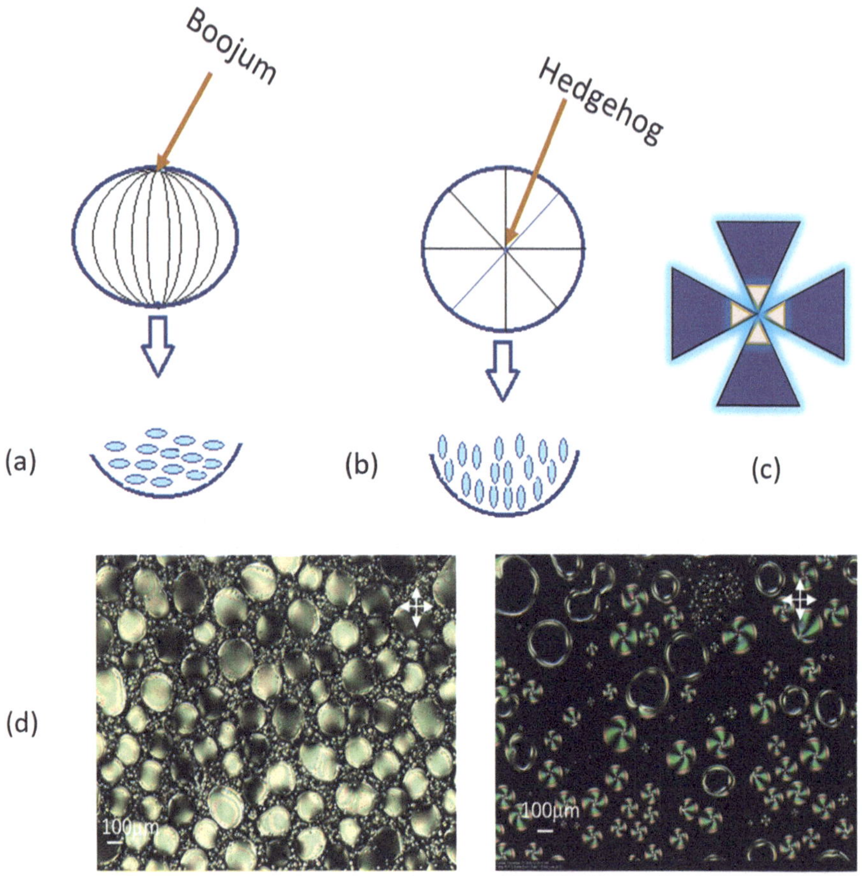

Fig. 5 (**a**) Bipolar configuration for planar anchoring in nematic LCs. (**b**) Redial configuration for homeotropic anchoring in nematic LCs. (**c**) Maltese-type pattern. (**d**) POM textures of typical PDLCs at 10 × magnification and crossed polarizers

regulated orientation of PDLC molecules allows for adjustable transparency [60]. Additionally, PDLCs are used in light shutters, gas sensors, wearable devices, biosensors, diffraction gratings, and modulators [61–66]. In this section, we will discuss some of the most promising applications of PDLCs. The applications of PDLCs are schematically shown in Fig. 6.

PDLCs as a Gas Sensor

Detecting toxic gases and volatile organic compounds (VOCs) is an emerging area of research with broad applications in military, household, medical, and various industrial sectors [60]. Gas sensors play a pivotal role across industries,

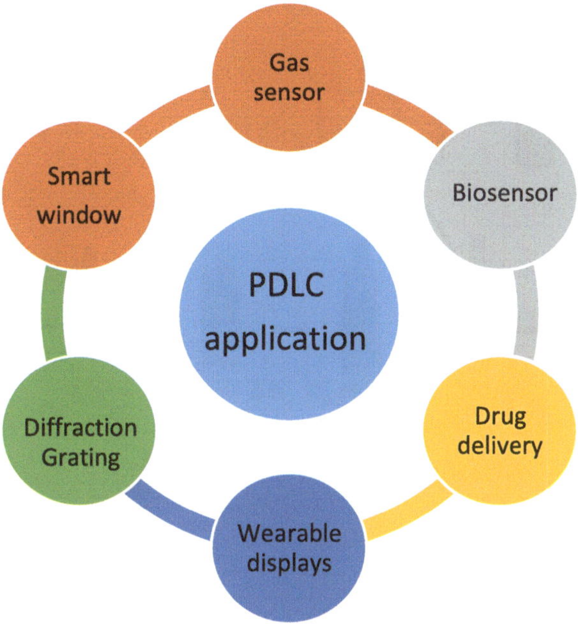

Fig. 6 Applications of PDLCs

from industrial production (such as methane detection in mines) and automotive emissions monitoring to medical diagnostics, air quality assessment (including carbon monoxide detection), and environmental surveillance [60–64]. Detection of chemicals and biological analytes using highly sensitive devices is crucial for various applications. Recent studies show that nematic LC systems can be employed as reliable, low-cost, portable, and highly sensitive sensors for the detection of chemicals and biological analytes due to their unique features [60]. Out of various gas sensors, PDLC-based gas sensors offer numerous advantages, including low power consumption, rapid response times, high sensitivity even at low analyte concentrations, reproducibility, and eco-friendliness [60, 61]. Conventional LC-based gas sensors have technical issues like the flow of LC and complex fabrication processes. However, PDLC gas sensors do not require expensive or complex instrumentation and operate effectively under ambient conditions.

In the future, these sensors could be developed as wearable gas sensors made of PDLC, as they do not require a set of polarizers or electric fields. Various research groups have observed reversible reorientation of LC molecules in the polymer matrix upon exposure to and removal of different gases or VOCs, resulting in changes in the order parameters and electrical and optical properties of the PDLC material [60, 65]. Incorporating functionalized nanoparticles or patterning PDLC films has enabled researchers to detect different gases remarkably [66–68]. For example, Pagidi et al. [66] doped functionalized carbon nanotubes (fCNTs) into

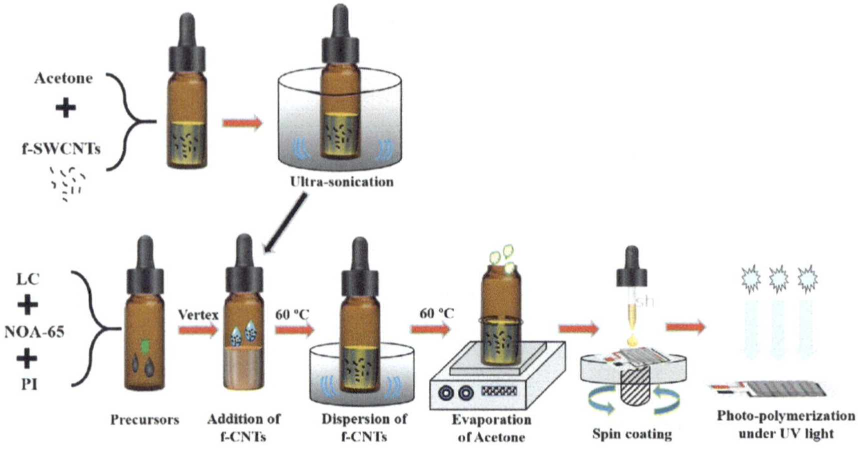

Fig. 7 Schematic picture for fabrication method of fCNT-PDLC-based gas sensor. (Reprinted with permission from [66])

PDLC to enhance the sensitivity of NO_2 gas. Figure 7 shows a schematic picture of the fabrication process of the fCNT-PDLC device. Initially, fCNTs were suspended in the PDLC mixture. After solvent evaporation, the prepared mixture was spin-coated over interdigitated electrodes (IDE), followed by a curing process for phase separation between the LC and polymer matrix. Their work demonstrates that the resistance of fCNT-PDLC devices increased linearly upon exposure to NO_2 gas, showing good sensitivity, reversibility, and reproducibility for NO_2 gas compared to NH_3, CO, H_2, and H_2S gases. This enhancement in sensing performance was attributed to the reorientation of LC molecules within the PDLC matrix, as depicted in the transient responses shown in Fig. 8a–d.

In another study, Zhang et al. [67] investigated the fast and reversible detection of toluene vapor using a PDLC thin film sensor prepared by the photo-induced phase separation of 5CB nematic LC and silicone adhesive (Platsil gel) on a glass substrate with decal transfer lithography [67]. The LC phase transitioned from nematic to isotropic is easily observable through optical appearance under a polarized optical microscope (POM). This change is likely due to the reorientation of LC droplets caused by exposure to toluene vapor. The relative intensity ($RI\% = [(I_0 - I_t)/I_0] \times 100$), where I_0 and I_t are the intensities of the LC droplets after exposure to toluene vapor at 0 and t seconds, respectively) decreased with increasing the toluene concentration and exposure time, indicating high sensitivity of the PDLC to toluene vapor. Moreover, Huang et al. [68] fabricated a pillar-patterned PDLC sensor using photolithography. The PDLC system, composed of trifunctional monomers and LCs, was designed to detect solvents such as ethanol and acetone. The authors observed a change in the orientation of LC droplets from an ordered (bright) to a disordered state (dark) in response to vapor, which caused the scattering of incident light. They concluded that the PDLC responded better to acetone vapor

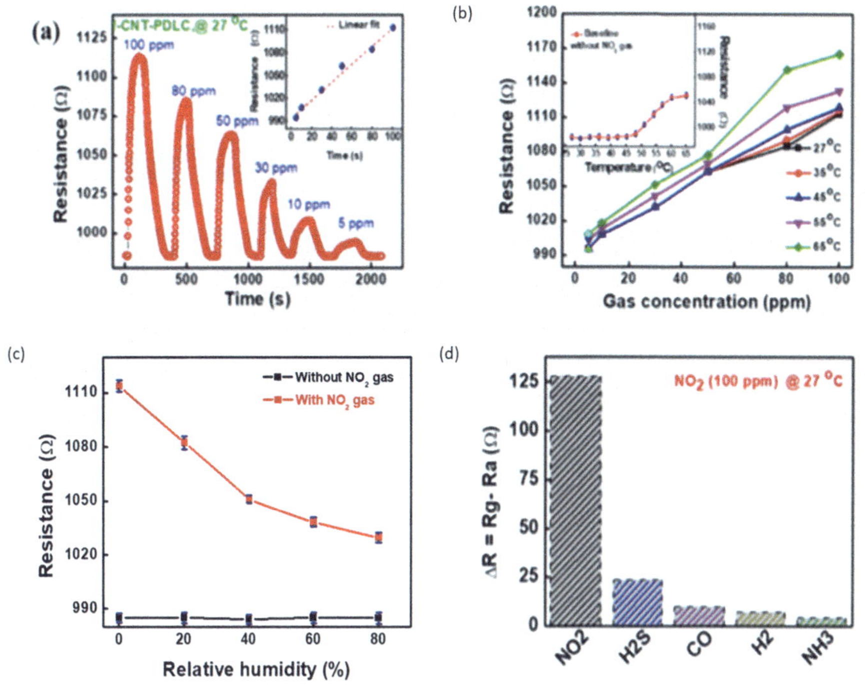

Fig. 8 (**a**) Transient responses of fCNT-PDLC sensor at 27 °C with different concentrations (5–100 ppm) of NO_2 gas. (**b**) Temperature-dependent sensing response of the device concerning gas concentration. (**c**) Resistance response of the device for relative humidity ranging from 0 to 80% with and without the exposure of NO_2 gas (100 ppm) at 27 °C. (**d**) Selectivity test for several gases NO_2, H_2S, CO, H_2, and NH_3 of 100 ppm at 27 °C. (Reprinted with permission from [66])

than ethanol, and the patterned film enhanced the sensor's response compared to a normal PDLC film [68]. In 2020, Liu et al. [69] observed that the concentration of zinc oxide (ZnO) nanotubes increased the droplet size of PDLC to the microscale. The presence of ZnO nanotubes in the PDLC system shifted the absorption peak toward higher energy and decreased absorbance, increasing permittivity, dielectric loss, and dielectric strength of the PDLC. When ethanol vapor was introduced into the ZnO-doped PDLC sample, the dielectric constant and dielectric loss increased while the impedance decreased. Wang et al. [70] prepared electrospun fiber mats of polylactic acid (PLA) and 5CB for the detection of toluene and acetone vapors. The peeling of the fiber mat and macroscopic appearance are shown in Fig. 9a, b, respectively. The vapors altered the morphology of the fiber mats due to absorption by the LC core while passing through the polymer sheath of the fibers, as can be depicted in Fig. 9c.

Similarly, Song et al. [71] demonstrated the innovation of ultra-flexible PDLC-based OFETs (organic field-effect transistors) incorporating poly(methyl methacrylate) (PMMA)-dispersed 5CB. This multifunctional sensor is capable of detecting

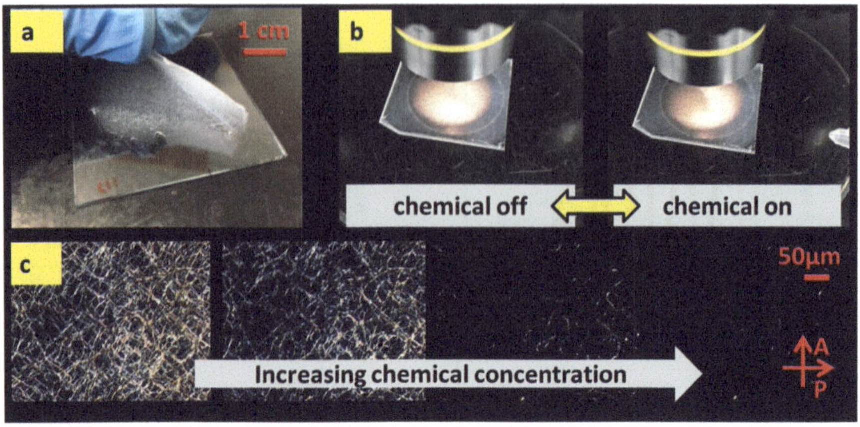

Fig. 9 (**a**) Illustration of the peeling of the fiber mat. (**b**) Macroscopic appearance of the fiber mat without and with toluene vapor. (**c**) Polarizing microscopic textures of a LC/polymer fiber mat responding to increasing concentration of toluene vapor. (Reprinted with permission from [70])

four distinct stimuli—weak air (gas) flow, strong physical force (touch), light, and heat—allowing for the simultaneous sensing of two different stimulations. The sensor's response was quantified by observing changes in drain current corresponding to gas flow, touch load, and temperature variations. Notably, the device exhibited the ability to detect extremely weak N_2 gas flow (0.3 sccm) and reliably responded to direct physical touches (0.6 g). The results indicate that these stimulations change the alignment of 5CB molecules.

Vein, Bag et al. [72] developed four chemi-resistive sensors for the detection of methanol (MeOH) vapor in ambient environments by introducing conductive nanofillers like carbon black, multi-walled carbon nanotubes, and reduced graphene oxide into a sulfonated poly(etherimide) (PEI)/liquid crystal polymer (LCP) composite (sPEI-LCP) with high thermal stability over a wide temperature range (25–507 °C). The DC electrical resistance of the films changed upon adsorption of MeOH vapor, and the sensor response increased with vapor concentration (300–1200 ppm). Reyes et al. [73] also fabricated mat fibers surrounded by a polyvinylpyrrolidone sheath to detect toluene gas. In the presence of toluene vapor, the optical properties of the LC fibers changed due to the reorientation of LC droplets and changes in anchoring energy at the PVP-5CB interface. Furthermore, Lai et al. [74] developed a reproducible gas sensor based on carbon nanotube (CNT)-doped PDLC (NLC E7 + U401 polymer) dispensed on IDE patterned over a silicon substrate, as shown in Fig. 10a–d. A linear sensor response was observed by measuring changes in resistance after exposing the sensor to acetone gas concentrations ranging from 100 to 1500 ppm, with an increase in response from 0.65 to 20.15% and a quick response time (>32 s). The increased sensor response time with higher LC content in the PDLC and higher sensitivity was observed for

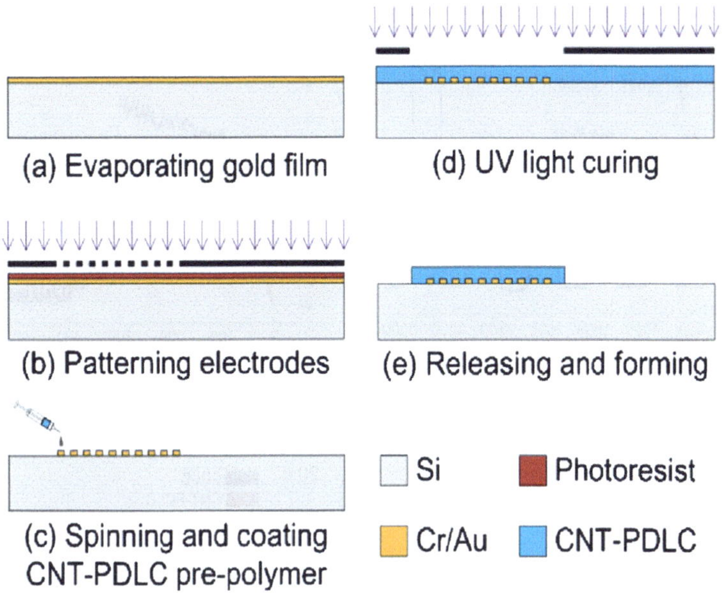

Fig. 10 (**a**–**e**) Fabrication process of CNT-/PDLC-based chemi-resistive sensor. (Reprinted with permission from [74])

acetone compared to ethanol, toluene, and hexane. CNT-PDLC also showed higher sensitivity than undoped PDLC, as confirmed by the results depicted in Fig. 11a–d.

Kirchner et al. [75] observed that LC devices containing cholesteryl nonanoate, cholesteryl oleyl carbonate, cholesteryl chloride, and polymers (PMMA, EG80A) can detect toxic gaseous analytes such as aliphatic amines (1-butylamine, diethylamine, and trimethylamine). These devices demonstrated significantly improved shelf life and exhibited a colorimetric effect upon exposure to vapors. The reflectance of the LC films increased with the concentration of vapors. The authors also observed that PMMA produced better results than the hydrophilic EG80A and that primary amines (such as 1-butylamine) interact more readily with the trifluoroacetyl group, forming a hemiaminal more easily compared to diethylamine [75].

Parmar et al. [76] observed the reorientation of the director upon applying shear stress to the top surface of a PDLC thin film. The microdroplets of partially exposed PDLC became brighter with the flow of gas. The authors also noted an increase in transmitted light intensity.

Nanomaterial-PDLC Composites: Enhancing Smart Window Performance

In recent years, nanomaterial-PDLC composites have also emerged as a fascinating area within LCs due to their exceptional properties and technological applications.

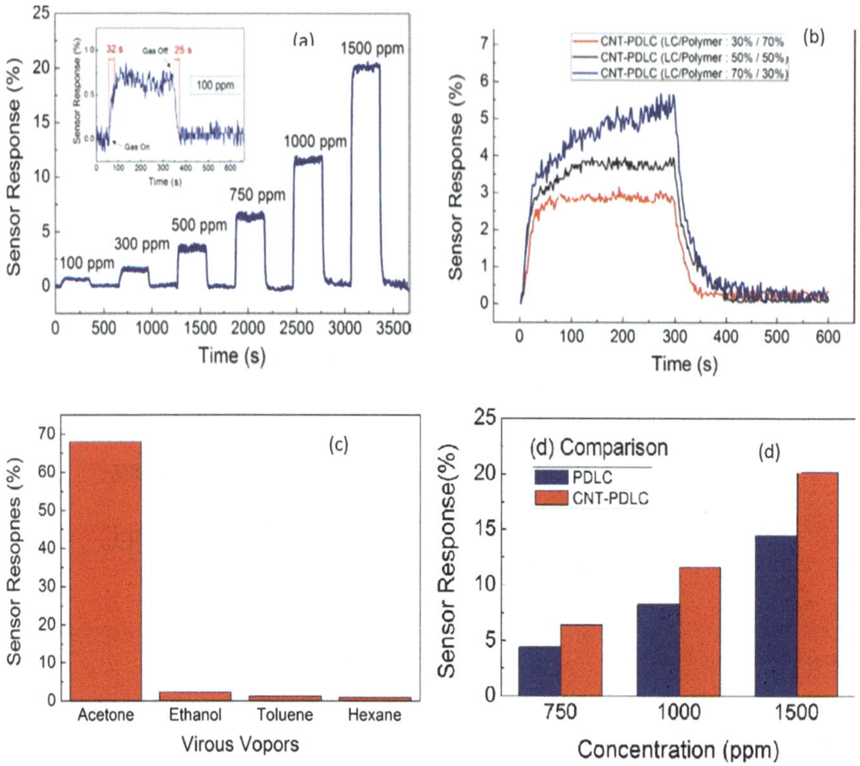

Fig. 11 (**a**) Sensor response of the CNT-PDLC sensor for varying concentrations of acetone (inset shows the response at 100 ppm). (**b**) Sensor response dependence on the composition of LC and polymer. (**c**) Response of the CNT-PDLC sensors on the exposure of acetone vapor and other common gases at 5000 ppm concentrations. (**d**) Comparative response of the pure PDLC and CNT-PDLC sensors for varying acetone concentrations. (Reprinted with permission from [75])

PDLCs are being used in both indoor and outdoor environments to manage privacy (such as in conference rooms, intensive care units, bathrooms, shower enclosures, and fitting rooms) or serve as temporary surfaces for projections [77–79]. The selection of nanomaterials and their size, shape, structure, and physical properties play a pivotal role in tailoring the performance of PDLCs for advanced applications, including smart windows [78]. Traditional PDLCs typically involve mixing of LC, monomer, and initiator as main components. The dispersion of nanoparticles (NPs) into PDLC systems has successfully enhanced the optical, thermal, and mechanical stability of the polymer matrix.

Other inorganic NPs such as ZnO, MgO, CuO, BTO, $BaTiO_3$, Fe_3O_4, and TiO_2 were doped into PDLCs, and a decrease in driving voltage was observed [14, 18–20]. NPs primarily influence the dielectric constant, refractive index, and anchoring forces at the interface between the polymer and LC [77, 78]. In this series, incorporation of anisotropic alumina nanowires (ANWs) into PDLC composites

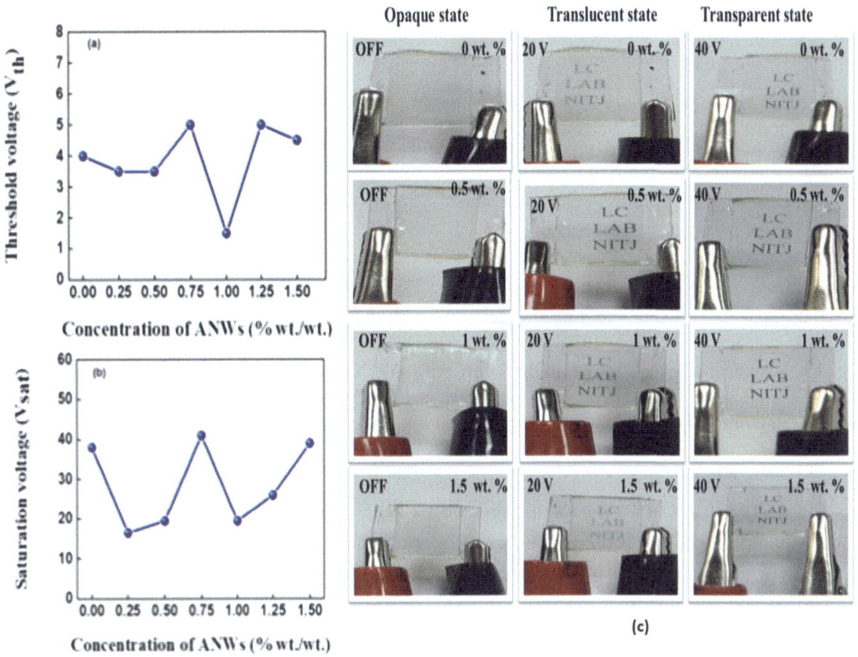

Fig. 12 Concentration of ANW dependency on (**a**) V_{th} and (**b**) V_{Sat}. (**c**) Operation mechanism of pure PDLC and ANW/PDLC composites with doping concentration of 0.5, 1.0, and 1.5 wt% at varying electric fields of 0, 20, and 40 V. (Reprinted with permission from [79])

holds promise for the development of smart windows with enhanced optical, electro-optical, and dielectric properties. In 2023, our group demonstrated a threshold voltage (V_{th}) of 1.5 V and a saturation voltage (V_{sat}) of 16.5 V for 1.0 and 0.25 wt% of ANWs in PDLC, as depicted in Fig. 12a, b [79]. In the absence of an applied electric field, incident light is scattered, and the PDLC film appears opaque. However, in the application of controlled electric field ($E > V_{th}$), the LC's director reorients, and as a result of matching of refractive index of LC with polymer, PDLC film appears transparent. The transformation of PDLC at different applied voltage is clearly shown in Fig. 12c [79].

In another study, Park et al. [80] fabricated PDLC-based stretchable smart windows containing mixed ink of Ag NW/PEDOT/PSS coated over hybrid transparent and stretchable electrodes (TSEs). They conducted bending and stretching assessments to validate the remarkable mechanical characteristics of the Ag NW and PEDOT/PSS hybrid TSE. The stretchable PDLC-based smart window displayed an on-state transmittance of 56% at 80 V and an off-state transmittance of 2% at 0 V, as illustrated in Fig. 13. In contrast to ITO-based PDLC smart windows, which are prone to breakage under stretching, the Ag NW/PEDOT/PSS hybrid electrode-based PDLC smart window endured stretching up to 30%.

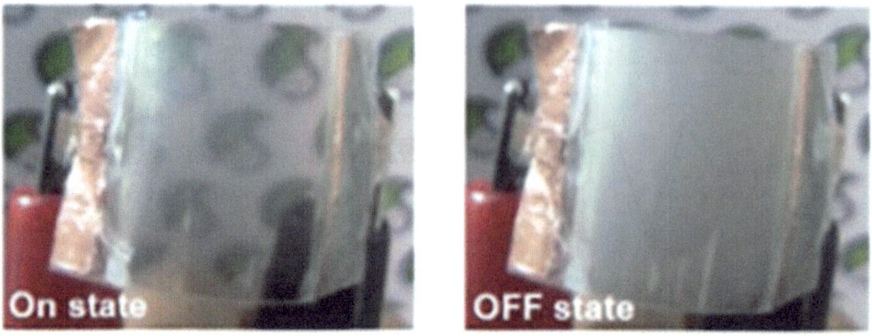

Fig. 13 Visual characteristics of PDLC film with Ag NW/PEDOT/PSS hybrid TSEs. A transparent on-state and off-state opaque picture of a PDLC-based smart window. (Reprinted with permission from [80])

In our recent work, we proposed GO-PDLC composites, which enhance our understanding of how GO influences the properties of PDLC, opening up possibilities for their application in energy-efficient smart windows with low voltage operation and increased transparency [81]. Notably, with just 0.005 wt% GO flakes, the saturation voltage (V_{sat}) dropped by 90.8%, from 125 V in PDLC to 11.5 V in GO-PDLC composites, as depicted from Fig. 14b. The GO flakes may align along the electric field direction in the on-state, aiding the alignment of LC molecules along the field, as shown in Fig. 14a. The improved dielectric properties observed with GO dispersion are likely due to strong π–π stacking interactions between GO and LCs and the higher polarizability of GO, attributed to its oxygen-containing functional groups [81]. Nasir et al. [82] demonstrated the light scattering behavior of PDLC film under different voltages at a frequency of 60 Hz. The homogenously filled PDLC film was opaque in the absence of an electric field due to light scattering, as depicted in Fig. 15a. However, upon application of an electric field, PDLC film became transparent, as shown in Fig. 15b. The authors also observed ~70% change in transmittance. Voltage versus transmission curves for the glass-based PDLC (g-PDLC) were measured in comparison with the PET-based PDLC (p-PDLC) and show the typical behavior of general PDLCs.

Figure 16 illustrates that the transparency of PDLC film can be controlled by adjusting the applied voltage. Lower voltages lead to partial alignment of LC molecules in the electric field, causing light scattering and limited transparency. As the voltage increases, the number of unaligned LC molecules decreases, resulting in reduced light scattering and increased transparency. Despite the capability for variable transparency, most PDLC devices in the market, offer only binary on-/off-states. Commercially, these devices are available as adhesive smart film rolls, allowing for easy customization to fit existing windows [83].

A recent study by Ellahi et al. [84] developed a technique for manufacturing thin PDLC films tailored for smart glasses. The method involved utilizing a nematic LC system, epoxy resin monomers, hardeners, and epoxy resin. The refinement of smart

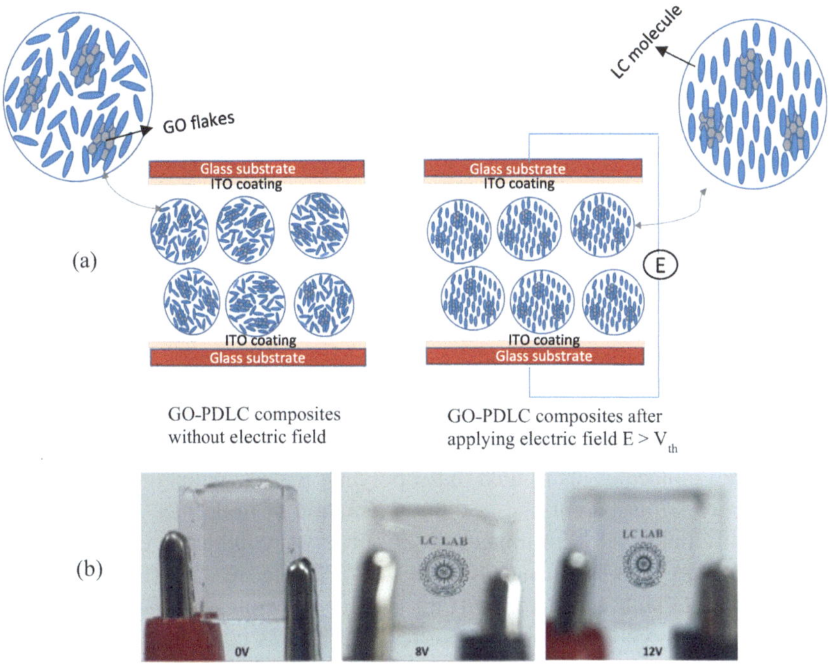

Fig. 14 (**a**) Schematic illustration of GO-PDLC composites in off-state ($E = 0$ V) and in on-state ($E \neq 0$ V). (**b**) Visual characterization of 0.005 wt% GO-PDLC composite at different voltages, 0 V (opaque state), 8 V (translucent state), and 12 V (transparent state). (Reprinted with permission from [81])

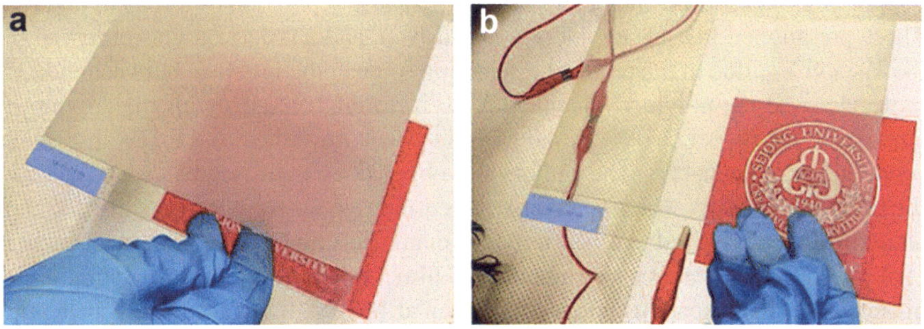

Fig. 15 The PDLC film (**a**) in the absence of an electric field (opaque state), (**b**) in the presence of an electric field (transparent state). (Reprinted with permission from [82])

PDLC films is crucial for enhancing the performance of optical and electro-optical display panels. This research employed the PIPS method, employing thermosetting resin 3-(diethylamino) propylamine as the primary functional heat-curing agent and hardener. Cross-linking agents such as poly(propylene glycol) diglycidyl ether,

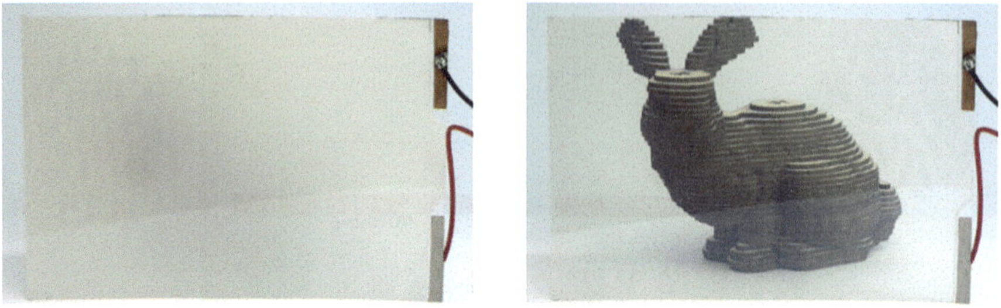

Fig. 16 Off- and transparent state of PDLC. (Reprinted with permission from [83])

trifunctional trimethylolpropane triglycidyl ether, and tetrafunctional pentaerythritol tetra glycidyl ether epoxy resin were also utilized.

In another study, shielded fog of PDLC film was observed in the off-state due to disordered molecules distributed in the polymer matrix. When the power of film is on, it appears transparent due to the reorientation of molecules in the direction of the electric field [85]. Shivaraja et al. [86] observed completely opaque states of PDLC and ODA-SWCNT nanocomposites of 5PCH in off-state (Fig. 17a–d) and transparent state with electric field (Fig. 17e–h). An improved transmittance was noticed for ODA-SWCNT nanocomposites of 5PCH as compared to pure PDLC. The sharpness of the character "LC," the derivative of the images in on-state, was obtained using Sobel, a mathematical operator for edge detection. Figure 17i–l illustrates that the sharpness of the word "LC" increased with doping of ODA-SWCNT, and 0.015 wt% of ODA-SWCNT is optimized concentration.

Hea et al. [87] demonstrated that applying voltage during UV curing results in the pre-alignment of LC molecules before the phase separation of LC and polymer. These pre-aligned PDLCs are more easily driven, and a reduction was observed in the V_{th} and V_{sat} due to the reduced interaction force at the interface between the LC molecules and the polymer. V_{th} and V_{sat} were further reduced with the doping of zinc oxide whiskers (ZnOW) in the PDLC system, attributed to the UV-absorbing properties of the whiskers. The authors also observed an increase in droplet size with the doping of ZnOW, which were distributed throughout the polymer network and altered the refractive index of the polymer matrix, leading to a decreased CR. Choi et al. [88] also fabricated a sky-blue PDLC film, and a direct visual operation from opaque state to transparent state is easily seen at 100 V DC voltage, also depicted in Fig. 18.

Liang et al. [89] formulated a PDLC film using methyl-2-propenoic acid cyclohexyl ester (CHMA) as a diluent and employed fractional phase UV light ($\lambda = 365.0$ nm) to initiate polymerization. The incorporation of NPs into the PDLC significantly reduced the V_{th} from 39.5 to 21.1 V and the driving voltage (V_d) from 71.0 to 45.0 V, compared to pure PDLC, thereby enhancing electro-optical performance with minimal contrast loss. The transition from opaque (off-state) to transparent (on-state) was notably improved. The authors also observed that the transmittance of PDLC doped with 10% CHMA slightly surpassed that of pure

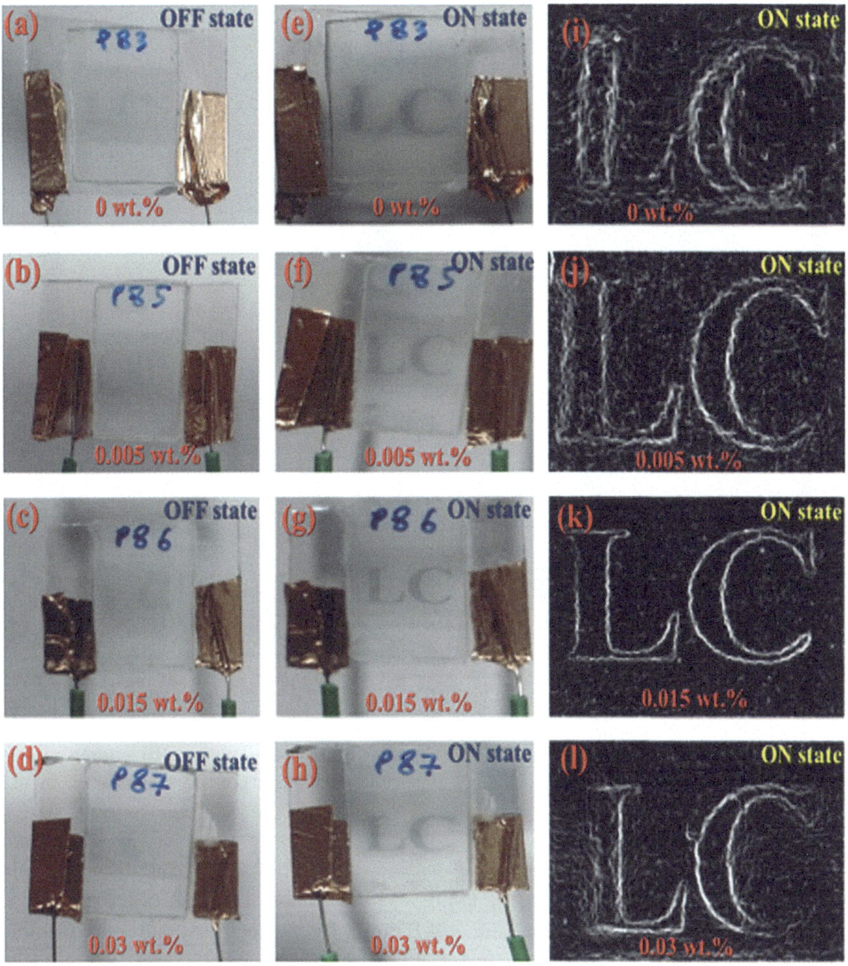

Fig. 17 PDLC films are in (**a–d**) an opaque state and (**e–h**) a transparent state. (**i–l**) Derivative pictures of films from the Sobel operator. (Reprinted with permission from [86])

PDLC. Kumari et al. [90] fabricated a $BaTiO_3$-doped PDLC smart window with 0.5 wt% concentration that works at 20 V. The study reported a decrease in both V_{th} and V_{sat} with increasing $BaTiO_3$ NPs up to 0.4 wt%, resulting in a 50% reduction in both parameters. Beyond this concentration, however, V_{th} and V_{sat} began to rise again, though they remained lower than those of pure PDLC. Choi et al. [91] developed a flexible, large-area PDLC device using spray-coated Ag NW-GO electrodes on a PET substrate, instead of the traditional glass substrate [92, 93]. The fabricated PDLC device demonstrated effective performance without damage, as illustrated in Fig. 19.

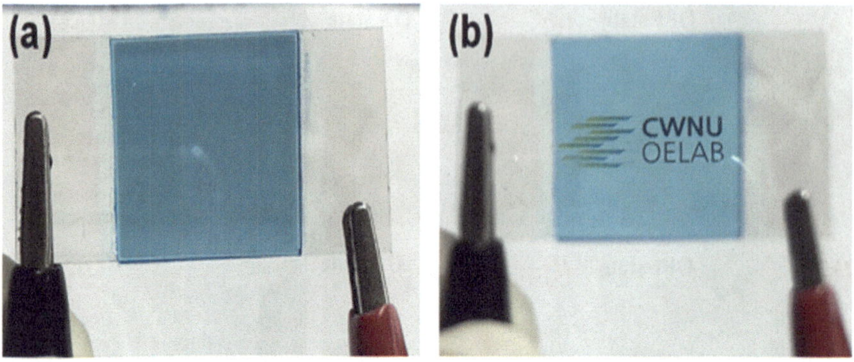

Fig. 18 (**a**) Off-state (**b**) on-state of deep sky-blue PDLC film. (Reprinted with permission from [86])

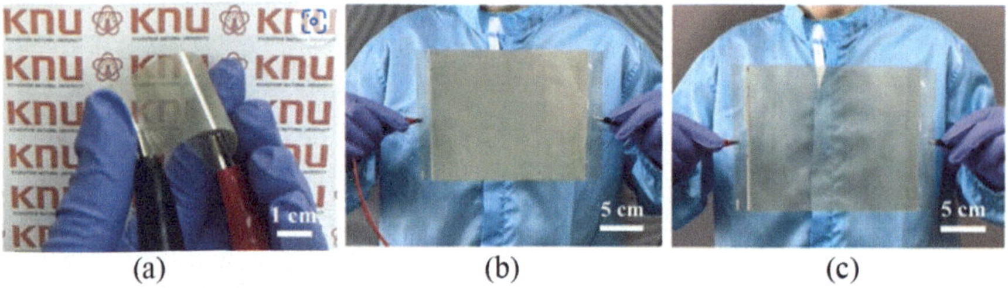

Fig. 19 Pictures of the fabricated PDLC device: (**a**) flexibility in the on-state, (**b**) without applied voltage, (**c**) with applied voltage. (Reprinted with permission from [91])

Kim et al. [94] developed highly transparent electrodes with excellent durable and extremely foldable properties, using nylon 6 nanofiber-reinforced cellulose acetate (NF-r-CA) film. In the presence of E-field, LC molecules inside the droplets align along applied field direction, and PDLC films appear transparent, as shown in Fig. 20a. NF-r-CA PDLC cell works well even under bending, as depicted in Fig. 20b. The film appeared to be more or less equally dark level to that of ITO PDLC film at V_{on}, as shown in Fig. 20c.

PDLC for Biosensing Application

Application of PDLC-based materials is not limited to displays, holographic PDLCs, and gas sensors. However, PDLC-based biosensors have emerged as a promising platform for label-free, real-time detection of biomolecules and cells. The response of LC droplets of PDLCs to electric field and its impact on optical properties can be exploited for biosensing applications by functionalizing the PDLC surface with receptor molecules that selectively bind to the target [95–97]. Several studies have demonstrated the potential of PDLC-based biosensors

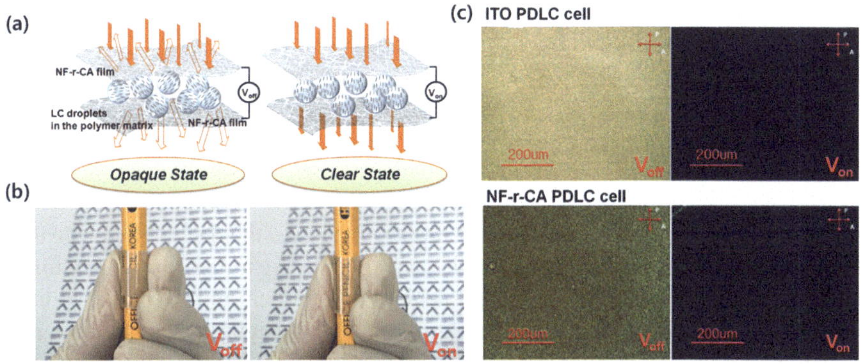

Fig. 20 (**a**) Schematic of opaque and transparent states of the NF45-r-CA-based PDLC film. (**b**) Photographs of the NF45-r-CA PDLC film rolled up a pencil at V_{off} (applied voltage ~0 V_{rms}) and V_{on} (applied voltage ~24.24 V_{rms}). (**c**) Comparison of POM images for the ITO and NF-r-CA PDLC-based PDLC films at V_{off} and V_{on}, respectively. (Reprinted with permission from [94])

for the detection of various biomolecules, including proteins, nucleic acids, and cells. These bioagents induce a change in the orientation of the LC molecules leading to a measurable change in the optical properties of the PDLC film. In addition to protein and nucleic acid detection, PDLC-based biosensors have also been used for cell-based applications. The unique optical properties of PDLCs enable sensitive and selective detection of target analytes, while the ease of fabrication and functionalization makes them a versatile platform for a wide range of biosensing applications. Perju et al. [95] developed a PDLC film consisting of a polysulfone matrix and 5CB. The contact angle of the fabricated film indicates a balanced hydrophilic/hydrophobic property, with moderate wettability making it promising for biocompatible applications. Additionally, the increased surface free energy suggests its potential use as a sensor. Ailincai et al. [96] developed PDLC-based biosensors using cholesteryl acetate and polyvinyl alcohol boric acid for the selective detection of blood analytes such as D-glucose, D-galactose, D-alanine, L-arginine, and salmon sperm DNA. The presence of these analytes significantly alters the optical properties (POM textures) of the PDLC film, causing a shift from planar to homeotropic anchoring of the LC molecules. The preferential interactions between the LC molecules or polymer chains and the analytes drive this reorientation. Lee et al. [97] introduced an innovative signal amplification method for label-free LC-based biosensing by employing a UV-cured LC-prepolymer film on a single glass substrate. A dark-to-bright transition in the optical textures of the composite film was observed with increasing BSA concentrations. It was determined that UV exposure did not affect the brightness of the optical texture. This LC-photopolymer composite film was also used in an immunoassay for the cancer biomarker CA125, showing a dark-to-bright transition at 10^{-7} g mL^{-1} concentration of an anti-CA125 for pristine AY40-006 and 10^{-9} g mL^{-1} for AY40-006/NOA65. This single-substrate platform can be more easily integrated with microarray technologies for high-throughput screening and analysis than traditional LC-based biosensing techniques using two glass substrates.

2 Conclusion

PDLC films exhibit a wide range of properties that make them suitable for diverse applications across various fields. Display technologies are utilized in privacy films, smart windows, and switchable optical devices due to their ability to control light transmission and scattering in response to external stimuli such as voltage and temperature. PDLCs also play a crucial role in sensing applications, particularly gas sensing, offering high sensitivity, selectivity, and rapid response times. By incorporating functionalized nanoparticles (NPs) or patterning PDLC films, researchers have achieved significant advancements in detecting gases like NO_2, toluene, acetone, ethanol, and methanol, promising for environmental monitoring, industrial safety, and healthcare diagnostics. Furthermore, PDLCs contribute to energy-efficient technologies, such as smart glass, which can regulate building temperatures and reduce energy consumption. In the biomedical field, PDLC materials show potential in applications such as drug delivery systems, tissue engineering scaffolds, and biosensors. Overall, PDLCs are at the forefront of innovation across multiple sectors, addressing a range of challenges and driving advancements in science and technology.

Acknowledgments This research was funded by the DST Delhi's CRG scheme (project CRG/2022/007362) at Dr. B. R. Ambedkar National Institute of Technology, Jalandhar.

References

1. Collings, P.J., Hird, M.: Introduction to liquid crystals chemistry and physics. Taylor and Francis Ltd, London (1997)
2. Reinitzer, F.: Wiener Monatscher Chemiie. **9**, 421–441 (1888)
3. Lehman, O.: On flowing crystals. Z. Phys. Chem. **4**, 462–472 (1889)
4. Lagerwall, J.P.F., Scalia, G.: A new era for liquid crystal research: applications of liquid crystals in soft matter nano, bio and micro technology. Curr. Appl. Phys. **12**, 1387–1412 (2012)
5. Bahadur, B.: Liquid crystals: applications and uses. World Scientific, Singapore (1991)
6. Demus, D., Goodby, J., Gray, G.W., Spiess, H.-W., Vill, V.: Handbook of liquid crystals, pp. 1–3. Wiley-VCH, Weinheim (1998)
7. Friedel, G.: The mesomorphic states of matter. Ann. Phys. **18**, 273–474 (1922)
8. Singh, S., Dunmur, D.A.: Liquid crystals fundamentals, p. 221. World Scientific Publishing, London (2002)
9. Brown, G.H.: Liquid crystals and some of their applications in chemistry. Anal. Chem. **41**, 26A–39A (1969)
10. Brown, G.H.: Properties and applications of liquid crystals. J. Electron. Mater. **2**, 403–430 (1973)

11. Brown, G.H.: Structure, properties, and some applications of liquid crystals. J. Opt. Soc. Am. A. **63**, 1505–1514 (1973)
12. Chandrasekhar, S.: Liquid crystals, 2nd edn. Cambridge University Press, Cambridge (1992)
13. Palffy-Muhoray, P.: The diverse world of liquid crystals. Phys. Today. **60**, 54–60 (2007)
14. Singh, A.K., Malik, P.: Analysis of physical properties of nanomaterials-polymer dispersed liquid crystal composites. NIT, Jalandhar (2024)
15. Dierking, I., Al-Zangana, S.: Lyotropic liquid crystal phases from anisotropic nanomaterials. Nano. **7**(10), 305 (2017)
16. Shen, Y., Dierking, I.: Perspectives in liquid-crystal-aided nanotechnology and nanoscience. Appl Sci. **9**, 2512 (2019). https://doi.org/10.3390/app9122512
17. Kumar, A., Singh, D.P., Singh, G.: Recent progress and future perspectives on carbon-nanomaterial-dispersed liquid crystal composites. J. Phys. D. Appl. Phys. **55**, 38 (2022). https://doi.org/10.1088/1361-6463/ac2ced
18. Tripathi, P.K., Misra, A.K., Manohar, S., Gupta, S.K., Manohar, R.: Improved dielectric and electro-optical parameters of ZnO nanoparticle (8% Cu^{2+}) doped nematic liquid crystal. J. Mol. Struct. **1035**, 371–377 (2013). https://doi.org/10.1016/j.molstruc.2012.10.052
19. Singh, U.B., Dhar, R., Dabrowski, R., Pandey, M.B.: Enhanced electro-optical properties of a nematic liquid crystals in presence of $BaTiO_3$ nanoparticles. Liquid Crystal. **41**, 953–959 (2014). https://doi.org/10.1080/02678292.2014.894209
20. Garbovskiy, Y., Glushchenko, A.: Ferroelectric nanoparticles in liquid crystals: recent progress and current challenges. Nano. **7**, 361 (2017). https://doi.org/10.3390/nano7110361
21. Urbanski, M.: On the impact of nanoparticle doping on the electro-optic response of nematic hosts. Liquid Crystal Today. **24**, 102–115 (2015)
22. Basu, R., Iannacchione, G.: Carbon nanotube dispersed liquid crystal: a nano electromechanical system. Appl Phys Lett. **93**, 183105-1-3 (2008)
23. Baik, I.S., Jeon, S.Y., Lee, S.H., Park, K.A., Jeong, S.H., An, K.H., Lee, Y.H.: Electrical-field effect on carbon nanotubes in a twisted nematic liquid crystal cell. Appl Phys Lett. **87**, 263110-1-3 (2005)
24. Kaur, S., Singh, S.P., Biradara, A.M., Choudhary, A., Sreenivas, K.: Enhanced electro-optical properties in gold nanoparticles doped ferroelectric liquid crystals. Appl Phys Lett. **91**, 023120-1-3 (2007)
25. Drazic, P.S.: Liquid crystal dispersion. World Scientific, Singapore (1995)
26. Doane, J.W., Golemme, A., West, J.L., Whitehead, J.B., Wu, B.G.: Polymer dispersed liquid crystals for display application. Mol Cryst Liq Cryst. **165**(1), 511–532 (1988). https://doi.org/10.1080/00268948808082211
27. Simoni, F.: Non-linear optical properties of liquid crystals and polymer dispersed liquid crystals. World Scientific, Singapore (1995)
28. Zhong, T., Mandle, R.J., Goodby, J.W., Zhang, L., Zhang, C.: Comparative studies of polymer-dispersed liquid crystal films via a thiol-ene click reaction. Polym. Adv. Technol. **30**, 2781–2789 (2019)
29. Hu, G., Chen, H., Liu, Z., Zhang, S., Zhou, Y., Zhu, B., Gu, H.: Tailoring structure and properties of polymer-dispersed liquid crystal by quenching process. Liq Cryst. **46**, 1–9 (2020)
30. Malik, P., Raina, K.K., Gathania, A.K.: Effects of polymer viscosity on the polymerization switching and electro-optical properties of unaligned liquid crystal/UV curable polymer composites. Thin Solid Films. **519**(3), 1047–1051 (2010)
31. Li, C., Chen, M., Shen, W., Chen, G., Zhang, L., Yang, H.: A study on the polymer structures and electro-optical properties of epoxy-mercaptan-based polymer dispersed liquid crystal films. Liq Cryst. **46**, 1718–1726 (2019)
32. Ghosh, A., Sundaram, S., Mallick, T.K.: Evaluation of thermal performance for a smart switchable adaptive polymer dispersed liquid crystal (PDLC) glazing. Sol. Energy. **195**, 185–193 (2020)
33. Malik, P., Raina, K.K.: Droplet orientation and optical properties of polymer dispersed liquid crystal composite films. Opt. Mater. **27**(3), 613–617 (2004)

34. Fahad, M., Ye, H., Jeon, S., Hong, J.H., Hong, S.K.: Enhancement of reflective optical properties using photoluminescence polymer-dispersed liquid crystal with added chiral dopant. Sci. Adv. Mater. **8**, 1745–1751 (2016)
35. Zhou, L., Chen, G., Shen, W., Zhang, C., Lanying, Z.: Effect of functionality of thiol on the optical properties of liquid crystals/polymer composite films. Liq Cryst. **5**, 1–9 (2020)
36. Jyoti, D., Khushboo, P.M., Arshdeep, S.: Effect of polymer concentration on morphology, dielectric and optical properties in a polymer dispersed ferroelectric liquid crystal. Liq. Cryst. **43**(5), 623–631 (2016)
37. Jain, A.K., Deshmukh, R.: An overview of polymer-dispersed liquid crystals composite films and their applications. In: Liquid crystal and display technology, pp. 1–68. IntechOpen (2020)
38. Shibaev, V.: Liquid crystalline polymers. Polym Sci Compr Ref. **1**, 259–285 (2012)
39. Rumi, M., Bunning, T.J., Sio, L.D.: Polymer dispersed liquid crystals. Soft Matter Series. **8**, 61–104 (2019). https://doi.org/10.1039/9781788013321-00061
40. Singh, S., Srivastava, J.K., Singh, R.K.: Polymer dispersed liquid crystals. Liq Cryst Polym. **7**, 195–250 (2016). https://doi.org/10.1007/978-3-319-22894-5_7
41. Bronnikov, S., Kostromin, S., Zuev, V.: Polymer-dispersed liquid crystals: progress in preparation, investigation, and application. J Macromol Sci Part B. **52**, 1718–1735 (2013)
42. Ellahi, M., Liu, F., Song, P., Gao, P.Y., Rafique, M.Y., Khan, D.F., Cao, H., Yang, H.: Characterization and morphology of polymer-dispersed liquid crystal films. Soft Mater. **12**, 339–345 (2014)
43. Pozhidaev, E.P., Kaznacheev, A.V., Torgova, S.I., Kesaev, V.V., Barbashov, V.A.: Polymer dispersed liquid crystals with electrically controlled light scattering in the visible and near-infrared ranges. Opt Mater Express. **10**(12), 3030–3040 (2020)
44. Mucha, M.: Polymer as an important component of blends and composites with liquid crystals. Prog. Polym. Sci. **28**(5), 837–873 (2003)
45. Raina, K.K., Chaudhary, A., Malik, P., Shukla, R., Mehra, R.: Polymer dispersed liquid crystals devices: role of photopolymerization to control defect orientation, optical and electro-optical properties. Liq Cryst. **50**, 977 (2023)
46. Ibragimov, T.D., Bayramov, G.M., Imamaliyev, A.R.: Design and electro-optic behaviour of novel polymer-liquid crystalline composites. J. Mol. Liq. **221**, 1151–1154 (2016)
47. Coates, D.: Polymer-dispersed liquid crystals. J. Mater. Chem. **5**(12), 2063–2072 (1995)
48. Malik, P., Raina, K.K.: Dichroic dye-dependent studies in guest–host polymer-dispersed liquid crystal films. Phys. B Condens. Matter. **405**(1), 161–166 (2010)
49. Islam, M.S., Chan, K.Y., Thien, G.S., Low, P.L., Lee, C.L., Wong, S.K., Noor, E.E., Au, B.W., Ng, Z.N.: Performances of polymer-dispersed liquid crystal films for smart glass applications. Polymers. **15**(16), 3420 (2023)
50. Shen, W., Wang, L., Zhong, T., Chen, G., Li, C., Chen, M., et al.: Electrically switchable light transmittance of epoxy-mercaptan polymer/nematic liquid crystal composites with controllable microstructures. Polymer. **160**, 53–64 (2019)
51. Serbutoviez, C., Kloosterboer, J.G., Boots, H.M.J., Touwslager, F.J.: Polymerization-induced phase separation. 2. Morphology of polymer-dispersed liquid crystal thin films. Macromolecules. **29**(24), 7690–7698 (1996)
52. Singh, A.K., Malik, P.: Textural, electro-optical, dielectric and fluorescence studies of citrate buffer stabilized gold nanoparticles doped in polymer-dispersed liquid crystals composites. Liq. Cryst. **49**(6), 864–874 (2022). https://doi.org/10.1080/02678292.2022.2027532
53. Raina, K.K., Kumar, P., Malik, P.: Morphological control and polarization switching in polymer dispersed liquid crystal materials and devices. Bull. Mater. Sci. **29**, 599–603 (2006)
54. Mani, S., Patwardhan, S., Hadkar, S., Mishra, K., Sarawade, P.: Effect of polymer concentration on optical and electrical properties of liquid crystals for photonic applications. Mater Today Proc. **62**(13), 7035–7039 (2022)
55. Kumar, P., Sharma, V., Jaggi, C., Malik, P., Raina, K.K.: Orientational control of liquid crystal molecules via carbon nanotubes and dichroic dye in polymer dispersed liquid crystal. Liq. Cryst. **44**, 843 (2016)

56. Saeed, M.H., Zhang, S., Cao, Y., Zhou, L., Hu, J., Muhammad, I., et al.: Recent advances in the polymer dispersed liquid crystal composite and its applications. Molecules. **25**(23), 5510 (2020)
57. Singh, A.K., Malik, P., Chauhan, G., Hegde, G., Malik, P.: Synthesis and characterization of bio waste-based porous carbon nanoparticle-polymer dispersed ferroelectric liquid crystal composites. J. Mol. Liq. **390**, 123024 (2023)
58. Perju, E., Paslaru, E., Marin, L.: Polymer-dispersed liquid crystal composites for bio-applications: thermotropic, surface and optical properties. Liq Cryst. **42**(3), 370–382 (2015)
59. Bouteiller, L., Barny, P.L.: Polymer-dispersed liquid crystals: preparation, operation and application. Liq Cryst. **21**(2), 157–174 (1996)
60. Esteves, C., Ramou, E., Porteira, A.R.P., Barbosa, A.J.M., Roque, A.C.A.: Seeing the unseen: the role of liquid crystals in gas-sensing technologies. Adv Opt. Mater. **8**, 1902117 (2020). https://doi.org/10.1002/adom.201902117
61. Chehri, A., Farjow, H.T., Mouftah, Fernando, X.: Canadian Conference on Electrical and Computer Engineering (CCECE). Institute of Electrical and Electronics Engineering, p. 001532 (2011)
62. Manna, S., Bhunia, S.S., Mukherjee, N.: Vehicular pollution monitoring using IoT. International conference on recent advances and innovations in engineering (ICRAIE) Institute of Electrical and Electronics Engineering, p. 1 (2014)
63. Van Hooren, M.R.A., Leunis, N., Brandsma, D.S., Dingemans, A.M.C., Kremer, B., Kross, K.W.: Differentiating head and neck carcinoma from lung carcinoma with an electronic nose: a proof of concept study. Eur. Arch. Otorrinolaringol. **273**, 3897–3903 (2016)
64. Zampolli, S., Elmi, I., Ahmed, F., Passini, M., Cardinali, G.C., Nicoletti, S., Dori, L.: An electronic nose based on solid state sensor arrays for low-cost indoor air quality monitoring applications. Sens Actuators B Chem. **101**, 39 (2004). https://doi.org/10.1016/j.snb.2004.02.024
65. Khan, A., Schaefer, D., Tao, L., Miller, D.J., Sun, K., Zondlo, M.A., Harrison, W.A., Roscoe, B., Lary, D.J.: Low power greenhouse gas sensors for unmanned aerial vehicles. Remote Sens. **4**, 1355–1368 (2012)
66. Pagidi, S., Pasupuleti, K.S., Reddeppa, M., Ahn, S., Kim, Y., Kim, J.H., Kim, M.D., Lee, S.H., Jeon, M.Y.: Resistive type NO_2 gas sensing in polymer-dispersed liquid crystals with functionalized-carbon nanotubes dopant at room temperature. Sens Actuators B Chem. **370**, 132482 (2022). https://doi.org/10.1016/j.snb.2022.132482
67. Zhang, W., Fan, D., Gong, Y., Liang, W.: Polymer dispersed liquid crystal thin films for sensing toluene. Liq. Cryst. **48**(12), 1643–1649 (2021). https://doi.org/10.1080/02678292.2021.1897893
68. Huang, C.Y., Lin, S.H.: Organic solvent sensors using polymer-dispersed liquid crystal films with a pillar pattern. Polymers. **13**(17), 2906 (2021). https://doi.org/10.3390/polym13172906
69. Liu, Y., Zheng, J., Jiang, Z., Zhu, Q., Chen, Q., Zhuang, S.: Optical and dielectric analysis of ZnO nano-rods doped polymer dispersed liquid crystal and ethanol gas sensing investigation. Liq. Cryst. **47**(14–15), 2247–2256 (2020). https://doi.org/10.1080/02678292.2020.1766133
70. Wang, J., Jakli, A., West, J.L.: Liquid crystal/polymer fibber mats as sensitive chemical sensors. J. Mol. Liq. **467**, 490–495 (2018). https://doi.org/10.1016/j.molliq.2018.01.051
71. Song, M., Seo, J., Kim, H., Kim, Y.: Ultrasensitive multi-functional flexible sensors based on organic field-effect transistors with polymer-dispersed liquid crystal sensing layers. Sci. Rep. **7**, 2630 (2017). https://doi.org/10.1038/s41598-017-02160-x
72. Bag, S., Rathi, K., Pal, K.: Effect of carbon derivatives in sulfonated poly(etherimide)—liquid crystal polymer composite for methanol vapor sensing. Nanotechnology. **28**(20), 205501 (2017). https://doi.org/10.1088/1361-6528/aa66b6
73. Reyes, C.G., Sharma, A., Lagerwall, J.P.F.: Non-electronic gas sensors from electrospun mats of liquid crystal core fibres for detecting volatile organic compounds at room temperature. Liq. Cryst. **43**(13–15), 1986–2001 (2016). https://doi.org/10.1080/02678292.2016.1212287

74. Lai, Y.T., Kuo, J.C., Yang, Y.J.: A novel gas sensor using polymer-dispersed liquid crystal doped with carbon nanotubes. Sens Actuators A. **215**, 83–88 (2014). https://doi.org/10.1016/j.sna.2013.12.021
75. Kirchner, N., Zedler, L., Mayerhofer, T.G., Mohr, G.J.: Functional liquid crystal films selectively recognize amine vapours and simultaneously change their colour. Chem. Commun. **14**, 1512–1514 (2006). https://doi.org/10.1039/b517768e
76. Parmar, D.S., Singh, J.J.: Gas flow-field induced director alignment in polymer dispersed liquid crystal microdroplets deposited on a glass substrate. Liq. Cryst. **14**(2), 361–369 (1993). https://doi.org/10.1080/02678299308027651
77. Sharma, V., Kumar, P., Chinky, Raina, K.K., Malik, P.: Preparation and electro-optic study of reverse mode polymer dispersed liquid crystal—performance augmentation with the doping of nanoparticles and dichroic dye. J. Appl. Polym. Sci. **137**(22), 48745 (2019)
78. Jayoti, D., Malik, P., Singh, A.: Analysis of morphological behaviour and electro-optical properties of silica nanoparticles doped polymer dispersed liquid crystal composites. J. Mol. Liq. **225**, 456–461 (2017)
79. Malik, P., Singh, A.K.: Metal oxide alumina nanowire-induced polymer-dispersed liquid crystal composites for low power consumption smart windows. J. Mol. Liq. **378**, 121573 (2023)
80. Park, J.Y., Kim, H.K.: Highly stretchable polymer-dispersed liquid crystal-based smart windows with transparent and stretchable hybrid electrodes. RSC Adv. **8**, 36549 (2018)
81. Kaur, M., Malik, P.: New graphene oxide doped polymer dispersed liquid crystal nanocomposites targeted to eco-friendly and energy-efficient smart windows. J. Mol. Liq. **410**, 125565 (2024)
82. Nasir, N., Hong, H., Rehman, M.A., Kumar, S., Seo, Y.: Polymer-dispersed liquid-crystal-based switchable glazing fabricated via vacuum glass coupling. RSC Adv. **10**, 32225–32231 (2020)
83. Kretzer, M.: Polymer dispersed liquid crystal. Materiability Research Group (2024)
84. Ellahi, M., Taimur, S., Baloch, N., Bhayo, A.M., Sarwar, A., Qadar, F.: Study of polymer-dispersed liquid crystal (PDLC) thin film technology for smart electronic devices. J. Electron. Mater. **53**, 1094–1104 (2024)
85. Wu, Y.L., Wu, C.M.: Effect of doped SiO_2 microspheres on the high temperature properties of PDLC films. Liq. Cryst. **51**(1), 120–127 (2024)
86. Shivaraja, S.J., Gupta, R.K., Manjuladevi, V.: Faster switching polymer dispersed liquid crystal devices incorporated with functionalized SWCNTs. J. Mol. Liq. **354**, 118905 (2022)
87. Hea, Z., Lia, J., Zhanga, W., Gaoa, J., Maa, C., Yaoa, R., Yaob, X., Miaoc, Z., Wang, D.: The regulation of electro-optical properties of the polymer dispersed liquid crystal films doped ZnO whiskers under applied pre-orientation voltage. Liq. Cryst. **49**, 1964–1973 (2022)
88. Choi, S.H., Kim, J.A., Heo, G.S., Park, H.G.: Electro-optical characteristics of polymer-dispersed liquid crystal containing copper(II) phthalocyanine as a function of UV irradiation time. J. Mol. Liq. **363**, 119821 (2022)
89. Lianga, Z., Zhaob, Y., Gaoc, H., Wanga, D., Miaob, Z., Caoa, H., Yanga, Z., He, W.: Influence of ZnO NPs on morphological and electro-optical properties of polymer-dispersed liquid crystals. Liq. Cryst. **48**, 1699–1708 (2021)
90. Kumari, A., Sinha, A.: Role of $BaTiO_3$ nanoparticles on electro-optic performance of epoxy-resin-based PDLC devices. Liq. Cryst. **48**(1), 23–24 (2021)
91. Choi, Y., Kim, C.S., Jo, S.: Spray deposition of Ag nanowire–graphene oxide hybrid electrodes for flexible polymer dispersed liquid crystal displays. Materials. **11**, 2231 (2018)
92. Singh, A., Salmi, Z., Joshi, N., Jha, P., Decorse, P., Lecoq, H., Truong, S.L., Jouini, M., Aswal, D., Chehimi, M.: Electrochemical investigation of free-standing polypyrrole–silver nanocomposite films: a substrate free electrode material for supercapacitors. RSC Adv. **3**, 24567–24575 (2013)
93. Kim, Y., Hong, J., Lee, S.: Fabrication of a highly bendable LCD with an elastomer substrate by using a replica-molding method. J. Soc. Inf. Disp. **14**, 1091–1095 (2006). https://doi.org/10.1889/1.2408391

94. Kim, I., Kim, T., Lee, S., Kim, B.: Extremely foldable and highly transparent nanofiber-based electrodes for liquid crystal smart device. Sci. Rep. **8**, 11517 (2018)
95. Perju, E., Stoleru, E., Marin, L.: Polymer-dispersed liquid crystal composites for bio-applications: thermotropic, surface and optical properties. Liq. Cryst. **42**(3), 370–382 (2015). https://doi.org/10.1080/02678292.2014.992055
96. Lee, M.J., Duan, F.F., Wu, P.C., Lee, W.: Liquid crystal–photopolymer composite films for label-free single-substrate protein quantitation and immunoassay. Biomed. Opt. Express. **11**(9), 4915–4927 (2020). https://doi.org/10.1364/BOE.398858
97. Ailincai, D., Farcău, C., Stoleru, E., Marin, L.: PDLC composites based on polyvinyl boric acid matrix—a promising pathway towards biomedical engineering. Liq. Cryst. **43**(13), 1–13 (2016). https://doi.org/10.1080/02678292.2016.1172353

Soft Material–Based Photo Actuators in Microfluidics

Surya Narayana Sangitra, Piyush Sahu, and Ravi Kumar Pujala

Abstract This chapter delves into photo actuators, examining their classifications and photo-responsive behaviors across materials like hydrogels, polymers, and liquid crystals. It shows the uses of these actuators in industries including soft robotics, micropumps, and nanodevices and investigates the principles underlying them, including photothermal, photochemical, and photoelectric effects. The chapter also discusses advancements like light-driven bimorph actuators and carbon nanotube-based systems, emphasizing their potential in modern technologies. It outlines future directions focused on improving efficiency, response times, and durability, positioning photo actuators as key innovations in material science and engineering.

Keywords Photo actuator · Light-responsive · Microfluidics · Actuator mechanism · Structure manipulation

1 Introduction

Microfluidic devices deal with the characterization and manipulation of small amounts of liquids in multidisciplinary engineering and scientific research areas, including fluid physics, chemical analysis, biological assays, droplet microreactors, food safety, photonic applications, 3D printing, and various biomedical and tissue engineering applications [1–3]. The microfluidic device typically operates based on mechanical, hydraulic, electrokinetic, pneumatic, and electric and magnetic force, which is only possible by attaching external devices such as valves, pumps, or electrodes [1]. Inspired by semiconductor-based electronic chips, higher-performance microfluidic chips have been developed for various valves, sensing structures, and

S. N. Sangitra · P. Sahu · R. K. Pujala (✉)
Soft and Active Matter Group, Department of Physics, Indian Institute of Science Education and Research (IISER) Tirupati, Tirupati, Andhra Pradesh, India
e-mail: pujalaravikumar@iisertirupati.ac.in

V. Kumar, Y. K. Mishra (eds.), *Soft Materials for Functional Applications*, Advanced Structured Materials 225, https://doi.org/10.1007/978-981-97-9468-3_8

actuators [2]. The different actuation mechanisms based on microfluidic valves are reviewed by Qian et al. [4]. They have discussed electric, magnetic, gas (pneumatic), material, and biological actuation. In the traditional actuation method, problems exist like low control accuracy, fluid leakage, high cost, high energy consumption, and poor reliability. While dealing with biological samples, there is also a chance of contamination. Therefore, the light-actuated (also called photo-actuated) method plays an important role to overcome the drawbacks. The light-actuated methods are clean, contactless, environment friendly, low cost, and consume less energy than all other methods [5].

Conventional solid materials are rigid, limiting their application when interacting with natural systems; therefore, soft materials with a multi-responsive nature have been used for developing new-generation actuators. Liquid crystals, gels, and polymers are examples of stimuli-responsive soft materials that have a major role in developing next-generation actuators [5]. These materials combine various functionalities onto a single chip by reacting to minute changes in external factors like pH, temperature, pressure, electric and magnetic fields, and light. Therefore, lab-on-chips (LOC) is another name for contemporary generation microfluidic devices [1]. Whitesides et al. discussed an excellent review of the origin and future of microfluidics [6]. This is a general review paper, but our interest is mainly focused on studying the integration of microfluidic and light (optic/photonic). In literature, the fusion of microfluidic and light has emerged as a new field of research known as "optofluidic." Monat et al. reported a detailed review of integrated optofluidics and how optics controls and monitors fluidic manipulations [7]. Since polydimethylsiloxane (PDMS) has been a necessary component of most microfluidic devices for the last two decades or more, light has typically been utilized to extract information from microfluidic chips, frequently employing the transparency of PDMS [8]. A more advanced approach involves illuminating a chip with structured light (maybe highly intense laser light), serving the dual purpose of powering and regulating fluidic operations. This method eliminates the need for external pumps and valves, capitalizing on our precise light control to steer intricate reactions [1, 5, 9]. Remarkably, light can pass through a chip's thickness, eliminating the need for internal regulating logic and getting around the complexity barrier of packed boundaries. Even if the optics may initially use bulky external lasers and cameras, it is conceivable to integrate and miniaturize them, opening the door for genuinely tiny devices [1, 5, 7]. In this chapter, we first discussed the classification of photo actuators based on different manipulated methods and interactions. Then, we discussed the different soft materials and actuation mechanisms used to develop photo-responsive actuation in microfluidic. Then, we discussed the applications of photo actuators in microfluidics. Finally, we summarized the report after discussing the future direction of photo actuation in the field of microfluidic.

2 Classification of Photo Actuators

Photo actuators can be classified by the method of manipulation and the interaction involved.

2.1 *Based on Manipulated Method*

Two main approaches have emerged to make this concept feasible. Light-induced surface tension gradients for droplet manipulation or optical tweezers for cell sorting is a method that uses light to manipulate the fluid or its constituents directly. Alternatively, the interaction is facilitated when a photo-responsive substance absorbs light [4].

Direct Manipulation

Damien Baigl [5] thoroughly discusses the direct approach, which discusses many techniques for directly manipulating light and explains how it might be used to influence fluids. However, there are some disadvantages as well. For example, a specific optical setup is required, as illustrated in Fig. 1, and local heating and potential contamination cannot be avoided. As a result, the indirect method is better suited to address these issues [5].

Indirect Method

Photoactive soft materials that work as actuators are frequently used in microfluidics to facilitate light-mediated interactions with fluids. Advanced lab-on-chip (LOC) systems that perform functions like mixing, valving, and pumping inside microscale

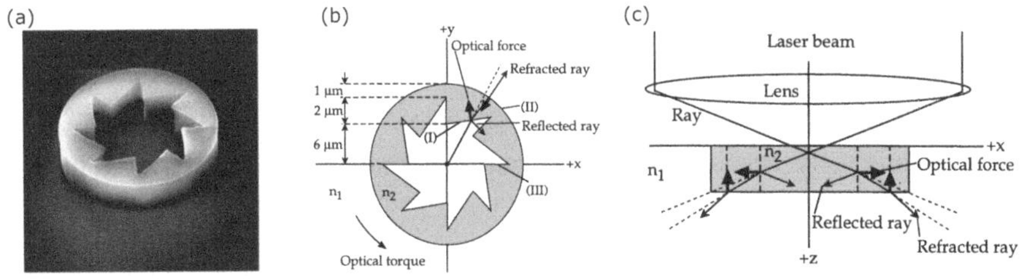

Fig. 1 (**a**) SEM micrographs showing fabricated micro-objects used for optical trapping (**b**) and (**c**) schematic representation for optical trapping, from side view and top view, respectively. (Reprinted with permission from Higurashi et al. [10]. Copyright 1997, American Institute of Physics)

channels are equipped with these materials. Contrasted with direct light–fluid interactions, indirect methods utilizing soft photo actuators offer broader prospects, including fluid motion control, beyond the scope of conventional approaches [1].

2.2 Based on Interaction

Some photo-responsive materials comprise photochromic molecules that absorb light signals and change them into various characteristics [11]. These compounds may find their way into gel-like substances, polymer compounds, and fluids, among other soft actuators. In this regard, we classify photo-stimulated actuators based on the light bandwidth needed to activate them, differentiating between visible and near-infrared (NIR) light, thus defining their separate uses.

Visible Radiation Driven

Using sunlight to activate soft robot actuators reduces energy consumption, allowing them to operate efficiently in natural environments. Present-day photo-stimulated actuators, primarily made of materials based on carbon or liquid crystal networks of polymers (LCNs), frequently need strong light, but their deformation capabilities are restricted. For example, Gelebart et al. presented a light-triggered plastic mill rotating continuously in response to high-intensity light [12]. Dong et al. eliminated extra device requirements by introducing sustained flapping-wing actuation in sunlight using a metal–polymer bilayer [13]. Deng et al. utilized laser-induced graphene (LIG) structures to create self-morphing sensitive actuators [14]. Liu et al. achieved sequential self-folding in printed 2D shapes using visible light. Biological cells, like contractile cells such as skeletal muscle and cardiomyocytes, are integrated into soft actuators, offering precise control over motion through optogenetic modification [15]. Williams et al. demonstrated artificial flagellar swimmers and synthetic batoid-like fish, showcasing controlled motion through optical actuation [16].

Near-Infrared (NIR)

Near-infrared (NIR) light-triggered soft actuators have seen increasing interest recently, primarily because NIR light can penetrate biomaterials with minimal energy loss, making it a preferred method for wireless actuation in biomedical applications [17]. However, the advancement of NIR-driven actuators has been constrained by slow actuation speeds and diminished mechanical properties, often due to suboptimal fabrication processes. These actuators, including liquid crystal elastomers (LCEs) [18] and various polymer-based systems [19], are typically fabricated by incorporating photothermal agents like carbon nanotubes, metal nanoparticles, or graphene into the polymer matrix [20]. These agents absorb NIR light and convert it into heat, causing changes in the polymer, such as phase transitions or swelling. Nonetheless, the poor solubility of these agents within the polymers has hindered the speed and mechanical effectiveness of the actuators. To overcome these challenges, Liu et al. developed single-axially aligned main-chain LCEs with chemically bonded NIR-absorbing cross-linkers, resulting in an actuator that quickly heats up and exhibits significant mechanical strength, enabling it to lift objects much heavier than its own weight [21]. Similarly, Tian et al. introduced a poly-dopamine (PDA)-coated LCE-based elastomer, which demonstrated fast responses to NIR light, including contraction, rolling, and bending. This innovation paves the way for swimmer robotic systems that can be precisely controlled using NIR laser-induced movements [22].

3 Photo-Responsive Soft Materials

Innovative, sensitive soft materials, like polymers, gel-like substances, and liquid crystals, have recently advanced to the point where they respond to minute changes in their surroundings. These materials provide a different strategy and significantly accelerate the development of the next generation of Lab-on-a-Chip (LOC) devices. These devices combine many functions onto one chip, including multiple actuators with different mechanical characteristics and tiny channels.

3.1 Hydrogels

Hydrogel actuators that are optically responsive alter their structure in response to light, transferring energy without needing physical interaction. They can respond just to particular light wavelengths [23]. There are three types of these actuators, each with a unique mechanism. Photo-switchable molecules, such as azobenzene and spiropyran, exhibit bidirectional isomerization in response to particular light wavelengths. For example, spiropyran dissociates in the presence of UV light and returns to a closed ring form in visible light [3]. Changes in hydrophobic properties brought upon by this bidirectional isomerization cause volumetric changes in the hydrogel. Hydrogel-based actuators with spiropyran for microfluidics can serve as bidirectional light-responsive valves [24]. In an acidic medium, the hydrogel transforms, forming a stable hydrophilic isomer called merocyanine-H^+ (McH^+). Upon exposure to light, this McH^+ isomer can be converted into the ring-closed hydrophobic spiropyran form. This change in molecular structure fundamentally alters the properties of the hydrogel, shifting it from a predominantly hydrophilic state to one that is mainly hydrophobic. Consequently, the hydrogel demonstrates reversible behavior, expanding and contracting as it absorbs and releases water molecules, as illustrated in Fig. 2.

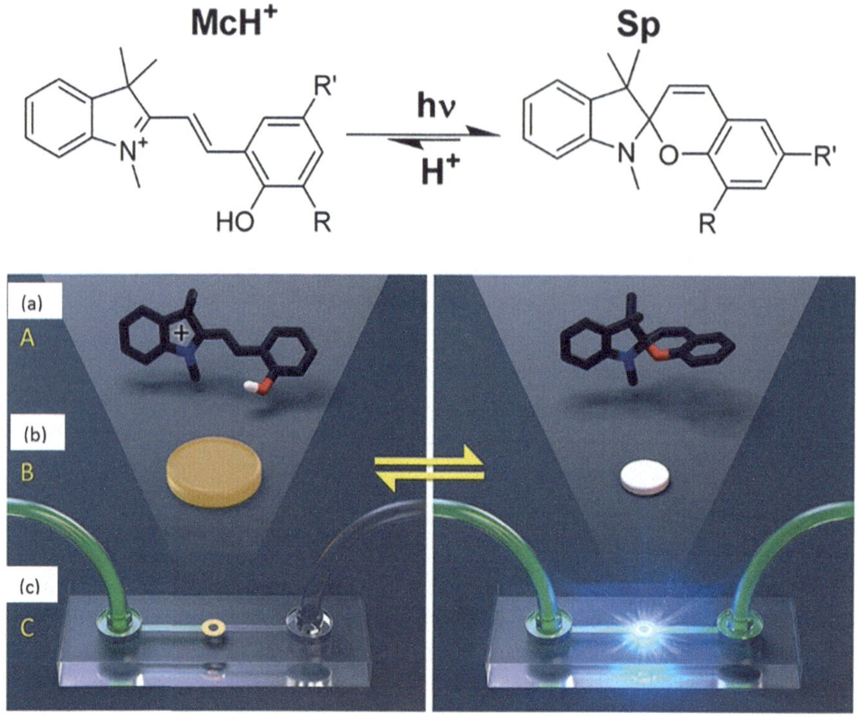

Fig. 2 Transformation of a protonated merocyanine (McH^+) into the spiropyran (Sp) form (**a**), illustrating the consequent impact on the dimensions of a hydrogel under light irradiation (**b**). This process is integrated as a light-responsive valve within microfluidic systems (**c**). (Reprinted with permission from ter Schiphorst et al. [24])

3.2 Liquid Crystals

Liquid crystal polymers (LCPs), derived from chemically bonding small-molecule liquid crystals, possess unique properties like light-responsiveness and flexibility due to their high molecular weight and liquid crystal order [25]. Deformations in LCPs, often caused by changes in molecular shape and orientation, have garnered attention, leading to the development of new LCP materials. For instance, Jiang utilized anthracene-containing liquid crystal networks (LCNs) and UV-induced photolysis to control actuation behavior through spatially controlled optical cross-linking [26]. Liquid crystalline elastomers (LCEs) are materials that display asymmetric contraction upon heating and expansion upon cooling. This combination of features allows for the creation of a wide range of light-responsive LCE materials [27]. Surface alignment techniques can sometimes be used to minimize surface energy by taking advantage of the tendency of liquid crystals to form

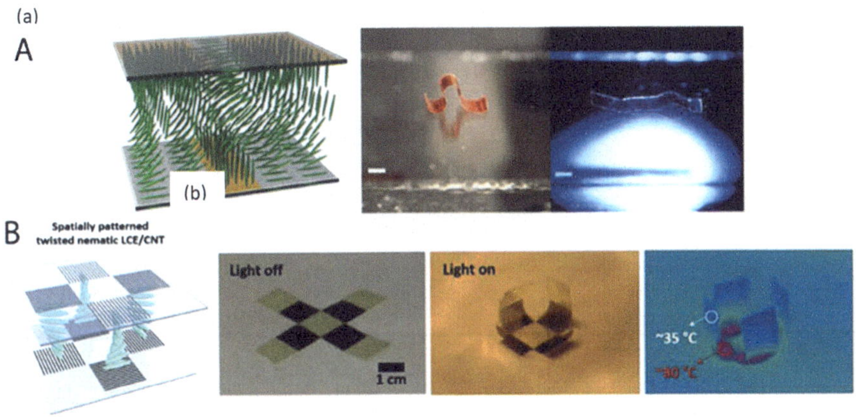

Fig. 3 (**a**) Molecular reorientation of liquid crystal elastomers (LCE) from a nematic to a homeotropic orientation both inside and outside the plane is shown in diagram (**b**) When exposed to light, this transition produces crawling actuation that is photothermally produced. Furthermore, the orientation of the carbon nanotube forest causes twist alignment to occur throughout the thickness of a liquid crystal elastomer (LCE) cell. This alignment makes it possible to actuate photothermally induced bending. (Reprinted with permission from Ambulo et al. [25])

organized monodomains on directed surfaces. Surface tethering is the process by which mesogens align along the preset direction when an LCE precursor is filled within a pre-built cell. Furthermore, because of the misalignment energy penalty, the intended orientation diffuses to neighboring layers [25]. One can do substrate programming by rubbing substrates coated with polymers, lining up photo-responsive molecules with polarized radiation, or employing lithography to shape topographical characteristics, as illustrated in Fig. 3.

3.3 *Polymers*

The field of shape memory polymers (SMPs) is expanding quickly due to its advantages, which include elastic deformation and possible biocompatibility [28]. SMPs were first primarily employed for temperature response, but they can also undergo a simpler form change in response to light-induced heat using a clear light source. Reversible variations in chain mobility during heat transitions are the cause of this phenomena [29–31]. As seen in Fig. 4, supramolecular shape memory polyurethane complexes comprising azobenzene- and pyridine-containing polyurethanes, for example, have good light-responsive form memory [29, 30].

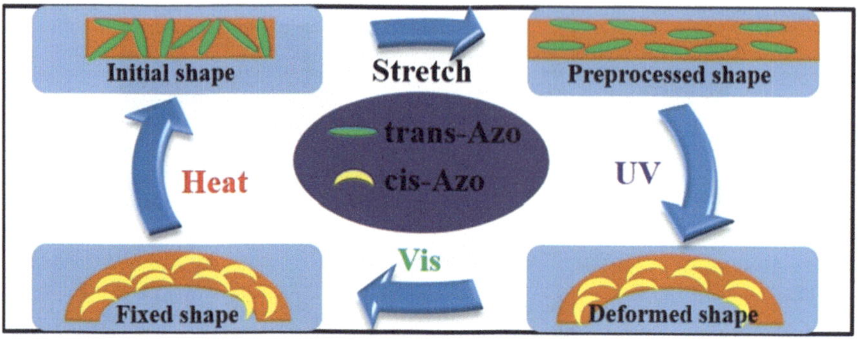

Fig. 4 Diagram illustrating the molecular process behind the photothermal staged-responsive shape memory characteristics. The initial shape stretched to form a trans-azo, then illuminated with UV radiation to deform the shape, making it a cis-azo. In contrast, treatment with visible radiation makes fixed shapes cis-azo, which upon heat regains its initial stage. (Reprinted with permission from Li et al. [30])

4 Actuation Mechanism for Photo Actuators

4.1 Photothermal

Three main mechanisms are involved in photothermal actuation: phase transition distortion, expansion deformation, and guest molecular deformation [32].

Expansion Deformation Through Thermal Photoresponse

The bilayer model is a widely used design approach for thermal photoresponse actuators. Here, a polymer with a high coefficient of thermal expansion (CTE) and a high degree of flexibility usually makes up the first layer. In contrast, a composite of polymer and other materials with a lower CTE often makes up the second layer. The degree of deformation is influenced by temperature, modulus, thickness, and CTE of both layers. More specifically, at the same temperature, more deformations are caused by greater variations in the characteristics between the layers. For thermal expansion actuators, this idea directs material selection and design [29].

Deformation of Guest Molecules Under Thermal Photoresponse

Beyond expansion deformation, thermal photoresponse also induces guest molecule deformation, primarily driven by volume changes in hydrophilic molecules upon light exposure. This phenomenon is notable in flexible materials rich in hydrophilic groups like poly-dopamine (PDA) and spiropyran. Through the use of this process, scientists have combined the effects of nanomaterials with volume changes in other soft organic materials to create light-responsive actuators. They used PDA-modified reduced graphene oxide's (PDA-RGO) photothermal conversion and molecular desorption capabilities, for example, to create a light-driven, double-layer actuator that can bend or not bend in the face of periodic near-infrared radiation [29].

Phase Transition Deformation Under Thermal Photoresponse

When exposed to environmental stimuli like temperature and light, certain flexible materials, such as paraffin wax and shape memory polymers (SMPs), experience unique changes in their physical characteristics. Phase transition deformation under thermal photoresponse is the collective term for these transitions, which include changes from liquid to solid states and glass transitions. Researchers have added phase-change materials and light-responsive nanomaterials to homogenous composite materials to speed up the phase-change temperature and shorten the reaction time to light stimuli [29].

4.2 *Photochemical*

Light-induced chemical transformation of photochemical materials frequently results in modifying their molecular structure [33]. Among these photochemical materials, spiropyran and azobenzene have become popular alternatives, especially when combined with hydrogels or liquid crystal materials; azobenzene is preferred because of its quick light response [34]. Numerous processes, including cis isomerization, ring opening and closure, cycloadditions, and bond exchange, are involved in the photochemical deformation process [35]. For example, under UV light, azo compounds display both trans-cis and cis-trans isomerization, mainly because of the structure of the –N=N– bond between the benzene rings [36]. At room temperature, these molecules usually exist in a trans-configuration that is thermodynamically stable. But when exposed to UV light, the azo compound's electrons absorb energy and transform to a higher energy state, which causes the trans-isomer to become the cis-isomer [37].

4.3 *Photoelectric*

Carbon materials at the nano/micron size cause transitions in the electrons inside the sp2 hybridized orbital of carbon nanoscale material when exposed to particular light wavelengths. This shift in turn causes the carbon nanolattice to vibrate more intensely, producing heat and enabling optical actuation using light and magnetism [38, 39]. For instance, scientists developed a brand-new 3D-linked graphene

Table 1 List of materials following different actuation mechanisms

Sl. no.	Materials	Actuation mechanism
1	Azobenzene and its derivatives	Photoelectric conversion actuation
2	Bi1-xSbx thin films	Photoelectric conversion actuation
3	Carbon nanoparticles	Photothermal and photoelectric conversion actuation
4	Cross-linked polymer structures	Photochemical conversion actuation
5	Divinylidene	Photochemical conversion actuation
6	Liquid crystal polymers and elastomers	Photochemical conversion actuation
7	Metal nanoparticles	Radiation and thermal energy coupled actuation
8	Poly-dopamine	Radiation and thermal energy coupled actuation
9	Poly-pyrrole	Radiation and thermal energy coupled actuation
10	Semiconductor nanostructures	Radiation and thermal energy coupled actuation
11	Shape memory polymers	Radiation and thermal energy coupled actuation
12	Silicone	Photoelectric conversion actuation
13	Spiropyan	Photochemical conversion actuation
14	Transition metal carbides and nitrides	Photothermal conversion actuation

sponge and investigated its capacity to activate light [40]. In addition to carbon nanomaterials, other materials that show notable deformation in response to light stimulation include silicone [41] and two-dimensional Bi1-xSbx thin films [42]. Table 1 provides a comprehensive overview of the state-of-the-art advancements in these studies.

5 Applications

5.1 Micropumps and Mixtures

Light-Driven Micropumps

Light-driven micropumps offer a promising avenue for achieving fluid manipulation within microfluidic systems. The use of light as a power source for micropumps opens up new possibilities for accurate and adaptable fluid handling in systems that are smaller. The promise of light-driven processes in microfluidic applications was demonstrated in a study by Luo et al. that assessed a variety of micropumps, including bubble-heat micropumps and light-driven pumps based on light-thermal transfer [43]. Agarwal has investigated programmable autonomous micropumps and micromixers, demonstrating how these devices were made possible by fusing microfluidic tectonics with MEMS [44]. These developments highlight the potential of light-driven micropump technology to aid in the creation of sophisticated microfluidic systems. In order to demonstrate the development and possible applications of light-driven micromixers in molecular diagnostics and biotechnology, Au et al. carried out an extensive investigation of the advancements made in microvalves, micropumps, and micromixers within PCR microfluidic chips during the previous 10 years [45]. Additionally, a study by Lam et al. showed how light-driven micromixers may be used to achieve accurate fluid handling and control by presenting an integrated digitally programmable microfluidic system for continuous solution supply with real-time concentration control [46].

5.2 *Emulsification and De-emulsification*

Application of Light-Driven Emulsification

Light-driven emulsification has found applications in various fields, including pharmaceuticals, the food industry, and materials science. It has enabled the creation of well-defined microemulsions and has facilitated the development of novel drug delivery systems and functional materials. Understanding the stabilization mechanisms of oil–water emulsions and the processes of de-emulsification forms the basis for developing light-driven approaches for targeted and efficient separation of emulsions. Furthermore, light-responsive Pickering emulsions have been reported by Li et al. to exhibit reversible transitions between emulsification and de-emulsification, triggered by alternating light stimuli, showcasing the potential for dynamic and controllable emulsion systems [47]. Dagkesamanskaya et al. concentrated on how light irradiation affected emulsions made with C7 azo C7, a photo-responsive gemini surfactant. This comprises measuring the equilibrated and dynamic interfacial tensions, analyzing the changes in emulsion droplets with light irradiation, and assessing the impact of light irradiation on emulsions [48]. The report also delves into a discussion of prior research on photo-induced demulsification and the crucial role played by azobenzene groups in the demulsification process [48]. Biswas et al. discuss the dependence of light intensity and the photo kinetics of destabilization, which has applications in efficient phase separation for various uses like photocatalysis and drug delivery [49].

5.3 *Optical Switch*

Tang et al. developed a novel microfluidic chip design that reduces instrumentation complexity and cost by employing optical switches for sequential signal readout across multiple detection channels. The chip, fabricated via two-photon lithography, offers high stability, customizable switching frequencies, and rapid response times

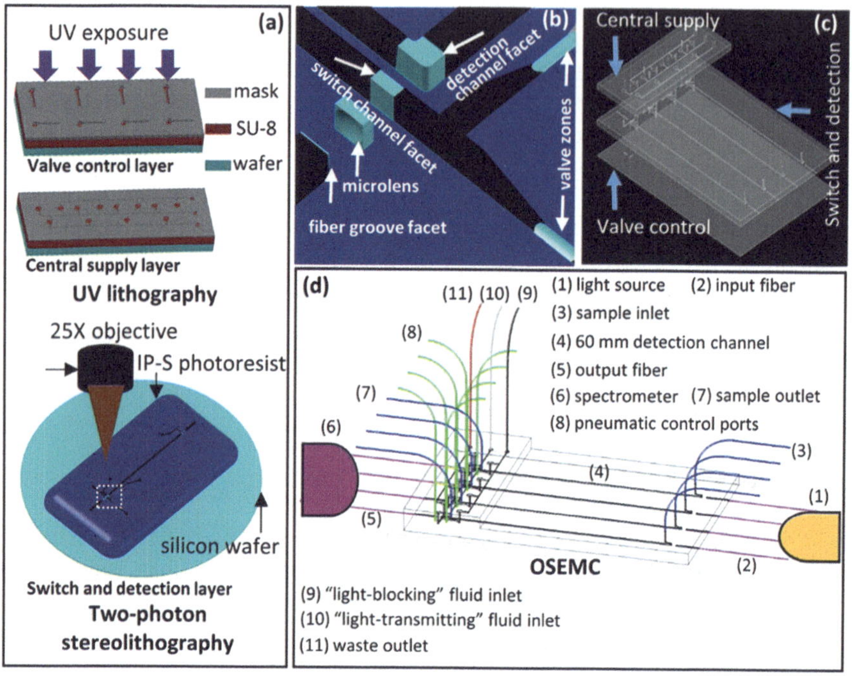

Fig. 5 The optically stimulated electrofluidic microchip (OSEMC) system: setup and operation. (**a**) Two-photon stereo lithography is used to build molds of the switch and detection layers, and UV lithography is used to create molds with 2D planar structures for the valve control layer and central supply layer. (**b**) Close-up of the region corresponding to the dashed white box in (**a**), showing the microvalves and key optical structures created utilizing the hierarchical printing technique (highlighted in blue). (**c**) An exploded diagram showing the three-layered layers that make up the OSEMC. (**d**) Descriptive schematic showing the OSEMC system for multichannel colorimetric analysis, complete with in-image descriptive annotations. (Reprinted with permission from Tang et al. [50])

[50]. Figure 5 illustrates the setup and operation of the Optically Stimulated Electrofluidic Microchip (OSEMC) system.

5.4 *Nonreciprocal Self-Oscillator*

Soft robots made of sensitive materials that can move on their own as their surroundings change are described by Deng et al. Achieving a self-oscillating material construct with self-sustaining movements produced by mechanisms like self-shadowing, mechanical zero-elastic-energy mode, Belousov–Zhabotinsky reaction, etc. is crucial. This material construct must vibrate mechanically under a continuous stimulus field. Although the application of self-oscillating materials in

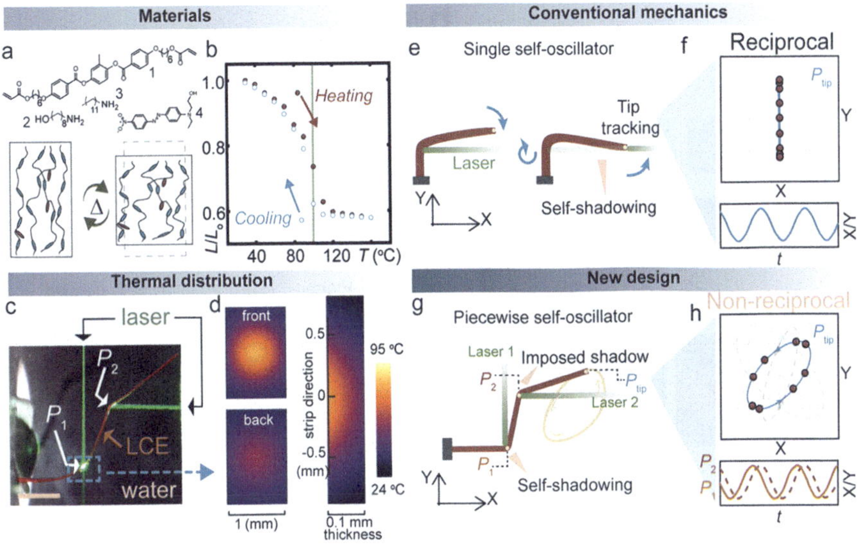

Fig. 6 Illustration of the divided light-driven self-oscillator conceptually. (**a**) Top: Elements used in manufacturing LCEs or light-responsive liquid crystal elastomers. (1) Liquid crystal crosslinker, (2) extenders for chains, (3) dye that absorbs light; (4) bottom: diagram showing anisotropic deformation of a planar-aligned LCE upon thermal stimulation. (**b**) Deformation brought on by heat over a single thermal cycle. The green bar shows the temperature needed to reach 50% of maximal contraction. (**c**) A picture showing how an LCE strip bends in response to two water-based laser beams. P1 and P2 indicate the orthogonal laser excitation points on the LCE. Laser specs: 1 mm point size, 50 mW power, 532 nm wavelength. Sample: 0.1 mm thick. Five millimeters is the scale bar. (**d**) A simulation showing the heat gradient across the sample thickness (right) and the photothermal distribution on the front and rear surfaces of the LCE (left). Spot size: 1 mm; laser power: 80 mW. Sample: 0.1 mm thick. (**e**) Schematic illustrating how self-shadowing causes photo-oscillation when excited by a single laser beam; (**f**) associated tip displacement trajectory. (**g**) Diagram showing the segmented photo-oscillator triggered by two laser beams oriented perpendicularly. While laser 2 applies a secondary oscillation onto the oscillating cantilever from an orthogonal direction, laser 1 excites the strip's bottom part. (**h**) The nonreciprocal stroke pattern that corresponds. (Reprinted with permission from Deng et al. [51])

microfluidics is rarely described, the paper offers examples of dispersed robotic systems that have proven self-sustained walking, swimming, item movement, and electric power generation [51]. The conceptual representation of the segmented light-driven self-oscillator is summarized in Fig. 6.

5.5 *Peristaltic Pumping*

Dradrach et al. discovered an innovative optofluidic method, leveraging the peristaltic deformations of a light-controlled liquid crystal gel (LCG) strip for efficient

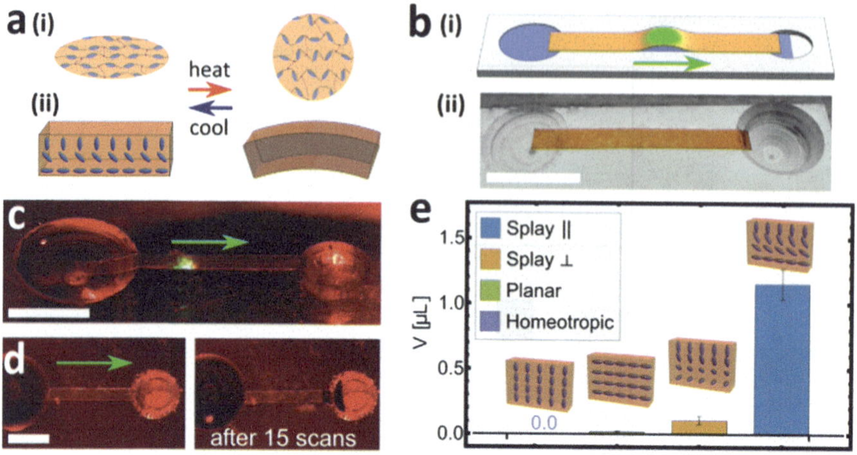

Fig. 7 An experimental demonstration and illustration of an LCG photo-peristaltic pump. Diagram of an LCE showing contraction along the alignment direction (**a**) (i). (ii) An LCG sheet bent with a splay-bend director as a result of expansion on the top after heating and contraction on the bottom after heating. (**b**) (i) Diagram illustrating the experimental configuration, with the direction of fluid transport and scanning indicated by an arrow. (ii) The system's unstimulated image. (**c**) A photo of the experimental setup that shows the peristaltic motion caused by light. (**d**) Before and after images (black-colored for clarity) showing the movement of water under 15 laser scans. (**e**) Comparing the amount of liquid moved for each of the four LC orientations in a single scan. Every strip was 1.7 mm broad and subjected to 12.8 W/cm^2 of peak light intensity. Every image has a 1 cm scale bar. (Reprinted with permission from Dradrach et al. [52])

microfluidic transport. Unlike conventional systems, this approach eliminates the need for valves, external mechanical connections, or intricate fabrication processes. The pump's simplicity lies in its design, where scanning illumination induces peristalsis in the LCG strip, providing adaptability and ease of adjustment during tasks. The authors propose its potential as a universal "lab-on-a-chip," offering versatility akin to electro-wetting droplet approaches. It is suitable for fast prototypes, useful applications, and post hoc microfluidic function additions. While acknowledging limitations, such as the inverse relationship between pumping rate and strip length, the study suggests future advancements, including employing 2D sheets of actuating material for enhanced mechanical strength and miniaturization. Figure 7 shows an illustration and experimental demonstration of an LCG photo-peristaltic pump. This research highlights the promising avenues of the proposed optofluidic approach in advancing microfluidics [52].

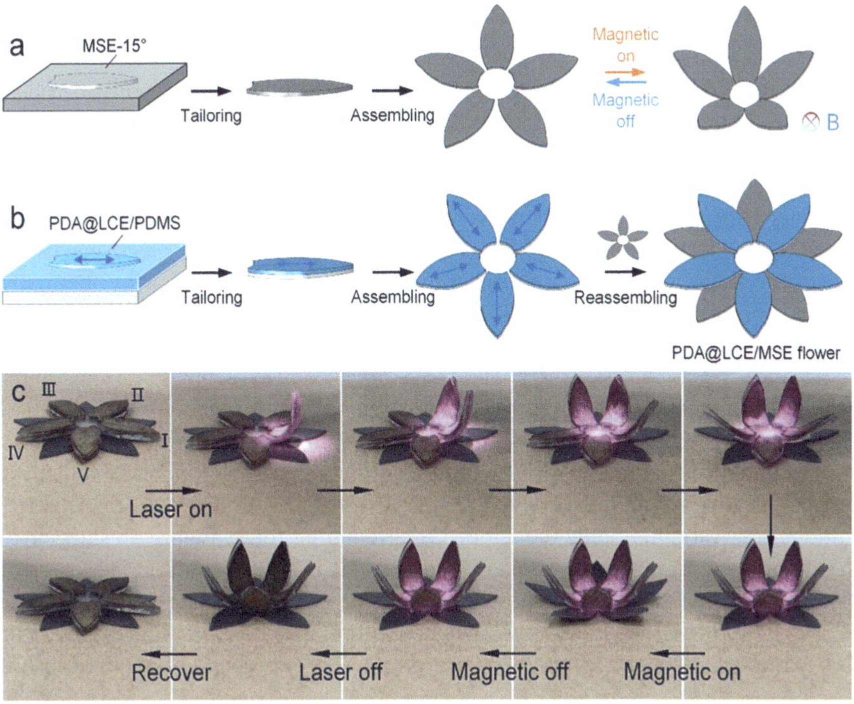

Fig. 8 Manufacturing procedure for (**a**) MSE flower and (**b**) PDA@LCE/MSE flowerlike structure. (**c**) The flowerlike PDA@LCE/MSE's multimode deformation in response to infrared laser and magnetic field control. (Reprinted with permission from Zhang et al. [53])

5.6 *Programmable Shape Deformation*

A multi-responsive soft actuator that combines a liquid crystal elastomer (LCE) and a magnetically sensitive elastomer (MSE) was introduced by Zhang et al., showcasing the extraordinary potential for applications in the field of soft photo actuators in microfluidics. The actuator shows sophisticated and controllable deformations, such as bidirectional bending, shrinkage/bending, rolling/bending, and twisting/bending, by modulating stimulus strength, geometrical dimensions, and material properties in concert. The fabrication method of the MSE flower and the flowerlike PDA@LCE/MSE is shown in Fig. 8. Most importantly, the actuator's photosensitive properties allow for both sequential and localized shape-morphing manipulation, offering a significant increase in degrees of freedom (DoFs). The potential benefits of this capacity include improved precision and adaptability of soft photo actuators in microfluidic devices, as well as opportunities for bioengineering, soft manipulators, and flexible electronics in microfluidic environments [53].

5.7 Underwater Soft Robotics

Chujun et al. discuss the development of high-speed underwater hydrogel robots in their recent article [54]. They introduce a novel approach to achieving rapid and programmable actuation in thermally responsive hydrogels. Traditionally, hydrogels undergo actuation through stimuli-induced swelling and deswelling, constrained by water diffusion kinetics. However, the authors propose a method involving dynamic photoactivated disulfide bond exchange to create spatial-selective network anisotropy. This modification allows for actuation primarily driven by thermally induced conformation changes in locally aligned polymer chains. By incorporating photothermal fillers, the researchers demonstrate light-powered oscillations with frequencies reaching up to 1.7 Hz. This advancement enables various high-speed movements such as continuous swimming, stepwise walking, and rotation The process of photomechanical programming introduces remarkable flexibility in customizing actuation modes by adjusting the external deformation force. As depicted in Fig. 9a, an identical sample can exhibit two distinct actuation modes, with the sole variation being the applied programming force. Figure 9b showcases chronological programming, where initially, a flat hydrogel palm adopts an "ok" gesture with only two fingers showing actuation. The remaining three fingers are then programmed as well, creating a whole fist form that allows for the actuation of all five fingers. Figure 9c shows how the spatio-selective property of photo exposure

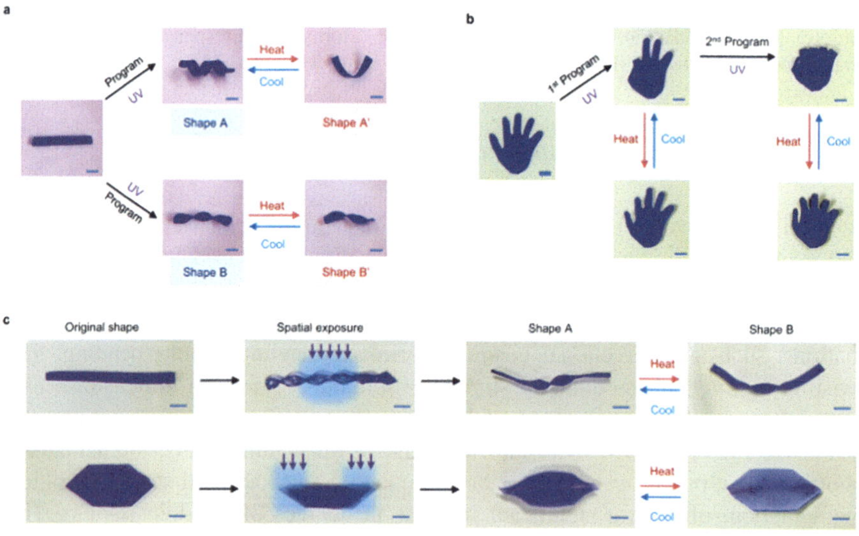

Fig. 9 Dynamic responses of the photomechanically programmed hydrogel. (**a**) Two distinct actuation modes facilitated by applying varying programming forces on identical samples. (**b**) Temporal progression of programming. (**c**) Spatially selective programming of hydrogels for region-specific actuations. All scale bars measure 5 mm. (Reprinted with permission from Ni et al. [54])

may be used to precisely designate the actuation zones of a uniformly deformed sample. This emphasizes how flexible and liberating reconfigurable actuations are [54].

5.8 Photo Actuators for Integrated Microfluidic Pumps

Efficient fluid management is crucial as lab-on-a-chip systems strive to integrate intricate, multistep microfluidic operations. In order to create integrated fluidic elements, Schmidleithner et al. use a splay-aligned liquid crystal network (LCN) photo actuator that deforms when exposed to blue light. An actuator material that is precisely defined is generated through manufacturing parameter optimization. A $3 \times 8 \times 0.05$ mm LCN film curves by 380 μm under 455 nm LED illumination at 41 mW/cm^2, producing forces of 1.1 mN. The size of the LCN film, temperature, and light intensity can all be used to modify the actuator's responsiveness, which remains constant over time. We demonstrate its integration into a microfluidic chip by combining this actuator with a polydimethylsiloxane (PDMS) membrane, achieving fluid movement of 77 nL in a 2-s stroke. Two passive PDMS check valves complete the integrated micropump, and the LCN film functions as a pumping membrane in a pumping configuration, attaining a constant pumping rate of 0.1 μL/min. The main benefit of this micropump is its noncontact actuation mechanism, which makes it simple to integrate into microfluidic chips and eliminates the need for connections to external devices. Figure 10 displays an examination of the fluid flow within a channel. Furthermore, the low operating power and simple manufacturing method satisfy the essential needs for point-of-care applications [55].

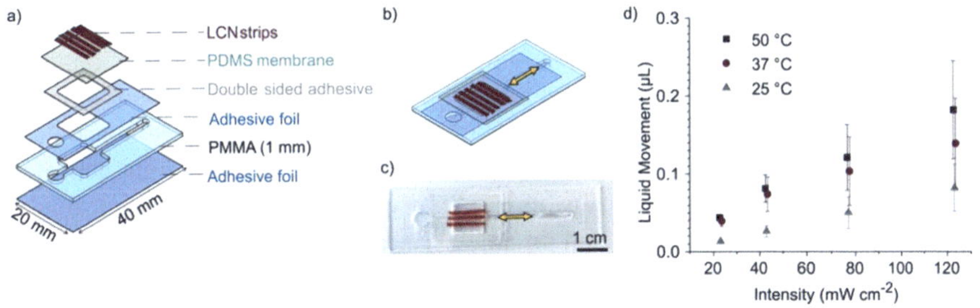

Fig. 10 Examination of a channel's internal fluid flow. (**a**) Microfluidic chip design diagram. (**b**) The test channel setup was assembled. The fluid movement's direction is shown by the yellow arrow. Fluid is forced into the channel when the light is on, and it exits the channel when the LED is off. (**c**) The test chip's image. Three LCN strips, each measuring 10 mm in length and 1 mm in width, are part of the active unit. (**d**) Fluid movement for 2-s light pulses at different ambient temperatures as a function of intensity ($n = 3$, mean ± standard deviation). For every pulse, higher temperatures and intensities facilitate more fluid flow inside the channel. (Reprinted with permission from Schmidleithner et al. [55])

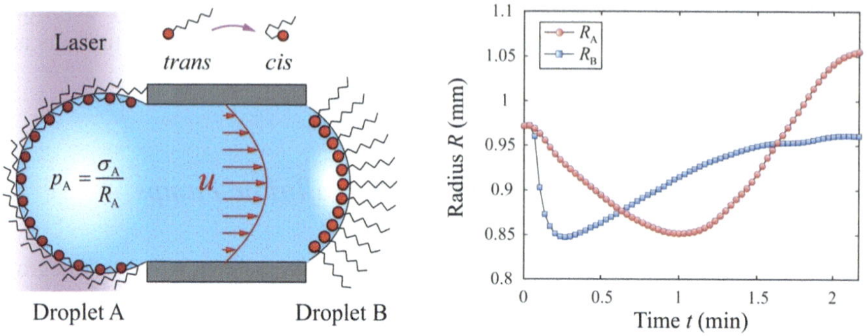

Fig. 11 Explanation of how light-actuated surface tension can control liquid movement and droplet switching. When the laser strikes the trans-configuration, it changes to the cis-configuration. As time increases, the radius on both sides increases, with initial drops in radius. (Reprinted with permission from Zhao et al. [56])

5.9 Manipulation of Liquid Transport and Droplet Switch

Manipulation of microfluidic systems is essential in many disciplines, including biology, chemistry, and materials science. Conventional approaches to controlling these systems frequently need complex configurations or rely on outside factors, which might limit their ability to be integrated and miniaturized. Zhao et al. present a unique method of light-activated photosensitive azobenzene surfactant control for microfluidic devices. This light-driven method offers a simpler and more practical way to control microfluidic devices. They employ their approach to overcome the surface energy barrier of a double droplet system (DDS) and regulate the transition between two bifurcations, demonstrating the efficacy of their strategy. Using this light-driven method, the experimental results successfully illustrate transitions between the DDS bifurcations. They create a dynamic model during experiments that accounts for the surfactant absorption process in order to better comprehend the changeover mechanism. Furthermore, they demonstrate a real-world implementation of our suggested methodology using a light-activated adjustable liquid lens, emphasizing its possible wider application in microfluidic systems, with results summarized in Fig. 11. The goal of this research is to progress the creation of novel light-driven manipulation techniques for microfluidic applications [56].

6 Future Directions

The integration of soft material–based photo actuators with microfluidic systems has unlocked a myriad of possibilities in various fields, ranging from biomedical applications to materials science and beyond. As we move forward, several promising avenues emerge for further exploration and development in this burgeoning field.

6.1 Enhanced Actuation Efficiency

Future research should be directed toward enhancing the efficiency and performance of photo actuators, with a particular focus on optimizing actuation speed, responsiveness, and energy consumption. These factors are crucial for advancing the capabilities of photo actuation technologies, especially in applications that demand high precision and rapid response, such as in microfluidic systems used for biochemical analysis, diagnostics, and other lab-on-a-chip technologies.

Improvements in material design are essential for achieving these enhancements. Developing advanced photo-responsive materials that offer faster actuation speeds and more excellent responsiveness to light stimuli is a crucial area of research. Materials with improved photomechanical properties, such as higher photo-induced strain and quicker relaxation times, can significantly enhance actuator performance. Exploring novel materials, including polymers, hydrogels, and composite materials, that combine desirable attributes such as light sensitivity and mechanical robustness could lead to substantial performance gains.

Advances in fabrication techniques also play a critical role in optimizing photo actuator performance. Precision fabrication methods, including advanced lithography, micro-manufacturing, and additive manufacturing, are necessary to produce actuators with high structural integrity and uniformity. These techniques enable the creation of intricate actuator designs with precise control over their mechanical properties, which is vital for achieving rapid and reliable actuation. Furthermore, innovations in fabrication processes that reduce material waste and improve energy efficiency during production are also important considerations.

In addition to material and fabrication advancements, the optimization of opto-mechanical coupling mechanisms is crucial. Effective opto-mechanical coupling involves the efficient conversion of light energy into mechanical motion. Research efforts should focus on refining these mechanisms to minimize energy losses and enhance the overall efficiency of photo actuators. Techniques such as the integration of plasmonic structures or advanced light focusing methods can improve light absorption and distribution, leading to more efficient actuation.

In material design, fabrication techniques and opto-mechanical coupling mechanisms hold the potential to achieve unprecedented levels of control and precision in microfluidic manipulation. By addressing these critical aspects, future research can lead to the development of photo actuators that are faster, more responsive, and energy-efficient, thus broadening the scope and capabilities of microfluidic technologies.

6.2 Multifunctional Integration

There is an increasing interest in integrating multiple functionalities onto a single microfluidic chip, encompassing capabilities such as pumping, mixing, sorting, and sensing. This trend reflects the need for more sophisticated and compact systems to perform complex tasks with enhanced efficiency and reduced operational complexity. Future research and development efforts will likely focus on designing and implementing multifunctional photo-responsive materials that can concurrently support various functions within microfluidic systems.

Integrating multifunctional photo-responsive materials holds the promise of significantly advancing microfluidic technology by enabling the concurrent execution of diverse tasks. For instance, materials that exhibit multiple photo-responsive behaviors could be engineered to simultaneously drive fluid flow, facilitate mixing, and enable sorting or sensing functions. Such materials would simplify the overall architecture of microfluidic devices by reducing the need for separate components dedicated to each function. This integration could lead to more streamlined device designs, reduced fabrication complexity, and lower overall system costs.

Advancements in material science will be pivotal in realizing these multifunctional capabilities. Researchers are exploring a range of photo-responsive materials, including smart polymers, hydrogels, and composite materials, that can undergo multiple, distinct transformations in response to light. These materials could be engineered to exhibit controlled swelling, contraction, or phase transitions, allowing them to perform various tasks such as pumping fluids, mixing reagents, or sorting particles based on size or type.

The development of advanced fabrication techniques will be crucial for integrating these multifunctional materials into microfluidic devices. Techniques such as micro-manufacturing and layer-by-layer assembly will enable the precise patterning and incorporation of photo-responsive materials into microfluidic channels and components. These methods must accommodate the complex interactions between the different functionalities to ensure seamless operation within the device. The successful implementation of multifunctional photo-responsive materials will necessitate the development of sophisticated control strategies. These strategies should allow for precise modulation of the photo-responsive properties in real time, enabling dynamic adaptation to varying experimental conditions or operational requirements. Innovations in optical control and feedback mechanisms will be essential to fully harness the capabilities of these advanced materials.

6.3 Biocompatibility and Biodegradability

As the applications of photo actuators continue to expand within the biomedical and healthcare domains, there is an escalating emphasis on the development of actuators utilizing biocompatible and biodegradable materials. These materials are critical in ensuring that photo actuation technologies can be safely employed in biological environments without eliciting adverse reactions or long-term environmental harm. Integrating such materials not only aligns with the requirements of medical applications but also addresses broader concerns regarding sustainability and environmental impact.

Future research must therefore prioritize the identification and characterization of sustainable materials that can perform effectively as photo actuators while minimizing their ecological footprint. This includes exploring novel polymers, composites, and other material systems that exhibit the necessary photo-responsive properties while being nontoxic, degradable, and easily assimilated or excreted by biological systems. The challenge lies in balancing the functional performance of these materials with their biocompatibility and biodegradability, ensuring that they maintain their efficacy throughout their operational lifespan and degrade safely post-use.

The development of such materials is integral to the advancement of eco-friendly microfluidic technologies. By incorporating biocompatible and biodegradable photo actuators into microfluidic platforms, researchers can design systems that meet the rigorous demands of medical and clinical applications and adhere to principles of environmental stewardship. These systems could be employed in various biomedical applications, including drug delivery, tissue engineering, and diagnostics, where the safety and environmental impact of the materials used are paramount.

The safe disposal of these materials after their functional life is a critical consideration. Using biodegradable photo actuators ensures that microfluidic devices can be disposed of without contributing to environmental pollution, thus supporting the broader goal of developing sustainable healthcare technologies. This focus on sustainability could drive innovation in the field, creating new materials and technologies that harmonize with human health and environmental preservation.

6.4 Miniaturization and Integration

Ongoing research and development efforts are crucial for miniaturizing and integrating photo actuation systems into compact and portable microfluidic platforms. Achieving these objectives is expected to significantly expand the applicability and functionality of such systems across various scientific and industrial domains.

Central to this endeavor is the advancement of microfabrication techniques. Additive manufacturing, commonly known as 3D printing, has become a pivotal technology. Its ability to produce highly precise, microscale structures facilitates the creation of complex, multi-material architectures essential for embedding photo actuation mechanisms within the constrained dimensions of microfluidic devices. The versatility of additive manufacturing enables the exploration of novel geometries and material combinations, optimizing system performance while maintaining a minimal spatial footprint.

Microscale patterning techniques, including photolithography and soft lithography, are indispensable in achieving the precise spatial organization of components within microfluidic systems. These methodologies allow for the accurate placement of photo-responsive materials, ensuring that the actuation mechanisms function as intended at the microscale. As these patterning techniques continue to evolve, they are expected to enable the fabrication of increasingly sophisticated and efficient photo actuation systems, thereby enhancing the performance and scalability of these devices. The synergy of these microfabrication advancements is anticipated to yield significant improvements in the functionality, reliability, and accessibility of miniaturized photo actuation systems. Integrating such systems into portable microfluidic platforms could revolutionize applications requiring mobility, such as point-of-care diagnostics, environmental sensing, and wearable technologies. Additionally, the scalability inherent in advanced microfabrication techniques could facilitate the mass production of these devices, rendering them more cost-effective and broadly accessible.

6.5 Dynamic Control and Feedback Mechanisms

Future photo actuation systems hold the promise of significant advancements through the incorporation of dynamic control and feedback mechanisms, which would enable real-time adjustments and optimization of fluidic operations. These systems could drastically improve the precision and efficiency of processes by allowing continuous monitoring and fine-tuning of actuation responses as conditions change.

The integration of advanced sensing technologies, such as microfluidic sensors and high-resolution imaging systems, would be a key component in achieving this level of control. By providing detailed, real-time data on various parameters like flow rates, fluid composition, and environmental factors, these sensors would enable the development of sophisticated closed-loop control systems. Such systems could autonomously adapt to fluctuations in the operating environment, ensuring optimal performance without the need for constant human intervention.

For instance, combining photo actuation with integrated sensors in microfluidic applications could allow for precise manipulation of tiny volumes of fluids, which is crucial for chemical analysis, drug delivery, and biological research applications. The feedback from these sensors could adjust the intensity, duration, and wavelength of light exposure in real time, optimizing the actuation process to maintain the desired fluidic behavior.

Moreover, this dynamic control could extend to more complex applications, such as soft robotics and wearable devices, where adaptive systems are essential for responding to unpredictable or changing environments. In these contexts, photo-responsive actuators equipped with integrated sensing and feedback could enable robots and wearable technologies to adjust their behavior autonomously, enhancing their functionality and resilience.

6.6 Exploration of Novel Applications

The versatility of soft material–based photo actuators presents a wealth of opportunities for innovative applications across diverse fields. These actuators, which harness the responsive properties of soft materials to convert light into mechanical movement, have already demonstrated significant potential in microfluidics. However, their full range of capabilities is far from being realized.

Future research should extend beyond the traditional confines of microfluidics, delving into soft robotics, where these materials could create highly flexible, adaptive, and lightweight robotic systems. These robots could perform tasks in environments that are too delicate or complex for conventional rigid robots, such as biomedical procedures or exploring unstructured terrains. Additionally, in wearable devices, photo-responsive materials could lead to the development of smart textiles that change shape or function in response to light, enhancing user comfort, safety, and interaction. For instance, clothing could automatically adjust its insulation properties based on environmental light conditions, or wearable sensors could provide real-time health monitoring by reacting to light signals. Environmental monitoring is another promising field where these actuators could be utilized. Devices equipped with photo-responsive materials could be deployed in remote or hazardous locations, autonomously responding to changes in light levels, thereby providing continuous data collection without the need for human intervention. In smart agriculture, the integration of photo-responsive actuators could revolutionize farming practices. For example, they could be used to create self-regulating systems for crop management, where light-controlled devices optimize plant exposure to sunlight, regulate water usage, or even deter pests, all while minimizing human labor and resource consumption.

7 Conclusion and Challenges

7.1 Conclusion

Light is an excellent external energy source and stimulant that can be applied remotely. Advanced light-response technology has been developed as a result of light exploration. Simple processing, quick reaction, and flexibility are benefits of flexible light-responsive actuators, which offer several opportunities for use in photon-energy accumulation systems. This chapter examines the advancement of current work on soft actuator materials that respond to light and their actuation methods of manufacture, modes of deformation, performance, and their uses. Thus far, numerous actuators made of different light-responsive materials have been produced. Various actuation techniques have been applied in biomimetic microfluidics and pliable robots.

This chapter shows that the light-responsive actuator is currently only being used for simple tasks such as gripping and simulating the movement of certain animals when they crawl and jump. There will be no immediate release of this application. Using the grasping device in the article as an example, in real life, these actuators typically produce very small forces that are insufficient to lift objects or be topical. As a result, their use is limited to controlling microscopic objects, even though they could be used as soft grippers. Consequently, in order to increase the likelihood of an application, output force can be increased by it harder. For instance, the thickness, it is possible to increase the flexible photo-responsive actuator's stiffness while making sure the actuator performs, certain photo-responsive nanoparticles may be included. These nanoparticles will improve the photo-responsive polymer's synthesis and photo-responsive performance while lowering the material's flexibility and bolstering output force and rigidity.

The main focus of flexible robotics research has been on driving techniques and intelligent materials. But light driving has emerged as a more practical and economical driving style in recent years. Current studies, however, indicate that most of their answers rely on temporary or permanent light source illumination to ensure the robot's continued mobility, as the robot would rapidly stop moving and waste energy if the light is switched off.

Integrating soft material–based photo actuators with microfluidic systems holds tremendous potential for revolutionizing diverse industries and scientific disciplines. By harnessing the synergistic capabilities of advanced materials, photonics, and microfluidics, researchers can continue to push the boundaries of innovation, paving the way for transformative advancements in healthcare, biotechnology, materials science, and beyond. As we embark on this exciting journey, collaboration across multidisciplinary fields and a commitment to sustainability and ethical practices will be essential in shaping the future landscape of photo-driven microfluidics.

7.2 *Challenges and Viewpoint*

Despite extensive research and application of light-responsive actuators, several challenges remain, particularly for high-performance and multifunctional uses. These actuators show promise for roles such as grippers in flexible robotics and controllers in microfluidic systems, yet their application is often confined to laboratory settings. Therefore, advancing the performance of these actuators and transitioning them from the lab to practical use is crucial. Additionally, while many light-responsive actuators rely on artificial light sources, there is a pressing need to develop devices that can be activated by natural light. Although some current models employ sophisticated technologies that closely mimic real-world applications, they still face material choice and design limitations. Selecting the optimal materials remains complex due to their diverse advantages and disadvantages. Light-sensitive actuators are assessed based on parameters like bending angle, curvature, or displacement. A standardized evaluation criterion should be established to enhance comparability across research efforts, which would facilitate better alignment and accelerate progress in actuator development. Large-area actuators, in particular, must achieve high performance and environmental resilience. Integrating inorganic materials, polymers, and crystals into actuator design could boost both performance and mechanical strength, paving the way for new applications. Additionally, coupling these actuators with smart sensor systems can enhance their multifunctionality. Expanding their application range to address modern needs will also benefit from combining various stimuli-responsive and photosensitive materials.

Acknowledgements RKP acknowledges the Department of Science and Technology for an INSPIRE Faculty Award Grant [DST/INSPIRE/04/2016/002370] and the Core Research Grants [CRG/2020/006281, CRG/2021/004759] from SERB, Government of India. The authors acknowledge IISER Tirupati for funding and research facilities. RKP acknowledges the CNRS for the award of a Visiting Professorship at the Laboratoire de Physiques des Solides in 2023.

References

1. Chen, L.-J., Li, Q.: Soft photoactuators in microfluidics. In: Photoactive Functional Soft Materials, pp. 167–196. John Wiley & Sons Ltd. (2019)
2. El-Atab, N., Mishra, R.B., Al-Modaf, F., et al.: Soft actuators for soft robotic applications: a review. Adv. Intell. Syst. **2**, 2000128 (2020). https://doi.org/10.1002/aisy.202000128
3. Ding, M., Jing, L., Yang, H., et al.: Multifunctional soft machines based on stimuli-responsive hydrogels: from freestanding hydrogels to smart integrated systems. Mater. Today Adv. **8**, 100088 (2020). https://doi.org/10.1016/j.mtadv.2020.100088
4. Qian, J.-Y., Hou, C.-W., Li, X.-J., Jin, Z.-J.: Actuation mechanism of microvalves: a review. Micromachines. **11**, 172 (2020). https://doi.org/10.3390/mi11020172
5. Baigl, D.: Photo-actuation of liquids for light-driven microfluidics: state of the art and perspectives. Lab Chip. **12**, 3637–3653 (2012). https://doi.org/10.1039/C2LC40596B
6. Whitesides, G.M.: The origins and the future of microfluidics. Nature. **442**, 368–373 (2006). https://doi.org/10.1038/nature05058
7. Monat, C., Domachuk, P., Eggleton, B.J.: Integrated optofluidics: a new river of light. Nat. Photon. **1**, 106–114 (2007). https://doi.org/10.1038/nphoton.2006.96
8. Raj, M.K., Chakraborty, S.: PDMS microfluidics: a mini review. J. Appl. Polym. Sci. **137**, 48958 (2020). https://doi.org/10.1002/app.48958
9. Hao, Y., Zhang, S., Fang, B., et al.: A review of smart materials for the boost of soft actuators, soft sensors, and robotics applications. Chin. J. Mech. Eng. **35**, 37 (2022). https://doi.org/10.1186/s10033-022-00707-2
10. Higurashi, E., Ohguchi, O., Tamamura, T., et al.: Optically induced rotation of dissymmetrically shaped fluorinated polyimide micro-objects in optical traps. J. Appl. Phys. **82**, 2773–2779 (1997). https://doi.org/10.1063/1.366163
11. Wang, L., Li, Q.: Photochromism into nanosystems: towards lighting up the future nanoworld. Chem. Soc. Rev. **47**, 1044–1097 (2018). https://doi.org/10.1039/C7CS00630F
12. Gelebart, A.H., Jan Mulder, D., Varga, M., et al.: Making waves in a photoactive polymer film. Nature. **546**, 632–636 (2017). https://doi.org/10.1038/nature22987
13. Dong, X., Xu, J., Xu, X., et al.: Sunlight-driven continuous flapping-wing motion. ACS Appl. Mater. Interfaces. **12**, 6460–6470 (2020). https://doi.org/10.1021/acsami.9b20250
14. Deng, H., Zhang, C., Su, J.-W., et al.: Bioinspired multi-responsive soft actuators controlled by laser tailored graphene structures. J. Mater. Chem. B. **6**, 5415–5423 (2018). https://doi.org/10.1039/C8TB01285G

15. Liu, Y., Shaw, B., Dickey, M.D., Genzer, J.: Sequential self-folding of polymer sheets. Sci. Adv. **3**, e1602417 (2017). https://doi.org/10.1126/sciadv.1602417
16. Williams, B.J., Anand, S.V., Rajagopalan, J., Saif, M.T.A.: A self-propelled biohybrid swimmer at low Reynolds number. Nat. Commun. **5**, 3081 (2014). https://doi.org/10.1038/ncomms4081
17. Ma, M., Guo, L., Anderson, D.G., Langer, R.: Bio-inspired polymer composite actuator and generator driven by water gradients. Science. **339**, 186–189 (2013). https://doi.org/10.1126/science.1230262
18. Ko, H., Javey, A.: Smart actuators and adhesives for reconfigurable matter. Acc. Chem. Res. **50**, 691–702 (2017). https://doi.org/10.1021/acs.accounts.6b00612
19. Yang, Y., Zhan, W., Peng, R., et al.: Graphene-enabled superior and tunable photomechanical actuation in liquid crystalline elastomer nanocomposites. Adv. Mater. **27**, 6376–6381 (2015). https://doi.org/10.1002/adma.201503680
20. Liu, X., Wang, X., Liu, T., Keller, P.: Gold nanoparticles incorporated nematic gel micropillars capable of laser actuation at room temperature. Macromolecules. **49**, 8322–8331 (2016). https://doi.org/10.1021/acs.macromol.6b01930
21. Liu, L., Liu, M.-H., Deng, L.-L., et al.: Near-infrared chromophore functionalized soft actuator with ultrafast photoresponsive speed and superior mechanical property. J. Am. Chem. Soc. **139**, 11333–11336 (2017). https://doi.org/10.1021/jacs.7b06410
22. Chen, S., Wu, R., Li, P., et al.: Acid-interface engineering of carbon nanotube/elastomers with enhanced sensitivity for stretchable strain sensors. ACS Appl. Mater. Interfaces. **10**, 37760–37766 (2018). https://doi.org/10.1021/acsami.8b16591
23. Lee, Y., Song, W.J., Sun, J.-Y.: Hydrogel soft robotics. Mater. Today Phys. **15**, 100258 (2020). https://doi.org/10.1016/j.mtphys.2020.100258
24. ter Schiphorst, J., Coleman, S., Stumpel, J.E., et al.: Molecular design of light-responsive hydrogels, for in situ generation of fast and reversible valves for microfluidic applications. Chem. Mater. **27**, 5925–5931 (2015). https://doi.org/10.1021/acs.chemmater.5b01860
25. Ambulo, C.P., Tasmim, S., Wang, S., et al.: Processing advances in liquid crystal elastomers provide a path to biomedical applications. J. Appl. Phys. **128**, 140901 (2020). https://doi.org/10.1063/5.0021143
26. Jiang, Z.-C., Xiao, Y.-Y., Tong, X., Zhao, Y.: Selective decrosslinking in liquid crystal polymer actuators for optical reconfiguration of origami and light-fueled locomotion. Angew. Chem. **131**, 5386–5391 (2019). https://doi.org/10.1002/ange.201900470
27. Davidson, E.C., Kotikian, A., Li, S., et al.: 3D printable and reconfigurable liquid crystal elastomers with light-induced shape memory via dynamic bond exchange. Adv. Mater. **32**, 1905682 (2020). https://doi.org/10.1002/adma.201905682
28. Chen, H.-M., Wang, L., Zhou, S.-B.: Recent progress in shape memory polymers for biomedical applications. Chin. J. Polym. Sci. **36**, 905–917 (2018). https://doi.org/10.1007/s10118-018-2118-7
29. Huang, Y., Yu, Q., Su, C., et al.: Light-responsive soft actuators: mechanism, materials, fabrication, and applications. Actuators. **10**, 298 (2021). https://doi.org/10.3390/act10110298
30. Li, Y., Zhuo, H., Chen, H., Chen, S.: Novel photo-thermal staged-responsive supramolecular shape memory polyurethane complex. Polymer. **179**, 121671 (2019). https://doi.org/10.1016/j.polymer.2019.121671
31. Wang, Y., Pan, Y., Zheng, Z., Ding, X.: Reconfigurable and reprocessable thermoset shape memory polymer with synergetic triple dynamic covalent bonds. Macromol. Rapid Commun. **39**, 1800128 (2018). https://doi.org/10.1002/marc.201800128
32. Yang, Y., Shen, Y.: Light-driven carbon-based soft materials: principle, robotization, and application. Adv. Opt. Mater. **9**, 2100035 (2021). https://doi.org/10.1002/adom.202100035
33. Mauro, M.: Gel-based soft actuators driven by light. J. Mater. Chem. B. **7**, 4234–4242 (2019). https://doi.org/10.1039/C8TB01893F
34. Pang, X., Lv, J., Zhu, C., et al.: Photodeformable azobenzene-containing liquid crystal polymers and soft actuators. Adv. Mater. **31**, 1904224 (2019). https://doi.org/10.1002/adma.201904224

35. Kitagawa, D., Tsujioka, H., Tong, F., et al.: Control of photomechanical crystal twisting by illumination direction. J. Am. Chem. Soc. **140**, 4208–4212 (2018). https://doi.org/10.1021/jacs.7b13605
36. Huang, C., Lv, J., Tian, X., et al.: Miniaturized swimming soft robot with complex movement actuated and controlled by remote light signals. Sci. Rep. **5**, 17414 (2015). https://doi.org/10.1038/srep17414
37. Taniguchi, T., Blanc, L., Asahi, T., et al.: Statistical modeling of photo-bending actuation of hybrid silicones mixed with azobenzene powder. Actuators. **8**, 68 (2019). https://doi.org/10.3390/act8040068
38. Kobayashi, M., Abe, J.: Optical motion control of maglev graphite. J. Am. Chem. Soc. **134**, 20593–20596 (2012). https://doi.org/10.1021/ja310365k
39. Maragó, O.M., Bonaccorso, F., Saija, R., et al.: Brownian motion of graphene. ACS Nano. **4**, 7515–7523 (2010). https://doi.org/10.1021/nn1018126
40. Zhang, T., Chang, H., Wu, Y., et al.: Macroscopic and direct light propulsion of bulk graphene material. Nat. Photon. **9**, 471–476 (2015). https://doi.org/10.1038/nphoton.2015.105
41. Vogt, P., De Padova, P., Quaresima, C., et al.: Silicene: compelling experimental evidence for graphenelike two-dimensional silicon. Phys. Rev. Lett. **108**, 155501 (2012). https://doi.org/10.1103/PhysRevLett.108.155501
42. Tang, S., Dresselhaus, M.S.: Constructing anisotropic single-Dirac-cones in Bi1–xSbx thin films. Nano Lett. **12**, 2021–2026 (2012). https://doi.org/10.1021/nl300064d
43. Luo, X., Yang, L., Cui, Y.: Micropumps: mechanisms, fabrication, and biomedical applications. Sensors Actuators A Phys. **363**, 114732 (2023). https://doi.org/10.1016/j.sna.2023.114732
44. Agarwal, A.K., Sridharamurthy, S.S., Beebe, D.J., Jiang, H.: Programmable autonomous micromixers and micropumps. J. Microelectromech. Syst. **14**, 1409–1421 (2005). https://doi.org/10.1109/JMEMS.2005.859101
45. Au, A.K., Lai, H., Utela, B.R., Folch, A.: Microvalves and micropumps for BioMEMS. Micromachines. **2**, 179–220 (2011). https://doi.org/10.3390/mi2020179
46. Lam, R.H.W., Li, W.J.: A digitally controllable polymer-based microfluidic mixing module array. Micromachines. **3**, 279–294 (2012). https://doi.org/10.3390/mi3020279
47. Li, Z., Shi, Y., Zhu, A., et al.: Light-responsive, reversible emulsification and demulsification of oil-in-water pickering emulsions for catalysis. Angew. Chem. Int. Ed. **60**, 3928–3933 (2021). https://doi.org/10.1002/anie.202010750
48. Dagkesamanskaya, A., Langer, K., Tauzin, A., et al.: Use of photoswitchable fluorescent proteins for droplet-based microfluidic screening. J. Microbiol. Methods. **147**, 59–65 (2018). https://doi.org/10.1016/j.mimet.2018.03.001
49. Biswas, S., Karishma, S., Ramesh, B., et al.: Light-induced destabilisation of oil-in-water emulsions using light-active bolaform surfactants. Soft Matter. **19**, 199–207 (2023). https://doi.org/10.1039/D2SM01207C
50. Tang, J., Cao, X., Qiu, G., et al.: Optical-switch-enabled microfluidics for sensitive multichannel colorimetric analysis. Anal. Chem. **93**, 6784–6791 (2021). https://doi.org/10.1021/acs.analchem.1c00674
51. Deng, Z., Zhang, H., Priimagi, A., Zeng, H.: Light-fueled nonreciprocal self-oscillators for fluidic transportation and coupling. Adv. Mater. **36**, e2209683 (2023). https://doi.org/10.1002/adma.202209683
52. Dradrach, K., Zmyślony, M., Deng, Z., et al.: Light-driven peristaltic pumping by an actuating splay-bend strip. Nat. Commun. **14**, 1877 (2023). https://doi.org/10.1038/s41467-023-37445-5
53. Zhang, J., Wang, Y., Sun, Y., et al.: Multi-stimuli responsive soft actuator with locally controllable and programmable complex shape deformations. ACS Appl. Polym. Mater. **5**, 6199–6211 (2023). https://doi.org/10.1021/acsapm.3c00858
54. Ni, C., Chen, D., Wen, X., et al.: High speed underwater hydrogel robots with programmable motions powered by light. Nat. Commun. **14**, 7672 (2023). https://doi.org/10.1038/s41467-023-43576-6

55. Schmidleithner, C., You, Y., Staniszewski, F., et al.: Splay aligned liquid crystal network photoactuators for integrated microfluidic pumps. Sensors Actuators B Chem. **412**, 135820 (2024). https://doi.org/10.1016/j.snb.2024.135820
56. Zhao, Y., Zhang, B., Lv, C.: Manipulation of liquid transport and droplet switch using light-actuated surface tension. Colloids Surf. A Physicochem. Eng. Asp. **687**, 133473 (2024). https://doi.org/10.1016/j.colsurfa.2024.133473

Soft Material–Based Photonic Crystal

Sagar Kumar Pathak and Ravi Kumar Pujala

Abstract Taking inspiration from natural photonic crystals, the need for developing "soft materials" that can change color in response to external stimuli is increasing as photonic crystal technology progresses. These materials can be used to design optical switches or chemical and biological sensors as those are capable of detecting temperature, humidity, pressure, electric and magnetic fields, ions and solvents, biological materials, and hazardous materials. Apart from sensing applications, other device, display, and security applications are also discussed. This review summarizes the recent developments in soft material–based photonic crystals, including classification based on dimension and material, principles for design and fabrication, mechanism of color tuning, and their tunable structural color in responsiveness to various stimuli. Various techniques for creating soft material–based photonic crystals have been reviewed, including top-down and bottom-up approaches. Most of these techniques depend on self-assembly processes to accomplish ordering. Additionally, the present restrictions and prospective future paths in the domains of sensing and self-assemblies are also explored.

Keywords Soft materials · Photonic crystals · External stimuli · Sensing

1 Introduction

Knowledge of the properties of natural materials helped humans upgrade themselves from the Old Stone Age to the present twenty-first century. Initially, the upgrade aimed to find physically hard materials, but later on, with the advancement of science and technology, people started exploring the inner properties of materials. In this century, we deal with materials' electrical, mechanical, and optical properties. The topic of this book chapter is a gift given by advancing those three topics.

S. K. Pathak · R. K. Pujala (✉)
Soft and Active Matter Group, Department of Physics, Indian Institute of Science Education and Research (IISER) Tirupati, Tirupati, Andhra Pradesh, India
e-mail: pujalaravikumar@iisertirupati.ac.in

V. Kumar, Y. K. Mishra (eds.), *Soft Materials for Functional Applications*,
Advanced Structured Materials 225, https://doi.org/10.1007/978-981-97-9468-3_9

With the advancement of mechanical properties of the material, we can produce some fantastic artificial materials like ceramic, plastics, etc. After lots of satisfying applications from the electronics regime, from transistors to diodes, researchers are now exploring the optical properties of the materials. Massive technological development and advancement could be possible using the optical properties of the material, such as high-speed computing, optical fiber, laser medium, and telecommunication and sensors [1–4].

Crystals have periodic atoms or molecules in an ordered pattern. That imaginary pattern is called a lattice; the atoms or molecules are called the basis. The periodicity of atoms implies a periodic potential across each atom, which gives the concept to explain the electronic properties inside the crystal [5]. Using quantum mechanics and solid-state physics, we learn about the electronic energy bandgap and the propagation of electron waves [6]. A photonic crystal structure features a regularly varying refractive index, where atoms and molecules are replaced by periodic arranged dielectric materials of nanoscale, and periodic potential is returned by the dielectric function. Generally, the dielectric function is equivalent to the refractive index of that medium [7].

Initially, the researcher focused on photonic crystals made of "hard materials" with little color tuning. They are employed in the commercial sector as high-efficiency thin-film optics that reflect specific light frequencies, such as optical filters [8] or dielectric mirrors [9]. As photonic crystal technology advances, there is a growing need to create "soft materials" that can change color in response to outside stimuli [10]. Soft material–based photonic crystals deal with colloidal particle self-assembly [11], physically or chemically crossed-linked polymer gels [12], polymer-based thin films [13], etc. A photonic gel is one soft polymeric matrix that can incorporate photonic bandgap devices. Gels are polymer networks with enough solvents cross-linked chemically or mechanically [14]. They are readily malleable and experience reversible volume changes in response to variations in the solvent's pH [15], light intensity [16], mechanical force [17], temperature [18], and other environmental factors [19, 20]. Because of the stimuli-responsive properties of soft material–based photonic crystals, these materials offer substantially enhanced color quality, sensitivity, reaction time, durability, and selectivity, making them highly promising for use in optical switches and sensors [21]. Sensing of temperature and humidity [22], pressure [23], electric and magnetic field [24, 25], ions and solvents [26], biological materials (e.g., glucose, carbohydrates, drugs, and macromolecules) [27], and hazardous materials [28] (e.g., environmental pollutants, explosives) has already been observed for different soft material photonic crystals [29].

2 Optics of Soft Materials

Bragg's law gives the idea about the mechanism of photonic crystals, the intensity, and the wavelength of the reflective optical wave. This equation given by Braggs applies to any medium that exhibits a periodic arrangement of material characteristics that vary according to the scale of the incident wavelength of any kind of excitation, such as electrical, optical, or acoustic. These properties include mass density, atomic potential, and refractive index [21]. Bragg diffraction was initially observed in crystalline materials whose atomic spacing was similar to the wavelength of the incoming X-ray. This phenomenon led to the specular reflection of specific wavelengths at particular angles [30].

If two planes have separation d, the after-interference of the reflected wave wavelength will be

$$2d\sin\theta = m\lambda$$

where d is the interplanar spacing, λ is the wavelength of the light source, and θ is the angle between incident and reflected rays [31].

Using Snell's and Bragg's laws for two different refractive mediums, this equation will be

$$2(n_1d_1 + n_2d_2)\sin\theta = m\lambda$$

Here, n_1 and n_2 are the refractive indexes for two different mediums [32].

Photonic crystals are typically considered periodic dielectric structures with bandgaps relevant to photons or quanta of light. By the previous equation, we can state that for a soft matter system made of polymer layers that can reflect a particular wavelength of lambda, we can tune their interplanar separation and refractive index of the dielectric medium.

So, the propagation of light through two layers will depend on the optical thickness of the two layers. Considering this principle, a lot of photonic crystals have been constructed. The chemical structure or composition of the materials can be used to adjust the reflective index, n, and the fabrication parameters can be used to alter the layer distance (= d). More structural color tuning may result from a layer distance change than a refractive index change. The majority of one-dimensional (1D), tunable photonic crystals rely on altering the thickness of the layer in conjunction with a shift in the refractive index.

The diffraction wavelength intensity (R) is given by

$$R = \frac{(1-Y)^2}{(1+Y)^2} \times 100\%$$

where Y is

$$Y = \frac{n_1 n_s}{n_2 n_o} \times 2N$$

n_s is the substrate's refractive index, n_o is the refractive index of the surrounding medium, and N represents the no. of bilayers. Now, if the value of the total number of bilayers or the ratio of contrast increases, the diffracted light intensity will also increase [33].

3 Classification of Photonic Crystals

Depending on the dimensional order, photonic crystals can be classified into one-dimensional photonic crystals, two-dimensional (2D) photonic crystals, and three-dimensional (3D) photonic crystals (Fig. 1).

3.1 One-Dimensional Photonic Crystal

Many species in the natural world have stunning structural hues; some can alter their color in response to external cues. Adaptive color changes are typically associated with signaling, reproduction, and camouflage. The design of artificially tunable one-dimensional photonic crystals will benefit from the knowledge and inspiration of the color-tuning behavior found in natural photonic crystals, particularly in one-dimensional photonic crystals. Neon tetra (a type of tropical fish) is a typical illustration of color shift from a natural one-dimensional photonic crystal (Fig. 2) [34]. The lateral stripes of neon tetra exhibit a vivid blue hue during the day, which transitions to a deep violet hue after dusk. It can change color swiftly under pressure or when the fish is aroused [35].

A few stacks of periodically arranged light-reflecting platelets are present in an iridophore of the neon tetra, and they can produce multilayer optical interference events. As illustrated in Fig. 2. Nagaishi et al. [35] suggested a model known as the Venetian blind model. According to this hypothesis, platelet tilt angles are thought to

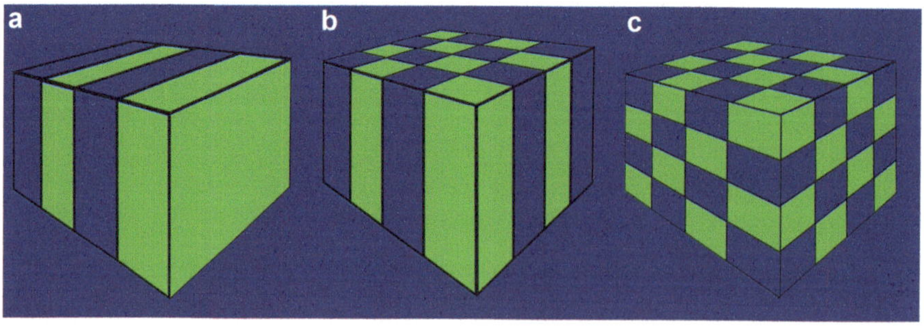

Fig. 1 Represent different types of photonic crystals by the periodicity of the dielectric medium. (**a**) One-dimensional photonic crystal, (**b**) two-dimensional photonic crystal, and (**c**) three-dimensional photonic crystal. (https://docs.google.com/spreadsheets/d/1LR4mpnL6OCiIgvrlHg20vzZHGH_jYLs9GZWldvodJJU/edit?usp=sharing)

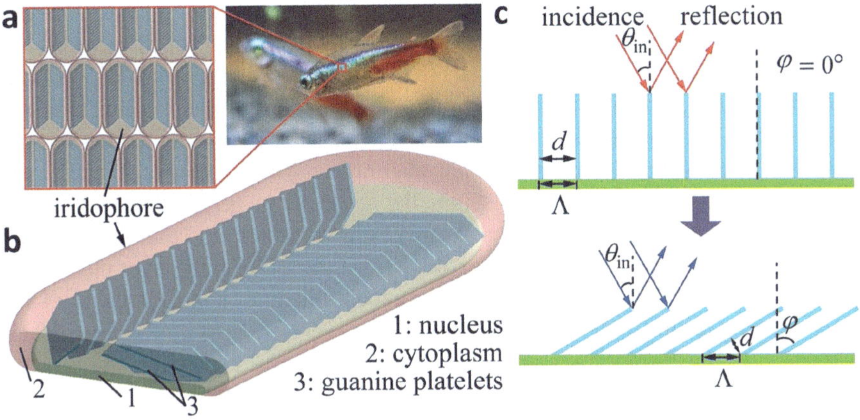

Fig. 2 One-dimensional natural photonic crystal of neon tetra. (**a**) Schematic picture of neon tetra fish with iridophore. (**b**) Detail diagram of iridophore consisting of nucleus, cytoplasm, and guanine platelets. (**c**) Schematic of "Venetian blind," mechanism of one-dimensional photonic crystal. (Adapted with permission from Chih-Hao Chang et al. [34] Copyright 2019 American Chemical Society) (https://docs.google.com/spreadsheets/d/1LR4mpnL6OCiIgvrlHg20vzZHGH_jYLs9GZWldvodJJU/edit?usp=sharing)

be physiologically regulated. Motor proteins attached to filaments connect platelets to microtubules, where they move to alter the tilt angle.

Several examples are the wings of Morpho butterflies and the neck feathers of domestic pigeons, which are also found to be in the regime of 1D photonic crystals.

The most straightforward photonic crystal shape, known as a one-dimensional photonic crystal or Bragg mirror, has a periodicity in a single direction. Because of constructive interference, the layers, which alternate between high and low refractive index layers, provide an amplified reflection. The number and thickness of layers, the angle at which light is incident, and the refractive index contrast all affect the optical characteristics of 1D photonic crystals. The strain can adjust the thickness and, thus, the reflectance of particular wavelengths.

Also, by adding a single layer to the stack, the simplest type of photonic crystal, a 1D multilayer stack with dielectric constant variation along only one direction, can function as a notch filter or a filter that lets through a specific range of input light wavelengths.

Hydrogels are a promising self-assembled system that displays 1D photonic crystal behavior. Anisotropic hydrogel developed by Haque et al. [17], these hydrogels are 1D photonic crystals whose photonic structure is defined by the assembly of hydrophobic poly(dodecyl glyceryl itaconate) (PDGI) bilayers embedded in a polyacrylamide matrix. The periodic stacking of bilayers made of self-assembled DGI monomers created lamellar liquid crystals in water. This structure held firm even with a cross-linker and initiator present in an aqueous acrylamide monomer solution. After polymerization, the polymeric-DGI (PDGI) lamellar structure became immobile inside the PAAm network structure. However, the resulting gel was isotropic on a macroscopic level as the self-assembled lamellar domains were oriented arbitrarily in the precursor solution [36]. The produced platelike bulk hydrogel by Haque et al. [37] exhibits perfect anisotropic swelling behavior, meaning that it only swells in the direction perpendicular to the lamellar bilayers and does not expand along the lamellar bilayers. It comprises around 4000 PDGI bilayers with unidirectional alignment. With an elastic modulus ten times higher in the direction of the lamellar layers than perpendicular to them, the gel exhibits substantial mechanical anisotropy. Furthermore, the hydrogel has outstanding visual color because Bragg reflects on the PDGI multilayer planes. By releasing a mechanical tension or strain, the exceptional mechanical strength of the gel enables reversible tuning of its structural color over a wide spectrum range. They also observed anisotropic swelling of gels concerning surfactant concentration. The swelling ratio (m/m), parallel and perpendicular to the lamellar layers, saturates to 2.4 and 1.0, respectively, for the dodecyl glyceryl itaconate concentration ratio from 0 to 0.2 mol/L. As the strain, ε, increased from 0 to 0.64, the wavelength maximum (λmax) fell from approximately 600 nm to 415 nm. Up to a strain value of 45%, a linear relationship between λmax and ε was found, with a corresponding peak shift of about 150 nm. The large color change occurred due to a decrease in the lamellar distance. Because the gel had a structure composed of two alternating layers, almost all the compressive deformation was localized in the soft PAAm layer. This caused a decrease in the lamellar distance, with no compressive deformation in the hard PDGI lamellar bilayer except for the lateral expansion. For the same system, Jian Ping Gong et al. [38] showed the effect of co-surfactants and the tuning of optical properties for those hydrogels. The chemical structure of the co-surfactants determines the concentration at which 1D swelling may be maintained: for sodium dodecyl sulfate (SDS), up to $R = 1{:}250$; for sodium stearate (SST), up to $R = 1{:}500$; and for dodecyl triethylamine chloride (DTAC), up to $R = 1{:}4000$. In contrast, NBA (*N*-butylamine) showed 1D swelling even at the highest concentration tested ($R = 1{:}64$). R defines the molar ratio of co-surfactant to DGI (Fig. 3).

Increased head-to-head repulsion, particularly for charged co-surfactants, causes self-assembled DGI structures to diverge from lamellar to curved ones (hexagonal, micellar) when co-surfactant concentration rises, as observed by Haque et al. [37] in another article. Therefore, when the concentration of co-surfactant increases, more flaws are generated. More specifically, it has been demonstrated lately that swelling PDGI/PAAm hydrogels in SDS solution tends to cause 3D swelling and converts the lamellar bilayers into micelles (above the critical micellar concentration of SDS).

Since the number of bilayers and the contrast of refractive index make a perfect tunable 1D crystal, researchers design the system in such a way that they can make a good mismatch in refractive index. Using various organic/inorganic polymer materials [39], creating void space [40] in the layer, and including multiple nanoparticles [41] are integral to this design.

Bai Yang et al. [39] prepared organic–inorganic hybrid films through alternating thin films of polydimethylaminoethyl methacrylate-co-ethylene glycol dimethacrylate (PDMAEMA-co-PEGDMA) and titania sol by spin-coating. Since the refractive index of the titania layer (1.831) is higher than the polymer layer (1.506), an

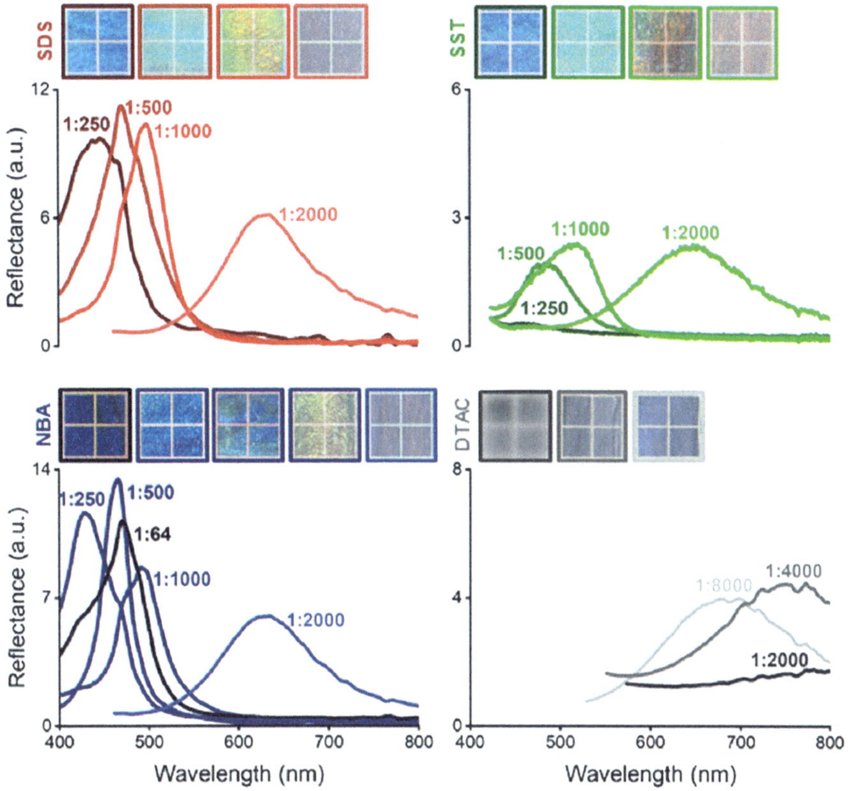

Fig. 3 Images of the hydrogels arranged on a black ground (each square measuring 5 mm by 5 mm), and the hydrogels' reflectance intensity spectra show a rising *R* from left to right. (Adapted with permission from Jian Ping Gong et al. [38] Copyright 2023 American Chemical Society) (https://docs.google.com/spreadsheets/d/1LR4mpnL6OCiIgvrlHg20vzZHGH_jYLs9GZWldvodJJU/edit?usp=sharing)

obvious photonic stopband can be easily obtained in several layers. Meanwhile, the materials take on evident color as the photonic stopband falls into the visible region. From the reflective spectra, they observed the stopband is at 352 nm in air and shifts to 766 nm with color changing from blue-green to wine as it emerges into 0.1 mol/L citric acid, which is due to the swelling of the polymer layer and the decrease of its refractive index. As the concentration of the SCN− changes from 10^{-6} mol/L to 10^{-1} mol/L, the photonic stopband changes from 750 nm to 490 nm with color changes from red to blue. The last byproduct of cyanide compound detoxification is thiocyanate (SCN−), which is eliminated in serum, saliva, and urine. Saliva, urine, and blood serum are particularly useful for measuring SCN since they are thought to be biomarkers that differentiate smokers from nonsmokers [42].

Another example of a mismatch of refractive index is Kleine et al. [43], who reported the highest reflectance of around 90–97%, with a significant mismatch of the refractive index of $\Delta n \approx 0.5$, creating around 22 layers with chalcogenide hybrid inorganic/organic polymers. Chalcogenide hybrid inorganic/organic polymers (CHIPs), solution-processable polymers with ultrahigh refractive index, made it possible to fabricate these extremely reflective sheets quickly. When low n polymers, like cellulose acetate ($n = 1.47$), are used to manufacture narrowband IR Bragg reflectors, the high refractive index (n) of CHIP materials ($n = 1.75$–2.10) permits the creation of high refractive index contrast ($n \sim 0.5$). With an all-polymeric Bragg mirror, this is the largest refractive index contrast ($n \sim 0.5$) yet shown, immediately enabling high reflectivity from films with 22 layers or fewer. A novel path to IR Bragg reflectors and other reflective coatings with potential uses for IR photonics, commercial sensing, and LIDAR is made possible by easy access to modular, thin, highly reflective films from affordable CHIP materials.

Fabricating 1D photonic crystals is more complex than 2D or 3D pcs. Maintaining stability in a typical chemical environment is the most challenging job. The group of Mark Robertson [44] fabricated one type of chemically stable 1D photonic crystals with the help of only two materials named poly(vinylidene fluoride-co-chlorotrifluoroethylene) (PVDF-CTFE) and phenolic resin (resol) through a dip coating process, whereby the substrate moving velocity during the dip coating process may be easily adjusted to tune the reflected wavelengths of the 1D-PC. Due to their naturally high refractive index contrast ($\Delta n \approx 0.24$), the two layers may withstand exposure to a wide range of organic solvents and basic and acidic aqueous solutions with remarkable stability. To achieve varied thicknesses from the same solution composition, the dip coating technique offered a rather easy way to modify the film thickness by adjusting the substrate moving velocity. The film thickness of PVDF-CTFE can be tuned by substrate withdrawal velocity and solution concentration. Greater film thicknesses are the outcome of raising the substrate withdrawal velocity and the PVDF-CTFE concentration in solution. A broad range of film thickness (from 175 to 375 nm) can be achieved for an 8 weight percent PVDF-CTFE solution in MEK by adjusting the substrate withdrawal rate, which ranges from 0.36 to 1.43 mm/s. This conclusion implies that the reflectance peak can be easily identified because the wavelengths reflected by the 1D photonic crystals depend directly on the thicknesses of the alternating layers.

Various top-down and bottom-up approaches have been considered in this chapter to develop 1D photonic crystals. For example, Baer et al. [45] used one of the top-down approaches (i.e., scale layer-multiplying coextrusion technology) to fabricate elastomeric 1D photonic crystals with almost 100% reflection of narrow reflection band with polystyrene and poly(methyl methacrylate) layers following one another layer. This method gave them up to 128 layers in the crystal.

Colloidal particles are a kind of soft material–based photonic crystals created when monodisperse micro/nanoparticles self-assemble [46–50]. Hernán Míguez et al. [51] provided a quick and dependable process for creating eight layers of 1D photonic crystals based on nanoparticles. This process involves alternating different kinds of nanoparticles and adjusting the porosity of each layer to provide a periodic modulation of the refractive index. To be used as precursor dispersions for a spin-coating technique, TiO_2 nanocrystals and SiO_2 nanoparticles—used to create highly uniform oxide films—were suspended in a mixture of water and methanol.

Yadong Yin [52] reports that assembling homogeneous nonmagnetic colloidal particles in fluids using external magnetic fields leads to the fast development of photonic crystal formations. Due to their minimal magnetic dipole moment, nonmagnetic particles as small as a few hundred nanometers have proven difficult to manipulate magnetically. These particles are ideal building blocks for creating photonic crystals with bandgaps in the visible regime. The instabilities that lead to aggregation at large concentrations or in strong magnetic fields have hindered the capacity to increase the dipole moment of magnetic holes. In this paper, we have successfully assembled 185 nm nonmagnetic polymer beads into photonic crystal structures, from 1D chains to 3D assemblies, as determined by the interplay of magnetic dipole force and packing force. This is achieved by utilizing the superior stability of highly surface-charged magnetite nanocrystal-based ferrofluids. This general strategy holds promise for rapidly creating large-area photonic crystals using nonmagnetic particles as building blocks. 3D photonic crystals with a high reflectance (83%) in the visible range can be produced in a strong magnetic field with a large field gradient within a few minutes.

3.2 Two-Dimensional Photonic Crystal

Like 1D photonic crystals, we also have various 2D photonic crystals in our beautiful nature. Iridescent setae from polychaete worms and barbules of male peacocks are such examples of natural 2D photonic crystals.

Jian Zi, Xindi Yu, and Yizhou Li [53] reported how color is produced in peacock feathers. They discovered that the pigmentation of differently colored barbules is attributed to the cortex, which has a 2D photonic crystal structure. According to simulations, the photonic crystal structure has a partial photonic bandgap along the direction average to the cortex surface for frequencies where light is highly reflected. Different hues are produced by regulating the number of periods in the photonic crystal structure and the lattice constant. When the number of periods decreases, more colors are added, resulting in a mixed coloration (Fig. 4).

In recent soft matter research, two-dimensional photonic crystals constructed by monodisperse colloidal particles have gained massive attention [55, 56]. For example, spherical particles, when rapidly dispersed over a surface, self-assemble into randomly oriented, hexagonally arranged 2D array micro-crystallites. Many fabrication methods have been defined to produce this type of photonic crystals; the main ingredients are colloidal particles of micron size, and to stabilize them on a two-dimensional system, researchers use polymers or surfactants.

Using a needle tip flow technique, the Asher group presented a quick way to produce large-area 2D colloidal crystals at the air/water interface [57]. They developed a 2D array monolayer of polystyrene in hydrogel film made of poly(*N*-isopropyl acrylamide) (PNIPAAm). They were implanted with a close-packed for

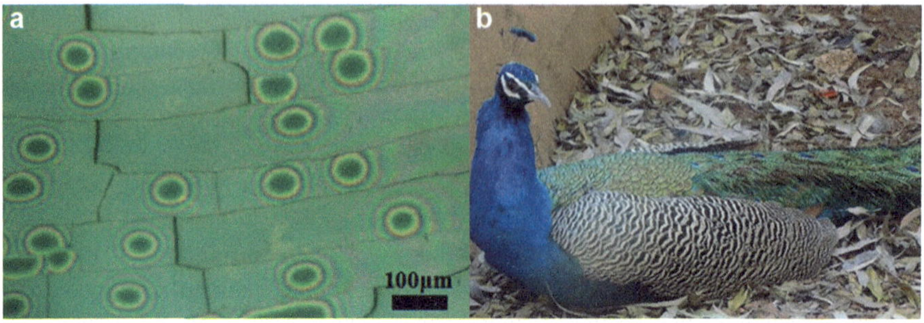

Fig. 4 (**a**) Optical microscope images show plenty of eye patterns similar to the peacock tail feathers, 2D photonic crystal (Adapted with permission from Jingxia Wang et al. [54]). (**b**) Schematic of peacock taken at Horsley Hills Zoo (Andhra Pradesh, India) (https://docs.google.com/spreadsheets/d/1LR4mpnL6OCiIgvrlHg20vzZHGH_jYLs9GZWldvodJJU/edit?usp=sharing)

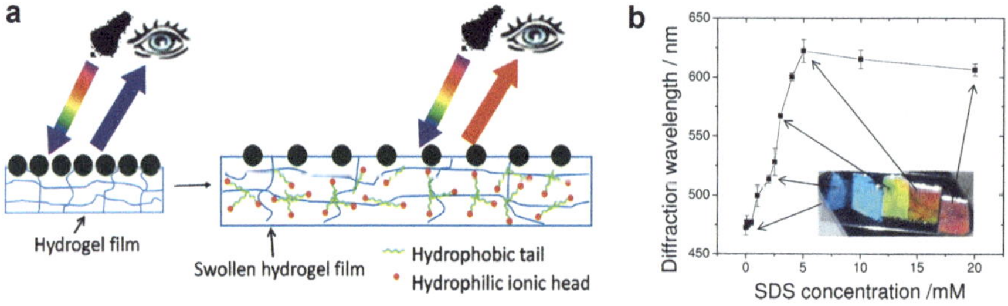

Fig. 5 (**a**) Diagram showing how a PNIPAAm hydrogel network is polymerized onto a 2D array of 490 nm PS particles on a liquid Hg surface to create a 2D photonic crystal sensor. The PNIPAAm hydrogel expands when molecules of the surfactant bond to it. When the PNIPAAm hydrogel swells, the 2D particle spacing increases, red-shifting the diffracted wavelength. (**b**) Wavelength of diffraction against SDS content. Images obtained near the Littrow configuration are displayed in the inset at a 28° angle from the 2D array regular to the source and camera. (Adapted with permission from Sanford A. Asher et al. [57] Copyright 2012 American Chemical Society) (https://docs.google.com/spreadsheets/d/1LR4mpnL6OCiIgvrlHg20vzZHGH_jYLs9GZWldvodJJU/edit?usp=sharing)

visually detecting amphiphilic molecules in water. They constructed a photonic red-shift crystal using polystyrene particles of size 490 nm. They used PNIPPAm (a thermos-responsive polymer with LCST 32°) and a Hg surface to stabilize the structure. Later on, they show the most exciting feature of using soft materials. Using different cationic and anionic surfactant (e.g., SDS, CTAB) concentrations, they got light diffraction of various wavelengths, from red to blue (Fig. 5).

Another facile approach of robust polymer-based two-dimensional (2D) colloidal crystal (CC) layers was presented by the Zhigang Zhu group [58]. This technology enables the convenient fabrication of an anti-counterfeiting coating with a polymeric 2D colloidal photonic crystal, allowing the fast preparation of functional polymer photonic materials. Briefly, a 2D CC was prepared by self-assembling polystyrene (PS) submicron spheres (size 5 µm), which were then transferred to substrates by the adhesive polymeric solution of PVA into DMSO via classic horizon evaporation method [59] and cured to form a photonic film. Such photonic films strongly and angle-dependently diffract visible light, and the photonic layers' high transparency ensures the substrate's readout. Such 2D colloidal photonic crystal–PVA layers could be prepared onto different material surfaces, and the 2D crystal structure was proved by Debye diffraction by forming a Debye ring.

A similar type of photonic crystal material based on monolayered colloidal crystals (MCCs) was created by Jun Zhou et al. [60] for glucose-sensing. A 4-boronobenzaldehyde-functionalized poly(vinyl alcohol) hydrogel was used to cover the two-dimensional (2D) template created by the self-assembly of polystyrene nanoparticles of size 600 nm into a highly organized MCC. The physiological ranges of tears and blood are covered by this sensor, which effectively diffracts visible light, whose structural color may shift from red to yellow to green when the glucose content varies from 0 to 20 mM. With its high sensitivity and accuracy in determining the glucose concentration in tears, the sensor also exhibits a quick response of 180 s at each glucose titration. This intelligent sensing material offers a promising future for point-of-care glucose monitoring.

3.3 Three-Dimensional Photonic Crystal

The significance of three-dimensional (3D) photonic crystals is growing due to the quick advancement of photonic integrated circuits [61, 62]. The precision and accuracy needed for these crystals are far higher than that of 1D and 2D photonic crystals, not to mention the need for optical activity and bandgaps in the UV and visible spectra. As a result, most 3D photonic crystals in use today are only helpful at microwave or infrared wavelengths and are challenging to manufacture [63]. Liquid photonic crystals, or blue-phase liquid crystals (PLCs), are self-assembled into three dimensions. In a specific type of chiral nematics mesophases known as BPs, the director axes self-assemble into cylinders of double-twist helices. The phases are commonly referred to as BPI, BPII, and BPIII, in sequence of emergence with increasing temperature. Double-twist cylinders (DTCs) produce a cubic crystal lattice in the first two BPs by stacking three dimensions across the bulk. BPII is a simple cubic, while BPI is a body-centered cubic [64–66]. The photonic bandgaps, regions in the visible spectrum where light is prohibited, have a few hundred nanometers lattice spacing. The estimation of the reflection wavelength λ, which corresponds to the lattice plane (h k l), can be achieved by satisfying Bragg's law. The resulting Bragg reflections are circularly polarized due to the chiral character of BP. The twist orientation of the structure and the bandgaps' spectral location (from ultraviolet through visible to near-infrared) may be altered by varying the handedness and mixing ratio of the chiral ingredients. Moreover, topological defect structures found in BPLCs yield photonic crystalline metamaterials by enabling the three-dimensional periodic confinement of nanoparticles.

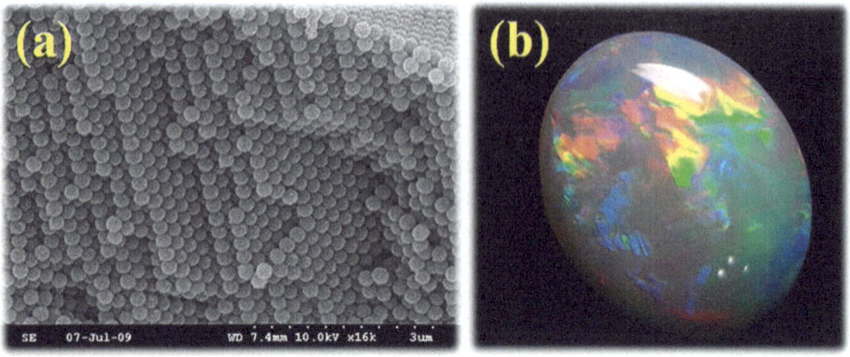

Fig. 6 (**a**) Typical SEM image of colloidal crystals; (**b**) a natural opal. (Adapted with permission from Zhongze Gu et al. [67] Copyright 2014 American Chemical Society) (https://docs.google.com/spreadsheets/d/1LR4mpnL6OCiIgvrlHg20vzZHGH_jYLs9GZWldvodJJU/edit?usp=sharing)

Various colloidal photonic crystal compounds with a macroscopic spherical form have been synthesized recently. The photonic bandgaps of spherical colloidal photonic crystals are independent of rotation under surface illumination at a given incidence angle of light, extending the range of possible applications due to their spherical symmetry. Depending on the specific optical features and applications, many forms of spherical colloidal PhCs may be created, including close-packed, non-close-packed, inverse opal, biphasic or multiphasic Janus structured, and core-shell structured geometries (Fig. 6).

It is possible to create inverse opal beads using co-assembly or template replication. The components of the final inverted opal specimen are inserted into the empty spaces in the templates throughout the replication process. These ingredients comprised hydrogel prepolymer solution, metal oxide precursors, ultrafine nanoparticles, and so on for the inverse opal film construction process. To get a complete filling and prevent the hybrid or inverse opal beads from being released from the rigid materials, the colloidal crystal bead templates must be implanted entirely in the rigid materials during the production of the inverse opal beads. On the other hand, the hybrid beads could be detached from the hydrogel even after they were fully embedded because soft hydrogel materials could be ripped without damaging the colloidal crystal bead templates. After etching the nanoparticles, it is possible to create inverse opal hydrogel beads with a perfect spherical shape [68–70].

The techniques utilized for 2D construction may be modified and used to create 3D crystals. Colvin et al. [71] have shown that 3D lattices with precisely regulated layer counts may be constructed by applying lateral attractive capillary forces. They produced 3D photonic crystals using the close-packed arrays formed when monodisperse silica spheres spontaneously crystallized. Thick samples with thicknesses up to 50 μm may be created using a vertical deposition approach to pack the spherical colloids into closely packed silica–air arrays. They also showed that different spheres with diameters 307.2, 353.0, and 383.7 nm, respectively, show different colors in white light. Lopez and colleagues [72] developed the electrophoretic deposition technique to arrange silica colloids into three-dimensional opaline patterns. Anyone can use sedimentation, repulsive electrostatic interactions, and physical confinement methods to construct 3D photonic crystals based on colloidal particles.

Cross-linked polymer networks forming three-dimensional hydrogels are hydrophilic structures that can hold onto water molecules and alter their phase and volume in response to external stimuli. It is also one of the recent soft matter–based photonic crystals. Hydrogels with optical clarity and good processability have been demonstrated in recent work to create a variety of micro- and nanostructures that are appealing for biological and environmental sensing applications based on photonics. Furthermore, hydrogel-based photonic devices of biocompatible and biodegradable polymers can stimulate, diagnose, and treat living tissues by providing light [67].

Nanosphere self-assembly in the presence of polymers plays a vital role in 3D photonic crystals. In the report of Wang et al. [73], *N*-isopropyl acrylamide, *N*-test-butyl acrylamide, and functional monomer acrylic acid were copolymerized to create the nano hydrogel colloids. With a maximum red shift of 24 nm, the nano hydrogel colloidal array's reflection spectrum allowed for quickly detecting relative humidity ranging from 20% to 100%. Using the nano hydrogel colloidal solution, which was diluted to 3.0 mg/mL to prepare the array with a specific thickness, a series of 3D nano hydrogel colloidal were self-assembled using the vertical deposition method [74]. A gravimetric analysis following lyophilization confirmed the solution concentration as 6.0–8.1 mg/mL. It is a simple, economical, and successful method of getting NHCA (nano hydrogel colloidal array) from nano hydrogel colloidal solution—the consistent microstructure of the as-prepared NHCA combined with the NHCs' uniform size and fine spherical shape. The organized array also implies the excellent uniform quality of the NHCs. Typically, as-prepared NHCA has around 20 layers and a thickness of three micrometers. With such a narrow array, the NHCA and the target have a more extensive interface and may respond more quickly [75].

Qin et al. [76] fabricated PS-based (98 ± 1.0 nm) hydrogel to detect mercury in seawater through the cleavage of S-S bonds in grafted *N,N′*-cystamine bisacrylamide molecules; spatially dispersed -SH groups functionalized the hydrogel. It may preferentially interact with Hg^{2+} ions to form S-Hg-S bridge links in saltwater. A wavelength change in the light diffracted by the hydrogel is correlated with the hydrogel's volume shrinking due to the Hg^{2+} binding. With a portable spectrometer, the amount of bound Hg^{2+} ions may be quantitatively evaluated, with a detection limit of 10^{-9} M in seawater. This is made possible by the wavelength shift.

4 Design and Fabrication of Photonic Crystals

By both top-down and bottom-up approaches, researchers can construct periodic photonic crystals [77].

4.1 Top-Down

The bulk material is broken down into nanoscale structures or particles using a top-down method. Top-down techniques are more straightforward and rely on miniaturizing bulk production procedures or dividing or removing bulk material to produce the desired structure with the right attributes. These approaches frequently employ conventional microfabrication processes, such as photolithography and etching, to create microstructures from bulk materials with the required size, shape, and order.

Holographic Photopolymerization

Combining holography and polymer-dispersed liquid crystal, holographic photopolymerization offers a highly adaptable method for generating polymeric photonic bandgap structures [78]. Holographic photopolymerization involves recording pure laser interference to produce passive diffractive optical elements [79]. Photopolymer materials are used in the process; these polymers are light-sensitive and react chemically when they come into contact with light. A liquid crystal cell containing polymer-dispersed liquid crystal materials is used to build the holographic structures. The interference pattern produced by two or more laser beams, whose intensity and angles can be adjusted, is recorded. Light can be diffracted in the cell when a periodic refractive index profile is generated due to a reasonably good phase separation between liquid crystals and polymers [80].

For example, Urbas et al. [81] produced functional photoresponsive optical materials by utilizing liquid crystal mixes from azobenzene for photo-optic purposes inside these material systems. When exposed to specific wavelengths, azoLCs undergo a reversible trans-cis photoisomerization process. The azoLC's conformational shift alters the ensemble's optical characteristics when blended with a host nematic LC by upsetting the order of the surrounding LCs [82].

Layer-Multiplying Coextrusion

Using a small number of extruders, layer multiplication coextrusion (LMCE) has been frequently employed to produce large numbers (100–1000 s) of very thin layers. This method has been demonstrated to improve property performance in

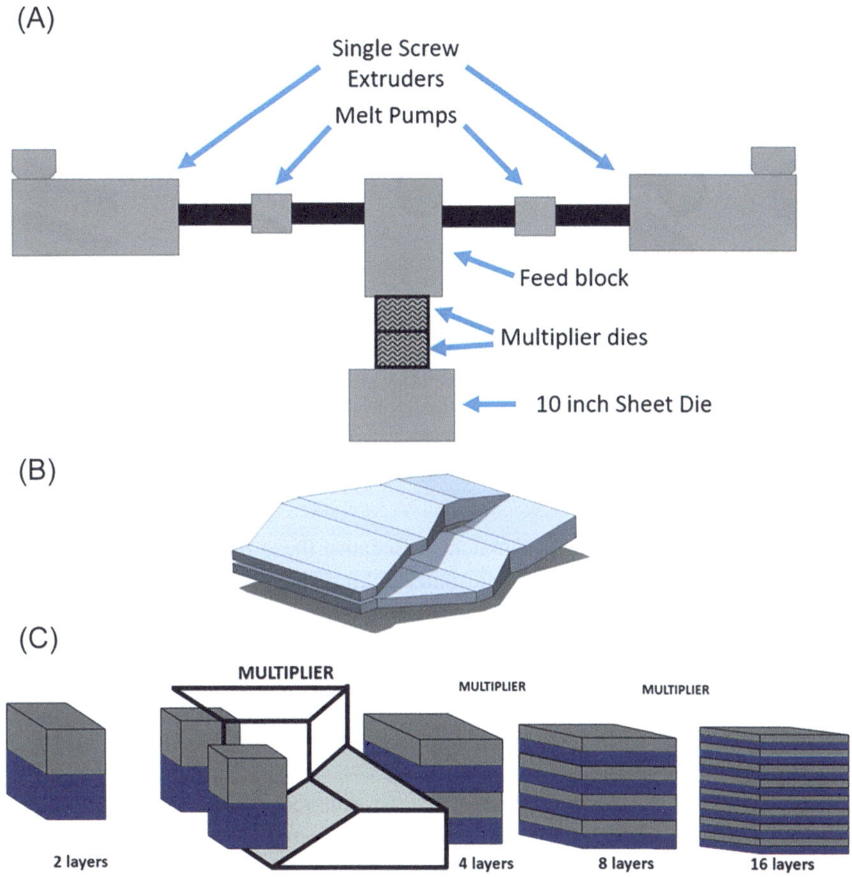

Fig. 7 Multilayer coextrusion (**a**) schematic of coextrusion setup, (**b**) HAR multiplier flow path, and (**c**) diagram of multiplier function. (Adapted with permission from João Maia et al. [90] Copyright 2023 Wiley Periodicals LLC) (https://docs.google.com/spreadsheets/d/1LR4mpnL6OCiIgvrlHg20vzZHGH_jYLs9GZWldvodJJU/edit?usp=sharing)

the following areas: barrier [83], insulation [84], mechanical [85], and optical [86]. The coextrusion of polymers into tubing with a high layer count has been added to the technology [87–89]. Studies investigating filled polymer systems as layer constituents have been carried out (Fig. 7).

A single screw extruder and melt gear pumps comprised the coextrusion line schematically depicted in Fig. 8a. Multilayer sheets were produced using a feed block, two high aspect ratio (HAR) multipliers with the flow route shown in Fig. 8a few-inch sheet die, and a chill roller with variable speed control. To get layer thicknesses above, around, and below the particle sizes of the microspheres, the chill roller speed was adjusted to alter the total sheet thickness [90]. Compared to conventional multipliers, the HAR multipliers utilized in this study operate better

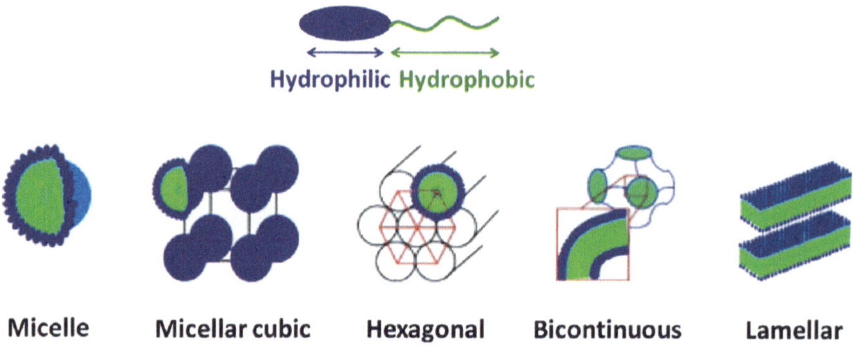

Fig. 8 Diagrammatic illustration of an amphiphilic molecule and self-assembled surfactants' normal and reverse forms. (Adapted with permission from Jonathan P. Hill et al. [91] Copyright 2014 Molecules) (https://docs.google.com/spreadsheets/d/1LR4mpnL6OCiIgvrlHg20vzZHGH_jYLs9GZWldvodJJU/edit?usp=sharing)

with materials with viscosity mismatches. Because these multipliers maintain the cross-sectional area during the splitting and stacking, the melt contracts and expands simultaneously, reducing viscous encapsulation and better stability [92–94].

Some iridescent microlayer films of polystyrene (PS) and poly(methyl methacrylate) (PMMA) were the first coextruded films to exhibit the properties of 1D photonic crystals. Coextruded plastic films have been produced with individual layer thicknesses ranging from 0.6 to 0.1 μm, consisting of 231 alternating layers of polystyrene and polymethylmethacrylate [95].

It is proven that layer-multiplying coextrusion technology at the laboratory scale can create narrowband, 1D photonic crystals from polymeric materials. Films that selectively filter various visible electromagnetic spectrum regions are used to illustrate how the photonic bandgap is tuned. By comparing the obtained UV-vis transmission spectra with model simulations, the layer homogeneity of the photonic crystals is assessed, and atomic force microscopy is used to confirm it. Polystyrene and poly(methyl methacrylate) alternating layer assemblies show a small reflection band with about 100% reflection, which is in good accord with the uniform layer prediction. The creation of elastomeric 1D photonic crystals demonstrates the process's adaptability. This technology is expected to make it possible to realize novel polymers quickly and easily [45].

Photolithography

Photolithography is a process that involves the transfer of a pattern onto a substrate by selective exposure to light. The process starts with simply putting a wafer in a photolithography system where the substrate is covered with active material. After this, a photoresist is applied to the wafer. Therefore, photo masking is applied to the area exposed to UV radiation. The system is then exposed to UV radiation.

Next, the UV-irradiated photoresist is developed and removed, and the UV-exposed active material is etched. The residual photoresist is finally released to obtain the wafer with the appropriate pattern for the industrial-scale manufacture of MSCs, hazardous etchants, and complicated procedures must be revised.

Soft lithography is one type of photolithography that consists of the main components of smooth materials like polymers or gels. Jae-Hwang Lee et al. [96] reported the fabrication of metallic photonic crystals by depositing gold on a polymer substrate. The photonic crystals behave like complete metallic structures with a photonic band edge at a wavelength of 3.5 μm. The stages of the upgradation of lithography technique are shown in the chart.

Sl. No.	Lithographic technique	Year
1	E-beam lithography	1995
2	Deep X-ray lithography	1997
3	Holographic lithography	2000
4	Deep UV lithography	2002
5	Nanoimprint lithography	2007

4.2 *Bottom-Up (Self-Assembly)*

We can get periodic photonic crystals by self-assembly method from a small building block of copolymer nanoparticles or colloidal particles. This method produces an ordered structure by minimizing the system's free energy. As part of the crystallization process, some defects could form, and hence, the free energy would increase; the process chose to reduce defects and their free energy. For example, the formation of 2D periodic patterns is achieved by electrostatically assembled closed-packed polystyrene particles [97, 98].

Colloidal spheres can have their properties further altered by applying a layer of variable shell thicknesses with a distinct chemical composition. The resulting core-shell colloids can preserve their narrow size distribution when the coating is used uniformly. The shells' structure, size, and makeup allow for a wide range of tuning of the spheres' optical [99], electrical [100], thermal [101], solvent [102], and stress/strain [23].

Self-Assembly of Block Copolymers

Since it may produce ordered structures in a variety of morphologies, such as spheres, cylinders, bicontinuous structures, lamellae, vesicles, and many more complicated or hierarchical assemblies, block copolymer (BCP) self-assembly has garnered significant attention for many years [91].

A block copolymer is a polymer made up of molecules with a linear arrangement of blocks; a block is defined as a segment of a polymer molecule where the monomeric units have at least one constitutional or configurational feature that is not present in the adjacent portions. The characteristic that sets apart a block copolymer is constitutional, meaning that each block is made up of units that are derived from a particular species of monomer; this definition does not include branched structures where the branches are made up of blocks, such as star or radical block copolymers. Those block copolymers are self-assembled and show the minimum free energy configuration, which could be a building block of photonic crystals [103].

The domain spacing needs to be improved for optical applications because BCPs with a typical molecular weight of ~50 kg/mol create microdomains on a typical length scale of ~20 nm. The average domain size must be increased to $\lambda/4n$, where λ is the wavelength of light and n is the corresponding block's refraction index, to create a bandgap at visible or near-IR wavelengths [14].

By combining symmetric poly(styrene-b-isoprene) (PS–PI) diblock copolymers, which have approximate molecular weights of 200 kg/mol per block, to produce a 60/20/20 ternary mix with the corresponding homopolymer of each block, the first example of a visible BCP photonic crystal was created. The systematic incorporation of homopolymers with low molecular weight into the host BCP offers a method for fine-tuning the photonic bandgap, or "stopband," throughout the visible wavelength range by only adjusting the mixed homopolymer content [104].

Self-Assembly of Surfactants

Surfactants have hydrophilic water attractive head along with one hydrophobic head compatible with oil. In solution, those surfactants can be assembled and create variously shaped structures.

Due to their unique hydrophilic and hydrophobic nature, they produce various types of aggregation bilayers (spherical, cylindrical, etc.) by self-assembly [105]. As roughly depicted in Fig. 8, the different structures might arise when a solute's concentration rises above the critical micelle concentration. The lamellar phase has a multi-bilayer structure where the hydrophilic head aligns with the hydrophobic tails at the water interface. In contrast, the hydrophobic tails align parallel to one another, hiding from the water.

Surfactants are classified based on the charge of the hydrophilic group. Surfactants can be nonionic or ionic (anionic and cationic). Anionic surfactants are frequently polar groups that include carboxylate, sulfate, phosphate, and sulfonate. Most cationic surfactants are based on nitrogen-containing quaternary compounds with positive charges. The most widely used cationic and anionic surfactants are cetyltrimethylammonium bromide (CTAB) and sodium dodecyl sulfate (SDS), respectively [106]. This chapter has already discussed the effect of surfactants in photonic crystals.

Using a lamellae-structured polystyrene-b-poly(2-vinylpyridine) (PS-b-P2VP) block copolymer (BCP), an ultrahigh-sensitive sensor is created by Yeo-Wan Chiang

[107] that can directly visualize the structural color change and detect surfactants that are commonly used, such as sodium methyl sulfate (SMS) and sodium dodecyl sulfate (SDS). Due to concentration-dependent compatibility between the quaternized P2VP (QP2VP) blockchains and SDS molecules, two distinct detecting mechanisms are found for low and high concentrations of the SDS surfactant, respectively. These mechanisms include unexpected blue-shifting and red-shifting reflectance wavelengths.

Other investigations were conducted into the effects of sodium oleate and sodium dodecyl sulfate by Wei Wang [108], two anionic surfactants, on the creation and characteristics of silica colloidal nanoparticles. Adsorption of anionic surfactants increased the size, monodispersity, and negative surface charge density of produced silica particles at a concentration of 0.001 M. Without much washing of the produced silica nanoparticles, colloidal photonic crystals easily self-assemble as homogeneity of particle size and particle–particle interactions rise. Bragg's law describes how photonic crystals deflect light in the visible spectrum. The organized colloidal particle arrays in dried thin films show a face-centered cubic shape. This work presented a novel method for synthesizing ordered colloidal silica thin films.

5 Stimuli-Responsive Photonic Crystals

Stimuli-responsive photonic crystals are periodic dielectric nanostructures that change their diffraction intensity or wavelength in response to external stimuli, such as chemical [109] or physical [110]. These materials' diffraction wavelengths can change due to their sensitivity to various stressors, such as temperature, solvent, and stress. Photonic hydrogels have the potential to be responsive photonic crystals because of their softness and ability to allow the diffusion of certain species, which allows for a quick drop in layer thickness without necessitating a change in volume. The photonic bandgap architectures must include certain responsive materials to create adjustable photonic crystals.

The photonic bandgap architectures must include certain responsive materials to create tunable photonic crystals. The design and production process includes selecting responsive materials, incorporating them into the crystals, and optimizing the performance of the photonic system. Good responsive photonic crystals have a large tuning range, reversible tuning, and high response rates and can be incorporated into existing electronics.

Those photonic crystals may be produced in two kinds. In the first case, responsive materials are directly prepared as building blocks that may be used to fabricate photonic crystals.

In the second scenario, the responsive materials are injected into the interstitial space after the periodic structures have been specified, creating a composite material that is both mechanically and optically stable.

5.1 Temperature-Sensitive Photonic Crystals

Most thermosensitive photonic crystals are made by embedding colloidal crystals in thermosensitive polymers like poly(*N*-isopropyl acrylamide) (PNIPAM), a thermosensitive polymer that undergoes a reversible volume phase transition (coil to globule) between a hydrated state and a dehydrated state around its lower critical solution temperature in water. The first example of responsive photonic crystals was likely the groundbreaking work on thermosensitive photonic crystals by Weissman et al. [111]. A blue shift in the diffraction occurs when the temperature rises, causing the polymer to release water and compress, reducing the inter-particle distance in the colloidal crystal. Consequently, photonic crystals containing PNIPAM polymer typically exhibit negative thermos-sensitivity, meaning they shrink at high temperatures and vice versa. This is accompanied by a blue shift in the photonic bandgap as the temperature rises. As PNIPAM has a lower critical solution temperature (LCST, 32.8 °C) in water [112, 113], changing the system's temperature from 10 to 35 °C shows the diffracted wavelength shift from 704 to 460 nm [12]. Few assemblies, like pure hydrogel colloids, show intensity change instead of wavelength shift.

Since the polymer network swells and contracts, 1D photonic crystals based on polymer gels may exhibit solvent sensitivity. The lamellar photonic gel is a typical example, which produces large (>1250 nm, 575%) reversible changes of the stopband for solvent swelling/deswelling by employing hydrophobic block–hydrophilic polyelectrolyte block polymer [12]. Youngjong Kang et al. [114] describe a polyelectrolyte block polymer that is hydrophobic and hydrophilic and forms a basic 1D periodic lamellar structure. As a result, a responsive photonic crystal is produced, which may be adjusted by applying pressure to a fluid reservoir, which causes the hydrophilic layers to swell. By forcing the hydrophilic layer to expand along the layer standard, the glassy hydrophobic layer produces variations in the index of refraction and layer thickness that result in incredibly excellent optical tunability.

5.2 Mechanical Photonic Crystals

Periodically structured soft materials having a photonic stopband that may be modified to reflect particular colors by mechanical stress are known as mechanical photonic crystals. These materials, which include rubbers and gels made of polymers, often have photonic crystal structures arranged in a soft matrix. Biaxial expansion or contraction along perpendicular directions occurs in tandem with the compression or stretching along one direction to maintain a constant volume. A photonic stopband shift is the outcome of the deformation, which also alters the lattice size or reflective index. Artificial mechano-chromic photonic crystals are a relatively new addition to the family of photonic crystals compared to their lengthy study history. This is likely because it is challenging to fabricate soft photonic crystals with a considerable photonic stopband shift under mechanical conditions [115].

Compression and stretching are examples of mechanical stress, the most basic stimulus to which 1DPCs respond. The first time elastomeric inverse opals (3DPCs) were used as mechano-chromic sensors was in 2006, according to the Ozin group [116].

A few examples have been reported; for instance, Steiner et al. showed that a PC made of polymethylsiloxane (PDMS) and PSPI (tri-block copolymer made of polyisoprene with a 22 weight percent minority polystyrene phase) shows a decrease in layer thickness when stretched. As a result, the photonic stopband is blue-shifted and can be tuned as a function of strain across the entire visible spectrum [117].

5.3 Photochromic Photonic Crystals

Photonic crystals can change radiation color by changing their refractive index contrast or lattice spacing.

By covalently joining azobenzene derivatives to polymerized crystalline colloidal arrays (PCCAs), which are then embedded in cross-linked hydrogels, the Asher group created a photochemically controlled 3D photonic gel. Visible excitation produces *trans*-azobenzene, whereas UV excitation is produced by *cis*-azobenzene. The hydrogel swells due to the more advantageous free energy of combining *cis*-azobenzene, which causes the photonic stopband to red-shift [118]. An optically switchable SiO_2/ITO 1D photonic crystal was created by釋Paternò et al. The optical characteristics of the device may be switched by tuning the indium tin oxide (ITO) plasmonic response in the near-infrared region and translating the effect to the visible light spectrum using a UV light photodoping method [119].

5.4 Electrochromic Photonic Crystals

Electric tunability is a superior option, as it is with current displays. In general, an active photonic crystal with a refractive index contrast and lattice dimension that can be quickly, continuously, and reversibly changed by an electrical input is needed to achieve this goal.

To convert an optical response from an electrical stimulus, many tactics have been used, which may be divided into four categories: (1) indirect electrically induced alteration of the photonic structures, (2) reorientation of liquid crystal molecules, (3) electrophoretic organization of colloidal suspension, and (4) electrochemical processes [120]. Ozin and Manners [116] provided the first example of tunable photonic crystals made possible by electrochemical processes. They showed that the photonic lattice could be electrochemically induced to expand or contract in an anisotropic manner, thereby allowing for precise tuning of the optical stopband of a colloidal photonic crystal concerning position, width, and intensity. By photopolymerizing a combination of (ethyl-methyl)sila-[1]-ferrocenophane and sila(cyclobutyl)-[1]-ferrocenophane infused into a self-assembled colloidal crystal of silica, the photonic structure was created.

The innovative polymer's active sites experience a charge change due to the applied voltage inside the coated glass, which causes the photonic structure to inflate and the reflected wavelength range for red-shifting.

According to Lu et al. [121], system adjustments at lower voltages may be accomplished using PS–P2VP copolymer multilayers using a water/ethanol combination as the solvent. Specifically, a voltage between 0 and 2.5 V is enough to induce the diffracted wavelength to change from blue to red.

5.5 Liquid Crystal–Based Electrical Photonic Crystals

Since liquid crystals' molecular structures and alignment are anisotropic, they exhibit a significant optical anisotropy. Because of this high sensitivity to electric fields, photonic crystals infused with liquid crystals have received a lot of attention as a potential material for electro-optical systems. A great deal of studies have focused on using LC RPCs in advanced optics and physics [75, 122–127].

Based on liquid crystal–infiltrated TiO_2 inverse opals, the group of Zhang, Ying, et al. [128] shows electrically switchable photonic crystals. Polystyrene opal templates create macroporous anatase TiO_2 inverse opals by sandwich vacuum backfilling and then calcining. The hybrid organic/inorganic structure's optical characteristics are determined by measuring the reflectance of the Bragg peak, whose location may be altered by applying an external electric field following liquid crystal infiltration. The applied electric field causes the liquid crystal molecules inside the spherical voids to reorient, which is the physical process underpinning this switchable behavior. This results in a considerable shift in the refractive index contrast between the liquid crystal and the TiO_2 inverse opal.

When the liquid crystal's long molecular axis is aligned parallel to the sphere surfaces, photonic crystals consisting of nematic liquid crystal intercalated into the space of densely packed silica spheres (synthetic porous opal) show a notable shift in the optical Bragg reflection peak caused by an electric field showed by D. Kang [129].

The inverse opal provides a greater empty volume (about 74%) for liquid crystal penetration than the opal structure. Therefore, the increased optical anisotropy is anticipated to produce more excellent tunability under electric modulation. A polymer inverse opal matrix infiltrated by 5CB was reported by Ozaki et al. [130] to exhibit a similar blue shift in reflection upon an increase in electric field.

6 Applications

6.1 Device

PhCs are used in several notable applications, including full-color paperlike displays, self-collimation [131], negative refractions [132], optical diodes [133], and light bending [134].

In this chapter on soft matter–based photonic crystals, we learn the tunable stimuli-responsive properties of soft matter–based photonic crystals. Realizing photonic crystals' tunability is crucial for their potential use in optical devices, including screens, lasers, and switching systems. M. Kolle and their group show how the transparent utility rubbers polydimethylsiloxane and polystyrene-polyisoprene may be assembled in a scalable layer assembly to create tunable distributed Bragg reflectors (DBRs) and resonant micro-cavities. An affine contraction of the layer thicknesses results from stretching the devices by more than 60%, which tunes both the cavity and DBR modes throughout the visible spectrum. These quickly and reversibly stretchable cavities have quantitative optical strain detection and tunable micro-laser applications [117]. In smart windows [135], multicolor circularly polarized luminescence [136] and anti-counterfeiting [137] it are being used.

It has already been used as an electro-optic switch. Lingling Ma et al. [138] reported sub-microsecond switching of chiral liquid crystals (LCs) using 1D photonic microstructures, in which, instead of changing the orientation, electric fields change the orientational order of molecules and quench their fluctuations. Consequently, the modified refractive indices cause the photonic bandgap to vary quickly roughly 100 ns four times quicker than traditional electro-optic switching in cholesterics.

6.2 Displays

A variety of colors might be produced using a photonic crystal. Reflective displays are incredibly effective for outdoor applications like billboard advertising and signage. In contrast, emissive display technologies like liquid crystal, light-emitting diode, and plasma displays struggle to overcome high production costs, high energy consumption, and glaring sunlight. Researchers can develop various display units using electrical [139] and magnetic [140] responsive properties of soft matter–based photonic crystals.

Dr. Cheng-Kai Liu [141] and their group can find the most efficient method to build polymer-network liquid crystal–based high-contrast and scattering-type liquid crystal displays. Their liquid crystal cell was compared with traditional ST-TRLCD cells. Comparisons reveal that to achieve better dark states in the transmissive and reflective modes of the reported ST-TRLCDs with the optimized configuration, CRs can be effectively enhanced by assembling a linear polarizer at the appropriate position. The main trade-off is the loss of brightness in the reflective modes. In this context, the PNLC cell, which functions as an electrically switchable polarizer, can be either homogeneous PNLC (H-PNLC) or PN-90° twisted nematic LC (PN-90° TNLC). A detailed comparison is made between the optoelectronic characteristics of H-PNLC cells and PN-90° TNLC cells. The findings show that the ST-TRLCD may achieve the maximum CR with an optimal H-PNLC cell design.

By adjusting photonic crystal orientation by the external magnetic field, magnetically controlled displays can also be achieved. The first synthesis of mesoscopic, monodisperse particles with nanoscopic ferromagnetic cobalt ferrite inclusions is presented by Sanford A. Asher et al. [142] and is a good choice for magnetically controlled display. The magnetically sensitive photonic crystals formed from these monodisperse ferromagnetic composite particles easily self-assemble and effectively Bragg diffract incoming light. The diffracted wavelength may be controlled by applying magnetic fields to the photonic crystal orientation.

6.3 Sensing

A different observation has already been made with the soft material–based system, which can be tuned as a sensing application for temperature [143], ion [144], biological object [145], and gas and humidity [146, 147] in the visible range.

Gas sensor technologies are gaining much attention (the Web of Science™ core database has 45,770 papers tagged with "gas sensor" between 2013 and 2021). As new applications are developed, new technological hurdles and financial needs arise. To create a composite MOF–silica colloidal crystal (MOF–SCC), Lu et al. [148] employed the layer-by-layer technique to produce the metal-organic framework (MOF) HKUST-1, a persistently microporous MOF with the empirical formula $Cu_2(BTC)_3$. Red shifts as large as 16 nm were seen in the MOF–SCC stopband when exposed to carbon disulfide (CS_2).

The range of measured amounts includes the concentrations of surfactants found in both industrial wastewater and municipal sewage discharges. 2D photonic crystal has applications in bio-sensing. Molecularly imprinted photonic crystals (PCs) have garnered significant interest due to their ability to detect biological and environmental chemicals using stimuli-sensitive hydrogels [149]. One metabolite of the Kynurenine pathway, L-Kynurenine (KYN), is a possible marker for cancer and immune suppressive diseases. Here, Rizvi et al. reported a two-dimensional (2D) photonic crystal (PC) hydrogel sensor that is molecularly imprinted to detect L-KYN in human serum. Methacrylic acid serves as the functional monomer in the polystyrene-based 2D PC colloidal arrays (2D PCCA) hydrogel used by the sensor, which allows it to imprint the L-KYN template by H-bonding [150]. Another stimuli-responsive colloidal hydrogel made of polystyrene particles is proposed by Fudouzi et al. [151]. Their techniques, which seem to have broad applicability, are based on the physical mechanisms of shear alignment for non–close-packed crystals and evaporation under silicone liquid for close-packed colloidal photonic crystals. For example, applications involving the sensing of color changes wouldn't need the material to have high optical quality. They examined basic applications that utilize the color of colloidal photonic crystals.

A noteworthy advancement has been made in the practical application with the help of gels or elastomers. They demonstrated an evolved gas pressure sensor through structural color changes or Bragg's diffractive peak tuning by successfully preparing thin films of superior materials.

Ciminelli [152] reported the design of a 2D photonic bandgap filter in the polymeric, which can be used in sensing applications. Teflon, PMMA, PVDF, and polystyrene are among the polymeric materials that have been studied. Polystyrene was selected because of its high refractive index value of 1.59, good thermal and electrical insulator qualities, optical resistance, and high stability. They compared hexagonal, square, and honeycomb lattices; out of them, square and hexagonal lattices are the most useful for realizing very compact dielectric mirrors.

Song et al. [147] created a humidity sensor by integrating water-sensitive PAAm into a P(St-MMAAA) photonic crystal framework. The photonic crystal hydrogel's hue may change reversibly in response to varying humidities. Given that the film displays various colors that span the visible spectrum in various humidity conditions, it might be a more affordable option for measuring humidity than conventional sensors.

6.4 Security

Counterfeiting is a major worldwide problem that harms the social economy and risks people's and communities' security. Therefore, there is a pressing need to create innovative anti-counterfeiting materials and technologies. Because of their distinct optical properties, photonic crystals (PCs) with vibrant structural hues are intriguing materials for improved security applications. In particular, it is anticipated that responsive PCs (RPCs), whose optical signals are easily modified by external stimuli, would increase security through optical encryption and information security. Additionally, RPCs may be utilized to verify the legitimacy of medications, papers, and currencies. The methods for creating responsive photonic crystals (RPCs) are outlined in a review done by W. Wang, which includes employing responsive materials as PC building blocks and packing responsive materials into the interstitial space of periodic photonic structures [153].

7 Future Directions and Conclusion

Many fabrication techniques are now available to realize photonic structures with different characteristics; even for fine-tuning the same parameter, various stimulating mechanisms can induce changes in photonic properties. PCs with a variety of functionalized approaches have developed quickly, leading to the construction of multiple types of PCs for the detection of a wide range of targets, such as temperature, humidity, pressure, electric field, magnetic field, ions, solvents, biological and clinical markers, and hazardous materials. However, while colloidal PC sensors respond instantly to electric fields and magnetic fields, their slow response when detecting biomolecules and other targets in aqueous solutions has severely hampered their practical applications in field detection, particularly for target molecules that demand rapid responses, like acute biological diagnosis and hazardous substance monitoring in the military. For this reason, there is an urgent need to find alternative colloidal PCs. That makes it an exciting field with tremendous opportunities in both fundamental and technological aspects. Understanding the system's self-assembly mechanism is exciting and can be done in any small laboratory without sophisticated equipment. Combining two different methods to fabricate a photonic system would be trending in the future. For example, a magnetic field can stabilize.

Colloidal particles form a stable two-dimensional closed-pack array during polymerization. Because of the affordability and reasonable response rate of those soft matter–based self-assembled photonic crystals, industrialists and start-up companies should consider bringing this technology to the marketplace in advertisement displays and sensors.

References

1. Goswami, D.: Optical computing. Resonance. **8**, 56–71 (2003). https://doi.org/10.1007/BF02837869
2. Addanki, S., et al.: Review of optical fibers—introduction and applications in fiber lasers. Results Phys. **10**, 743–750 (2018)
3. Reich, S., Goesmann, M., Heunoske, D., Schäffer, S., Lueck, M., Wickert, M., Osterholz, J.: Change of dominant material properties in laser perforation process with high-energy lasers up to 120 kilowatt. Sci. Rep. **13**(1), 21611 (2023). https://doi.org/10.1038/s41598-023-48511-9
4. Dorn, R., et al.: Nonlinear optical materials for integrated optics: telecommunications and sensors. Adv. Mater. **4**, 464–473 (1992)
5. Li, S.S.: Classification of solids and crystal structure. In: Li, S.S. (ed.) Semiconductor Physical Electronics. Springer, New York (2006). https://doi.org/10.1007/0-387-37766-2_1
6. Devanathan, V.: Electron energy bands in solids. J. Chennai Acad. Sci. **4**, 39–56 (2021)
7. Butt, M.A., Khonina, S.N., Kazanskiy, N.L.: Recent advances in photonic crystal optical devices: a review. Opt. Laser Technol. **142**, 107265 (2021)
8. Ivan Epstein, L.: The design of optical filters. J. Opt. Soc. Am. **42**, 806–810 (1952)

9. Bronnbauer, C., Gasparini, N., Brabec, C.J., Forberich, K.: Guideline for efficiency enhancement in semi-transparent thin-film organic photovoltaics with dielectric mirrors. Adv. Opt. Mater. **4**(7), 1098–1105 (2016). https://doi.org/10.1002/adom.201600080
10. Däntl, M., Jiménez-Solano, A., Lotsch, B.V.: Stimuli-responsive one-dimensional photonic crystals: design, fabrication and sensing. Mater. Adv. **3**(20), 7406–7424 (2022). https://doi.org/10.1039/D2MA00793B
11. Cai, Z., et al.: From colloidal particles to photonic crystals: advances in self-assembly and their emerging applications. Chem. Soc. Rev. **50**(10), 5898–5951 (2021)
12. Kang, Y., Walish, J.J., Gorishnyy, T., Thomas, E.L.: Broad-wavelength-range chemically tunable block-copolymer photonic gels. Nat. Mater. **6**, 957–960 (2007)
13. Bourezgui, A., Al-Hossainy, A.F., El Azab, I.H., Alresheedi, F., Mahmoud, S.A., Bassyouni, M., Abdel-Aziz, M.H., Zoromba, M.S.: Combined experimental and TDDFT computations for the structural and optical properties for poly(ortho phenylene diamine) thin film with different surfactants. J. Mater. Sci. Mater. Electron. **32**(5), 5489–5503 (2021). https://doi.org/10.1007/s10854-021-05271-4
14. Chan, E.P., Walish, J.J., Urbas, A.M., Thomas, E.L.: Mechanochromic photonic gels. Adv. Mater. **25**(29), 3934–3947 (2013). https://doi.org/10.1002/adma.201300692
15. Kang, N.G., Cho, B., Kang, B.G., Song, S., Lee, T., Lee, J.S.: Structural and electrical characterization of a block copolymer-based unipolar nonvolatile memory device. Adv Mater. **24**(3), 385–390 (2012). https://doi.org/10.1002/adma.201103862. Epub 2011 Dec 12. PMID: 22162078
16. Kamenjicki Maurer, M., Lednev, I.K., Asher, S.A.: Photoswitchable spirobenzopyran-based photochemically controlled photonic crystals. Adv. Funct. Mater. **15**, 1401–1406 (2005)
17. Haque, M.A., Kamita, G., Kurokawa, T., Tsujii, K., Gong, J.P.: Unidirectional alignment of lamellar bilayer in hydrogel: one-dimensional swelling, anisotropic modulus, and stress/strain tunable structural color. Adv. Mater. **22**(45), 5110–5114 (2010). https://doi.org/10.1002/adma.201002509
18. Chiappelli, M.C., Hayward, R.C.: Photonic multilayer sensors from photo-crosslinkable polymer films. Adv. Mater. **24**(45), 6100–6104 (2012)
19. Qin, H., Liu, P., Chen, C., Cong, H.-P., Yu, S.-H.: A multi-responsive healable supercapacitor. Nat. Commun. **12**, 4297 (2021). https://doi.org/10.1038/s41467-021-24568-w
20. Li, C., Lau, G.C., Yuan, H., Aggarwal, A., Dominguez, V.L., Liu, S., Sai, H., Palmer, L.C., Sather, N.A., Pearson, T.J., Freedman, D.E., Amiri, P.K., de la Cruz, M.O., Stupp, S.I.: Fast and programmable locomotion of hydrogel-metal hybrids under light and magnetic fields. Sci. Robot. **5**, eabb9822 (2020). https://doi.org/10.1126/scirobotics.abb9822
21. Yue, Y., Gong, J.P.: Tunable one-dimensional photonic crystals from soft materials. J. Photochem. Photobiol. C Photochem. Rev. **23**, 45–67 (2015)
22. Yan, D., Lu, W., Qiu, L.L., Meng, Z.H., Qiao, Y.: Thermal and stress tension dual-responsive photonic crystal nanocomposite hydrogels. RSC Adv. **9**, 21202–21205 (2019)
23. Yue, Y.F., Kurokawa, T., Haque, M.A., Nakajima, T., Nonoyama, T., Li, X.F., Kajiwara, I., Gong, J.P.: Mechano-actuated ultrafast full-colour switching in layered photonic hydrogels. Nat. Commun. **5**, 4659 (2014)
24. Liu, J., Mao, Y.W., Ge, J.P.: Electric field tuning of magnetically assembled photonic crystals. J. Mater. Chem. **C1**, 6129–6135 (2013)
25. Chen, K., Fu, Q.Q., Ye, S.Y., Ge, J.P.: Multicolor printing using electric-field-responsive and photocurable photonic crystals. Adv. Funct. Mater. **27**, 1702825 (2017)
26. Hong, W., Hu, X.B., Zhao, B.Y., Zhang, F., Zhang, D.: Tunable photonic polyelectrolyte colorimetric sensing for anions, cations and zwitterions. Adv. Mater. **22**, 5043–5047 (2010)
27. Cai, Z.Y., Sasmal, A., Liu, X.Y., Asher, S.A.: Responsive photonic crystal carbohydrate hydrogel sensor materials for selective and sensitive lectin protein detection. ACS Sens. **2**, 1474–1481 (2017)
28. Li, H., Wang, J.X., Pan, Z.L., Cui, L.Y., Xu, L.A., Wang, R.M., Song, Y.L., Jiang, L., Qi, F., et al.: Amplifying fluorescence sensing based on inverse opal photonic crystal toward trace TNT detection. J. Mater. Chem. **21**, 1730e1735 (2011)

29. Qi, F., et al.: Anal. Chim. Acta. **1123**, 91–112 (2020)
30. Bragg, W.H., Bragg, W.L.: The reflection of X-rays by crystals. Proc. R. Soc. Lond. Ser. A. **88**(605), 428–438 (1913)
31. González, A.: 1.5 X-ray crystallography: data collection strategies and resources. Compr. Biophys. **1**, 64–91 (2011). https://doi.org/10.1016/B978-0-12-374920-8.00106-5
32. Mayonado, G., et al.: Investigation of the Bragg-Snell law in photonic crystals. BFY Proc., 60–63 (2015)
33. Sözüer, H.S., Haus, J.W., Inguva, R.: Photonic bands: convergence problems with the plane-wave method. Phys. Rev. B. **45**(24), 13962 (1992)
34. Luo, Z., Evans, B.A., Chang, C.-H.: Magnetically actuated dynamic iridescence inspired by the neon tetra. ACS Nano. **13**(4), 4657–4666 (2019). https://doi.org/10.1021/acsnano.9b00822
35. Nagaishi, H., Oshima, N., Fujii, R.: Light-reflecting properties of the iridophores of the neon tetra, Paracheirodon innesi. Comp. Biochem. Physiol. **95A**, 337–341 (1990)
36. Hayakawa, M., Onda, T., Tanaka, T., Tsujii, K.: Hydrogels containing immobilized bilayer membranes. Langmuir. **13**(14), 3595–3597 (1997)
37. Haque, M.A., Kurokawa, T., Nakajima, T., Kamita, G., Gong, J.P., Fatema, Z.: Surfactant induced bilayer-micelle transition for emergence of functions in anisotropic hydrogel. J. Mater. Chem. B. **10**, 8386–8397 (2022). https://doi.org/10.1039/d2tb00172a19
38. Lama, M., Gong, J.P.: Mechanical reinforcement of lamellar bilayer hydrogels by small amounts of co-surfactants. ACS Omega. **8**(28), 25185–25194 (2023)
39. Wang, Z., Zhang, J., Tian, Z., Wang, Z., Li, Y., Liang, S., Cui, L., Zhang, L., Zhang, H., Yang, B.: Organic inorganic hybrid photonic hydrogels as a colorful platform for visual detection of SCN. Chem. Commun. **46**, 8636–8638 (2010)
40. López, C.: Materials aspects of photonic crystals. Adv. Mater. **15**, 1679–1704 (2003). https://doi.org/10.1002/adma.20030038624
41. Song, D.-P., Li, C., Li, W., Watkins, J.J.: Block copolymer nanocomposites with high refractive index contrast for one-step photonics. ACS Nano. **10**, 1216–1223 (2016). https://doi.org/10.1021/acsnano.5b0652522
42. Guien, Z., Bin, L., Jing, F., Suling, F.: Determination of trace thiocyanate in body fluids by a kinetic fluorimetric method. Talanta. **44**(7), 1141–1147 (1997)
43. Kleine, T.S., Diaz, L.R., Konopka, K.M., Anderson, L.E., Pavlopolous, N.G., Lyons, N.P., Kim, E.T., Kim, Y., Glass, R.S., Char, K., Norwood, R.A., Pyun, J.: One dimensional photonic crystals using ultrahigh refractive index chalcogenide hybrid inorganic/organic polymers. ACS Macro Lett. **7**, 875–880 (2018)
44. Robertson, M., Sanford, A., Qiang, Z.: Easy to pattern, chemically resistant one-dimensional polymer photonic crystals. ACS Appl. Polym. Mater. **3**(5), 2626–2634 (2021)
45. Kazmierczak, T., Song, H., Hiltner, A., Baer, E.: Polymeric one-dimensional photonic crystals by continuous coextrusion. Macromol. Rapid Commun. **28**, 2210–2216 (2007)
46. Cong, H., Yu, B., Tang, J., Li, Z., Liu, X.: Current status and future developments in preparation and application of colloidal crystals. Chem. Soc. Rev. **42**(19), 7774–7800 (2013)
47. Meseguer, F.: Colloidal crystals as photonic crystals. Colloids Surf. Physicochem. Eng. Aspects. **270**, 1–7 (2005)
48. Vlasov, Y.A., Bo, X.Z., Sturm, J.C., Norris, D.J.: On-chip natural assembly of silicon photonic bandgap crystals. Nature. **414**(6861), 289–293 (2001)
49. Xia, Y., Gates, B., Yin, Y., Lu, Y.: Monodispersed colloidal spheres: old materials with new applications. Adv. Mater. **12**(10), 693–713 (2000)
50. Wang, J., Zhang, Y., Wang, S., Song, Y., Jiang, L.: Bioinspired colloidal photonic crystals with controllable wettability. Acc. Chem. Res. **44**(6), 405–415 (2011)
51. Colodrero, S., Ocaña, M., Míguez, H.: Nanoparticle-based one-dimensional photonic crystals. Langmuir. **24**(9), 4430–4434 (2008)
52. He, L., Hu, Y., Kim, H., Ge, J., Kwon, S., Yin, Y.: Magnetic assembly of nonmagnetic particles into photonic crystal structures. Nano Lett. **10**(11), 4708–4714 (2010). https://doi.org/10.1021/nl103008v

53. Zi, J., Yu, X., Li, Y., Hu, X., Xu, C., Wang, X., Liu, X., Fu, R.: Coloration strategies in peacock feathers. Proc. Natl. Acad. Sci. **100**(22), 12576–12578 (2003)
54. Wang, M., Meng, F., Wu, H., Wang, J.: Photonic crystals with an eye pattern similar to peacock tail feathers. Crystals. **6**, 99 (2016). https://doi.org/10.3390/cryst6080099
55. Tikhonov, A., Kornienko, N., Zhang, J.T., Wang, L., Asher, S.A.: J. Nanophoton. **6**, 063509–1–9 (2012)
56. Zhang, J.T., Wang, L., Luo, J., Tikhonov, A., Kornienko, N., Asher, S.A.: J. Am. Chem. Soc. **133**, 9152–9155 (2011)
57. Zhang, J.-T., Smith, N., Asher, S.A.: Two-dimensional photonic crystal surfactant detection. Anal. Chem. **84**(15), 6416–6420 (2012)
58. Chen, C., Wang, X.H., Dong, Z.Q., Chen, G., Han, L.X., Zhu, Z.G.: Anti-counterfeiting layer of 2D colloidal crystal based photonic material. Mater. Sci. Forum. (2019). https://doi.org/10.4028/www.scientific.net/msf.972.185
59. Prevo, B.G., Velev, O.: Controlled, rapid deposition of structured coatings from micro-and nanoparticle suspensions. Langmuir. **20**(6), 2099–2107 (2004)
60. Chen, C., Dong, Z.-Q., Shen, J.-H., Chen, H.-W., Zhu, Y.-H., Zhu, Z.-G.: 2D photonic crystal hydrogel sensor for tear glucose monitoring. ACS Omega. **3**(3), 3211–3217 (2018). https://doi.org/10.1021/acsomega.7b02046
61. Nagarajan, R., Joyner, C.H., Schneider, R.P., Bostak, J.S., Butrie, T., Dentai, A.G., et al.: Large-scale photonic integrated circuits. IEEE J. Selected Top. Quantum Electron. **11**(1), 50–65 (2005)
62. Carroll, L., Lee, J.S., Scarcella, C., Gradkowski, K., Duperron, M., Lu, H., et al.: Photonic packaging: transforming silicon photonic integrated circuits into photonic devices. Appl. Sci. **6**(12), 426 (2016)
63. Lin, S.Y., Fleming, J.G., Hetherington, D.L., Smith, B.K., Biswas, R., Ho, K.M., Sigalas, M.M., Zubrzycki, W., Kurtz, S.R., Bur, J.: A three-dimensional photonic crystal operating at infrared wavelengths. Nature. **394**(6690), 251–253 (1998)
64. Gandhi, S.S., Chien, L.C.: Unraveling the mystery of the blue fog: structure, properties, and applications of amorphous blue phase III. Adv. Mater. **29**, 1704296 (2017). https://doi.org/10.1002/adma.201704296
65. Castles, F., et al.: Stretchable liquid-crystal blue-phase gels. Nat. Mater. **13**, 817–821 (2014). https://doi.org/10.1038/nmat3993
66. Kikuchi, H.: Liquid crystalline blue phases. In: Kato, T. (ed.) Liquid Crystalline Functional Assemblies and Their Supramolecular Structures, pp. 99–117. Springer, Berlin (2007)
67. Zhao, Y., et al.: Spherical colloidal photonic crystals. Acc. Chem. Res. **47**(12), 3632–3642 (2014)
68. Zhao, Y.J., Zhao, X.W., Tang, B.C., Xu, W.Y., Li, J., Hu, J., Gu, Z.Z.: Quantum-dot-tagged bioresponsive hydrogel suspension array for multiplex label-free DNA detection. Adv. Funct. Mater. **20**, 976–982 (2010)
69. Kim, Y.C., Cho, C.Y., Kang, J.H., Cho, Y.S., Moon, J.H.: Synthesis of porous carbon balls from spherical colloidal crystal templates. Langmuir. **28**, 10543–10550 (2012)
70. Wang, J.Y., Hu, Y.D., Deng, R., Liang, R.J., Li, W.K., Liu, S.Q., Zhu, J.T.: Multiresponsive hydrogel photonic crystal microparticles with inverse-opal structure. Langmuir. **29**, 8825–8834 (2013)
71. Jiang, P., et al.: Single-crystal colloidal multilayers of controlled thickness. Chem. Mater. **11**(8), 2132–2140 (1999)
72. Holgado, M., et al.: Electrophoretic deposition to control artificial opal growth. Langmuir. **15**(14), 4701–4704 (1999)
73. Ali, I., et al.: Recent advances in syntheses, properties and applications of TiO_2 nanostructures. RSC Adv. **8**(53), 30125–30147 (2018)
74. Reculusa, S., Masse, P., Ravaine, S.: Three-dimensional colloidal crystals with a well-defined architecture. J. Colloid Interface Sci. **279**, 471–478 (2004)
75. Ge, J., Yin, Y.: Responsive photonic crystals. Angew. Chem. Int. Ed. **50**(7), 1492–1522 (2011). https://doi.org/10.1002/anie.200907091

76. Qin, J., Dong, B., Li, X., Han, J., Gao, R., Su, G., Gao, L., Wang, W.: Fabrication of intelligent photonic crystal hydrogel sensors for selective detection of trace mercury ions in seawater. J. Mater. Chem. C. **5**(3), 8482–8488 (2017)
77. Belda-Lois, J., et al.: Rehabilitation of gait after stroke: a review towards a top-down approach. J. NeuroEng. Rehabil. **8**, 66 (2011)
78. Weiser, M.-S., et al.: Self-processing, diffusion-based photopolymers for holographic applications. Macromol. Symp. **296**, 133–137 (2010)
79. Bunning, T.J., Natarajan, L.V., Sutherland, R.L., Tondiglia, V.P.: Holographic polymer-dispersed liquid crystals (H-PDLCs). Annu. Rev. Mater. Sci. **30**(1), 83–115 (2000)
80. Urbas, A., Klosterman, J., Tondiglia, V., Natarajan, L., Sutherland, R., Tsutsumi, O., Ikeda, T., Bunning, T.: Optically switchable Bragg reflectors. Adv. Mater. **16**, 1453–1456 (2004)
81. Urbas, A., Tondiglia, V., Natarajan, L., Sutherland, R., Yu, H., Li, J.-H., Bunning, T.: Optically switchable liquid crystal photonic structures. J. Am. Chem. Soc. **126**, 13580–13581 (2004)
82. Ikeda, T., Tsutsumi, O.: Optical switching and image storage by means of azobenzene liquid-crystal films. Science. **268**(5219), 1873–1875 (1995)
83. Feng, J., Li, Z., Olah, A., Baer, E.: High oxygen barrier multilayer EVOH/LDPE film/foam. J. Appl. Polym. Sci. **46425**, 1–9 (2018)
84. Rahman, A., Wu, H., Zhang, Z., Baer, E.: Enhanced properties of oriented multilayer polypropylene film/foams. SPE ANTEC, 1806–1810 (2016)
85. Lee, S., Zhu, L., Maia, J.: The effect of strain-hardening on the morphology and mechanical and dielectric properties of multi-layered PP foam/PP film. Polymer. **70**, 173–182 (2015). https://doi.org/10.1016/j.polymer.2015.06.029
86. Messin, T., Marais, S., Follain, N., et al.: Impact of water and thermal induced crystallizations in a PC/MXD6 multilayer film on barrier properties. Eur. Polym. J. **111**, 152–160 (2019). https://doi.org/10.1016/j.eurpolymj.2018.12.021
87. Schneider, T., Colton, M.F., Mccauley, K.M.: Microlayer and nanolayer tubing and piping via layer multiplication coextrusion. II. Rheologically mismatched systems. J. Appl Polym. **137**, 48684 (2020). https://doi.org/10.1002/app.48684
88. Schneider, T.: Multi-layered tubing and piping: technology development and transfer leading to new dimensions in annular layered structures. Dissertation (2020)
89. Schneider, T, Maia J: Studying layer alignment in flows of multi-layered systems through dies with a geometric transition
90. Steinmetz, E., Maia, J.: Layer multiplication co-extrusion of cross-linked polymer microsphere-filled systems. Polym. Eng. Sci. **64**(3), 1045–1056 (2024). https://doi.org/10.1002/pen.26595
91. Hill, J.P., Shrestha, L.K., Ishihara, S., Ji, Q., Ariga, K.: Self-assembly: from amphiphiles to chromophores and beyond. Molecules. **19**, 8589–8609 (2014)
92. Harris, P.J., Patz, J., Huntington, B.A., Bonnecaze, R.T., Meltzer, D., Maia, J.: Improved interfacial surface generator for the coextrusion of micro- and nanolayered polymers. Polym. Eng. Sci. **54**, 636–645 (2014). https://doi.org/10.1002/pen.23597
93. Huang, R.: Multilayer Co-Extrusion and Twin Screw Compounding of Polymeric Elastomer Systems. Case Western Reserve University (2014)
94. Huang, R., Chari, P., Tseng, J., Zhang, G., Cox, M., Maia, J.M.: Microconfinement effect on gas barrier and mechanical properties of multilayer rigid/soft thermoplastic polyurethane films. J. Appl. Polym. Sci. **41849**, 1–11 (2015). https://doi.org/10.1002/app.41849
95. Radford, J.A., Alfrey Jr., T., Schrenk, W.J.: Reflectivity of iridescent coextruded multilayered plastic films. Polym. Eng. Sci. **13**(3), 216–221 (1973)
96. Yoon, J., Lee, W., Thomas, E.L.: Self-assembly of block copolymers for photonic-bandgap materials. MRS Bull. **30**(10), 721–726 (2005)
97. Grzybowski, B.A., Winkleman, A., Wiles, J.A., Brumer, Y., Whitesides, G.M.: Electrostatic self-assembly of macroscopic crystals using contact electrification. Nat. Mater. **2**(4), 241–245 (2003). https://doi.org/10.1038/nmat860
98. McCarty, L.S., Winkleman, A., Whitesides, G.M.: Electrostatic self-assembly of polystyrene microspheres by using chemically directed contact electrification. Angew. Chem. Int. Ed. **46**(1–2), 206 (2007)

99. Kim, S.H., Jeon, S.J., Yi, G.R., Heo, C.J., Choi, J.H., Yang, S.M.: Optofluidic assembly of colloidal photonic crystals with controlled sizes, shapes, and structures. Adv. Mater. **20**, 1649–1655 (2008)
100. Walish, J.J., Kang, Y., Mickiewicz, R.A., Thomas, E.L.: Bioinspired electrochemically tunable block copolymer full color pixels. Adv. Mater. **21**, 3078–3081 (2009)
101. Valkama, S., Kosonen, H., Ruokolainen, J., Haatainen, T., Torkkeli, M., Serimaa, R., ten Brinke, G., Ikkala, O.: Self-assembled polymeric solid films with temperature-induced large and reversible photonic-bandgap switching. Nat. Mater. **3**, 872–876 (2004)
102. Lee, W., Yoon, J., Thomas, E.L., Lee, H.: Dynamic changes in structural color of a lamellar block copolymer photonic gel during solvent evaporation. Macromolecules. **46**, 6528–6532 (2013)
103. Liu, S., et al.: Recent progress in responsive photonic crystals of block copolymers. J. Mater. Chem. C. **8**(47), 16633–16647 (2020)
104. Fink, Y., Urbas, A.M., Bawendi, M.G., Joannopoulos, J.D., Thomas, E.L.: J. Lightwave Technol. **17**, 1963 (1999)
105. Nagarajan, R., Ruckenstein, E.: Theory of surfactant self-assembly: a predictive molecular thermodynamic approach. Langmuir. **7**, 2934–2969 (1991)
106. Holmberg, K., Jönsson, B., Kronberg, B., Lindman, B.: Surfactants and Polymers in Aqueous Solution. John Wiley & Sons, New York (1998)
107. Zheng, S., Lin, Y.L., Chang, C.C., Lee, M.J., Chen, Y.F., Lee, L.R., Chang, M.H., Chen, J.T.: Boosting ion conductivities: light-modulated azobenzene-based ionic liquids in vertical nanochannels. ACS Appl. Mater. Interfaces. **15**(38), 45418–45425 (2023). https://doi.org/10.1021/acsami.3c08792
108. Wang, W., Gu, B., Liang, L.: Effect of anionic surfactants on synthesis and self-assembly of silica colloidal nanoparticles. J. Colloid Interface Sci. **313**(1), 169–173 (2007). https://doi.org/10.1016/j.jcis.2007.04.042
109. Chiasera, A., Scotognella, F., Criante, L., Varas, S., Della Valle, G., Ramponi, R., Ferrari, M.: Disorder in photonic structures induced by random layer thickness. Sci. Adv. Mater. **7**, 1207–1212 (2015)
110. Bellingeri, M., Chiasera, A., Kriegel, I., Scotognella, F.: Optical properties of periodic, quasi-periodic, and disordered one-dimensional photonic structures. Opt. Mater. **72**, 403–421 (2017)
111. Weissman, J.M., Sunkara, H.B., Albert, S.T., Asher, S.A.: Thermally switchable periodicities and diffraction from mesoscopically ordered materials. Science. **274**, 959–963 (1996)
112. Schild, H.: Poly(N-isopropylacrylamide): experiment, theory and application. Prog. Polym. Sci. **17**, 163–249 (1992)
113. Shibayama, M., Tanaka, T.: Poly(N-isopropylacrylamide): experiment, theory and application. Adv. Polym. Sci. **109** (1993)
114. Arsenault, A.C., Clark, T.J., von Freymann, G., Cademartiri, L., Sapienza, R., Bertolotti, J., Vekris, E., Wong, S., Kitaev, V., Manners, I., Wang, R.Z., John, S., Wiersma, D., Ozin, G.A.: Nat. Mater. **5**, 179 (2006)
115. Park, W., Lee, J.-B.: Mechanically tunable photonic crystal structure. Appl. Phys. Lett. **85**(21), 4845–4847 (2004)
116. Arsenault, A.C., et al.: Vapor swellable colloidal photonic crystals with pressure tunability. J. Mater. Chem. **15**(1), 133–138 (2005)
117. Kolle, M., Zheng, B., Gibbons, N., Baumberg, J.J., Steiner, U.: Stretch-tuneable dielectric mirrors and optical microcavities. Opt. Express. **18**, 4356–4364 (2010)
118. Kamenjicki, M., Lednev, I.K., Mikhonin, A., Kesavamoorthy, R., Asher, S.A.: Photochemically controlled photonic crystals. Adv. Funct. Mater. **13**, 774–780 (2003)
119. Paternò, G.M., et al.: Solution processable and optically switchable 1D photonic structures. Sci. Rep. **8**(1), 3517 (2018)
120. Nucara, L., Greco, F., Mattoli, V.: Electrically responsive photonic crystals: a review. J. Mater. Chem. C. **3**(3), 8449–8467 (2015)

121. Lu, Y., Xia, H., Zhang, G., Wu, C.: Electrically tunable block copolymer photonic crystals with a full color display. J. Mater. Chem. **19**(33), 5952–5955 (2009)
122. Wilk, R., Vieweg, N., Kopschinski, O., Koch, M.: Liquid crystal based electrically switchable Bragg structure for THz waves. Opt. Exp. **17**(9), 7377–7382 (2009)
123. Arriaga, J., Dobrzynski, L., Djafari-Rouhani, B.: Electrotunable band gaps of one-and two-dimensional photonic crystal structures based on silicon and liquid crystals. J. Appl. Phys. **104**(6), 063108 (2008)
124. Tolmachev, V.A., Perova, T.S., Grudinkin, S.A., Melnikov, V.A., Astrova, E.V., Zharova, Y.A.: Appl. Phys. Lett. **90**, 011908 (2007)
125. Haurylau, M., et al.: Electrically tunable silicon 2-D photonic bandgap structures. IEEE J. Sel. Top. Quantum Electron. **12**(6), 1527–1533 (2006)
126. Song, M.H., et al.: Electrotunable non-reciprocal laser emission from a liquid-crystal photonic device. Adv. Funct. Mater. **16**(14), 1793–1798 (2006)
127. Maune, B., et al.: Liquid-crystal electric tuning of a photonic crystal laser. Appl. Phys. Lett. **85**(3), 360–362 (2004)
128. Zhang, Y., et al.: Electrically switchable photonic crystals based on liquid-crystal-infiltrated TiO_2-inverse opals. Opt. Exp. **27**(11), 15391–15398 (2019)
129. Kang, D., Maclennan, J.E., Clark, N.A., Zakhidov, A.A., Baughman, R.H.: Electro-optic behavior of liquid-crystal-filled silica opal photonic crystals: effect of liquid-crystal alignment. Phys. Rev. Lett. **86**(18), 4052–4055 (2001). https://doi.org/10.1103/PhysRevLett.86.4052
130. Ozaki, M., Shimoda, Y., Kasano, M., Yoshino, K.: Electric field tuning of the stop band in a liquid-crystal-infiltrated polymer inverse opal. Adv. Mater. **14**(7), 514–518 (2002). https://doi.org/10.1002/1521-4095(20020404)14:7<514::aid-adma514>3.0.co,2-4
131. Noori, M., Soroosh, M., Baghban, H.: Self-collimation in photonic crystals: applications and opportunities. Ann. Der Phys. **530**, 1700049 (2018)
132. Dutta, H.S., Goyal, A.K., Srivastava, V., Pal, S.: Coupling light in photonic crystal waveguides: a review. Photon. Nanostruct. Fundam. Appl. **20**, 41–58 (2016)
133. Liu, B., Liu, Y.-F., Jia, C., He, X.-D.: All-optical diode structure based on asymmetrical coupling by a micro-cavity and FP cavity at two sides of photonic crystal waveguide. AIP Adv. **6**, 065316 (2016)
134. Zhang, Y., Li, B.: Photonic crystal-based bending waveguides for optical interconnections. Opt. Exp. **14**, 5723–5732 (2006)
135. Zaky, Z.A., Aly, A.H.: Novel smart window using photonic crystal for energy saving. Sci. Rep. **12**(1), 1–9 (2022). https://doi.org/10.1038/s41598-022-14196-9
136. Shi, Y., Han, J., Li, C., Zhao, T., Jin, X., Duan, P.: Recyclable soft photonic crystal film with overall improved circularly polarized luminescence. Nat. Commun. **14**(1), 1–9 (2023). https://doi.org/10.1038/s41467-023-41884-5
137. Pan, M., Wang, L., Dou, S., Zhao, J., Xu, H., Wang, B., Zhang, L., Li, X., Pan, L., Li, Y.: Recent advances in colloidal photonic crystal-based anti-counterfeiting materials. Crystals. **9**(8), 417 (2019). https://doi.org/10.3390/cryst9080417
138. Ma, L., Li, C., Sun, L., Song, Z., Yanqing, L., Li, B.: Submicrosecond electro-optical switching of one-dimensional soft photonic crystals. Photon. Res. **10**, 786–792 (2022)
139. Arsenault, A.C., Puzzo, D.P., Manners, I., Ozin, G.A.: Nat. Photon. **1**, 468 (2007)
140. Ge, J., Yin, Y.: Magnetically tunable colloidal photonic structures in alkanol solutions. Adv. Mater. **20**(18), 3485–3491 (2008)
141. Liu, C.-K., Chen, W.-H., Li, C.-Y., Cheng, K.-T.: High-contrast and scattering-type transflective liquid crystal displays based on polymer-network liquid crystals. Polymers. **12**, 739 (2020)
142. Xu, X., Majetich, S.A., Asher, S.A.: Mesoscopic monodisperse ferromagnetic colloids enable magnetically controlled photonic crystals. J. Am. Chem. Soc. **124**(46), 13864–13868 (2002). https://doi.org/10.1021/ja026901k
143. Kanai, T., Yano, H., Kobayashi, N., Sawada, T.: Enhancement of thermosensitivity of gel-immobilized tunable colloidal photonic crystals with anisotropic contraction. ACS Macro Lett. **6**, 1196–1200 (2017)

144. Saito, H., Takeoka, Y., Watanabe, M.: Simple and precision design of porous gel as a visible indicator for ionic species and concentration. Chem. Commun. **17**, 2126–2127 (2003)
145. Tian, T., Li, X., Cui, J., Li, J., Lan, Y., Wang, C., Zhang, M., Wang, H., Li, G.: Highly sensitive assay for acetylcholinesterase activity and inhibition based on a specifically reactive photonic nanostructure. ACS Appl. Mater. Interfaces. **6**, 15456–15465 (2014)
146. Kobler, J., Lotsch, B.V., Ozin, G.A., Bein, T.: Vapor-sensitive bragg mirrors and optical isotherms from mesoporous nanoparticle suspensions. ACS Nano. **3**(7), 1669–1676 (2009)
147. Tian, E.T., Wang, J.X., Zheng, Y.M., Song, Y.L., Jiang, L., Zhu, D.B.: Colorful humidity sensitive photonic crystal hydrogel. J. Mater. Chem. **18**(10), 1116–1122 (2008)
148. Lu, G., Farha, O.K., Kreno, L.E., Schoenecker, P.M., Walton, K.S., Van Duyne, R.P., Hupp, J.T.: Fabrication of metal-organic framework-containing silica-colloidal crystals for vapor sensing. Adv. Mater. **23**(38), 4449–4452 (2011)
149. Liu, X.Y., Fang, H.X., Yu, L.P.: Molecularly imprinted photonic polymer based on β-cyclodextrin for amino acid sensing. Talanta. **116**, 283–289 (2013). https://doi.org/10.1016/j.talanta.2013.05.003
150. Rizvi, A.S., Murtaza, G., Yan, D., Irfan, M., Xue, M., Meng, Z.H., Qu, F.: Development of molecularly imprinted 2D photonic crystal hydrogel sensor for detection of L-Kynurenine in human serum. Talanta. **208**, 120403 (2019)
151. Fudouzi, H., Sawada, T.: Colloidal photonic crystals made of soft materials: gels and elastomers. In: Meyrueis, P., Van de Voorde, M., Sakoda, K. (eds.) Micro- and Nanophotonic Technologies (2017)
152. Ciminelli, C., Armenise, M.: Modeling and design of a 2D photonic crystal microcavity on polymer material for sensing applications. Proc. SPIE. **6619**, 738607 (2007). https://doi.org/10.1117/12.738607
153. Wang, W., Zhou, Y., Yang, L., Yang, X., Yao, Y., Meng, Y., Tang, B.: Stimulus-responsive photonic crystals for advanced security. Adv. Funct. Mater. **32**(40), 2204744 (2022). https://doi.org/10.1002/adfm.202204744

Soft Material–Based Flexible Devices

Amrita De Adhikari

Abstract Soft material–based flexible devices are electronic devices made using soft and flexible materials. These materials allow the devices to be bent, stretched, twisted, and molded into different shapes. Unlike conventional electronics that use rigid materials like silicon, soft material–based flexible devices use materials that can be easily deformed. This allows the development of flexible electronics with a wide range of applications in various fields, such as wearable technology, medical devices, and flexible displays. Flexible devices made of soft materials are an exciting and cutting-edge electronics area. They provide endless possibilities for creating innovative and adaptable electronic systems. With ongoing research and technological advancements, the capabilities and applications of these flexible devices are expected to expand further.

Keywords Soft · Flexible · Wearable · Stretch · Molded

1 Introduction

Soft electronic devices can adapt freely to the surface contour and easily stretch with distortions. Flexible devices are integrated systems that integrate various functional materials on flexible substrates [1]. Developing soft material–based flexible devices has emerged as a pioneering area in the electronics industry, offering many opportunities to create novel and highly adaptable electronic systems. The unique properties of these devices allow them to bend, twist, and flex, making them ideal for use in various applications such as wearable electronics, healthcare monitoring systems, and soft robotics [2, 3]. This helps construct soft sensors, actuators, interconnects, etc. Such soft devices possess a high degree of freedom with conductive patterns. The advances in the field of flexible electronics open

A. De Adhikari (✉)
Department of Chemical Engineering, Indian Institute of Technology Kanpur, Kanpur, India

V. Kumar, Y. K. Mishra (eds.), *Soft Materials for Functional Applications*,
Advanced Structured Materials 225, https://doi.org/10.1007/978-981-97-9468-3_10

up noteworthy possibilities for research that can have a bigger impact on human life. With continuous research and technological advancements, the capabilities and potential applications of these flexible devices are expected to expand even further, paving the way for a new generation of electronic devices that are not only highly functional but also comfortable and safe to use. Fruitful upshots can help to visualize the connectivity of all with a collection of Internet of Things (IOTs) leading to next-gen electronic advancements.

Analysts estimated the growth of flexible electronics will be approximately 30% every year till 2028, making integrating solar cells, transistors, actuators, sensors, and other functional electronic devices important [2]. Commercially available wearables, like watches, belts, wristbands, and glasses, are composed of rigid or planar materials. This is the shortcoming of the growing wearable market. For this, the bulky and rigid materials must be replaced with thin, soft, skin-like materials for functionality and wearability. Soft and rigid materials exhibit distinct mechanical properties and behaviors, and their differences lie in factors such as flexibility, deformation, and response to external forces. The highlighted challenges to such soft, flexible devices include (1) achieving high conductivity, (2) parameters for obtaining optimal polymer matrix, (3) the nature of the conductive inks, and (4) various fabrication processes [4].

This chapter focuses on the latest developments in flexible devices based on soft materials, which are expected to bring about the upcoming generation of soft electronics. Firstly, we will examine the various methods used to create soft materials. Then, we will delve into the use of soft materials in manufacturing functional constituents such as semiconductors; conductors; insulators; devices such as transistors, sensors, and energy devices; as well as integrated arrangements. Lastly, this chapter will deliberate the difficulties inherent in emerging real-world presentations using flexible devices constructed on soft materials.

2 History of Flexible Devices

The origin of flexible displays can be pinpointed to the Xerox Palo Alto Research Centre in 1974, where employee Nicholas K. Sheridan conceived the Gyricon. This marked the inception of the first flexible e-paper display, replicating the tactile sensation of paper while capable of showcasing dynamic images [5]. Subsequently, flexible displays have found their way into numerous consumer electronics such as smartphones, tablets, TVs, and e-readers. Flexible devices, such as transistors and transducers, became industrially prominent in the late 1990s. These devices were crafted from organic or inorganic thin films or plastic substrates. During the period spanning from the 1990s to the early 2000s, light-emitting diodes and flexible transistors were predominantly designed on a bendable substrate utilizing organic materials [6–8]. This preference for organic materials stemmed from their inherent mechanical flexibility compared to inorganic films. These organic components could be formed through evaporation or solution-based processes on plastic amorphous substrates.

Nevertheless, a significant development occurred with introducing an inorganic thin film relocation method from a bulk wafer to an alternative substrate. This technique, pioneered by Yablonovitch et al. and Konagai et al., facilitated the creation of inorganic-based flexible devices since 2000 [9, 10]. In the early phases of flexible electronics development, a comprehensive exploration of the essential individualities of devices like sensors and transistors was conducted. This exploration encompassed considerations of inorganic and organic materials, electrical traits, machine-driven dependability, and bendability. While ongoing research continues to delve into fundamental properties using existing and novel material systems, the current focus on flexible and stretchable electronics is directed toward creating applications for flexible and/or stretchable devices [11–16].

As part of the evolution in device integration on elastic substrates, Rogers et al. achieved a milestone in 2001 by developing a paperlike electronic demonstration utilizing an organic material–based dynamic matrix backplane [17]. This demonstration is widely considered one of the earliest instances of material incorporation for flexible device uses. Subsequent advancements followed this breakthrough, with Someya et al. presenting a synthetic electronic skin (e-skin) in 2004, capable of detecting pressure distributions akin to human skin [18]. In 2005, their group further enhanced this technology by creating a device for sensing pressure and temperature distributions using organic-based thin film transistors (TFTs) on the active matrix backplane [19]. Concurrently, thin film Si-TFTs were successfully replaced by a flexible substrate in place of a silicon-on-insulator (SOI) wafer, enabling the functioning of TFTs with high flexibility and constancy in open air.

Through the 2000s, a significant hurdle for flexible electronics utilizing organic resources as the important constituent was the low and unstable performance in the open air. Nonetheless, with the advent of inorganic-based TFTs, new possibilities emerged for inorganic materials to play a role in flexible electronics [11]. Given the matured technique of Si-based TFTs as manufacturing standard for integrated circuits (ICs) made of Si, flexible Si-TFTs exhibited remarkable performance characteristics, boasting a 180 $cm^2/V\ s$ of field-effect mobility on a bendable substrate. Rogers et al., in the year 2008, achieved a significant milestone by successfully constructing an integrated digital circuit, specifically a four-bit row decoder, employing networks of carbon nanotubes as p-type transistors [20]. Following numerous advancements in the formation of inorganic materials, device structures, and the introduction of novel materials, flexible transistors utilizing both inorganic and organic materials have exhibited considerable prospective for flexible electronics, marked by enhanced performance and stability. Someya et al. demonstrated an active matrix organic light-emitting diode (OLED) flexible display utilizing the podiums of OLEDs and organic transistors [21]. Simultaneously, Javey et al. and Bao et al. reported two distinct e-skin device demonstrations [22, 23]. Javey et al.'s device featured an array of inorganic nanowire active matrix backplanes, showcasing high performance and elasticity achieved through a nanowire printing technique. At the same time, Bao et al. projected an exclusive configuration designed for tactile pressure detection with extraordinary sensitivity. The application of stretchable and flexible devices expanded into therapeutic and neuroscience arenas for in vivo experiments or implantation, leveraging their plasticity [24–26]. An innovative approach involved using materials like silk dissolved in water, enabling sensors and curative devices used for implantation without post-surgery removal, thereby reducing medical costs and patient burden [26].

Diverse stretchable and flexible device uses have emerged following these breakthroughs, including ultrathin flexible devices, self-healing sensors, and photovoltaic devices. Moreover, recent advancements include printed, wide variety, cheap, wearable soft healthcare appliances that assimilate numerous sensors for healthcare monitoring and human activity. Despite these achievements, challenges persist, particularly in realizing wireless systems, signal processing circuits, and batteries. Flexible transistors face limitations in constructing complex circuits despite substantial improvements. In 2016, Javey et al. addressed this challenge by demonstrating a hybrid system that integrated flexible chemical sensors with conventional inflexible integrated circuits (ICs) for signal processing and wireless capabilities on a flexible printed circuit film [27]. Additionally, Rogers et al. established a battery-free fusion system for wireless health monitoring, incorporating flexible and stretchable antennas and electrical chips [28].

Nevertheless, foldable smartphones are the most prominent manifestation of flexible display electronics. Only in the relatively recent past have these devices transitioned from conceptual prototypes to tangible products. The inaugural commercially available foldable smartphone, the Royole FlexPai, debuted in December 2018 and was crafted by the Chinese company Royole. Flexible displays are discreetly integrated into numerous widely manufactured consumer electronics, presenting themselves in the subtle curvature observed at the edges of smartphone screens and smartwatches. However, in recent times, technology companies have embraced more ambitious ventures concerning flexible displays. This includes the creation of foldable smartphones featuring flexible displays and the development of adjustable, flexible display TVs. Flexible display TVs have been available for approximately a decade, with both LG and Samsung asserting their positions as pioneers in introducing the first curved OLED displays [29].

3 Difference Between Soft and Rigid Materials

Choosing the right material for a system or device requires understanding the advantages and limitations of both soft and rigid materials. Here are some key differences between soft and rigid materials [30–32].

3.1 Flexibility

Soft Materials: Soft materials like elastomers, polymers, gels, and certain foams are deformable and can easily undergo large deformations without permanent damage. They are also characterized by their low stiffness, which makes them easy to bend, twist, and stretch.

Rigid Materials: Rigid materials like metals, ceramics, and hard plastics are known for their high stiffness and resistance to deformation. They can maintain their shape and structure even under applied forces, making them ideal for structural applications and heavy-duty machinery.

3.2 Deformation Behavior

Soft Materials: Soft materials can easily deformation under stress and can restore their initial shape after deformation. They often display nonlinear and time-dependent deformation behaviors.

Rigid Materials: Rigid materials can deform elastically up to a certain point, reaching their yield strength, beyond which permanent deformation (plastic deformation) occurs. Unlike soft materials, rigid materials do not easily restore their original shape once they undergo significant deformation.

3.3 Mechanical Modulus

Soft Materials: Soft materials exhibit low stiffness, measured by the elastic modulus that gauges a material's resistance to deformation under stress.

Rigid Materials: Rigid materials are stiff and require considerable force to deform due to their high elastic modulus.

3.4 Response to External Forces

Soft Materials: Soft materials are suitable for applications that require flexibility and impact absorption because they absorb and dissipate energy. They can also conform to irregular shapes.

Rigid Materials: Rigid materials are suitable for applications where structural integrity and stability are important as they transmit and reflect energy while maintaining their shape under external forces.

3.5 Manufacturability

Soft Materials: Soft materials can be shaped into intricate forms using manufacturing processes such as 3D printing, casting, and molding.

Rigid Materials: It may be difficult to shape metals or materials that require higher forces into intricate designs.

3.6 Cost and Availability

Soft Materials: Soft materials, such as polymers and elastomers, are affordable to produce. They are readily available and can be synthesized or produced at low cost.

Rigid Materials: Rigid materials, such as metals and ceramics, can be more expensive to produce due to specific processing techniques required, such as casting and machining.

4 Advantages of Soft Material–Based Flexible Devices

Flexible soft material devices have several advantages, leading to their increasing popularity and widespread use in various applications. Here are some key advantages [33–36].

1. **Conformability**:
   Flexibility: Flexible materials allow for the fabrication of devices that can bend, twist, and stretch, conforming to irregular surfaces such as the human body, improving comfort and wearability.
2. **Wearable Technology**:
   Comfort: Wearable technology benefits from flexible devices, which provide a comfortable and nonintrusive user experience. This is especially important for smartwatches, fitness trackers, and health monitors.
3. **Biocompatibility**:
   Soft and Biocompatible Materials: Soft materials with biocompatibility can be used in healthcare without adverse reactions.
4. **Lightweight**:
   Low Weight: Soft materials are usually lightweight and ideal for developing portable and wearable devices. This is mostly beneficial in applications where weight is acute, such as in portable medical devices or wearable electronics.
5. **Robustness to Mechanical Stress**:
   Durability: Soft devices can be more durable than rigid ones when subjected to mechanical stress, such as bending or twisting. This makes them ideal for applications that involve repeated deformation.

6. **Large-Area Coverage**:
   Scalability: Manufacturing processes for soft materials, such as roll-to-roll processing and printing technologies, enable the cost-effective production of devices over large areas. This scalability is advantageous for applications like flexible displays.
7. **Enhanced Safety in Human Interaction**:
   Soft Robotics: Soft materials are crucial in developing soft robotics, where safety during human–robot interactions is essential. Soft robots can adapt to human environments more safely than traditional rigid robots.
8. **Energy Efficiency**:
   Low Power Consumption: Integrating soft materials with low-power electronic components is important for energy-efficient wearable and portable electronics.
9. **Customizable and Versatile Designs**:
   Adaptability: Soft materials can be tailored for specific applications, making them ideal for devices ranging from medical sensors to flexible displays.
10. **Printability and Manufacturing Ease**:
    Printed Electronics: Printing technologies, like inkjet and screen printing, make it possible to manufacture soft devices on flexible substrates quickly and cost-effectively.
11. **Impact Absorption**:
    Softness and Elasticity: Soft materials are often designed to absorb impact energy, making them suitable for applications where shock absorption is essential, such as in protective gear or wearable sensors.
12. **Innovative Form Factors**:
    Unique Designs: Soft materials allow for unconventional and innovative device designs, expanding the potential for functionality across industries.

Soft material–based flexible devices possess advantages that make them highly versatile and applicable in various fields like electronics, healthcare, robotics, and consumer products. Ongoing research and advancements in materials science are continuously improving the performance and capabilities of these soft devices. This progress opens up new opportunities for innovation.

5 Development of Soft Material–Based Flexible Devices

The production of flexible devices using soft materials involves various processes that aim to create devices that can bend and flex without breaking. Common techniques include soft lithography, inkjet printing, and laser ablation. Soft lithography uses a mold to create a pattern on a substrate, while inkjet printing involves depositing small droplets of material onto a surface. Laser ablation, however, uses a laser to remove material from a substrate to create a pattern. Ultimately, these techniques are used to create devices such as flexible sensors, displays, and wearable electronics that can adapt to the shape of the human body and provide a comfortable and seamless user experience.

5.1 Printing Technologies

Printing technologies play a crucial role in manufacturing soft material–based flexible devices. Inkjet printing, screen printing, and gravure printing are some of the most commonly used printing technologies enabling functional material deposition onto flexible substrates (Fig. 1).

Inkjet Printing

Inkjet printing is a versatile and precise printing technique widely used in manufacturing soft material–based flexible devices. It involves depositing functional materials, such as conductive inks, onto a flexible substrate using a printer that ejects tiny droplets of material through a nozzle onto the substrate. Inkjet printing offers several advantages over other printing technologies, including high resolution, low material waste, and the ability to print a wide range of materials. Moreover, inkjet printing can be easily integrated with other manufacturing processes, such as soft lithography and laser ablation, to create complex patterns and structures. This makes inkjet printing an ideal choice for the production of flexible sensors, displays, and wearable electronics that require high precision and customization.

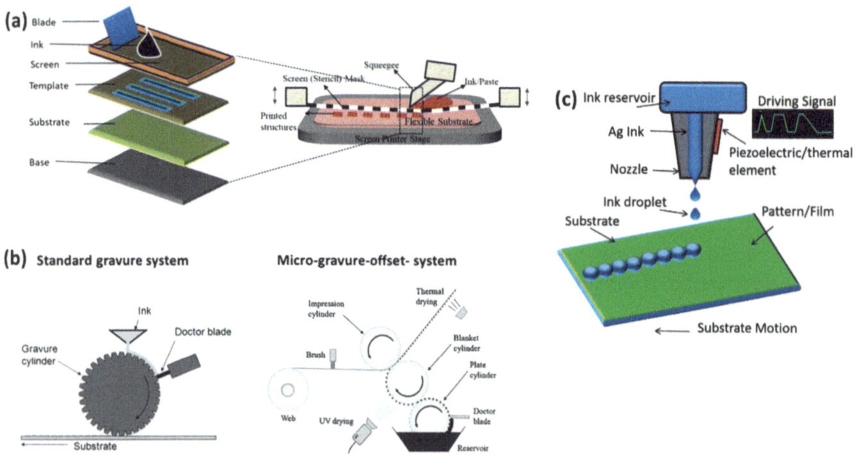

Fig. 1 Printing technologies to develop soft flexible devices. (**a**) Screen printing. (**b**) Gravure printing. (**c**) Inkjet printing [37]

Screen Printing

Screen printing is a widely used printing technology in manufacturing soft material–based flexible devices. It involves pushing ink or paste through a stencil, called a screen, onto a flexible substrate to create a pattern. Screen printing is a versatile technology that can deposit various functional materials, such as conductive inks, onto flexible substrates. It offers several benefits over other printing technologies, like low cost, high output, and the ability to print on various substrates. Screen printing can also be easily scaled up for mass production, making it an ideal choice for producing flexible sensors, displays, and wearable electronics. However, screen printing has limitations in terms of resolution and precision, which can affect the quality and performance of the printed devices.

Gravure Printing

Gravure printing is a high-speed printing technology commonly used to manufacture soft material–based flexible devices. It involves the transfer of ink or paste onto a flexible substrate using a rotary printing press with tiny cells engraved on its surface. The cells are filled with ink or paste, and the excess is scraped off, leaving ink or paste only in the cells. The ink or paste is transferred onto the substrate as the press rotates, creating a pattern. Gravure printing offers several advantages over other printing technologies, including high precision, high throughput, and the ability to print a wide range of materials. It can also achieve high-resolution images and patterns, making it suitable for producing flexible sensors, displays, and wearable electronics that require high precision and quality. However, gravure printing is generally more expensive than other printing technologies and requires a longer setup time, which can limit its use for small-scale production.

5.2 Soft Lithography

Soft lithography is a manufacturing technique widely used to produce soft material–based flexible devices. It involves using a mold made of an elastomeric material such as polydimethylsiloxane (PDMS) to create a pattern on a substrate. The mold is created by pouring liquid PDMS onto a master mold with the desired pattern and then curing it. Once the PDMS is cured, it is peeled off the master mold and used as a stamp to transfer the pattern onto the substrate. Soft lithography offers several advantages over other patterning techniques, including high resolution, high aspect ratio, and the ability to create complex patterns and structures. It can also be easily integrated with other manufacturing processes, such as inkjet printing and laser ablation, to create functional devices such as flexible sensors, displays, and wearable electronics. However, soft lithography has limitations in terms of scalability and reproducibility, which can affect its use for large-scale production. Soft material–

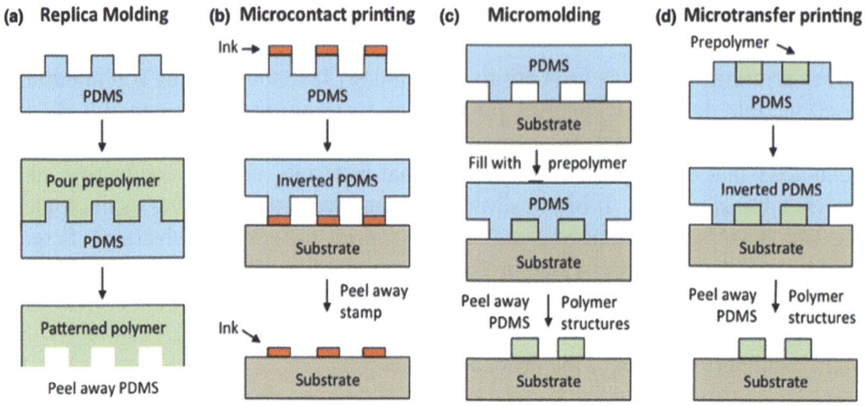

Fig. 2 Different soft lithography techniques [38]

based flexible devices are manufactured using various soft lithography techniques such as microcontact printing, replica molding, and nanoimprinting (Fig. 2).

Microcontact Printing

Microcontact printing involves the transfer of an ink or a functional material onto a substrate using a stamp made of an elastomeric material such as PDMS. The stamp is patterned with the desired pattern by a master mold, and the ink or functional material is transferred onto the substrate by bringing the stamp in contact with it.

Replica Molding

This technique involves creating a mold by pouring a liquid elastomer material onto a master mold with a desired pattern. After the elastomer is cured, it can be peeled off the master mold and used as a stamp to transfer the pattern onto the substrate.

Nanoimprinting

This process involves transferring a pattern onto a substrate using a mold made of a hard material like quartz. The mold is pressed onto the substrate with high pressure and temperature, which causes the substrate to flow into the mold cavities and replicate the pattern.

5.3 *Roll-to-Roll Manufacturing*

Roll-to-Roll (R2R) Processing

Roll-to-roll (R2R) manufacturing is a process used to produce flexible electronics, such as sensors, displays, and photovoltaics. It is a high-throughput manufacturing process, which means it can produce large quantities of items in less time. In this process, flexible substrates like plastic films, paper, and metal foils are continuously fed through processing steps, including printing, coating, and patterning. The result is the creation of large-area flexible electronic devices. R2R manufacturing has several benefits over other built-up processes, having high rates of manufacturing, low material waste, and the capacity to produce large-area devices at low cost.

The R2R (roll-to-roll) process is a manufacturing technique used to produce flexible electronics on a large scale. This process involves the use of printing technologies such as screen printing, inkjet printing, and gravure printing to deposit functional materials onto flexible substrates. Additionally, laser patterning and embossing can be used to create patterns and structures on the substrates. The R2R manufacturing technique has made flexible electronics more accessible and affordable, allowing for their use in a variety of uses comprising wearable devices, easy packing, and flexible displays (Fig. 3).

R2R manufacturing is a process that has some limitations, such as lower resolution compared to other manufacturing processes and the requirement for specialized equipment and processes. However, it still plays a crucial role in the flexible electronics industry, as it makes it possible to create flexible devices that can fit the contours of the human body and provide a comfortable and seamless user experience.

5.4 *Embossing and Molding*

1. **Hot Embossing**: Soft materials can be shaped into specific structures using hot embossing techniques. In this process, a patterned mold is pressed onto a soft

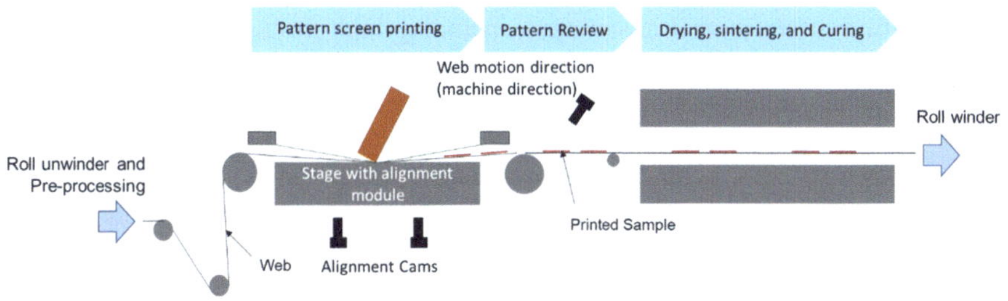

Fig. 3 Roll-to-roll manufacturing process [39]

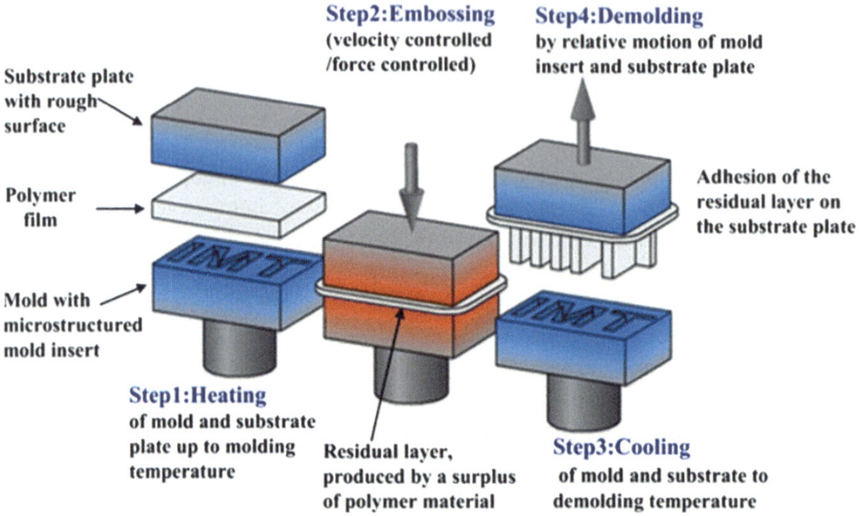

Fig. 4 Hot embossing process [40]

substrate. Another method is thermoforming where a soft material is heated until it is malleable and then molded into a specific shape. Both processes are used for shaping soft materials (Fig. 4).

2. **Thermoforming**: This comprises heating a soft material until it becomes malleable and then forming it into a specific shape using a mold.

5.5 *Spray Coating*

Spray coating is a highly effective and versatile technique used in the manufacturing of soft material–based flexible devices. It involves depositing functional materials such as conductive inks and adhesives onto a flexible substrate using a spray gun. The spray gun atomizes the material into tiny droplets that are propelled toward the substrate by a stream of air or gas. Compared to other coating technologies, spray coating offers several advantages, including high throughput, low cost, and the ability to coat large substrate areas. Spray coating is a flexible and versatile technique that can deposit a wide range of materials onto flexible substrates. These materials include ceramics, metals, and polymers. This technique is easily combinable with other manufacturing processes, such as inkjet printing and laser ablation, creating intricate patterns and structures. Spray coating is widely used in producing wearable electronics, flexible sensors, and displays, producing conformal and uniform coatings on substrates. However, it has some limitations, including lower resolution than other coating technologies and the need for precise control of spray parameters to achieve the desired coating properties [41, 42].

5.6 *Solution Processing*

Solution Casting

Soft materials are dissolved in a solvent to form a solution. The solution is then cast onto a substrate, and after the solvent evaporates, the soft material remains and forms the desired structure [43].

5.7 *Spin Coating*

This technique involves depositing a liquid precursor onto a rotating substrate, which spreads the material uniformly. After curing or drying, the desired thin film is obtained [44].

5.8 *3D Printing*

Additive Manufacturing

To create complex and customized structures, soft and flexible materials can be 3D printed using fused deposition modeling (FDM) or stereolithography (SLA) (Fig. 5).

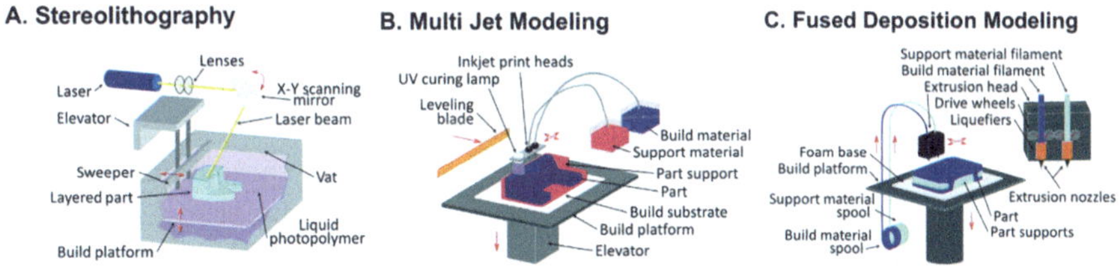

Fig. 5 3D printing technologies [45]

5.9 *Hybrid Integration*

Hybrid Integration with Rigid Components

In certain cases, both soft and flexible components are combined with rigid ones to reap the benefits of both. Techniques like hybrid printing or assembly are used for this purpose.

5.10 *Encapsulation Techniques*

Encapsulation with Soft Materials

Soft devices may require encapsulation to protect them from environmental factors. Encapsulation can be done using soft materials such as flexible polymers, which can be laminated or coated.

6 Characterization of Soft Flexible Materials

Testing soft flexible materials involves a combination of mechanical, thermal, chemical, and other analyses to understand their properties and behavior. Here are some common testing methods for soft flexible materials [46–51].

1. **Mechanical Testing**:
   (a) **Tensile Testing**: Measures the response of the material to stretching forces. It provides information regarding elongation at break, tensile strength, and Young's modulus.
   (b) **Compression Testing**: Evaluates the material's response to compressive forces, providing data on compressive strength and stiffness.
   (c) **Flexural Testing**: Assesses the material's bending properties, relevant for applications undergoing bending or flexing.
   (d) **Shear Testing**: Evaluate the material's response to shear forces. It measures the maximum stress the material can withstand in shear.
   (e) **Peel Testing**: Determines the force essential to peel two bonded materials apart. It measures the force per unit width necessary to peel the materials apart.
   (f) **Puncture Testing**: Assesses the material's resistance to puncture or penetrate. It determines the force required to puncture the material.
   (g) **Creep Testing**: Evaluate the material's deformation under a constant load over time. It measures the gradual deformation that occurs under a constant load.

2. **Rheological Testing**:
   (a) **Dynamic Mechanical Analysis (DMA)**: Determines the mechanical properties of materials as a function of time, temperature, or frequency. This is crucial for understanding viscoelastic behavior.
   (b) **Rotational Rheometry**: Evaluates soft materials' flow and deformation properties under controlled stress or strain. It measures the shear viscosity, the resistance to flow under shear stress, and shear rate dependence which indicates whether the material exhibits shear-thinning or shear-thickening behavior, yielding stress at which the material starts to flow.
   (c) **Shear Rheometry**: Evaluates the material's response to shear forces, providing insights into its flow and deformation characteristics.
   (d) **Capillary Rheometry**: Evaluates the flow behavior of soft materials through a capillary tube under different conditions. It measures the melt flow rate, i.e., the flowability of molten polymers.
   (e) **Extensional Rheometry**: Assesses the material's behavior under extensional forces, which is relevant for processes like extrusion. Measures the resistance to stretching or extension.
3. **Thixotropy Testing**: Measures the time-dependent recovery of a material's viscosity after shear thinning. Quantifies the extent of thixotropic behavior.
4. **Frequency Sweep Tests**: Evaluate how the material's rheological properties change with varying deformation frequencies. Storage and loss moduli are measured as a function of frequency.
5. **Temperature Sweep Tests**: Studies the effect of temperature on the material's rheological behavior. Changes in storage and loss moduli with temperature are measured.
6. **Thermal Analysis**:
   (a) **Thermogravimetric Analysis (TGA)**: Determines the material's weight loss as a function of temperature, helping to assess thermal stability and decomposition.
   (b) **Differential Scanning Calorimetry (DSC)**: Measures the flow of heat related with phase transitions, providing information about melting points, glass transition temperatures, and thermal stability.
7. **Chemical Analysis**:
   (a) **Nuclear Magnetic Resonance (NMR) Spectroscopy**: Analyzes the nuclear magnetic properties of atoms, providing detailed information about the molecular structure. Elucidation of molecular structures, identification of components, and quantification of different chemical species can be obtained.
   (b) **Fourier Transform Infrared Spectroscopy (FTIR)**: Identifies functional groups present in the material, aiding in characterizing its chemical composition.

(c) **Gas Chromatography–Mass Spectrometry (GC-MS)**: Separates and analyzes volatile compounds based on their mass and chemical structure. Detection and quantification of organic compounds and identification of additives or contaminants are also carried out.
(d) **X-Ray Fluorescence (XRF)**: Analyzes the elemental composition by measuring the fluorescent X-rays emitted when a sample is exposed to X-rays. The determination of elemental composition in various materials is also very easy.
(e) **Inductively Coupled Plasma Mass Spectrometry (ICP-MS)**: Determines elemental composition by ionizing the sample in an inductively coupled plasma source and measuring the resulting ions. Quantification of trace elements, heavy metals, and contaminants is possible.
(f) **Raman Spectroscopy**: Determines the scattering of monochromatic light via molecules, giving details about molecular vibrations. Identification of molecular structures and characterization of polymers and organic materials are carried out.
(g) **Energy-Dispersive X-Ray Spectroscopy (EDS)**: Analyzes the elemental composition of a material by identifying X-rays emitted when a sample is bombarded with electrons. Elemental analysis of materials and identification of impurities are performed.
(h) **Chemical Resistance Tests**: Evaluate how the material responds to exposure to various chemicals or environmental conditions.

8. **Durability and Fatigue Testing**: Cyclic Loading Tests: Assess the material's performance under repeated loading, helping to understand its fatigue behavior.
9. **Abrasion Resistance Testing**: Determines the material's resistance to wear and abrasion, which is important for applications where friction is a factor.
10. **Dielectric Strength Testing**: Measures the material's ability to withstand electric stress without breaking down.
11. **Electrical Conductivity Testing**: Determines the material's electrical conductivity or resistance.
12. **Biocompatibility Testing**:

    (a) **Cytotoxicity Tests**: Assess the material's impact on cell viability.
    (b) **Biological Reactivity Tests**: Evaluate the material's interaction with living tissues.
    (c) **Surface Roughness Measurement**: Assesses the roughness of the material's surface, which can be important in applications such as sealants or gaskets.

13. **Adhesion Testing**: Measures the material's adhesion to other surfaces relevant for bonding or coating applications.
14. **Processability Tests**: Evaluate how well the material can be processed using various manufacturing techniques like extrusion, injection molding, or 3D printing.

These tests are often performed in combination to provide a comprehensive understanding of the soft flexible material's properties and suitability for specific applications. It's important to tailor the testing approach based on the material's intended use and the critical factors influencing its performance.

7 Materials for Soft Flexible Devices

Soft and flexible devices require materials that can withstand frequent bending and stretching without breaking or losing their properties. Some of the commonly used materials for making soft and flexible devices are silicone, polyurethane, and different types of polymers. These materials are selected for their flexibility, durability, and biocompatibility, which make them ideal for medical implants and wearable electronic devices.

7.1 Organic Polymers

Organic polymers are commonly used in producing soft and flexible devices due to their excellent mechanical flexibility, biocompatibility, and optical transparency. One of the most popular polymers is polyimide, which is widely used in manufacturing flexible circuits and displays due to its high-temperature stability and excellent mechanical strength. Another frequently used polymer material is polyethylene terephthalate (PET), known for its excellent thermal stability, low dielectric constant, and high tensile strength. Polydimethylsiloxane (PDMS), on the other hand, is a highly elastic and transparent polymer often used to produce microfluidic devices and biosensors. Overall, organic polymers are an important class of materials that enable the development of soft and flexible devices with a wide range of applications [52–62] (Table 1).

7.2 Elastomers

Elastomers are a popular material choice for flexible devices due to their excellent stretchability and durability. They can be effortlessly molded into many forms and sizes, making them appropriate for a widespread applications such as wearable electronics, soft robotics, and sensors. Moreover, elastomers are biocompatible, which makes them ideal for medical devices and implants. In general, elastomers offer exceptional mechanical properties for soft devices, allowing them to withstand deformation and bending without compromising their functionality. Some useful elastomers are as follows:

Table 1 Organic polymers for soft flexible devices

Materials	Methods of fabrication	Substrates	Bending radius (mm)	Bending times	Application	Reference
DPP2T-TT	Bar coating	PEN	8	1000	Gas sensor	[52]
P3HT	Evaporation	PEN	8	100	Biosensor	[53]
P-29-DPP-SVS	Spin coating	PEN	8.3	1200	Gas sensor	[54]
DPP-DTT	Printing	6FDA-DABC	5	800	Flexible circuit	[55]
PDFDT	Bar coating	PEN	6	100	Gas sensor	[56]
PIDT-BT:TCNQ	Spin coating	PET	3	2000	Pressure sensor	[57]
PDPP3T/N2200	Printing	PET	6.2	800	Flexible circuit	[58]
PQT-12	Electro-spinning	PEN	6	100	Gas sensor	[59]
PTAA	Spin coating	Starch papers	0.79	1000	Flexible circuit	[60]
PQT-12/C [8]	Printing	PET	0.5	1000	Chemical sensor	[61]
PQDPP-2FT	Printing	PET	8	200	Logical circuit	[62]

1. Polydimethylsiloxane [63, 64]
2. Poly(ethylene glycol) methyl ether methacrylate/2-ureido-4[1*H*] pyrimidinone [65]
3. Polydimethylsiloxane/polycaprolactone [66]
4. Polyborosiloxane [67]
5. *N*-Isopropylacrylamide/2-methoxyethyl acrylate [68]
6. Carboxyl cellulose nanocrystals/chitosan [69]
7. Poly(ethylene glycol sebacate) [70]
8. Fluorocarbon elastomer [71]
9. Adamantanes/β-cyclodextrin nanogels [72]

7.3 Conductive Materials

Conductive materials play a crucial role in soft flexible devices by facilitating the transmission of electrical signals and power. Various types of conductive materials can be used in soft devices, including metals, carbon-based materials, and conductive polymers. Metals like gold, silver, and copper are excellent conductors and are commonly used in electrodes and interconnects. Carbon-based materials such as carbon nanotubes and graphene are also prevalent conductive materials owing to their unique mechanical and electrical properties. Conductive polymers such as poly(3,4-ethylenedioxythiophene) (PEDOT) and polyaniline are lightweight and flexible, making them ideal for applications in wearable electronics and soft robotics. Table 2 summarizes some conductive materials for soft flexible electronics. Ultimately, the choice of conductive material depends on the specific requirements of the soft flexible device, such as electrical conductivity, flexibility, and biocompatibility [73–102].

Table 2 Conductive materials for soft flexible electronics

Conductive materials	Applications	References
PEDOT:PSS	Sensor, energy store, electrode	[73–76]
Polyaniline	Biosensor, tissue engineering, drug delivery, bioelectrode, supercapacitor, strain sensor	[77–80]
Polypyrrole	Biosensor, tissue engineering, drug delivery, bioelectrode, supercapacitor, strain sensor	[81–84]
Graphene	Sensor, display, energy store and conversion, electrode, robot, lighting	[85–89]
Carbon nanotube	Tissue engineering, strain sensor, supercapacitor, biosensor, drug delivery	[90–93]
Metallic nanowires/nanoparticles	Tissue engineering, drug delivery, biosensor	[94–98]
MXenes	Energy store and conversion, sensor display, electrode, lighting, robot	[99–102]

Table 3 Natural materials for soft flexible devices

Natural materials (species)	Functional applications	References
Silk (fauna)	Optical functions, biodegradability, water solubility, lightweight, high adhesion, and elasticity	[103, 104]
Wood (flora)	Mechanically roust, lightweight, and optical	[105]
Pollen (flora)	Mechanical property, superhydrophobicity, lightweight, drag reduction, adhesion property, antimicrobial activity, and anti-radiation	[106–108]
Leaf (flora)	Low adhesion, superhydrophobicity, and self-cleaning	[109–111]
Chitin	High adhesion, biodegradability, lightweight, mechanical flexibility, and optical property	[112–114]
Butterfly (fauna)	Superhydrophobicity, structural color, directional adhesion, antireflection, chemical sensing capacity, optical purposes, lightweight, self-cleaning, and fluorescence emission functions	[115–117]
Skin (fauna)	Extensibility, self-healing, drag reduction superhydrophobicity, and mechanical flexibility	[118, 119]
Peptide	Biodegradation and antimicrobial activity	[120–122]
Cellulose	Lightweight, high linkage, mechanical property, low thermal expansion, and optical property	[123–125]
Fish	Superoleophilicity in air and water, structural color, optical functions, and drag reduction	[126–128]

7.4 *Natural Materials*

Natural resources and derivatives have also gained attention because of their widespread availability, nontoxicity, and biocompatibility. Some commonly used natural materials include biomass, wood, paper, silk, fibroin, etc. The various kinds of functional groups and their naturally occurring ordered structures meet the requirements for particular applications. Table 3 summarizes some natural materials that have been useful for soft flexible device applications [103–128].

8 Applications of Soft Material–Based Flexible Devices

Soft material–based flexible devices have a broad range of applications in various fields due to their high flexibility and stretchability. These devices are suitable for use in wearable electronics, biomedical devices, soft robotics, and many other areas. For example, flexible sensors made from soft materials can continuously monitor the vital signs and body movements of patients in real time. Soft robotic devices can mimic the movements of living organisms and be used in various applications, including surgical procedures, rehabilitation, and industrial automa-

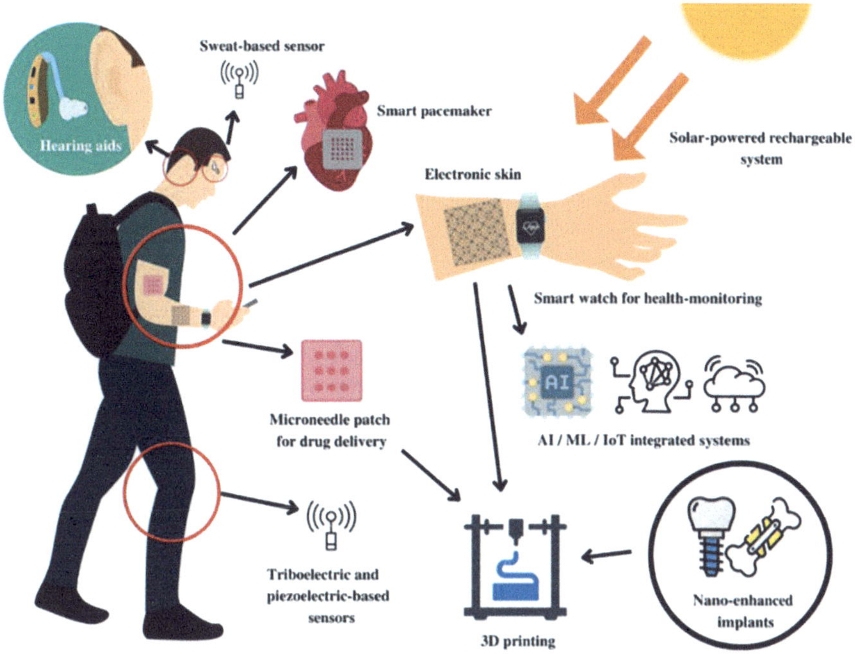

Fig. 6 Application of soft flexible materials integrating AI-ML/IOT systems [129]

tion. Soft material–based flexible devices also have great potential applications in the development of next-generation displays, such as flexible and foldable displays [129] (Fig. 6).

8.1 *Soft Materials for Wearable Electronics*

Soft and flexible materials are highly valued for wearable electronic devices because they can be integrated into clothing and worn comfortably for extended periods of time. Some commonly used soft materials for wearable electronics include silicone, polyurethane, and thermoplastic elastomers. These materials have unique properties, such as stretchability, durability, and biocompatibility, making them ideal for wearable electronics use. For example, soft materials can be used to create sensors that can detect various parameters, such as temperature, humidity, and pressure. They can also be used to create flexible batteries and energy-harvesting devices that can be integrated into clothing. In addition, soft materials can be used to create stretchable and conformable displays that can be worn on the skin. Overall, the use of soft materials in wearable electronics has opened up new possibilities for developing wearable devices that are comfortable, durable, and functional [130–135] (Table 4).

Table 4 Soft material–based wearable and flexible electronics

Device	Nanomaterials	Application	Reference
Moldable conducting nanocomposite	Networks of Ag nanowires have stacked polymeric structures	Integrating with wearable skin electronics that can improve breathability, flexibility, and electromechanical steadiness	[130]
Bioadhesive interface	Nanocomposites of graphene	Quick incorporation with damp tissues, unhindered bidirectional message at the interface of the device–tissue	[131]
Gas-penetrable skin electronics	Holey graphene and silicone-based elastomeric sponges	Quickens sweat evaporation, reduces the danger of swelling, and adds to the comfort of the user	[132]
Self-powered metamaterial backbone fusion cage implants	Triboelectric auxetic microstructures	Regulated observing for progress in bone healing	[133]
Bio-resorbable electronic stent	Au nanorod and nanoparticles of silica integrated with drug-encapsulated nanoparticles	Real flow and temperature watching	[134]
Lab-on-a-patch noninvasive perspiration biosensor	A microfluidics-assembled 3D Au nanostructure electrode	Immuno-detection of cortisol existing in perspiration	[135]

8.2 Soft Materials for Health Monitoring Devices

Soft material–based health-checking devices have gained noteworthy attention in recent years due to their flexibility, stretchability, and biocompatibility. These devices can be integrated into clothing or worn directly on the skin to monitor numerous health factors such as blood pressure, heart rate, and temperature. Some commonly used soft materials for health monitoring devices include silicone, hydrogels, and elastomers. These materials can be used to create soft sensors that can follow the body's shape and deliver accurate readings. Additionally, soft materials can create wearable adhesive patches that can monitor the body's vital signs in real time. These patches are easy to use and can transmit data wirelessly to

Table 5 Soft flexible devices for health monitoring

Source of the sample	Physiological parameter	Sensing devices	References
Monitoring sleep	Body temperature, heart rate, and breathing	Smart rings, smartwatch, and a sleep mask	[136–140]
Activities during daytime	Blood pressure, heart rate, abnormalities, oxygen level in blood, and rate of respiration	Wearable vests, wristbands, and smartwatch	[141, 142]
Monitoring blood	Different biomarkers and thoracic impedance	Implants in arms	[143]
All kinds of activities	Heart rate, heart rate, blood oxygen level, body temperature, ECG, EEG, and respiration rate	Earphones and textile-based devices	[144, 145]

a smartphone or other electronic device for further analysis. Overall, the applications of soft materials in health monitoring devices have the potential to revolutionize the way we monitor our health, making it more comfortable, noninvasive, and accessible [136–145] (Table 5).

8.3 Soft Materials for Flexible Displays

Soft and flexible materials are excellent for creating flexible displays, which have become increasingly popular for research due to their unique properties, such as bendability, rollability, and foldability. Commonly used soft materials for flexible displays include organic polymers like polyethylene naphthalate (PEN), polyethylene terephthalate (PET), and polyimide (PI), as well as inorganic constituents such as metal foils and oxide thin films. These materials can be used to create displays that are flexible enough to be integrated into clothing or worn on the skin or even wrapped around objects. Additionally, soft materials can create transparent and stretchable displays that conform to different shapes and sizes. These displays have numerous applications, such as in developing flexible smartphones, smartwatches, and electronic paper. Overall, the use of soft materials in flexible displays has opened up new possibilities for developing lightweight, durable, and highly functional displays [146] (Fig. 7).

8.4 Soft Robotics

Soft robotics is a developing meadow that includes the usage of soft and flexible materials to create robotic devices that can interact with the environment more

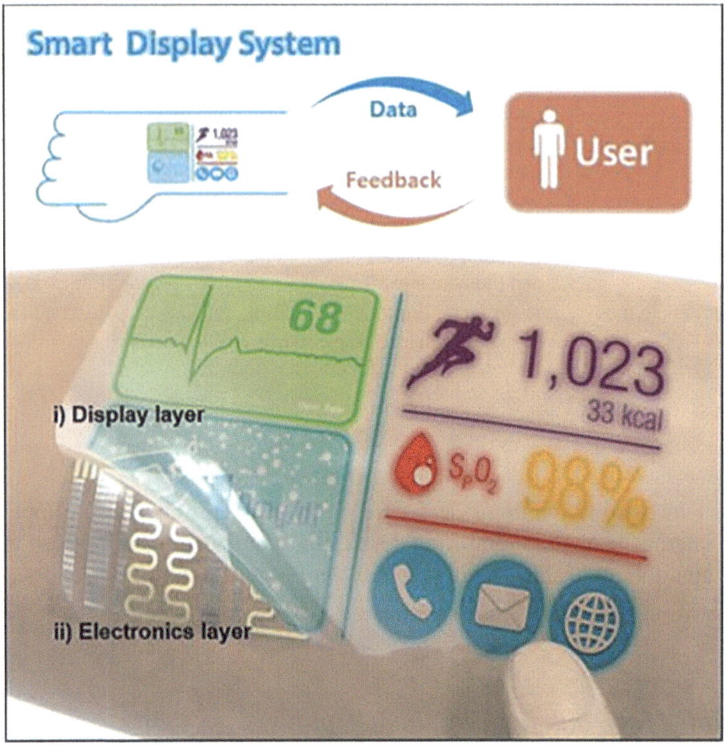

Fig. 7 Soft flexible displays [146]

naturally and safely than traditional rigid robots. These robots can be designed to mimic the movements of living organisms, such as octopuses and worms. They can be used in various applications, including surgical procedures, rehabilitation, and industrial automation. There are several types of soft robotics, each with their own unique characteristics and applications. Here are a few examples [147].

1. **Soft Grippers**: Soft grippers are robotic devices that use soft materials to grasp and manipulate objects. They can be designed to imitate the form of the body being gripped, which allows for a more secure grip.
2. **Soft Exosuits**: Soft exosuits are wearable devices that use soft materials to assist with movement. They can be used to augment the strength and endurance of the user, making it easier to perform physical tasks.
3. **Soft Sensors**: Soft sensors are devices that use soft and flexible materials to detect environmental changes. They can be used to monitor temperature, pressure, and other physical parameters.
4. **Soft Actuators**: Soft actuators use soft materials to move. They can be designed to bend, stretch, or twist in response to an external stimulus.

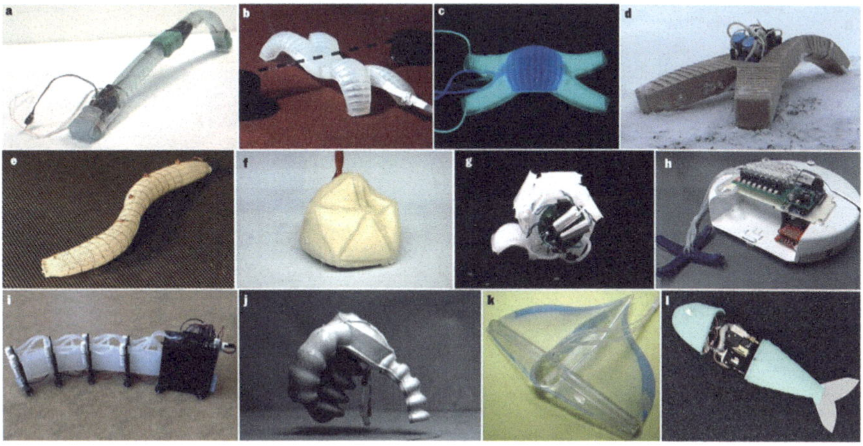

Fig. 8 Soft flexible robotics [147]

5. **Soft Robots for Medical Applications**: Soft robots can be used in medical applications, such as surgical procedures and rehabilitation. For example, soft robots can be used to assist with minimally invasive surgery or to help patients regain mobility after an injury (Fig. 8).

8.5 *Flexible and Stretchable Energy Harvesters*

Soft materials can be used to create energy harvesters that can be integrated into clothing or wearable devices, capturing energy from body movements or ambient sources.

1. **Flexible Solar Cells**: Advances in soft material technology may lead to the development of flexible and lightweight solar cells that can be integrated into various surfaces, expanding the possibilities for solar energy generation (Fig. 9).

8.6 *Human–Machine Interfaces*

1. **Soft Haptic Interfaces**: Soft materials can enhance the development of more natural and comfortable human–machine interfaces, such as soft touchscreens, haptic feedback devices, and interactive surfaces.
2. **Brain–Machine Interfaces**: Soft and flexible materials may create biocompatible interfaces for seamless integration with the human body, facilitating communication between the brain and external devices (Fig. 10).

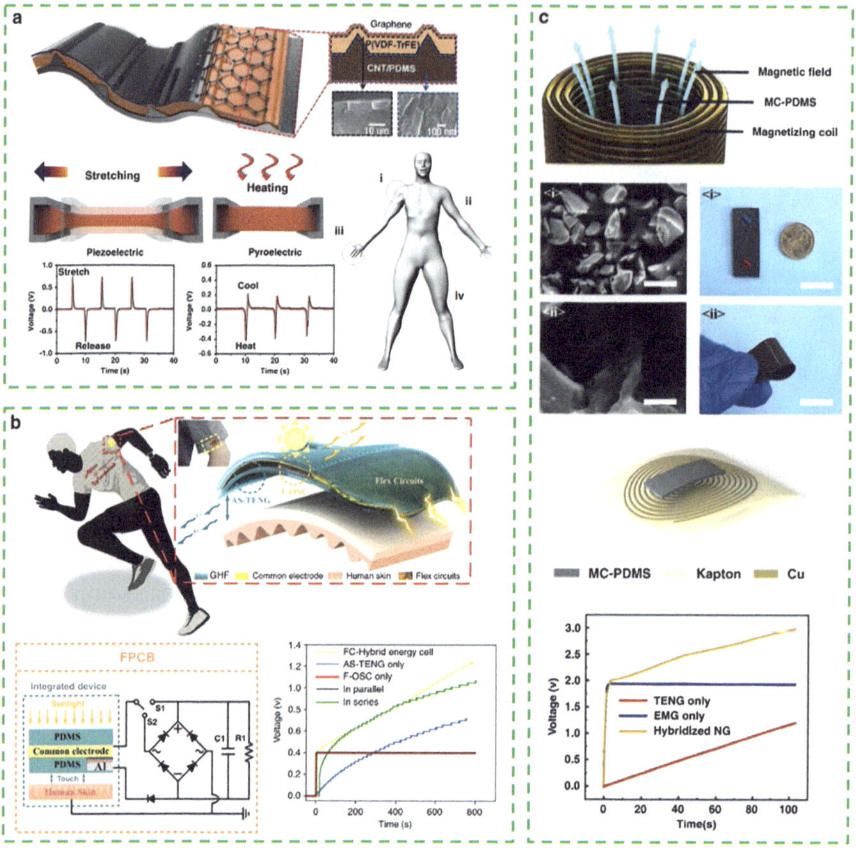

Fig. 9 Soft, stretchable, and flexible energy harvesters [148]

8.7 *Environmental Sensing*

Soft Sensors for Environmental Monitoring

Soft, flexible sensors can be employed for environmental monitoring applications, such as pollution detection, water quality assessment, and climate monitoring.

Therefore, soft flexible materials can be assigned as smart materials. There is no unanimity in stating "smart materials." For instance, NASA narrates them as "materials that remember different forms and can reconcile with particular stimuli" [129]. It can be termed as highly engineered constituents that react elegantly to their surroundings. Smart materials are placed in the class of improved materials that can sense and react/reply accordingly. Some exclusive potentials of smart materials are as follows [129]:

1. Immediacy: Quickly responding to environmentally friendly effects.

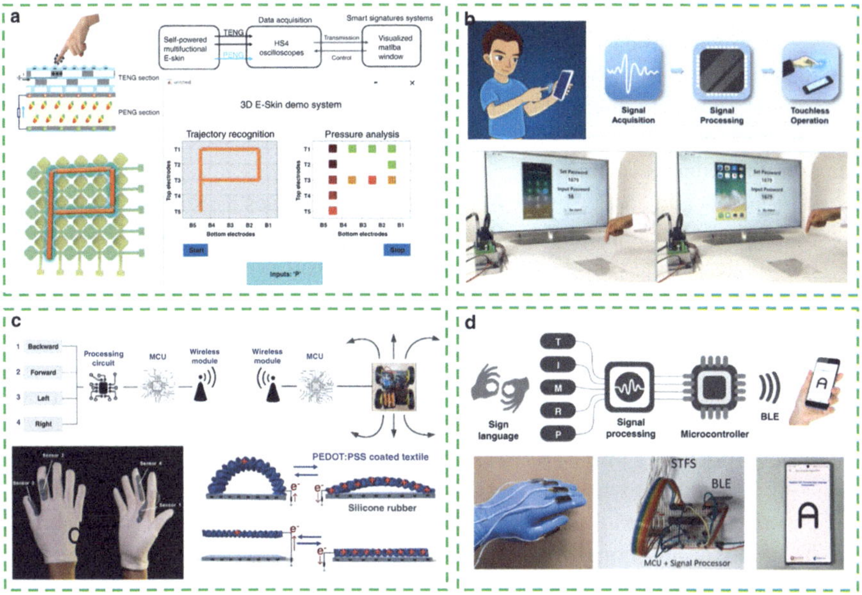

Fig. 10 Soft flexible devices as human–machine interfaces [148]

2. Transiency: Sensitive to the surroundings and able to exist in several states.
3. Selectivity: The reaction is very specific and anticipated.
4. Self-actuation: This aptitude is confined inside the material.
5. Directness: Action and reaction co-occur.

9 Challenges and Future of Soft Material–Based Flexible Devices

Soft material–based flexible devices have limitations when it comes to their durability and reliability. They may wear out over time due to repeated bending and stretching, leading to cracks or breaks in the material. In addition, the materials used in these devices may not be able to withstand harsh environmental conditions, such as extreme temperatures or exposure to chemicals. This can limit their effectiveness in certain applications. Furthermore, soft material–based flexible devices may not be as powerful or efficient as their rigid counterparts, which can limit their overall performance. While soft material–based flexible devices hold great promise for various applications, there are several challenges that researchers and engineers are actively addressing [149]. Understanding and overcoming these challenges will play a crucial role in shaping the future of soft materials in flexible devices. Here are some key challenges and considerations [150–154].

1. **Material Durability and Stability**:
   Soft materials can be more prone to wear and degradation compared to rigid materials. Ensuring the long-term durability and stability of soft devices is a critical challenge, particularly in applications where the devices are subjected to repeated bending, stretching, or other mechanical stresses.
2. **Manufacturing Techniques**:
   Scalable and cost-effective manufacturing processes for soft material–based devices are essential for their widespread adoption. Developing reliable methods for large-scale production, such as roll-to-roll printing and other flexible fabrication techniques, is an ongoing challenge.
3. **Integration with Traditional Electronics**:
   Integrating soft materials with traditional rigid electronics can be challenging. Ensuring seamless communication and compatibility between soft and rigid components is crucial for creating functional and reliable devices.
4. **Biocompatibility and Safety**:
   In applications where soft materials come into contact with the human body, ensuring biocompatibility and safety is critical. The materials used should be nontoxic and nonallergenic and meet regulatory standards for medical or wearable devices.
5. **Sensor Performance and Accuracy**:
   Soft sensors, while flexible and conformable, may face challenges in achieving the same level of performance and accuracy as traditional rigid sensors. Enhancing the sensitivity, stability, and reliability of soft sensors is an ongoing research focus.
6. **Energy Storage and Harvesting**:
   Soft materials for energy storage (e.g., soft batteries or supercapacitors) and energy harvesting (e.g., flexible solar cells) need to be developed with a focus on efficiency, energy density, and the ability to withstand repeated deformations.
7. **Environmental Impact**:
   The environmental impact of manufacturing soft materials, including the use of certain polymers and chemicals, is a concern. Researchers are exploring sustainable and eco-friendly alternatives to mitigate the environmental footprint of soft material–based devices.
8. **Standardization and Regulation**:
   Establishing standards and regulations for soft material–based devices is essential to ensure product quality, safety, and interoperability. Developing standardized testing methods for flexible devices will facilitate industry-wide adoption.
9. **User Experience and Design Challenges**:
   Designing user-friendly and aesthetically pleasing soft devices poses challenges, particularly in applications such as wearable technology. Balancing functionality, comfort, and style is crucial for user acceptance and adoption.

10. **Cost Considerations**:
    Soft materials can sometimes be more expensive than traditional rigid materials. Researchers and manufacturers must explore cost-effective alternatives and optimize processes to make soft material–based devices economically viable for mass production.
    Despite these challenges, ongoing research and innovation are actively addressing these issues, and the future of soft material–based flexible devices looks promising. As technology advances, it is likely that these challenges will be overcome, opening up new possibilities for applications in electronics, healthcare, robotics, and beyond.

10 Conclusion

The potential impact of soft material–based flexible devices is profound and spans various industries, offering transformative possibilities in electronics, healthcare, robotics, and beyond. Here are some final thoughts on the potential impact of these innovative devices.

Revolutionizing Electronics: Soft materials enable the creation of flexible and foldable electronics, paving the way for devices with novel form factors. This can revolutionize the design and functionality of smartphones, tablets, and other electronic gadgets, offering enhanced user experiences.

Wearable Technology Advancements: Soft materials are key to the development of comfortable and conformable wearable devices. This includes smart textiles, health monitors, and fitness trackers that seamlessly integrate into our daily lives, providing real-time data and insights for personal well-being.

Healthcare and Medical Applications: Soft materials are crucial in developing soft robotics for minimally invasive surgeries and medical procedures. Flexible and conformable sensors can be integrated into wearable health devices, improving patient monitoring and diagnostics.

Human–Machine Interfaces (HMIs): Soft materials contribute to the evolution of natural and intuitive human–machine interfaces. Soft touchscreens, haptic feedback devices, and interactive surfaces enhance the interaction between humans and technology, making interfaces more responsive and user-friendly.

Environmental Monitoring and Sensing: Soft sensors can be deployed for environmental monitoring, enabling data collection on pollution, water quality, and climate conditions. This has implications for sustainability and smart city initiatives.

Energy Harvesting and Storage: Soft materials facilitate the development of flexible energy-harvesting devices that can capture energy from body movements or ambient sources. This can contribute to the advancement of self-powered wearable devices and increase the efficiency of renewable energy technologies.

Innovation in Robotics: Soft robotics, made possible by compliant and flexible materials, opens up new possibilities in robotics. Soft robots can navigate complex environments, interact safely with humans, and find applications in healthcare, search and rescue, and exploration.

Accessibility and Inclusivity: The flexibility and adaptability of soft materials make it easier to design devices that are more accessible and inclusive. For example, devices that conform to different body shapes can benefit a broader range of users.

Sustainable Design Practices: As the field progresses, there's potential for developing environmentally friendly soft materials and manufacturing processes, contributing to more sustainable design practices in the electronics and materials industries.

Cross-disciplinary Collaborations: Developing soft material–based flexible devices requires collaboration across disciplines, including materials science, engineering, electronics, and healthcare. This interdisciplinary approach fosters innovation and accelerates the pace of technological advancement.

In conclusion, soft material–based flexible devices have the potential to reshape the way we interact with technology, monitor our health, and address societal challenges. While there are challenges to overcome, ongoing research and development are driving progress in this exciting field, and the future holds the promise of increasingly versatile, adaptable, and user-friendly devices.

References

1. Chen, X., Rogers, J.A., Lacour, S.P., Hu, W., Kim, D.H.: Materials chemistry in flexible electronics. Chem. Soc. Rev. **48**(6), 1431–1433 (2019)
2. Rogers, J.A., Chen, X., Feng, X.: Flexible hybrid electronics. Adv. Mater. (Deerfield Beach, Fla.). **32**(15), e1905590 (2020)
3. Kang, J., Tok, J.B.H., Bao, Z.: Self-healing soft electronics. Nat. Electron. **2**(4), 144–150 (2019)
4. Tan, J.M.R., Farraj, Y., Kamyshny, A., Magdassi, S.: Fabrication approaches of soft electronics. ACS Appl. Electron. Mater. **5**(3), 1376–1393 (2023)
5. Potu, A., Jayalakshmi, R., Umpathy, K.: Smart paper technology a review based on concepts of E-paper technology. IOSR J. Electron. Commun. Eng. **11**, 42–46 (2016)
6. Garnier, F., Horowitz, G., Peng, X., Fichou, D.: An all-organic "soft" thin film transistor with very high carrier mobility. Adv. Mater. **2**(12), 592–594 (1990)
7. Garnier, F., Hajlaoui, R., Yassar, A., Srivastava, P.: All-polymer field-effect transistor realized by printing techniques. Science. **265**(5179), 1684–1686 (1994)
8. Gustafsson, G., Cao, Y., Treacy, G.M., Klavetter, F., Colaneri, N., Heeger, A.J.: Flexible light-emitting diodes made from soluble conducting polymers. Nature. **357**(6378), 477–479 (1992)
9. Konagai, M., Sugimoto, M., Takahashi, K.: High efficiency GaAs thin film solar cells by peeled film technology. J. Cryst. Growth. **45**, 277–280 (1978)
10. Yablonovitch, E., Gmitter, T., Harbison, J.P., Bhat, R.: Extreme selectivity in the lift-off of epitaxial GaAs films. Appl. Phys. Lett. **51**(26), 2222–2224 (1987)

11. Menard, E., Lee, K.J., Khang, D.Y., Nuzzo, R.G., Rogers, J.A.: A printable form of silicon for high performance thin film transistors on plastic substrates. Appl. Phys. Lett. **84**(26), 5398–5400 (2004)
12. Hur, S.H., Khang, D.Y., Kocabas, C., Rogers, J.A.: Nanotransfer printing by use of noncovalent surface forces: applications to thin-film transistors that use single-walled carbon nanotube networks and semiconducting polymers. Appl. Phys. Lett. **85**(23), 5730–5732 (2004)
13. Ahn, J.H., Kim, H.S., Lee, K.J., Jeon, S., Kang, S.J., Sun, Y., Nuzzo, R.G., Rogers, J.A.: Heterogeneous three-dimensional electronics by use of printed semiconductor nanomaterials. Science. **314**(5806), 1754–1757 (2006)
14. Ko, H.C., Stoykovich, M.P., Song, J., Malyarchuk, V., Choi, W.M., Yu, C.J., Geddes III, J.B., Xiao, J., Wang, S., Huang, Y., Rogers, J.A.: A hemispherical electronic eye camera based on compressible silicon optoelectronics. Nature. **454**(7205), 748–753 (2008)
15. Sun, Y., Rogers, J.A.: Fabricating semiconductor nano/microwires and transfer printing ordered arrays of them onto plastic substrates. Nano Lett. **4**(10), 1953–1959 (2004)
16. Yoon, J., Baca, A.J., Park, S.I., Elvikis, P., Geddes III, J.B., Li, L., Kim, R.H., Xiao, J., Wang, S., Kim, T.H., Motala, M.J.: Ultrathin silicon solar microcells for semitransparent, mechanically flexible and microconcentrator module designs. Nat. Mater. **7**(11), 907–915 (2008)
17. Rogers, J.A., Bao, Z., Baldwin, K., Dodabalapur, A., Crone, B., Raju, V.R., Kuck, V., Katz, H., Amundson, K., Ewing, J., Drzaic, P.: Like electronic displays: large-area rubber-stamped plastic sheets of electronics and microencapsulated electrophoretic inks. Proc. Natl. Acad. Sci. **98**(9), 4835–4840 (2001)
18. Someya, T., Sekitani, T., Iba, S., Kato, Y., Kawaguchi, H., Sakurai, T.: A large-area, flexible pressure sensor matrix with organic field-effect transistors for artificial skin applications. Proc. Natl. Acad. Sci. **101**(27), 9966–9970 (2004)
19. Someya, T., Kato, Y., Sekitani, T., Iba, S., Noguchi, Y., Murase, Y., Kawaguchi, H., Sakurai, T.: Conformable, flexible, large-area networks of pressure and thermal sensors with organic transistor active matrixes. Proc. Natl. Acad. Sci. **102**(35), 12321–12325 (2005)
20. Cao, Q., Kim, H.S., Pimparkar, N., Kulkarni, J.P., Wang, C., Shim, M., Roy, K., Alam, M.A., Rogers, J.A.: Medium-scale carbon nanotube thin-film integrated circuits on flexible plastic substrates. Nature. **454**(7203), 495–500 (2008)
21. Sekitani, T., Nakajima, H., Maeda, H., Fukushima, T., Aida, T., Hata, K., Someya, T.: Stretchable active-matrix organic light-emitting diode display using printable elastic conductors. Nat. Mater. **8**(6), 494–499 (2009)
22. Takei, K., Takahashi, T., Ho, J.C., Ko, H., Gillies, A.G., Leu, P.W., Fearing, R.S., Javey, A.: Nanowire active-matrix circuitry for low-voltage macroscale artificial skin. Nat. Mater. **9**(10), 821–826 (2010)
23. Mannsfeld, S.C., Tee, B.C., Stoltenberg, R.M., Chen, C.V.H., Barman, S., Muir, B.V., Sokolov, A.N., Reese, C., Bao, Z.: Highly sensitive flexible pressure sensors with microstructured rubber dielectric layers. Nat. Mater. **9**(10), 859–864 (2010)
24. Kim, D.H., Lu, N., Ghaffari, R., Kim, Y.S., Lee, S.P., Xu, L., Wu, J., Kim, R.H., Song, J., Liu, Z., Viventi, J.: Materials for multifunctional balloon catheters with capabilities in cardiac electrophysiological mapping and ablation therapy. Nat. Mater. **10**(4), 316–323 (2011)
25. Tao, H., Hwang, S.W., Marelli, B., An, B., Moreau, J.E., Yang, M., Brenckle, M.A., Kim, S., Kaplan, D.L., Rogers, J.A., Omenetto, F.G.: Silk-based resorbable electronic devices for remotely controlled therapy and in vivo infection abatement. Proc. Natl. Acad. Sci. **111**(49), 17385–17389 (2014)
26. Hwang, S.W., Tao, H., Kim, D.H., Cheng, H., Song, J.K., Rill, E., Brenckle, M.A., Panilaitis, B., Won, S.M., Kim, Y.S., Song, Y.M.: A physically transient form of silicon electronics. Science. **337**(6102), 1640–1644 (2012)
27. Gao, W., Emaminejad, S., Nyein, H.Y.Y., Challa, S., Chen, K., Peck, A., Fahad, H.M., Ota, H., Shiraki, H., Kiriya, D., Lien, D.H.: Fully integrated wearable sensor arrays for multiplexed in situ perspiration analysis. Nature. **529**(7587), 509–514 (2016)

28. Kim, J., Salvatore, G.A., Araki, H., Chiarelli, A.M., Xie, Z., Banks, A., Sheng, X., Liu, Y., Lee, J.W., Jang, K.I., Heo, S.Y.: Battery-free, stretchable optoelectronic systems for wireless optical characterization of the skin. Sci. Adv. **2**(8), e1600418 (2016)
29. Takei, K.: History of flexible and stretchable devices. In: Flexible and Stretchable Medical Devices, pp. 1–6 (2018)
30. Chae, S.H., Lee, Y.H.: Carbon nanotubes and graphene towards soft electronics. Nano Converg. **1**, 1–26 (2014)
31. Byun, S.H., Sim, J.Y., Agno, K.C., Jeong, J.W.: Materials and manufacturing strategies for mechanically transformative electronics. Mater. Today Adv. **7**, 100089 (2020)
32. Rogers, J.A., Someya, T., Huang, Y.: Materials and mechanics for stretchable electronics. Science. **327**(5973), 1603–1607 (2010)
33. Lim, H.R., Kim, H.S., Qazi, R., Kwon, Y.T., Jeong, J.W., Yeo, W.H.: Advanced soft materials, sensor integrations, and applications of wearable flexible hybrid electronics in healthcare, energy, and environment. Adv. Mater. **32**(15), 1901924 (2020)
34. Wen, N., Zhang, L., Jiang, D., Wu, Z., Li, B., Sun, C., Guo, Z.: Emerging flexible sensors based on nanomaterials: recent status and applications. J. Mater. Chem. A. **8**(48), 25499–25527 (2020)
35. Sang, M., Kim, K., Shin, J., Yu, K.J.: Ultra-thin flexible encapsulating materials for soft bio-integrated electronics. Adv. Sci. **9**(30), 2202980 (2022)
36. Su, M., Song, Y.: Printable smart materials and devices: strategies and applications. Chem. Rev. **122**(5), 5144–5164 (2021)
37. Khan, S., Ali, S., Bermak, A.: Smart manufacturing technologies for printed electronics. In: Hybrid Nanomaterials-Flexible Electronics Materials, p. 18 (2019)
38. Lindquist, N.C., Nagpal, P., McPeak, K.M., Norris, D.J., Oh, S.H.: Engineering metallic nanostructures for plasmonics and nanophotonics. Rep. Prog. Phys. **75**(3), 036501 (2012)
39. Phung, T.H., Gafurov, A.N., Kim, I., Kim, S.Y., Kim, K.M., Lee, T.M.: IoT device fabrication using roll-to-roll printing process. Sci. Rep. **11**(1), 19982 (2021)
40. Worgull, M., Hétu, J.F., Kabanemi, K.K., Heckele, M.: Hot embossing of microstructures: characterization of friction during demolding. Microsyst. Technol. **14**, 767–773 (2008)
41. Govind, R.K., Mondal, I., Baishya, K., Ganesha, M.K., Walia, S., Singh, A.K., Kulkarni, G.U.: Large-area fabrication of high performing, flexible, transparent conducting electrodes using screen printing and spray coating techniques. Adv. Mater. Technol. **7**(6), 2101120 (2022)
42. Mariello, M., Qualtieri, A., Mele, G., De Vittorio, M.: Metal-free multilayer hybrid PENG based on soft electrospun/-sprayed membranes with cardanol additive for harvesting energy from surgical face masks. ACS Appl. Mater. Interfaces. **13**(17), 20606–20621 (2021)
43. Tan, P., Wang, H., Xiao, F., Lu, X., Shang, W., Deng, X., Song, H., Xu, Z., Cao, J., Gan, T., Wang, B.: Solution-processable, soft, self-adhesive, and conductive polymer composites for soft electronics. Nat. Commun. **13**(1), 358 (2022)
44. Lachaux, J., Salmon, H., Loisel, F., Arouche, N., Ochoa, I., Fernandez, L.L., Uzan, G., Mercier, O., Veres, T., Roy, E.: Soft thermoplastic elastomer for easy and rapid spin-coating fabrication of microfluidic devices with high hydrophilization and bonding performances. Adv. Mater. Technol. **4**(2), 1800308 (2019)
45. Bhattacharjee, N., Urrios, A., Kang, S., Folch, A.: The upcoming 3D-printing revolution in microfluidics. Lab Chip. **16**(10), 1720–1742 (2016)
46. De Mulatier, S., Ramuz, M., Coulon, D., Blayac, S., Delattre, R.: Mechanical characterization of soft substrates for wearable and washable electronic systems. APL Mater. **7**(3), 5063671 (2019)
47. Li, L., Xie, F., Wang, T., Wang, G., Tian, Y., Jin, T., Zhang, Q.: Stiffness-tunable soft gripper with soft-rigid hybrid actuation for versatile manipulations. Soft Robot. **9**(6), 1108–1119 (2022)
48. Sala, R., Regondi, S., Graziosi, S., Pugliese, R.: Insights into the printing parameters and characterization of thermoplastic polyurethane soft triply periodic minimal surface and honeycomb lattices for broadening material extrusion applicability. Addit. Manuf. **58**, 102976 (2022)

49. Xavier, M.S., Tawk, C.D., Yong, Y.K., Fleming, A.J.: 3D-printed omnidirectional soft pneumatic actuators: design, modeling and characterization. Sensors Actuators A Phys. **332**, 113199 (2021)
50. Gonzalez, D., Garcia, J., Voyles, R.M., Nawrocki, R.A., Newell, B.: Characterization of 3D printed pneumatic soft actuator. Sensors Actuators A Phys. **334**, 113337 (2022)
51. Wyss, H.M.: Rheology of soft materials. In: Fluids, Colloids and Soft Materials: An Introduction to Soft Matter Physics, pp. 149–163 (2016)
52. Zhang, S., Zhao, Y., Du, X., Chu, Y., Zhang, S., Huang, J.: Gas sensors based on nano/microstructured organic field-effect transistors. Small. **15**(12), 1805196 (2019)
53. Jeong, J., Essafi, M., Lee, C., Haoues, M., Diouani, M.F., Kim, H., Kim, Y.: Ultrasensitive detection of hazardous reactive oxygen species using flexible organic transistors with polyphenol-embedded conjugated polymer sensing layers. J. Hazard. Mater. **355**, 17–24 (2018)
54. Ryu, G.S., Park, K.H., Park, W.T., Kim, Y.H., Noh, Y.Y.: High-performance diketopyrrolopyrrole-based organic field-effect transistors for flexible gas sensors. Org. Electron. **23**, 76–81 (2015)
55. Park, H., Yoo, S., Won, J.C., Kim, Y.H.: Room-temperature, printed, low-voltage, flexible organic field-effect transistors using soluble polyimide gate dielectrics. APL Mater. **8**(1), 011112 (2020)
56. Nketia-Yawson, B., Jung, A.R., Noh, Y., Ryu, G.S., Tabi, G.D., Lee, K.K., Kim, B., Noh, Y.Y.: Highly sensitive flexible NH_3 sensors based on printed organic transistors with fluorinated conjugated polymers. ACS Appl. Mater. Interfaces. **9**(8), 7322–7330 (2017)
57. Yin, Z., Yin, M.J., Liu, Z., Zhang, Y., Zhang, A.P., Zheng, Q.: Solution-processed bilayer dielectrics for flexible low-voltage organic field-effect transistors in pressure-sensing applications. Adv. Sci. **5**(9), 1701041 (2018)
58. Angunawela, I., Nahid, M.M., Ghasemi, M., Amassian, A., Ade, H., Gadisa, A.: The critical role of materials' interaction in realizing organic field-effect transistors via high-dilution blending with insulating polymers. ACS Appl. Mater. Interfaces. **12**(23), 26239–26249 (2020)
59. Fan, H., Li, H., Han, J., McKeever, N., Yu, J., Katz, H.E.: A humid-air-operable, NO_2-responsive polymer transistor series circuit with improved signal-to-drift ratio based on polymer semiconductor oxidation. ACS Sens. **4**(12), 3240–3247 (2019)
60. Jeong, H., Baek, S., Han, S., Jang, H., Kim, S.H., Lee, H.S.: Novel eco-friendly starch paper for use in flexible, transparent, and disposable organic electronics. Adv. Funct. Mater. **28**(3), 1704433 (2018)
61. Kweon, O.Y., Lee, M.Y., Park, T., Jang, H., Jeong, A., Um, M.K., Oh, J.H.: Highly flexible chemical sensors based on polymer nanofiber field-effect transistors. J. Mater. Chem. C. **7**(6), 1525–1531 (2019)
62. Ni, Z., Dong, H., Wang, H., Ding, S., Zou, Y., Zhao, Q., Zhen, Y., Liu, F., Jiang, L., Hu, W.: Quinoline-flanked diketopyrrolopyrrole copolymers breaking through electron mobility over 6 $cm^2\ V^{-1}\ s^{-1}$ in flexible thin film devices. Adv. Mater. **30**(10), 1704843 (2018)
63. Yang, Z., Li, H., Zhang, L., Lai, X., Zeng, X.: Highly stretchable, transparent and room-temperature self-healable polydimethylsiloxane elastomer for bending sensor. J. Colloid Interface Sci. **570**, 1–10 (2020)
64. Yan, Q., Zhao, L., Cheng, Q., Zhang, T., Jiang, B., Song, Y., Huang, Y.: Self-healing polysiloxane elastomer based on integration of covalent and reversible networks. Ind. Eng. Chem. Res. **58**(47), 21504–21512 (2019)
65. Chen, J., Liu, J., Thundat, T., Zeng, H.: Polypyrrole-doped conductive supramolecular elastomer with stretchability, rapid self-healing, and adhesive property for flexible electronic sensors. ACS Appl. Mater. Interfaces. **11**(20), 18720–18729 (2019)
66. Liu, S., Lin, Y., Wei, Y., Chen, S., Zhu, J., Liu, L.: A high performance self-healing strain sensor with synergetic networks of poly (ε-caprolactone) microspheres, graphene and silver nanowires. Compos. Sci. Technol. **146**, 110–118 (2017)

67. Wu, T., Chen, B.: Synthesis of multiwalled carbon nanotube-reinforced polyborosiloxane nanocomposites with mechanically adaptive and self-healing capabilities for flexible conductors. ACS Appl. Mater. Interfaces. **8**(36), 24071–24078 (2016)
68. Yan, S., Zhang, G., Jiang, H., Li, F., Zhang, L., Xia, Y., Wang, Z., Wu, Y., Li, H.: Highly stretchable room-temperature self-healing conductors based on wrinkled graphene films for flexible electronics. ACS Appl. Mater. Interfaces. **11**(11), 10736–10744 (2019)
69. Cao, J., Lu, C., Zhuang, J., Liu, M., Zhang, X., Yu, Y., Tao, Q.: Multiple hydrogen bonding enables the self-healing of sensors for human–machine interactions. Angew. Chem. **129**(30), 8921–8926 (2017)
70. Günal, G., Okan, M., Gokcen, D., Caykara, T., Aydin, H.M.: Microwave-assisted synthesis of stretchable and transparent poly(ethyleneglycol-sebacate) elastomers with autonomous self-healing and capacitive properties. Soft Robot. **8**(3), 262–272 (2021)
71. Cao, Y., Tan, Y.J., Li, S., Lee, W.W., Guo, H., Cai, Y., Wang, C., Tee, B.C.K.: Self-healing electronic skins for aquatic environments. Nat. Electron. **2**(2), 75–82 (2019)
72. Li, X., Zhang, H., Zhang, P., Yu, Y.: A sunlight-degradable autonomous self-healing supramolecular elastomer for flexible electronic devices. Chem. Mater. **30**(11), 3752–3758 (2018)
73. Rong, Q., Lei, W., Chen, L., Yin, Y., Zhou, J., Liu, M.: Anti-freezing, conductive self-healing organohydrogels with stable strain-sensitivity at subzero temperatures. Angew. Chem. Int. Ed. **56**(45), 14159–14163 (2017)
74. Mano, N., Yoo, J.E., Tarver, J., Loo, Y.L., Heller, A.: An electron-conducting cross-linked polyaniline-based redox hydrogel, formed in one step at pH 7.2, wires glucose oxidase. J. Am. Chem. Soc. **129**(22), 7006–7007 (2007)
75. Dong, R., Zhao, X., Guo, B., Ma, P.X.: Self-healing conductive injectable hydrogels with antibacterial activity as cell delivery carrier for cardiac cell therapy. ACS Appl. Mater. Interfaces. **8**(27), 17138–17150 (2016)
76. Qu, J., Zhao, X., Ma, P.X., Guo, B.: Injectable antibacterial conductive hydrogels with dual response to an electric field and pH for localized "smart" drug release. Acta Biomater. **72**, 55–69 (2018)
77. Li, W., Gao, F., Wang, X., Zhang, N., Ma, M.: Strong and robust polyaniline-based supramolecular hydrogels for flexible supercapacitors. Angew. Chem. **128**(32), 9342–9347 (2016)
78. Chakraborty, P., Guterman, T., Adadi, N., Yadid, M., Brosh, T., Adler-Abramovich, L., Dvir, T., Gazit, E.: A self-healing, all-organic, conducting, composite peptide hydrogel as pressure sensor and electrogenic cell soft substrate. ACS Nano. **13**(1), 163–175 (2018)
79. Wang, Z., Zhou, H., Lai, J., Yan, B., Liu, H., Jin, X., Ma, A., Zhang, G., Zhao, W., Chen, W.: Extremely stretchable and electrically conductive hydrogels with dually synergistic networks for wearable strain sensors. J. Mater. Chem. C. **6**(34), 9200–9207 (2018)
80. Chen, J., Peng, Q., Thundat, T., Zeng, H.: Stretchable, injectable, and self-healing conductive hydrogel enabled by multiple hydrogen bonding toward wearable electronics. Chem. Mater. **31**(12), 4553–4563 (2019)
81. Hur, J., Im, K., Kim, S.W., Kim, J., Chung, D.Y., Kim, T.H., Jo, K.H., Hahn, J.H., Bao, Z., Hwang, S., Park, N.: Polypyrrole/agarose-based electronically conductive and reversibly restorable hydrogel. ACS Nano. **8**(10), 10066–10076 (2014)
82. Gan, D., Han, L., Wang, M., Xing, W., Xu, T., Zhang, H., Wang, K., Fang, L., Lu, X.: Conductive and tough hydrogels based on biopolymer molecular templates for controlling in situ formation of polypyrrole nanorods. ACS Appl. Mater. Interfaces. **10**(42), 36218–36228 (2018)
83. Shi, Y., Pan, L., Liu, B., Wang, Y., Cui, Y., Bao, Z., Yu, G.: Nanostructured conductive polypyrrole hydrogels as high-performance, flexible supercapacitor electrodes. J. Mater. Chem. A. **2**(17), 6086–6091 (2014)
84. Han, L., Yan, L., Wang, M., Wang, K., Fang, L., Zhou, J., Fang, J., Ren, F., Lu, X.: Transparent, adhesive, and conductive hydrogel for soft bioelectronics based on light-transmitting polydopamine-doped polypyrrole nanofibrils. Chem. Mater. **30**(16), 5561–5572 (2018)

85. Liu, Q., Zhang, M., Huang, L., Li, Y., Chen, J., Li, C., Shi, G.: High-quality graphene ribbons prepared from graphene oxide hydrogels and their application for strain sensors. ACS Nano. **9**(12), 12320–12326 (2015)
86. Xu, Y., Lin, Z., Huang, X., Wang, Y., Huang, Y., Duan, X.: Functionalized graphene hydrogel-based high-performance supercapacitors. Adv. Mater. **25**(40), 5779–5784 (2013)
87. Han, L., Lu, X., Wang, M., Gan, D., Deng, W., Wang, K., Fang, L., Liu, K., Chan, C.W., Tang, Y., Weng, L.T.: A mussel-inspired conductive, self-adhesive, and self-healable tough hydrogel as cell stimulators and implantable bioelectronics. Small. **13**(2), 1601916 (2017)
88. Jing, X., Mi, H.Y., Napiwocki, B.N., Peng, X.F., Turng, L.S.: Mussel-inspired electroactive chitosan/graphene oxide composite hydrogel with rapid self-healing and recovery behavior for tissue engineering. Carbon. **125**, 557–570 (2017)
89. Wang, Y., Huang, F., Chen, X., Wang, X.W., Zhang, W.B., Peng, J., Li, J., Zhai, M.: Stretchable, conductive, and self-healing hydrogel with super metal adhesion. Chem. Mater. **30**(13), 4289–4297 (2018)
90. Liang, Y., Zhao, X., Hu, T., Han, Y., Guo, B.: Mussel-inspired, antibacterial, conductive, antioxidant, injectable composite hydrogel wound dressing to promote the regeneration of infected skin. J. Colloid Interface Sci. **556**, 514–528 (2019)
91. Mottet, L., Le Cornec, D., Noël, J.M., Kanoufi, F., Delord, B., Poulin, P., Bibette, J., Bremond, N.: A conductive hydrogel based on alginate and carbon nanotubes for probing microbial electroactivity. Soft Matter. **14**(8), 1434–1441 (2018)
92. Han, L., Liu, K., Wang, M., Wang, K., Fang, L., Chen, H., Zhou, J., Lu, X.: Mussel-inspired adhesive and conductive hydrogel with long-lasting moisture and extreme temperature tolerance. Adv. Funct. Mater. **28**(3), 1704195 (2018)
93. Deng, Z., Hu, T., Lei, Q., He, J., Ma, P.X., Guo, B.: Stimuli-responsive conductive nanocomposite hydrogels with high stretchability, self-healing, adhesiveness, and 3D printability for human motion sensing. ACS Appl. Mater. Interfaces. **11**(7), 6796–6808 (2019)
94. Fantino, E., Chiappone, A., Roppolo, I., Manfredi, D., Bongiovanni, R., Pirri, C.F., Calignano, F.: 3D printing of conductive complex structures with in situ generation of silver nanoparticles. Adv. Mater. **28**(19), 3712–3717 (2016)
95. Navaei, A., Saini, H., Christenson, W., Sullivan, R.T., Ros, R., Nikkhah, M.: Gold nanorod-incorporated gelatin-based conductive hydrogels for engineering cardiac tissue constructs. Acta Biomater. **41**, 133–146 (2016)
96. Wei, Q.B., Luo, Y.L., Zhang, C.H., Fan, L.H., Chen, Y.S.: Assembly of Cu nanoparticles in a polyacrylamide grafted poly(vinyl alcohol) copolymer matrix and vapor-induced response. Sensors Actuators B Chem. **134**(1), 49–56 (2008)
97. Li, L., Wang, Y., Pan, L., Shi, Y., Cheng, W., Shi, Y., Yu, G.: A nanostructured conductive hydrogels-based biosensor platform for human metabolite detection. Nano Lett. **15**(2), 1146–1151 (2015)
98. Rong, Q., Han, H., Feng, F., Ma, Z.: Network nanostructured polypyrrole hydrogel/Au composites as enhanced electrochemical biosensing platform. Sci. Rep. **5**(1), 11440 (2015)
99. Cui, H., Mi, H., Ji, C., Guo, F., Chen, Y., Wu, D., Qiu, J., Xie, H.: A durable MXene-based zinc ion hybrid supercapacitor with sulfated polysaccharide reinforced hydrogel/electrolyte. J. Mater. Chem. A. **9**(42), 23941–23954 (2021)
100. Pandey, M., Deshmukh, K., Raman, A., Asok, A., Appukuttan, S., Suman, G.R.: Prospects of MXene and graphene for energy storage and conversion. Renew. Sust. Energ. Rev. **189**, 114030 (2024)
101. Li, Y., Huang, S., Peng, S., Jia, H., Pang, J., Ibarlucea, B., Hou, C., Cao, Y., Zhou, W., Liu, H., Cuniberti, G.: Toward smart sensing by MXene. Small. **19**(14), 2206126 (2023)
102. Iravani, S.: Role of MXenes in advancing soft robotics. Soft Matter. **19**(33), 6196–6212 (2023)
103. Zhu, B., Wang, H., Leow, W.R., Cai, Y., Loh, X.J., Han, M.Y., Chen, X.: Silk fibroin for flexible electronic devices. Adv. Mater. **28**(22), 4250–4265 (2016)

104. Kim, D.H., Viventi, J., Amsden, J.J., Xiao, J., Vigeland, L., Kim, Y.S., Blanco, J.A., Panilaitis, B., Frechette, E.S., Contreras, D., Kaplan, D.L.: Dissolvable films of silk fibroin for ultrathin conformal bio-integrated electronics. Nat. Mater. **9**(6), 511–517 (2010)
105. Zhu, H., Luo, W., Ciesielski, P.N., Fang, Z., Zhu, J.Y., Henriksson, G., Himmel, M.E., Hu, L.: Wood-derived materials for green electronics, biological devices, and energy applications. Chem. Rev. **116**(16), 9305–9374 (2016)
106. Brandon Goodwin, W., Gomez, I.J., Fang, Y., Meredith, J.C., Sandhage, K.H.: Conversion of pollen particles into three-dimensional ceramic replicas tailored for multimodal adhesion. Chem. Mater. **25**(22), 4529–4536 (2013)
107. Tzvetkov, G., Kaneva, N., Spassov, T.: Room-temperature fabrication of core-shell nano-ZnO/pollen grain biocomposite for adsorptive removal of organic dye from water. Appl. Surf. Sci. **400**, 481–491 (2017)
108. Mundargi, R.C., Potroz, M.G., Park, S., Shirahama, H., Lee, J.H., Seo, J., Cho, N.J.: Natural sunflower pollen as a drug delivery vehicle. Small. **12**(9), 1167–1173 (2016)
109. Bhushan, B., Jung, Y.C.: Micro- and nanoscale characterization of hydrophobic and hydrophilic leaf surfaces. Nanotechnology. **17**(11), 2758 (2006)
110. Ensikat, H.J., Ditsche-Kuru, P., Neinhuis, C., Barthlott, W.: Superhydrophobicity in perfection: the outstanding properties of the lotus leaf. Beilstein J. Nanotechnol. **2**(1), 152–161 (2011)
111. Latthe, S.S., Terashima, C., Nakata, K., Fujishima, A.: Superhydrophobic surfaces developed by mimicking hierarchical surface morphology of lotus leaf. Molecules. **19**(4), 4256–4283 (2014)
112. Jin, J., Lee, D., Im, H.G., Han, Y.C., Jeong, E.G., Rolandi, M., Choi, K.C., Bae, B.S.: Chitin nanofiber transparent paper for flexible green electronics. Adv. Mater. **28**(26), 5169–5175 (2016)
113. Jin, J., Reese, V., Coler, R., Carter, D., Rolandi, M.: Chitin microneedles for an easy-to-use tuberculosis skin test. Adv. Healthc. Mater. **3**(3), 349–353 (2014)
114. Wegst, U.G., Bai, H., Saiz, E., Tomsia, A.P., Ritchie, R.O.: Bioinspired structural materials. Nat. Mater. **14**(1), 23–36 (2015)
115. Perez Goodwyn, P., Maezono, Y., Hosoda, N., Fujisaki, K.: Waterproof and translucent wings at the same time: problems and solutions in butterflies. Naturwissenschaften. **96**, 781–787 (2009)
116. Vukusic, P., Sambles, J.R., Lawrence, C.R., Wootton, R.J.: Quantified interference and diffraction in single Morpho butterfly scales. Proc. R. Soc. Lond. B Biol. Sci. **266**(1427), 1403–1411 (1999)
117. Vukusic, P., Sambles, R., Lawrence, C., Wakely, G.: Sculpted-multilayer optical effects in two species of Papilio butterfly. Appl. Opt. **40**(7), 1116–1125 (2001)
118. Yuan, Y., Verma, R.: Measuring microelastic properties of stratum corneum. Colloids Surf. B Biointerfaces. **48**(1), 6–12 (2006)
119. Teyssier, J., Saenko, S.V., Van Der Marel, D., Milinkovitch, M.C.: Photonic crystals cause active colour change in chameleons. Nat. Commun. **6**(1), 6368 (2015)
120. Dai, B., Li, D., Xi, W., Luo, F., Zhang, X., Zou, M., Cao, M., Hu, J., Wang, W., Wei, G., Zhang, Y.: Tunable assembly of amyloid-forming peptides into nanosheets as a retrovirus carrier. Proc. Natl. Acad. Sci. **112**(10), 2996–3001 (2015)
121. Abbas, M., Zou, Q., Li, S., Yan, X.: Self-assembled peptide-and protein-based nanomaterials for antitumor photodynamic and photothermal therapy. Adv. Mater. **29**(12), 1605021 (2017)
122. Sopher, N.B., Abrams, Z.R., Reches, M., Gazit, E., Hanein, Y.: Integrating peptide nanotubes in micro-fabrication processes. J. Micromech. Microeng. **17**(11), 2360 (2007)
123. Malho, J.M., Arola, S., Laaksonen, P., Szilvay, G.R., Ikkala, O., Linder, M.B.: Modular architecture of protein binding units for designing properties of cellulose nanomaterials. Angew. Chem. **127**(41), 12193–12196 (2015)
124. Moon, R.J., Martini, A., Nairn, J., Simonsen, J., Youngblood, J.: Cellulose nanomaterials review: structure, properties and nanocomposites. Chem. Soc. Rev. **40**(7), 3941–3994 (2011)

125. Fratzl, P.: Cellulose and collagen: from fibres to tissues. Curr. Opin. Colloid Interface Sci. **8**(1), 32–39 (2003)
126. Fudouzi, H.: Tunable structural color in organisms and photonic materials for design of bioinspired materials. Sci. Technol. Adv. Mater. **12**, 064704 (2011)
127. Vernerey, F.J., Musiket, K., Barthelat, F.: Mechanics of fish skin: a computational approach for bio-inspired flexible composites. Int. J. Solids Struct. **51**(1), 274–283 (2014)
128. Vernerey, F.J., Barthelat, F.: Skin and scales of teleost fish: simple structure but high performance and multiple functions. J. Mech. Phys. Solids. **68**, 66–76 (2014)
129. Ghosh, A., Nag, S., Gomes, A., Gosavi, A., Ghule, G., Kundu, A., Purohit, B., Srivastava, R.: Applications of smart material sensors and soft electronics in healthcare wearables for better user compliance. Micromachines. **14**(1), 121 (2022)
130. Namkoong, M., Guo, H., Rahman, M.S., Wang, D., Pfeil, C.J., Hager, S., Tian, L.: Moldable and transferrable conductive nanocomposites for epidermal electronics. npj Flexible Electron. **6**(1), 41 (2022)
131. Rodriguez, N.R.M., Das, S., Kaufman, Y., Wei, W., Israelachvili, J.N., Waite, J.H.: Mussel adhesive protein provides cohesive matrix for collagen type-1α. Biomaterials. **51**, 51–57 (2015)
132. Li, Q., Chen, G., Cui, Y., Ji, S., Liu, Z., Wan, C., Liu, Y., Lu, Y., Wang, C., Zhang, N., Cheng, Y.: Highly thermal-wet comfortable and conformal silk-based electrodes for on-skin sensors with sweat tolerance. ACS Nano. **15**(6), 9955–9966 (2021)
133. Kabiri Ameri, S., Ho, R., Jang, H., Tao, L., Wang, Y., Wang, L., Schnyer, D.M., Akinwande, D., Lu, N.: Graphene electronic tattoo sensors. ACS Nano. **11**(8), 7634–7641 (2017)
134. Kim, J.H., Kim, S.R., Kil, H.J., Kim, Y.C., Park, J.W.: Highly conformable, transparent electrodes for epidermal electronics. Nano Lett. **18**(7), 4531–4540 (2018)
135. Tang, W., Zhou, Y., Chen, S., Yu, S., Yang, Y., Lin, J., Yin, S., Ma, Y., Hu, B.: Delamination-resistant imperceptible bioelectrode for robust electrophysiological signals monitoring. ACS Mater. Lett. **3**(9), 1385–1393 (2021)
136. Kamyshny, A., Magdassi, S.: Conductive nanomaterials for printed electronics. Small. **10**(17), 3515–3535 (2014)
137. Canovas, R., Padrell Sánchez, S., Parrilla, M., Cuartero, M., Crespo, G.A.: Cytotoxicity study of ionophore-based membranes: toward on-body and in vivo ion sensing. ACS Sens. **4**(9), 2524–2535 (2019)
138. Bellu, E., Medici, S., Coradduzza, D., Cruciani, S., Amler, E., Maioli, M.: Nanomaterials in skin regeneration and rejuvenation. Int. J. Mol. Sci. **22**(13), 7095 (2021)
139. Beard, R.B., Hung, B.N., Schmukler, R.: Biocompatibility considerations at stimulating electrode interfaces. Ann. Biomed. Eng. **20**, 395–410 (1992)
140. Drees, C., Makic, M.B., Case, K., Mancuso, M.P., Hill, A., Walczak, P., Limon, S., Biesecker, K., Frey, L.: Skin irritation during video-EEG monitoring. Neurodiagn. J. **56**(3), 139–150 (2016)
141. Hasan, M.M., Hossain, M.M.: Nanomaterials-patterned flexible electrodes for wearable health monitoring: a review. J. Mater. Sci. **56**(27), 14900–14942 (2021)
142. Bosetti, M., Massè, A., Tobin, E., Cannas, M.: Silver coated materials for external fixation devices: in vitro biocompatibility and genotoxicity. Biomaterials. **23**(3), 887–892 (2002)
143. Huigen, E., Peper, A., Grimbergen, C.A.: Investigation into the origin of the noise of surface electrodes. Med. Biol. Eng. Comput. **40**, 332–338 (2002)
144. Paladini, F., Sannino, A., Pollini, M.: In vivo testing of silver treated fibers for the evaluation of skin irritation effect and hypoallergenicity. J. Biomed. Mater. Res. B Appl. Biomater. **102**(5), 1031–1037 (2014)
145. Basova, T.V., Vikulova, E.S., Dorovskikh, S.I., Hassan, A., Morozova, N.B.: The use of noble metal coatings and nanoparticles for the modification of medical implant materials. Mater. Des. **204**, 109672 (2021)
146. Koo, J.H., Kim, D.C., Shim, H.J., Kim, T.H., Kim, D.H.: Flexible and stretchable smart display: materials, fabrication, device design, and system integration. Adv. Funct. Mater. **28**(35), 1801834 (2018)

147. Rus, D., Tolley, M.T.: Design, fabrication and control of soft robots. Nature. **521**(7553), 467–475 (2015)
148. Xu, C., Song, Y., Han, M., Zhang, H.: Portable and wearable self-powered systems based on emerging energy harvesting technology. Microsyst. Nanoeng. **7**(1), 25 (2021)
149. Gariya, N., Kumar, P.: A review on soft materials utilized for the manufacturing of soft robots. Materi. Today Proc. **46**, 11177–11181 (2021)
150. Yang, M., Yuan, Z., Liu, J., Fang, Z., Fang, L., Yu, D., Li, Q.: Photoresponsive actuators built from carbon-based soft materials. Adv. Opt. Mater. **7**(16), 1900069 (2019)
151. Appiah, C., Arndt, C., Siemsen, K., Heitmann, A., Staubitz, A., Selhuber-Unkel, C.: Living materials herald a new era in soft robotics. Adv. Mater. **31**(36), 1807747 (2019)
152. Tomé, L.C., Mecerreyes, D.: Emerging ionic soft materials based on deep eutectic solvents. J. Phys. Chem. B. **124**(39), 8465–8478 (2020)
153. Tamate, R.: Healable soft materials based on ionic liquids and block copolymer self-assembly. Polym. J. **53**(7), 789–798 (2021)
154. Wang, Z., Furukawa, S.: Pore-networked soft materials based on metal–organic polyhedra. Acc. Chem. Res. **57**, 2506–12563 (2024)

Soft Materials for Wearable Electronics

Arpita Roy and Aniruddha Adhikari

Abstract Recent advancements in soft materials and system integration technologies have paved the way for creating a diverse range of wearable flexible electronic devices capable of real-time monitoring of physiological signals and delivering therapeutics, thereby revolutionizing personalized healthcare. However, the differences in properties between rigid, inorganic electronic devices and soft, organic human tissues present significant challenges, including skin irritation, tissue damage, reduced signal quality, and limited device lifespan. Consequently, to mitigate these challenges, extensive research efforts have been directed on developing flexible and stretchable device configurations and novel soft materials. In this chapter, we provide a comprehensive summary of recent research and technological advances in soft materials, emphasizing the importance of various mechanical, physical, electrochemical, and biological properties essential for designing conformable and stretchable wearable devices. We discuss evolving classes of soft materials used in cutting-edge wearable and implantable sensor technologies. Furthermore, we argue emerging trends, limitations, and future prospects in the field, along with methods to overcome challenges such as signal reliability, biocompatibility, and comfort-of-wear. Collectively, this chapter offers valuable insights for researchers and stakeholders involved in the design, fabrication, and application of soft materials for next-generation wearable sensors, outlining the necessary requirements for material properties, sensor capabilities, electronics performance, and skin integration.

Keywords Flexible electronics · Stretchable electronics · Wearable sensor · Implantable device · Health monitoring · Advanced material

A. Roy (✉) · A. Adhikari
Department of Chemical and Biomolecular Engineering, University of California at Los Angeles (UCLA), Los Angeles, CA, USA

V. Kumar, Y. K. Mishra (eds.), *Soft Materials for Functional Applications*,
Advanced Structured Materials 225, https://doi.org/10.1007/978-981-97-9468-3_11

1 Introduction

In the heart of Gotham City, where shadows and secrets intertwine, emerges a figure cloaked in darkness. Armed with unparalleled intellect, and an arsenal of cutting-edge gadgets, the enigmatic vigilante prowls the city's gritty streets as a symbol of hope for the innocents and a chilling harbinger for the criminal echelons. As the night unfolds, so does the legend of the Dark Knight—a solitary avenger behind the bat mask, sworn to protect Gotham from its encroaching abyss.

You might be curious as to why, rather than delving into scientific intricacies, we chose to kick off this chapter with the tale of a fictional superhero first created in 1939 by Bob Kane and Bill Finger. Throughout history, fictional narratives have stretched the boundaries of imagination and consistently fueled scientific curiosity and innovation. For example, the depictions of submarines in Jules Verne's *Twenty Thousand Leagues Under the Sea* foreshadowed the development of actual combat submarines by John Philip Holland, and the futuristic "communicators" featured in *Star Trek* inspired the creation of flip phones by Martin Cooper. Similarly, at different junctures of life, we all are captivated and inspired by the extraordinary abilities of fictional superheroes—whether it's Superman's incredible strength, heat vision, and freeze breath, Spiderman's ability to crawl walls, or Wonder Woman's agility and truth-compelling lasso. Conversely, Batman, the enigmatic Dark Knight mentioned earlier, is somewhat different. His superpower does not stem from superhuman capabilities or extraterrestrial strengths but rather from the ingenious assortment of gadgets housed within his "utility belt." These tools include a grapple gun for enhanced mobility, a first aid kit for injury repair, a communicator for communication, a rebreather for breathing assistance, and more. Wearable flexible electronic devices are like the stealthy gadgets in Batman's utility belt that flexes and bends with the wearer offering a blend of functionality and comfort.

The research into wearable devices, once limited to rigid wristwatches and fitness trackers, has now significantly expanded to encompass a wide array of innovations like smart textiles, electronic skin (i.e., e-skin), and personalized healthcare monitoring systems. Nowadays, we envision garments that monitor vital signs in real time [1, 2], allowing us to detect health issues before they escalate, an injectable cardiac patch that continuously monitors electrocardiogram (ECG) [3, 4] and deliver medicine whenever needed [5, 6], or athletic wear that improves performance by analyzing biomechanics and muscle activity [7–9]. Such paradigm shift in wearable electronics research has taken place predominantly due to the development of novel mechanically stable but flexible soft materials and their seamless integration with modern electronics. The hybrid integration of biocompatible soft materials with miniaturized wireless electronic systems indisputably holds great promise, as achieving successful device performance hinges on high degree of mechanical flexibility, sensing accuracy, comfort-of-wear, and ease of use [10]. The structural flexibility and functional versatility of these materials have rendered them potential candidates for deployment in a wide array of applications, including display and informatics [11, 12], robotics [13–15], sports equipment

[16], artificial e-skin [17, 18], energy-harvesting and storage systems [19, 20], and healthcare devices, particularly for real-time monitoring of neural activity [21, 22], cardiovascular signals [23–26], prosthetic integrity [27–29], organ behavior [30, 31], movement [32–34], and chemical/biochemical profiles (metabolites, ions, drugs, and biomolecules) [35, 36]. Undoubtedly, the progressive integration of soft, flexible electronic devices into daily life is transforming our abilities to perceive and interact with the world, reminiscent of abilities once only seen in comic book superheroes.

In recent years, various types of soft materials (e.g., liquid metal, nanotubes and nanowires, conductive polymers, conductive hydrogels, and elastomeric nanocomposites) have been used for designing wearable bioelectronic sensors. The suitability of a material for the development of bioelectronic sensor is determined by a combination of several inherent chemical (e.g., to ensure specificity, selectivity, and resistance to chemical interference from other substances), mechanical (e.g., mechanical robustness, flexibility, and conformability to biological tissues to ensure comfortable wear), rheological (e.g., optimal viscosity, Young's modulus, and shear modulus), optical (e.g., transparency), electronic (e.g., appropriate electrical conductivity, impedance, or capacitance), and biological (e.g., biocompatibility, bioresorbability or biodegradability, low immunogenicity) properties. Therefore, to advance wearable technologies, it is crucial to understand the essential characteristics that a soft material must possess for potential use in bioelectronic devices, alongside fundamental design principles. This book chapter showcases the soft materials employed in wearable electronic sensors, elucidating the pivotal interplay between the inherent properties of these materials and cutting-edge bioelectronic technologies, unveiling a prospective research direction within this field.

2 Wearable Electronics: Definitions and Classification

A wearable device is any miniaturized electronic device with sensors that can be worn on the body or integrated into clothing or other body-worn accessories that are capable of longitudinal monitoring of vital physiological and biochemical parameters [37]. Nowadays, several wearable bioelectronic devices offer therapeutic functionalities as well [38]. Classically, wearable technology is defined as "hybrid, network-enabled devices that can be worn on or in the body, which are integrated with the user's everyday life and movements" [39]. One of the prerequisites of wearable technology is the "comfort-of-wear," which greatly influences the wearer's willingness to wear a wearable device. It can be defined as the device's ability to mitigate the disruptions arising from its physical existence during the day-to-day activities of the wearer [40]. In simpler terms, wearable devices keep a close watch on human data while allowing unrestricted movement and comfort of the wearer. This can only be made possible by incorporating lightweight, flexible, mechanically sturdy, and user-friendly components [18, 35, 41]. Wearable sensors can be worn at rest, as well as during walking and running, on the head, arm, finger, ear, chest, leg/foot, and anywhere on the body, or they can be implantable, such as smart pills or artificial organs [42]. Based on their placement on the body, wearable devices are generally classified into two categories: (1) epidermal electronics and (2) implantable devices (Fig. 1a). Epidermal electronics refer to wearable electronic devices designed to be worn directly on the skin's surface [18]. Examples include wearable patch for monitoring vital signs [43], smartwatch [44], noninvasive ECG or electromyogram (EMG) devices [45, 46], and contact lens [47]. On the other hand, an implantable device is a type of wearable electronic device designed to be placed inside the body, typically within or on organs or tissues [42]. Examples include ingestible smart pills for wireless monitoring heart rate, blood pressure, pH, and temperature [48]; implantable cardioverter defibrillators (ICDs) [49]; and pulse generator for deep brain stimulation [50]. Both categories of these devices or systems fundamentally share a common goal: to possess properties that closely mimic those of tissue surfaces, facilitating flexible and durable contact [54]. Examples of both types of wearable electronic devices are illustrated in Fig. 1a, b.

Two major approaches can be taken to make electronics stretchable and flexible [55]: (1) designing innovative materials that are flexible and stretchable in solitary or aggregated forms and (2) structural engineering, i.e., by transforming non-stretchable materials into particular configurations that are capable of absorbing strain without fracturing, e.g., by inducing various modes of mechanical deformation.

3 Why Soft Materials?

Despite remarkable advancement in wireless technologies, low-power electronics, the Internet of Things, and the performance and miniaturization of integrated circuits, mechanical design in wearable technology remains conceptually outdated: fragile components enclosed in rigid packaging, resulting in bulky devices that often need straps or adhesive tapes to be attached to the skin [54, 56, 57]. The mechanical mismatch at the bioelectronics–skin interface is a widely discussed topic and forms the most formidable challenge in the growth of next-generation, skin-integrated electronics [58, 59]. All established categories of high-performance electronics rely on single-crystal inorganic materials, such as silicon or gallium arsenide, typically in the form of rigid and flat semiconductor wafers. In contrast, the human body is soft, textured, stretchable, and curvilinear. This disparity in properties poses a challenge for creating devices that can seamlessly and conformably integrate with biological tissues, often constraining and frustrating the dynamic, natural motions of the organs [60]. Consequences include discomfort, irritation, tissue damage, severe foreign body reaction, and adverse immune responses [61–63]. Additionally, bioelectronic devices function at the interface between living systems and electronics by translating biological signals into electronic signals and vice versa [64]. Biological signals within living organism are primarily carried by ions rather than electrons and holes as in electronic semiconductor devices [65]. These ions exhibit significantly higher

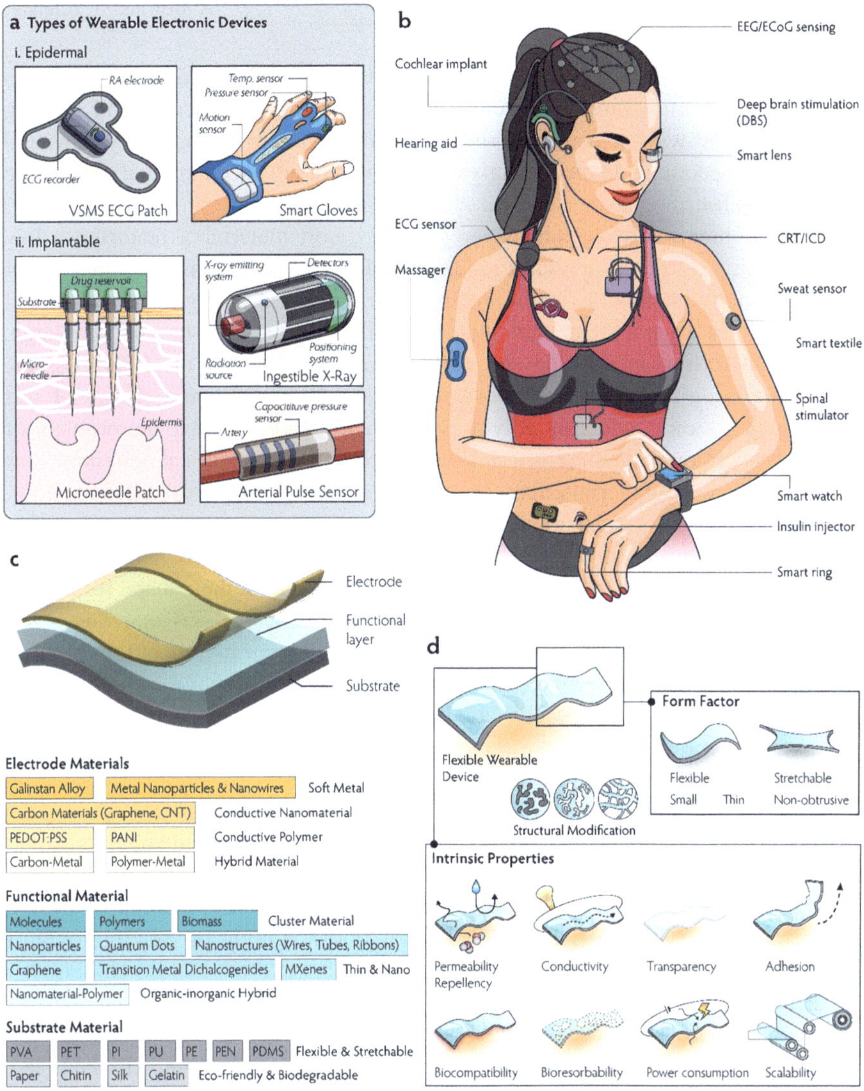

Fig. 1 Different classes of wearable devices, basic design, and essential properties. (**a**) Types of typical wearable electronic devices: epidermal (e.g., VSMS ECG patch, and smart gloves) and implantable (e.g., microneedle patch for drug delivery, ingestible X-ray, and arterial pulse sensor). *Abbreviations*: *RA* right arm, *VSMS* vital sign monitoring system, *ECG* electrocardiogram (*Ingestible X-ray* and *arterial pulse sensor* are redrawn with permission from Refs. [51, 52], Springer Nature Ltd). (**b**) Conventional wearable/implantable bioelectronic devices for healthcare, sports, and human machine interfaces. *Abbreviations*: *ECoG* electrocorticogram, *EEG* electroencephalogram, *CRT* cardiac resynchronization therapy, *ICD* implantable cardioverter defibrillators. (**c**) Basic architecture of a wearable electronic device and examples of intrinsically soft materials used for designing flexible electronic devices. *Abbreviations*: *CNT* carbon nanotube, *PEDOT:PSS* poly(3,4-ethylenedioxythiophene) polystyrene sulfonate, *PANI* polyaniline, *PVA* poly(vinyl alcohol), *PET* polyethylene terephthalate, *PI* polyimides, *PU* polyurethane, *PE* polyethylene, *PEN* poly(ethylene-2,6-naphthalenedicarboxylate), *PDMS* polydimethylsiloxane. (Adopted with permission from Ref. [53], Elsevier). (**d**) Schematic illustration of essential material properties for flexible wearable device (Adopted with permission from Ref. [10], Wiley-VCH)

conductivity in the water-rich biological milieu compared to electrons and holes [38]. Therefore, mitigating the discrepancy in mechanical properties and refining transduction pathways (electron to ion conversion) between biological systems and electrodes is becoming an increasingly crucial task for advancement in the wearable technologies [66].

In this context, soft elastomers, or gels (i.e., soft materials), featuring a low shear modulus (in the kilopascal-to-megapascal range, in contrast to conventional stiff materials with moduli in the gigapascal range), are emerging as functional components in wearable technologies [67]. Their capability to undergo large, reversible deformation matches the mechanical properties of the living tissues. Furthermore, ionic conductivity and water uptake ability of some classes of soft materials (e.g., ion-conducting polymers and hydrogels) can solve the problem of signal transduction at the bioelectronics interface [64]. In a classical definition, soft materials are those materials that can be easily deformed by external force [68]. The magnitude of the shear modulus of these materials is much smaller than that observed for true atomic crystals, and their mechanical behavior is intermediate between a solid and a liquid, in the sense that they maintain their shape for a time, but they eventually flow. The physicochemical properties of soft materials often resemble those of biological matter. As compared with their hard counterparts, soft materials have distinct advantages to be used in wearable electronics: they can be manufactured using simple designs, are less dense and less costly, can be processed at lower temperatures and in aqueous environments, and in many cases exhibit a higher level of biodegradability and bioresorbability [69]. A distinguishing quality of many soft materials is their stimulus-responsive characteristics, which make them attractive for biomedical applications. Several polymers and supramolecular architectures respond to external triggers by changing their macroscopic shape. These materials, capable of sensing and actuation, have the potential to exhibit behavioral responses. This feature is essential from the point of view not only of functionality but also of adaptability characteristics.

Therefore, in summary, the integration of soft electronics with nonplanar objects like living systems has numerous implications. Firstly, the close contact between the device and the nonplanar surface facilitates the gathering of high-quality data [18]. Contrary to rigid electronics, which creates air gaps at the interface, resulting in diminished contact area and potential signal disruptions (i.e., introduction of noise and artifacts), soft stretchable electronics ensure better signal quality. Secondly, flexible and compact devices enable mobile and widespread sensing applications, holding significant potential for Internet of Things technology [70]. Lastly, soft materials with mechanical properties like biological tissues and ionic conductivity can solve the long-lasting problems of biocompatibility and bioelectronic signal sensing in the domain of continuous healthcare monitoring [71, 72].

Essential components for building flexible electronic thin-film devices consist of substrates, electrodes, and functional layers (Fig. 1c). Lately, considerable attention has been directed toward developing intrinsically soft materials (Fig. 1c provides a comprehensive list), which represents the most intuitive and promising approach [53].

4 Essential Properties of Soft Materials for Wearable Electronics

A thorough understanding of fundamental material properties crucial for achieving optimal "comfort-of-wear" without compromising sensing capabilities can significantly drive advancements in the field of wearable electronics. Various material properties essential for developing advanced wearable electronic products are illustrated in Fig. 1d. As the primary objective of a wearable system is to attain flexibility and stretchability, it is first and foremost to carefully consider Young's modulus (i.e., a measurement of stiffness [73] or a material's ability to resist elastic deformation [74]) along with nonobtrusive form factors. Other essential physicochemical properties that enhance "comfort-of-wear" include transparency, adhesion property, gas/vapor permeability, and water repellency. Safe use of the wearable electronic device over long periods of time requires that the constituent material be both biocompatible and bioresorbable. Moreover, the integration of adjustable conductivity, high-power systems, and scalable processes along with coupling of wireless networks has facilitated the development of cost-effective and unobtrusive multiple sensory systems in the realm of wearable technology. In this section, we demonstrated a checklist of necessary material characteristics and current strategies to modulate them under five major categories: mechanical, physicochemical, biological, electrical, and device-level properties (summarized in Table 1).

4.1 Mechanical Properties

Biological systems, organs, and tissues are soft, elastic, and curved; silicon wafers are not. Addressing this fundamental disparity in mechanics and form between biological tissues and conventional rigid, planar integrated circuits presents the foremost challenge in seamless integration of electronics with the human body as wearable or implantable devices [121, 122]. In this context, two crucial mechanical properties that require consideration are softness, indicated by Young's modulus, and stretchability [123–126]. In recent years, the focus of research has primarily been on imparting stretchability to electronics [127, 128], with considerably less advancement in addressing the modulus mismatch with soft tissues [129, 130]. Although achieving stretchability often involves reducing Young's modulus, the moduli of these stretchable electronic materials and devices, including conductors and semiconductors, remain several orders of magnitude higher than most soft biotissues [131–133].

Table 1 Mechanical, physicochemical, biological, and electrical attributes necessary for the development of materials for next-generation wearable electronics, along with strategies to achieve them

Property	Requirement	Typical strategies	References
Young's modulus	Should be low and match the value of the target biological tissue or organ to ensure minimal mechanical stress at the interface between the human skin and the device (i.e., bioelectronics interface)	• Thinning of the inorganic or organic materials	[75, 76]
		• Using a porous substrate	
		• Creating "filamentary serpentine" structures of metals	[77]
		• Embedding inorganic materials inside or on top of the organic substrate	
		• Adding plasticizer into the polymer	[76]
		• Combining with liquid metal or ionic liquid	[78]
Stretchability	Should be high for secure attachment to the curved surfaces of human skin and maintaining mechanical strength to prevent device failure	• Reducing the dimensions of the electrode material	[79]
		• Designing metal electrodes to withstand strain	[79]
		• Utilizing a combination of ultrathin plastic with pre-strained wavy or buckled elastic substrates	[79]
		• Creating percolating networks of hybrid materials to form intrinsically stretchable composites	[79]
		• Incorporating intrinsically stretchable polymers, hydrogels, and ionic liquids into the device design	[80]
Form factor	Flexible, stretchable, small, thin, dry, and nonobtrusive form factors while achieving accurate sensing capabilities over prolonged period of time are essential to enhance usability and comfort-of-wear	• Reducing the size or thickness of each component, including electrodes, substrates, and chips	[75]
		• Creating components based on soft materials and implementing efficient encapsulation techniques	[81]
		• Designing band-aid patch, temporary tattoo	[82]
		• Gel-free, dry, and reusable adhesive	[83]

(continued)

Table 1 (continued)

Property	Requirement	Typical strategies	References
		• Transparent, lightweight, comfortable	[84]
		• Integrated devices with microfluidics, 3D scaffold forms, and fiber-based components	[85]
		• Incorporating wireless near-field communication protocols	[86]
Transparency	High optical transparency is required for visual optical monitoring, as well to make the device "invisible" and to improve the aesthetics of devices for daily usage	• Using transparent polymeric films	[87]
		• Restricting the width of metal electrodes	[88]
		• Utilizing transparent elastic films composed of carbon nanotubes, silicon nanowires, and metal nanowires	[89, 90]
		• Decreasing the thickness of the nanolayers made from papers	[91]
		• Using inherently transparent materials	[92]
Adhesion	Adequate or tunable bio-adhesion should be achieved through chemical, van der Waals, or electrostatic forces to ensure effective transfer of strain between two distinct surfaces at the bioelectronics interface	• Implementing wavy and suspended structures for electrodes and substrates	[93]
		• Engineering micropillar arrays with diverse tip geometries and aspect ratios	[94]
		• Employing dry adhesives to enhance signal accuracy and quality	[95]
		• Using kirigami-inspired geometries to adjust adhesion through variations in stiffness and actual contact width	[96]

(continued)

Table 1 (continued)

Property	Requirement	Typical strategies	References
Permeability	High permeability to gas (air, oxygen) and water is required to prevent the blockage of sweat glands and mitigate the risks of skin inflammation, even during prolonged wearing, tight adhesion, perspiration, and evaporation	• Using thin silicon substrates	[97]
		• Incorporating additional encapsulation layers to prevent degradation of device performance	[98]
		• Utilizing inorganic nanomeshes or nanofiber-based conductors	[99]
		• Engineering paperlike fibers or yarns from natural materials such as wood, silk, or cellulose	[100]
		• Using porous graphene on elastomer sponges	[101]
Water repellency	Should be unaffected by water or resist the ingress of water under specified conditions to enhance durability over long time	• Utilizing polymeric substrates with inherent water-repellent properties	[102]
		• Using hybrid nanofiller-shaped composites	[103]
		• Treating the surfaces with chemicals containing silane, carboxyl, and fluorine groups to impart repellency	[104]
Biocompatibility	Low toxicity or nontoxicity (i.e., low immunogenicity) is necessary for long-term usage of the wearable bioelectronics on the human skin or inside the body	• Encapsulation or functionalization with biocompatible materials	[41]
		• Using inherently nontoxic polymers and metal oxides	[81, 105]
		• Utilizing biodegradable porous nanostructures	[106]
		• Using gel-free dry adhesives	[107]
		• Using natural biomaterials	[108, 109]
Bioresorbability	Should be biodegradable for easy metabolism without triggering harmful physiological reactions and to tackle the problem of unmanageable electronic waste	• Using bioresorbable materials including Si, Mg, Zn, Mo, W, SiO_2, MgO, PLGA, PI, etc.	[110, 111]
		• Using biomaterials and bioinspired materials including proteins, enzymes, and DNA	[112]

(continued)

Table 1 (continued)

Property	Requirement	Typical strategies	References
Gauge factor	A high gauge factor across a wide strain range is essential for detecting the minute strains generated by subtle human expressions and bodily movements	• Using intrinsically high gauge factor materials such as semiconducting ZnO, $ZnSnO_3$, CNTs, and C-diamond	[113]
		• Introduction of cracks in a conductive thin film	[114]
		• Adopting ultrathin metal foils on pre-stretched dielectric materials for capacitive-type strain sensor	[115]
		• Designing hierarchical architecture with MXene–AgNW as the "brick" and PDA/Ni^{2+} as the "mortar"	[116]
		• Using 0D–1D–2D ternary fullerene–SiNWs–GO nanocomposites	[117]
Scalability	Large scalability and cost-effectiveness are attractive for future electronics	• Designing FET structure for thermistor-type temperature sensors	[41]
		• Microfabricating a solderable and stretchable sensing system, utilizing nickel–vanadium and gold pad metal layers, along with copper interconnection	[118]
		• Adopting large-scale, cost-effective manufacturing methods such as solution and printing processes	[119]
		• Incorporating 3D printability	[120]

Abbreviations: *PLGA* poly(lactic-*co*-glycolic acid), *PI* polyimide, *MXene* 2D titanium carbide $Ni_3C_2T_x$, *PDA* polydopamine, *GO* graphene oxide, *NW* nanowires, *FET* field-effect transistor

Young's Modulus

Young's modulus (E) or elastic modulus is defined as the ratio of principal stress in one direction (σ_x) to corresponding strain in the same direction (ε_x) during the linear elasticity regime of a uniaxial deformation (unit: Pa) (Eq. 1).

$$E_{(\mathrm{Pa})} = \frac{\sigma_x}{\varepsilon_x} = \frac{\mathrm{Stress}}{\mathrm{Strain}}. \tag{1}$$

Since its value broadly depends on atomic arrangement and bonding energy [134, 135], Young's modulus regulates both mechanical and electronic properties of a material that comprises integrated devices [136]. Achieving a tissue level elastic modulus in wearable electronic devices is critical for two reasons: first, to enhance the conformability of bio-integrated devices on nonzero Gaussian surfaces (i.e., human skin or tissue) particularly for enabling high-fidelity measurement (i.e., high signal-to-noise ratio) [137–139], and second, to suppress foreign body responses (**FBR**) (i.e., immune cell recruitment and inflammatory activation) and side effects (e.g., skin irritation, rashes, and allergic reaction) toward both epidermal and implanted devices [140, 141]. Unfortunately, the elastic modulus of different human organs and tissues (e.g., skin, heart, cartilage, brain, etc.) varies widely (0.5–100 kPa) [142], often being three to four orders of magnitude lower than that of the materials used in devices (e.g., silicon, gold, etc.) (Fig. 2). For instance, the elastic modulus of brain tissues ranges from 0.5 to 1.0 kPa [143, 144], while in cardiac tissues, it is 10–15 kPa [145, 146]. Of greater significance, human skin typically exhibits a wide elastic modulus ranging from 20 to 300 kPa [147, 148]. In contrast, silicon has an elastic modulus >130 GPa [149], and for gold, it is >40 GPa [150]. To decrease the high elastic modulus of conventional materials, several researchers have pursued a thinning strategy wherein the thickness of a material is reduced below a specific threshold. This principle relies on the fact that the elastic modulus

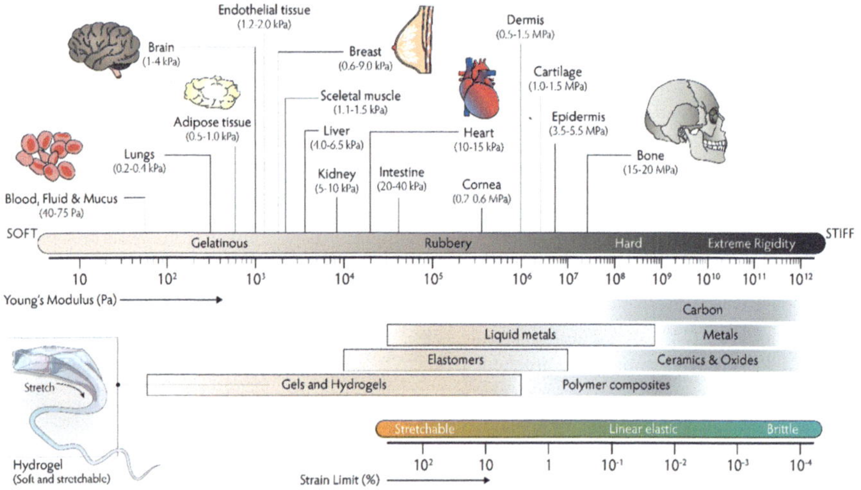

Fig. 2 Mismatch in mechanical properties between soft biological tissues and electronic device materials in terms of Young's modulus. Young's moduli for most of the biological tissues lie within 0.5–100 kPa range, while for conventional electronic materials, its >1 MPa. The lower scale depicts stretchability of the materials. Note that for use in wearable devices, materials should be stretchable. Inset shows hydrogel as a promising material for bioelectronics due to softness and stretchability

decreases inversely with the thickness of strained films at thickness values below 80 nm (Eq. 2) [136].

$$E = 54.872 \times h^{0.226} \tag{2}$$

where h is the film thickness. This simple approach led to the decrease in the bending stiffness (a quantitative measure of user comfort) of a wearable physiological sensor made up of gold (AU) film from 75.85 to 0.33 pN m^2 (comparable to human skin's uppermost layer), when the thickness was changed from 3 μm to <300 nm [151]. However, achieving this remains challenging due to the high moduli (above 100 MPa) of existing stretchable semiconductors and conductors, particularly on substrates with much lower moduli (e.g., elastomer, skin) [152, 153]. According to the fracture mechanics principles governing thin films on elastomeric substrates, a substantial modulus contrast across the interface would exacerbate stress concentrations at defect sites, which in turn would escalate the rate of cohesive energy release during stretching, leading to a heightened risk of crack propagation, and detachment from the substrate [154–156]. For example, during uniaxial tensile tests, copper (Cu) films with thickness <200 nm failed at very low strains as a result of intergranular fracture [157]. As an alternative, the concept of neutral mechanical plane has emerged [158]. This concept involves integrating a several-hundred-nanometer Si layer into stretchable or foldable circuit systems, with polyimide (PI) serving as the encapsulating layer. Interestingly, when considering electrode materials, the elastic stretchability of electronics improves as the modulus of a metal layer increases, both with and without the presence of strain, contrary to the trend observed in a substrate [159]. Examples include carbon-based materials (elastic modulus ranging from 0.5 to 1.8 TPa), such as monolayer or multilayered graphene and carbon nanotubes (CNTs), that show promise as electrode candidates [160–162]. It is worth noting here that, the stress endured by the packaged electrodes is proportional to Young's modulus of the polymeric substrate surrounding them, and this correlation enables the entire device to achieve high levels of stretchability [57, 85, 163, 164]. Therefore, the most crucial aspect for enhancing stretchability, particularly in terms of the elastic modulus, lies in selecting and optimizing substrate materials, structures, and processing methods. Such substrates must have a low modulus that will allow electrodes to deform out of plane, ensuring high stretchability post-integration without any detachment in the interface region [165]. A prime example is polydimethylsiloxane (PDMS) (E = 360–870 kPa), known for its mechanical compatibility with both skin and electrodes. Such mechanical compatibility has facilitated its use in a variety of form factors for wearable flexible electronics, including encapsulation layers, membrane films, porous platforms, functional composites, microfluidic substrates, and microneedles/pillars, among others [166, 167]. In addition, several inherently soft materials, such as poly(vinylpyrrolidone) (PVP), ethylene–vinyl acetate, Tegaderm, Ecoflex, and parylene, offer substantial reductions in weight and thickness, enhance mechanical compliance at the interface, and improve skin conformability (refer to Refs. [81, 104] for a detailed compilation of materials along with corresponding

E). Similarly, the controllable properties of naturally stretchable plastics have enabled adjustments in shape and structure, thereby influencing Young's modulus in these devices. Recent advancements in high-performance wearable devices heavily depend on the effective tuning of organic materials to lower the elastic modulus. Another way to address the problem of elastic modulus is to use intrinsically stretchable polymer, such as soft ionic conductors [168], or adoption of effective strategies to tune their structure. Elastic modulus of these polymers and nanofibrils (e.g., cellulose, chitin, silk, etc.) with unique functionalities like self-healing [169–171] and biodegradability [172] depends on factors such as ratio of units, structure, and orientation index rather than solely on chain entanglement or cross-linking [173, 174]. Thereby, these materials not only offer ease of processing but also enable the fine-tuning of Young's modulus, fracture strain, and fracture energy simply by altering the ratio of each unit in the polymer. Examples include rational design of a self-healing stretchable material with multi-strength hydrogen bonding interaction by varying the ratio of 4,4′-methylenebis(phenylurea) (MPU) units in the polymer [169] and tuning of E (from >10 GPa to 0.15 MPa) and stretchability (from <20% to >400%) in silk-based hydrogels by plasticization and control of the water content [175]. In many instances, plasticizers are employed to enhance both the toughness and flexibility of polymer films, thereby improving their formability for use in wearable applications [108, 176]. Given the dependency of the output response of a piezoelectric material on E (Eq. 3), precise modulation of Young's modulus is crucial for energy-harvesting wearable devices [177, 178].

$$D = d\sigma = dE\varepsilon \tag{3}$$

where D is the electrical charge separation and d is the piezoelectric coefficient [179]. In summary, controlling Young's modulus is essential for designing durable and comfortable wearable electronics that can bend, stretch, and even harvest energy or work in robots.

Stretchability

Stretchability plays a vital role in bio-integrated electronics due to two principal factors. First, most of the microelectronic materials are planar, necessitating sufficient stretchability for proper adaptation to the undevelopable surfaces (i.e., surfaces with nonzero Gaussian curvature) found in biological tissues and organs. Second, due to the soft and elastic nature of most bio-tissues, only devices capable of deforming together with these tissues can sustain their form and functionality while establishing a close and minimally stressful interface for bioelectronics. Consequently, numerous design approaches have been employed to attain stretchable bioelectronics utilizing materials with high elastic modulus, including out-of-plane buckling, island-plus-serpentine, or filamentary serpentine designs and fractal-, microcrack-, or kirigami-inspired designs. In this section, we discussed some of them with relevant examples.

A common technique for creating out-of-plane buckling structures involves initially applying a rigid thin-film device onto a pre-stretched flexible substrate and subsequently releasing the substrate to allow it to return to its original state. For example, when 100 nm gold electrodes deposited on 1 μm polyethylene terephthalate substrate are transferred onto a 200% pre-stretched acrylic elastomer and subsequently released, the gold electrodes undergo spontaneous wrinkling, enabling them to withstand tensile strains of up to 120% (in contrary, usual nanocrystalline gold thin films typically rupture at a tensile strain of 1% [180]) [31]. This strategy allowed the successful application of stretchable gold electrodes in monitoring the voiding function of a rat's bladder, with a volume change from 71% to 100% [31]. However, buckled ribbons or films that protrude out of the plane create a surface roughness issue, hindering the close contact between the device and the soft biological tissue. Therefore, as an alternative, the in-plane meandering ribbons (i.e., serpentines) became popular. Serpentines are essentially 2D springs that elongate through ribbon rotation and twisting, helping to reduce the intrinsic strains in a material [181, 182]. Stretchability of a material is amplified with the increasing hierarchy of self-similar serpentines [183, 184]. Example includes a multilayer wireless e-tattoo with incorporated ICs that showed one-time stretchability of 55% and cyclability of 10,000 times under 20% applied strain and was used for monitoring ECG, electrocardiogram (ECG), blood oxygen saturation level (SpO_2), heart rate (HR), and skin temperature [185]. While serpentines offer ample stretchability, their mechanical features do not accurately align with the nonlinear behavior exhibited by bio-tissues [186]. As a result, new designs utilizing microcrack and kirigami techniques have emerged to accommodate applied stretching and increase surface area coverage. For instance, soft ECG and EEG sensors made up of microcracked gold electrodes with stretchability up to 120% [187]. Another recent example is kirigami-inspired cuts on an elastomer with embedded aluminum foil with a stretchability of 400% and durability of 1000 cycles at 300% strain [188].

Other methods for creating composites with desired functionality and stretchability involve incorporating filler materials (that are stretchable on their own) into a plastic substrate, rather than using layered electrode and substrate structures. Nanomaterials, such as 1D [189–191] or hybrid [117, 192] forms, are utilized to enhance stretchability and conductivity, ensuring high performance even in the presence of cracks or notches. Hybridization of diverse materials into composites, including ultrathin 2D materials like graphene [193], molybdenum disulfide (MoS_2) [194], and MXene [116, 195], further enhances flexibility and transparency, making them suitable for wearable platforms. Additionally, novel concepts like ionic conductors and liquid metals (LM) show promise as "intrinsically stretchable" electronic materials. Ionic conductors, such as hydrogels and ionic conducting elastomers (ICEs), offer stretchability and additional functionalities like self-healing capabilities. Liquid metals, encapsulated in soft substrates, exhibit high stretchability and conductivity, making them suitable for various applications including triboelectric nanogenerator [196], glove-type strain sensor with meander resistors [197], and air quality monitoring system [198]. The other type of ionic conductor, comprising hydrogels [168, 170, 199, 200], plasticized silk proteins [175], and semiconducting polymers [10], also exhibits high stretchability and enhances the wearable device with additional features such as self-healing capabilities. Furthermore, the development of intrinsically stretchable arrays of transistors, utilizing materials like cross-linked styrene–ethylene–butadiene–styrene (SEBS) and conjugated polymer/elastomer phase separation induced elasticity (CONPHINE) film, along with carbon nanotubes (CNTs) for dielectric, semiconductor, and electrode components, enables the fabrication of semitransparent skin electronics without the need for hard–soft integration [201].

4.2 Physicochemical Properties

Small and Thin Form Factor

Form factor encompasses the size, shape, and physical arrangement of a device, collectively defining its physical dimensions and configuration. Mostly, in electronic packaging, consumer electronics, and nanomaterials fabrication, the term describes the physical size and shape of a material or device that represents a class of similarly sized components and prescribes a specific standard [202, 203]. For seamless integration with the body surface, the wearable devices must possess flexible/stretchable, small, thin, dry, and nonobtrusive form factors. To create functional electronics with a thickness as low as a few hundred nanometers, thin polymeric layers (i.e., substrate) are fabricated surrounding a functional electrode or device [87, 204, 205]. This method enables the device to be lightweight and thin and allows it to remain unobtrusive during human movement. To decrease the thickness of the substrate, several polymer-based materials are used, which include ultrathin organic [206], carbon-based composite [207], 3D cellular sensor array [208], and 3D monolithic types [209]. Such strategies have been employed to construct thin and light patch-type wearable vital sign monitoring devices including the electrochemical sweat sensor [82], interstitial fluid-based biochemical sensor [210], graphene-based physiological sensor [211], eutectic gallium–indium (EGaIn) liquid metal (LM) alloy-based electromyogram (EMG) sensor [212], commonly used Au-based sensor [213], and even the supercapacitor [214].

Transparency

In addition to electrical and mechanical properties, high optical transparency is crucial for wearable devices (particularly optoelectronic devices) and smart displays, ensuring a seamless integration with the human body. Optical transparency refers to the characteristic of a material that enables light to travel through it without significant obstruction or absorption [215]. Transmittance (T, %), the percentage of transmitted light in comparison to total incident light of a specific wavelength, is used to quantify optical transparency. It is of great importance to optimize the electrical conductivity and the optical transmittance for transparent electrodes as there is often a tradeoff between the two parameters [216]. For instance, while materials like PDMS offer flexibility and transparency, their conductivity and functionality fall short compared to newer materials like novel metals and carbon-based substances [217, 218]. To address this, recent studies have focused on enhancing transparency while adding functionality to wearable electronics. For instance, ultrathin organic field-effect transistors (OFETs) composed of thin Au metal films on PEN foil exhibit flexibility akin to paper when crumpled [87]. Another approach involves reducing the thickness of layered structures using micro- and nanopaper, resulting in highly transparent and gas-permeable films [100]. Additionally, composite materials comprising metals and polymers, like silver nanowires (AgNWs) embedded in polyamide nanofibers (NFs), offer high transparency and conductivity [99, 219]. Thus, composites with controlled scale or shape, comprising metals and polymers, exhibited significant transparency in the visible light spectrum. Examples include composites like Ag nanoparticles (NPs) embedded in polyurethane (PU) ($T = 77\%$) [220], AgNPs in PDMS ($T = 86\%$) [221], PDMS combined with Au nanowires (AuNW) and Ag nanowires (AgNW) ($T = 86\%$) [222], AgNWs integrated into PDMS ($T = 88.3\%$) [223], and Ag nanofibers (AgNF) along with AgNW in polyimide (PI) ($T = 89.05\%$) [88], all tailored for use in flexible strain and tactile sensors. Carbon-based materials such as graphene ($T > 90\%$) [224] and CNT/PDMS ($T > 85\%$) [225] also show promise due to their high transparency and conductivity. Films made from biomimetic materials like chitin and silk fibroin hybrids with AgNFs exhibit high transparency ($T > 90\%$) and mechanical strength [226, 227].

Adhesion

Adhesion is vital for soft wearable devices to ensure secure attachment, comfort, functionality, and durability particularly because direct contact between two surfaces with different structures tends to weaken over time, whether by design or by accident. While adhering to the skin can provide highly accurate signals, immediate challenges arise from the difficulty of detaching such devices from the skin. Moreover, medical adhesives (i.e., acrylic adhesives), commonly found in tapes, dressings, ECG electrodes, and medication patches, can cause allergic contact dermatitis [228]. These patches, during removal, often strip away superficial layers of the skin, compromising its barrier function and increasing the risk of infection, irritation, and subsequent delayed healing [229, 230]. Consequently, recent research has focused on dry features, specifically interfacial adhesion. Instead of relying on sticky adhesives, establishing a dry and continuous contact through van der Waals, electrostatic, or other non-covalent interactions is desired [94]. In contrast to commonly used acrylic medical bandages, dry adhesive offers several potential benefits. First, it can exhibit consistent and renewable adhesion due to its self-cleaning ability [231–233]. Second, it is less influenced by surface contaminants, oxidation, and other environmental factors because its adhesion originates from densely packed surface microstructures, rather than involving chemical components [234]. Third, it is expected to offer improved biocompatibility, as there is ample ventilation and minimal skin contact with potentially irritating chemicals [235]. Achieving conformal adhesion involves employing extremely thin electrodes, pattern-imprinted substrates, and self-adhesive materials to create flexible, stretchable form factors with dry contact between the skin and wearable flexible and stretchable electronic devices. For example, Kwak et al. [235] devised a method to achieve adhesion to dry skin without adhesive by patterning a rubbery, nontoxic substrate with micrometer-scale mushroom-shaped projections—a topology ideal for maximum adhesion to human skin [236]. Impressively, the substrate maintained strong adhesion for up to 30 cycles of attachment and removal, without inflicting considerable damage to the skin. This hybrid approach of incorporating micrometer-scale pillars in the adhesion area eliminates the need for glue while maintaining adhesion quality. Another example includes an ultrathin gold electrode (200 nm) partially encapsulated by parylene which has been shown to improve skin adhesion by reducing bending stiffness, enabling self-adhesion to any 3D biological surface [151]. By mimicking adhesion mechanisms found in nature, such as those used by geckos or octopuses, wearable devices can record vital signals even from wet skin, bypassing potential toxicity from conventional adhesives [95, 237]. For instance, stretchable electronics featuring octopus-like patterns imprinted on a carbon-based conductive polymer composite film exhibit high adhesive capabilities under both dry and wet conditions [237]. These bioinspired conductive suckers allow for high adhesion without leaving behind residues after detachment, even under water, thereby enabling the detection of vital biosignals like ECG and wrist bending motions. Other approaches include patterning PET films with laser cutting and coating them with adhesive silicon primer, as well as modifying adhesion strength using surfactants, such as Triton X, in PDMS substrates [238]. Hydrogels have also emerged as promising adhesives due to their composition and programmable changes in adhesion strength. These hydrogels demonstrate strong adhesion to the skin and various solid materials, high flexibility, and strain sensitivity, enabling the monitoring of subtle motions and physiological signals. Hydrogel is an excellent adhesive primarily due to its unique composition, consequent physical interactions, and chemical bonds, including hydrogen bonding, hydrophobic interactions, metal complexation, π–π stacking, cation–π interactions, and covalent bonding, introduced between the hydrogels and bio-tissues [239, 240]. For example, in situ polymerization of nucleobase-driven adhesive hydrogels onto conductive tough hydrogels results in robust adhesion to skin and various solid materials [199]. These hydrogels effectively block the current from the tough conductive layer, protecting the skin, and exhibit high flexibility and strain sensitivity, enabling detection of both large-range human motions and subtle physiological signals. They demonstrate strong bonding capabilities, attributed to high stretchability (up to 2100%) and adhesion strength even underwater (10–20 kPa), allowing for monitoring of delicate motions like swallowing and smiling with minimal resistance change [171]. Further, Han et al. [241] demonstrated that by synthesizing hydrogel with polydopamine (PDA)-doped polypyrrole nanofibrils within an elastic and transparent polyacrylamide network, they could achieve programmable changes in adhesion strength, reaching up to 30 kPa in maximum strength. Additionally, Oh et al. [242] showcased a temperature sensor with high thermal sensitivity (2.6% $°C^{-1}$ from 25 to 40 °C) by integrating functional hydrogel with a hierarchical surface design resembling octopus sucker-like microstructures. By coating PDMS surfaces with a thermoresponsive hydrogel layer (poly(*N*-isopropylacrylamide)-pNIPAM), they achieved strong adhesion, with or without the pNIPAM coating. The adhesion increased 14–23 times with a rim structure compared to a flat surface. At 40 °C, the pNIPAM de-swelled, causing the hydrogel to shrink and increase the concavity volume, enhancing the suction effect of the rim structure, resulting in even higher adhesion (13.4 kPa) and enabling easy detachment. These features offer reusability and high durability, making them suitable for skin-attached temperature sensors.

Permeability

Wearable electronic devices often operate in environments rich in gases, liquids, and molecules. This external environment can inevitably lead to the infiltration of chemical substances into the surfaces of wearable devices, causing a decline in sensor performance over time. Such environmental heterogeneity poses a significant challenge to achieve stability while maintaining adequate sensing performance. Additionally, conventional electronics are susceptible to moisture. Sweat formed under the device can damage the skin, leading to infection, discomfort, and delayed healing [229]. Therefore, it is crucial for wearable electronics to be selectively permeable or impermeable to gases like water vapor from the external environment and liquids containing salts, proteins, and other molecular substances from the biological environment.

Encapsulating wearable electronics with a low water-permeable or water-repellent layer enhances stability at the interfaces. For example, PDMS, while flexible and biocompatible, is permeable to water, organic solvents, and small lipophilic molecules, which prevents it from forming a hermetic seal [243]. However, this limitation can be mitigated by applying multilayer coatings with polymers like parylene or PI, which have lower permeability and higher durability [98, 244]. Inorganic encapsulation layers, such as thermal SiO_2 and silicon nitride bilayers, offer impermeability to water and ions while remaining flexible, making them attractive alternatives [245]. This approach has become essential for serpentine-patterned electrodes and other components employed in wearable devices to date. Furthermore, liquid metals (LMs), a rapidly growing material for flexible electronics, require encapsulation to prevent changes in surface oxide composition due to water exposure, which could weaken mechanical strength and breakdown voltage [246, 247]. Wang et al. [248] created Mg-GaIn electrode by blending gallium-based liquid metal with magnesium particles for use in cancer phototherapy after patterning them on a flexible paper substrate. The outcomes revealed a remarkable photothermal conversion efficiency of 61.5%. Moreover, these paper-based substrates exhibit special impermeability to liquid metals, streamlining the fabrication process of fully stretchable electronics [100, 249].

Breathability is another important aspect in respect to permeability of wearable electronics, ensuring long-term wearability and secure adhesion during the times of perspiration. Substrate materials with high permeability to water vapor and gas allow sweat to evaporate without blocking sweat gland ducts, reducing the risk of skin inflammation [90]. Recent research has demonstrated successful modulation of selective permeability by controlling the structure of substrate materials in terms of porosity and pore size [86, 250–252]. Electrodes made with impermeable graphene nanosheets achieve high water vapor permeability when combined with porous elastomer sponges, facilitating sweat transportation and evaporation to minimize inflammation risks [101]. Additionally, electrodes mimicking the porous structure of fabrics or textiles using nanowires [90, 253], vertically aligned 3D electrodes [254], or rugged substrates [93] also enhance water vapor permeability. For instance, Au nanomesh conductors exhibit excellent water vapor permeability by allowing water evaporation through their mesh-like structure, reducing the risk of skin irritation and inflammation during long-term wear [90].

Self-Healing

The concept of self-healing, once only seen in science fiction, has recently become a focal point of scientific inquiry. The idea is simple: to infuse electronics with the ability to heal themselves, much like our bodies do repair structural damage (i.e., scratches and cuts), allowing it to restore its mechanical and electrical properties [255]. The self-healing process in electronically active polymeric materials arises from the dynamic equilibrium between the cross-linking network and percolation pathways within the polymer system. Self-healing materials can be categorized into two main groups: autonomic and non-autonomic. Autonomic self-healing materials can recover without external intervention, while non-autonomic materials require external stimuli such as heat, light, or humidity to initiate healing [256]. Also, self-healing materials can be divided into intrinsic and extrinsic categories. Extrinsic self-healing materials involve encapsulation of a healing agent within capsules or vascular tubes within the material's structure [257]. Upon damage, these containers break, releasing the healing agent and recovery of the whole system. Intrinsic self-healing relies on reversible dynamic bonds, covalent or non-covalent. These reversible bonds break (usually being the weakest bonds) to dissipate energy when the material is damaged, subsequently reformed again under ambient conditions to restore material integrity [258]. Additional physical factors influencing intrinsic self-healing include polymer chain mobility, interfacial flow, and shape recovery properties [258]. Given the low conductivity of polymeric materials, strategies for designing self-healing and high-performance electronic materials often involve incorporating electronically active fillers into a dynamic polymer matrix. For example, nanostructured conductive networks based on one-dimensional wires, in contact with a cross-linked self-healing polymer matrix, can be rearranged dynamically to restore original percolation paths for electrical conduction after mechanical damage [259]. The size and surface properties of conductive fillers in a polymer matrix can affect both self-healing kinetics and electrical responses. However, it is essential to differentiate between mechanically tough self-healing materials and flowable viscoelastic materials. In highly flowable viscoelastic materials, healing primarily occurs through the mixing of materials at the reconnected interface. However, in self-healing materials, the repair process involves not only intermixing but also the restoration of broken cross-linking bonds. Therefore, flowable viscoelastic materials may undergo permanent plastic deformation even after mechanical force is applied and are often unable to fully restore their original properties. In contrast, combining strong and weak dynamic bonding can yield elastic yet tough self-healing materials.

Prior to the advent of self-healing electronics, researchers developed various self-healing polymers and elastomers [260, 261] that later found applications as both passive components, like substrates in sensing platforms and active components, such as dielectric layers in field-effect transistor-based sensors. For instance, hydrogels, known for their stretchability and conductivity due to their high water content and presence of small molecules, have been employed as substrates for wearable strain sensors by monitoring changes in resistance [170, 262]. Wang et al. [263] demonstrated a ternary polymer composite consisting of polyaniline (PANI), poly(acrylic acid), and phytic acid, exhibiting high stretchability ($\approx$500%) and approximately 99% self-healing efficiency within 24 h, facilitated by dynamic hydrogen bonding and electro-elastic interactions. However, self-healing hydrogels face challenges related to conductivity, toughness, water sensitivity, and low temperature resilience.

Several efforts have been made to enhance toughness [200], conductivity [264], and water insensitivity [169, 265] of hydrogels. For improving conductivity, graphene oxide (GO) sheets were combined with poly(sodium 4-vinyl-benzenesulfonate-*co*-*N*-(2-(methacryloyloxy)ethyl)-*N*,*N*-dimethylbutan-1-aminium bromide), resulting in high ionic conductivity (10.5 mS dm^{-1}) and efficient self-healing capabilities [264]. Additionally, water-insensitive elastomers were developed by Kang et al. [169], exhibiting unprecedented self-healing efficiency in both dry and wet conditions. Further research exploring the self-healing chemistry of reconfigurable electrode components has shown promise for fully healable multifunctional electronic devices, facilitating the development of integrated self-healing e-skin [259]. There, integrating conductive nanostructures, such as CNTs, into a dynamically cross-linked polymer network, PDMS–$MPU_{0.4}$–$IU_{0.6}$, enabled the restoration of electrical conductivity following cutting and damage within 12 h. These materials exhibit stretchability (up to 50% strain without crack propagation) and self-healing properties, making them suitable for use as electrodes in active electronic components. For instance, they were utilized in creating a fully healable multifunctional electronic skin system comprising a strain monitor, ECG sensor, and light-emitting capacitor display array. This system incorporates dielectric materials, AgNW networks, and Cu-doped zinc sulfide (ZnS) particles. For further information on self-healing electronics, the readers may turn to Refs. [255, 266].

4.3 Biological Properties

Biocompatibility

Ensuring the safety of wearable devices, both epidermal and implantable, is paramount, as they must not elicit any adverse reactions such as inflammation, irritation, foreign body rejection, or immune responses (e.g., irritation, allergy). Such reactions can range in severity, from minor immune responses characterized by the accumulation of defensive cells around the implant to more serious issues that can interfere with the proper functioning of nearby organs [123]. Therefore, biocompatibility (i.e., the ability of a material to interact appropriately with the host in a specific situation [267]) is a fundamental consideration for safety and comfort of the user. Intrinsic biocompatibility simplifies processing steps and eliminates the need for complex multilayered structures, addressing detachment issues commonly associated with wearable devices [268]. Organic thin-film transistors, for instance, leverage their inherent biocompatibility and ability to amplify vital signals from the body, bridging the gap between materials that are not fully biocompatible and wearable [268]. Ionic conducting polymers such as hydrogels and nanofibrils offer biocompatibility and biodegradability, expanding their applications in wearable health monitors. For example, tough conductive hydrogels composed of interpenetrating networks of PANI and poly(acrylamide-co-hydroxyethyl methacrylate) have been developed for the recurrent and sensitive detection of both large strains and subtle vibrations [269]. Other biomimicking nanofibrillar polymers, including silk, chitin, and cellulose, that exhibit highly ordered structures formed through the self-assembly of repetitive building blocks, stabilized by non-covalent interactions (i.e., resembles natural bio-repetitive sequences) are used in fabrication of wearable device due to their biocompatibility [173]. Recently, bacterial cellulose (BC), a flexible, biocompatible, and electrically conductive material, has been utilized in wearable health monitors owing to its unique nanochannel structure and high ionic conductivity (1.0×10^{-3} S cm^{-1}) [270]. Additionally, replacing conventionally used toxic materials (e.g., lead zirconate titanate) with biocompatible alternatives, such as group III nitride thin films, in piezoelectric generators enhances safety and performance [271].

Further, to reduce unwanted biomolecular interactions and improve biocompatibility, meticulous engineering of the chemical properties of the constituent materials, especially their surface chemistry, plays a vital role [272]. For example, to minimize protein interaction, researchers are focusing on the development of wearable electronic surfaces with uniform functional groups. It has been observed that the surfaces containing carboxyl (–COOH) or fluorocarbon ($-CF_x$) functional groups usually attenuate inflammatory/fibrotic reactions and cellular infiltration, while surfaces with amine ($-NH_2$) functional groups induce acute inflammatory response and cellular infiltration and promote fibrotic capsule formation [272]. Controlling the ion release and corrosion is another important element in reducing toxicity of the constituent materials [273].

Despite these advancements, regulatory testing primarily focuses on materials rather than fully assembled devices with passive and active components. Encapsulating strategies have been introduced to cover entire devices with biocompatible materials, ensuring both device and material compatibility with biological systems. Examples include LM/Ecoflex [197], LM/PU [274], or WFHE/PDMS [275]. In summary, the ultimate goal for the current wearable electronics is to use soft inherently bio-adhesive materials for gel-free attachment to the skin avoiding toxic or injurious effects that could trigger immune responses [94].

Bioresorbability

Most of the wearable devices are designed and manufactured with the aim of lasting longer and remaining functional for an indefinite amount of time [123]. However, in many instances, the presence of foreign devices in the body is neither desirable nor necessary. For example, cardiac pacemakers are intended to last lifelong, while monitoring of processes like wound healing, infection development, immune response, and tumor progression do not merit a permanent implant. The need for finite use wearables and implants has led to the development of bioresorbable electronics, i.e., a subset of transient electronics that can completely dissolute, disintegrate, or decompose to environmentally and biologically benign byproducts in aqueous and physiological milieu within a biologically relevant timeframe [276–278]. This kind of biodegradable electronics has two major benefits. First, they are typically biocompatible and safe for human use, as they can be metabolized in the body without triggering harmful physiological reactions. As a result, it eliminates the necessity for follow-up procedures and the potential risks linked with long-term implants, such as bacterial growth and ongoing device malfunctions and upkeep [279]. Second, they represent the most feasible and optimal solution to tackle the problem of unmanageable electronic waste (i.e., e-waste) [278].

The concept of "bioresorbable electronics" emerged in 2009 [280], where silk was utilized as a substrate for traditional silicon electronics, incorporating insoluble elements like gold and polyimide, providing both bioresorbability and conformal contact with the curved surfaces of target tissues and organs. In 2012, the first fully bioresorbable silicon electronic system was achieved [281], showcasing the bioresorbability of thin-film magnesium (Mg), magnesium oxide (MgO), and silicon nanomembranes (Si NMs) in a silicon-based complementary metal oxide semiconductor (CMOS) technology. Together with integrated sensors, actuators, power supply systems, and wireless control strategies, this implantable transient device has shown programmable nonantibiotic bactericidal effect. Subsequent research has explored a broader array of materials, including biodegradable polymers [282], dissolvable metals [283, 284], semiconductors [285–287], and dielectrics [288, 289], leading to the development of a wide range of bioresorbable electronic devices with diverse applications. The dissolution rate, which indicates the degree of bioresorbability, is expressed in terms of $\mu m\ h^{-1}$ or $mg\ h^{-1}$ in a representative biofluid (e.g., Hank's solution) [290] or deionized water at room temperature. This rate is influenced by the morphology and chemistry of materials, as well as the composition and temperature of the solution. Initial investigations in chemistry identified various bioresorbable materials suitable for substrates, encapsulation layers, and dielectrics, as well as providing multiple choices in organic and bio-organic semiconductors [290]. Bioresorbable materials can broadly be categorized into nanomembranes or porous silicon [106], metals (e.g., Mg, Zn, Fe, Tn, Mo, and their alloys), inorganic compounds (e.g., SiO_2, Si_3N_4, MgO), synthetic polymers (e.g., PVA, PVP, PLA, PLGA, polycaprolactone, poly(1,8-octanediol-*co*-citrate)), and natural/nature-derived polymers (e.g., silk, cellulose, sodium alginate) [110, 111]. Bioresorbable materials are currently used in the clinic for sutures [291], matrices for tissue engineering, carriers for drug delivery and release [292, 293], intravascular stents [294, 295], and plate–screw systems [296]. Over the past decade, researchers have created a range of bioresorbable electronic devices, including flexible CMOS circuits, wireless temperature sensors, and transient pressure sensors, utilizing materials that decompose or dissolve under controlled conditions. For example, Lei et al. [297] created fully disintegrable CMOS flexible circuits by integrating conjugated polymers with reversible imine bonds, biodegradable cellulose substrates, and Al_2O_3 and Fe electrodes. The degradation rate was controlled using decomposable encapsulation materials. This technology has been applied to real-time brain activity mapping by implanting silicon-based transient electronic devices. These devices sense intracranial pressure, temperature, fluid flow, and more [111, 298]. Additionally, silicon-based FETs offer the potential to monitor single molecule presence and kinetics, enabling online tracking of specific markers like inflammation or infections and predicting immune rejection and disease progression [299–301]. Further, Salvatore et al. [302] designed a wireless flexible temperature sensor using a bioresorbable magnesium electrode encapsulated by thin layers of Si_3N_4, SiO_2, and Ecoflex elastomer. Recently, a wearable transient pressure sensor was developed using porous MXene nanosheet paper sandwiched between biodegradable polylactic acid layers, exhibiting repeatable low detection limits (10.2 Pa) and power consumption [303]. Besides wearables, implantable devices represent an unquestionably compelling domain that leverages the benefits of biocompatibility. For instance, highly elastomeric and stretchable poly(citric acid-octanediol-PEG)-graphene (PCEG) nanocomposites with controlled biodegradation and electronic conductivity have shown to enhance myogenic differentiation and skeletal muscle regeneration [172]. Through the utilization of silicon nanomembranes, it becomes feasible to incorporate active, biodegradable components into fully degradable devices without compromising their functionality or confining the components to passive elements [281].

When designing a bioresorbable device, the initial consideration should be the intended duration of dissolution, which hinges on whether the device is envisioned for long-term or short-term monitoring. Regulation over dissolution rates is attained by selecting suitable materials for the device. The constituents of devices, including conductors, semiconductors, dielectrics, and substrate and encapsulation layers, are usually composed of different materials with varying degradation properties [123]. The substrate layer of bioresorbable devices can be fabricated from natural or synthetic polymers with tunable degradation properties. For instance, natural silk fibroin protein can serve as a substrate, breaking down into natural amino acids via enzymatic action in the body [304, 305]. Other alternatives such as synthetic polymers like poly(lactic-*co*-glycolic acid) (PLGA) degrade into their constituent monomers, glycolic and lactic acid [306]. Semiconducting and dielectric components such as silicon and silicon dioxide degrade into nontoxic $Si(OH)_4^-$ when exposed to physiological milieu [285]. Although the dissolution rate depends on thickness, the process typically occurs only on the surface of the layer and is thus limited to porous nanomembranes. Various metal conductors can be chosen based on the desired device lifespan; zinc and magnesium dissolve quickly [305], while tungsten, molybdenum, and iron have slower dissolution rates [304]. Although the degradation rates of conducting and semiconducting materials can be adjusted independently, the substrate layer often determines the device's lifespan since it forms the outermost layer [123].

4.4 Electrical Properties

Conductivity

The primary focus in the development of wearable flexible electronics revolves around enhancing their sensing capabilities and overall performance. It's crucial that these devices maintain reliable electrical conductivity even under high strain conditions over time, while seamlessly detecting vital signals without degradation from external factors, and show programmable change by structural deformation (i.e., a gauge factor) [10]. A major challenge in achieving this lies in thwarting the unpredictable fluctuations in mechanical and chemical properties associated with the irrevocable collapse of the nanocomposite structure. In this section, we delve into the electrical properties of recently developed materials and structures recognized for their potential as electrodes with high flexibility, stability, and conductivity. Numerous approaches have been taken by the researchers to design electronics that can bend and endure repeated high strains [307]. The use of conductive organic polymers like polypyrrole [308], polyaniline [309], and poly(3,4-ethylenedioxythiophene) (PEDOT) [310] as stretchable electrodes is the notable of them (Fig. 3). Due to biocompatible nature, these materials can be used in larger surface area resulting in greater charge transfer [311].

Liquid metals (LMs) have garnered significant interest due to their intrinsic liquid-like properties, high conductivity (even under strain), and scalable printability. Examples include screen-printed gallium (Ga) alloys (electrical conductivity ~3.8×10^6 S m^{-1} at strain 1000%) [274] and direct laser-written eutectic gallium–indium (GaIn) alloys (~5.7×10^5 S m^{-1} at strain >100%) [198], both exhibiting exceptional conductivity and strain resistance. Composites comprising of various functional nanomaterials and stretchable polymers have also demonstrated a wide range of conductivities depending on their architectures. For instance, kirigami-

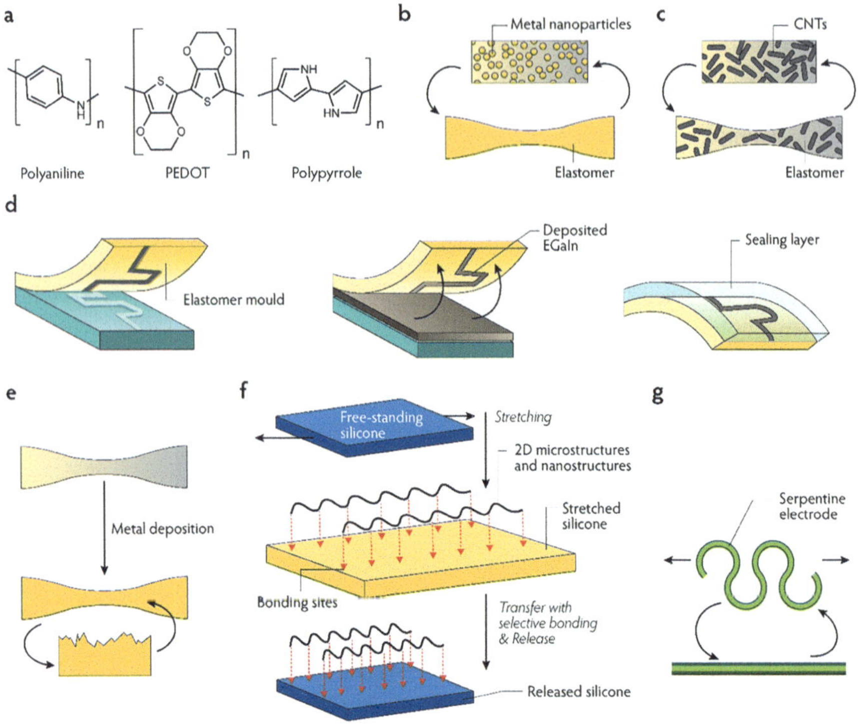

Fig. 3 Assembly of conductive and soft materials into complex three-dimensional architecture for wearable flexible electronic device fabrication. (**a**) Example of conductive polymers that can undergo deformation while retaining their conductivity. (**b**) Nanoparticles can be deposited onto or inside an elastomer prior to solidification, thereby making it conductive when the percolation threshold is surpassed. (**c**) Carbon nanotubes (CNTs) can be incorporated inside or on top of an elastomer. Due to high aspect ratio, nanotubes have the capability to bridge minor gaps formed during stretching. (**d**) Liquid metals can flow within microfluidic channels inside an elastomer, e.g., polydimethylsiloxane (PDMS). A eutectic gallium–indium (EGaIn) binary alloy is introduced into these channels and subsequently sealed with an additional layer of the elastomer. (**e**) Buckling of a metal conductor can be initiated by depositing it onto a pre-stretched elastomer. Upon relieving the strain, buckling of the metal layer occurs. Subsequently, the elastomer can be stretched and released while still maintaining conductivity. (**f**) 3D architectures can be fabricated by transfer printing a 2D structure onto a pre-stretched elastomer and defining specific binding sites. Upon relieving the strain from the elastomer, buckling of the structure occurs between the binding sites, yielding intricate architectures based on the chosen binding sites and 2D layout. (**g**) Fabrication of electrodes with specific architectures, i.e., a serpentine configuration enables deformation without triggering stress. *Abbreviation*: *PEDOT* poly(3,4-ethylenedioxythiophene). (Adopted with permission from Ref. [123], Springer Nature Ltd. Panels **d** and **f** are redrawn with permission from Ref. [312], Wiley-VCH, and Ref. [313], AAAS, respectively)

patterned AgNWs/parylene hybrid electrodes (sheet resistance 2.24 Ω sq^{-1}) have shown stability as wearable stretchable humidity sensors, due to a linear correlation between conductance and humidity [107]. Carbon-based composites, especially those incorporating graphene nanosheets, have shown enhanced conductivity compared to conventional materials. For example, due to its excellent electrical conductivity and transparency, graphene has been utilized alongside copper as a wearable lens, demonstrating impressive sheet resistances ranging from 850 to 1520 $\Omega\ sq^{-1}$ [314].

Conducting polymers, although inherently flexible, typically exhibit lower conductivity (resistivity 10^{-1} to 10^{4} Ω cm) compared to metals (bulk resistivity 10^{-6} Ω cm) [307, 315]. Therefore, to further enhance conductivity, novel metals or carbon-based nanoparticles can be incorporated to improve electrical transport, especially at interfaces [316]. For example, the use of a freestanding metal mesh for foldable supercapacitor electrodes demonstrated high conductivity (~4.0×10^{5} S m^{-1}) and areal capacitance (268 mF cm^{-2}), along with long cycling life (capacitive retention of 90.1%; 10,000 cycles) [317]. Strategies such as wrinkled structures [84] and doping [318] of conducting polymers (i.e., PEDOT:PSS) have been explored to increase conductivity for applications like wearable triboelectric nanogenerators. Another example is ionic conductive hydrogels. In contrast to liquid electrolytes, ionic conductive hydrogels integrate free ions with an elastic polymer network resulting in materials that are optically transparent, possess softness similar to tissues, and offer high stretchability [319]. The diffusion and transfer of ions can be efficiently managed by adjusting the size and arrangement of pores within the hydrogel. Additionally, by altering the chemical composition and structural design of hydrogels, it is possible to regulate the selectivity and sensitivity of ions that offer benefits for seamlessly transitioning between sensors and tissues at the bioelectronic interfaces [38]. The inclusion of salts like NaCl, LiCl, KCl, $CaCl_2$, and $FeCl_3$ in polymer solutions leads to the ionization of metal ions, allowing them to move in specific directions within the inner channels of the hydrogel when an external voltage is applied, thereby enhancing the conductivity of the hydrogel [320]. As the hydrogel deforms due to external forces, the resistance changes accordingly, and the speed of ion movement is affected by the electrostatic interaction within the hydrogel matrix [321]. Table 2 summarizes a representative list of wearable sensors made of conductive hydrogels. Biodegradable materials like silk fibroin have also shown promise, offering both low sheet resistance (10.5 $\Omega\ sq^{-1}$) and good extensibility (~60% strain), making them suitable for wearable applications [176].

Gauge Factor

Strain and pressure sensors are the most common wearable electronic devices for monitoring physiological signals. These sensors play a crucial role in detecting a user's vital signs such as heart rate, blood pressure, body motion, and other signals originating from the body. Regardless of whether the deformation is minimal or substantial, they typically operate based on resistive or capacitive mechanisms and are evaluated quantitatively using a parameter known as the gauge factor (GF). The GF is defined as the ratio between resistance and induced strain, providing

Table 2 Properties of different types of composite hydrogels and their application in wearable electronics

Type	Materials	NW struct.	Conductivity ($S\ m^{-1}$)	Toughness ($J\ m^{-3}$)	Advantage	Application	References
Conductive polymer	PEDOT:PSS	SN	4×10^4	–	Pure PEDOT:PSS hydrogels	Sensors	[322]
	PAA/$BaFe_{12}O_{19}$	SN	40	–	Nano barium ferrite as the cross-linker	Strain sensors	[323]
	PAAm/ALG/PPy	MN	2.16	5.5×10^6	Anti-swelling	Underwater sensors	[324]
	P(AAm-*co*-DMC)/PANI	DN	0.43	5.0×10^6	Semi-interpenetrating network endows with mechanical robustness	Strain sensor, e-skin	[325]
	HAPAA/PANI/glycerol	DN	2.23	1.1×10^7	Antifreezing, water-locking, and anti-heating	Pressure and strain sensor, e-skin	[326]
Metallic salts	Lysine/PAAm/LiCl	SN	73.6	–	Amino acid–mediated adhesion, transparency	Wearable strain sensors	[327]
	Hydroxypropyl cellulose/PVA	DN	3.4	5.85×10^6	3D printability	Artificial tissue	[328]
Metal material	P(AAm-*co*-MAAc)/EGaIn	DN	2.0×10^5		Self-shaping ability	Implantable electronics	[329]
	LS/PVA/Ag NPs	DN	7.1	5.07×10^7	Antibacterial, ultrahigh toughness	Wearable strain sensors	[330]
Micro–/nanomaterials	Agar/Borax/MXene	SN	81.4	–	Flexible, green sensor	Wearable strain sensor	[331]
	TA/SF/rGO	DN	1	13.9×10^6	Elongation	Bionics, soft robotics	[332]
	PVA-CA-MXene	DN	–	9.7×10^3	Oxidation resistance	Strain sensor	[333]
	ANFs/PVA	DN	1.66×10^4	–	Excellent Joule heating and electromagnetic interference shielding	Soft robots, artificial skin, bioelectronics	[334]

Abbreviations: *NW Struct.* network structure, *SN* single network, *DN* double network, *MN* multi-network, *PEDOT:PSS* poly(3,4-ethylenedioxythiophene) polystyrene sulfonate, *PAA* polyacrylic acid, *PAAm* polyacrylamide, *ALG* alginate, *PPy* polypyrrole, *DMC* [2-(methacryloyloxy)ethyl]trimethylammonium chloride, *PANI* polyaniline, *HAPAA* hydrophobic association poly(acrylic acid), *PVA* poly(vinyl alcohol), *LS* sodium lignosulfonate, *EGaIn* eutectic gallium–indium alloy, *MAAc* methacrylic acid, *NPs* nanoparticles, *MXene* 2D titanium carbide $Ni_3C_2T_x$, *CA* catechol, *SF* silk fibroin, *TA* tannic acid, *rGO* reduced graphene oxide, *ANFs* aramid nanofibers insight into the relationship between mechanical and electrical properties crucial for wearable sensor performance. The GF associated with piezoresistive mechanism can be expressed by the following equation:

$$\mathrm{GF} = \frac{\Delta R/R}{\varepsilon} = (1 + 2\nu) + \frac{\Delta\rho/\rho}{\varepsilon} \tag{4}$$

where R and ΔR are the initial resistance and the change in resistance, respectively, ε is the applied strain, ν is Poisson's ratio, and ρ and $\Delta\rho$ are the initial resistivity and the change in resistivity, respectively. Considering that the resistivity effect ($\Delta\rho/\rho$) in semiconductors is significantly greater than in metals (~100 times higher), semiconducting materials hold greater potential compared to thin-film metals (GF $\approx$ 2) [113]. Therefore, achieving both high stretchability and sensitivity simultaneously presents a significant challenge, as they often operate in different strain ranges. High stretchability demands that the wearable device maintains its structural integrity under large strains while high sensitivity requires the ability to detect even the smallest strains. Finding a balance between these conflicting requirements remains a key focus for researchers, as achieving both high stretchability ($\varepsilon \geq 100\%$) and high sensitivity (GF $\geq$ 50) is currently challenging.

Composite materials comprising conductive components within a soft substrate are commonly used for large pressure or strain applications. In this scenario, the alteration in resistance is primarily impacted by the carrier conduction occurring between the materials rather than the inherent properties of the material itself. Nanomaterial-based sensors, such as those composed of gold nanowires (AuNWs) and small-sized silver nanowires (AgNWs), have shown high sensitivity and cyclic stability [222]. Binary nanocomposites incorporating strain-sensitive materials, particularly 2D materials like MXene, have also demonstrated impressive performance in terms of both strain range and gauge factor [335, 336].

Micro- and nanoscale engineering of soft sensing materials, often inspired by biological structures, can significantly enhance the gauge factor and repeatability of strain sensors. Architectures like interlocked arrays of polymeric nanofibers or bioinspired structures resembling epidermis–dermis layers [337] or sea urchin-shaped microparticles [338] have shown incredible promise in this regard. Additionally, fiber-shaped micrometer fibers modified with intrinsic microbeads (250–550 μm) have been developed, demonstrating high scalability (5000 strain cycles at 30% strain) and sensitivity (GF 9–25, 120% linear strain) for stretchable strain-sensing applications [206].

5 Novel Materials for Soft Bioelectronics

5.1 *Liquid Metal*

Liquid metal, which remains in a liquid state at room temperature, offers high conductivity and flexibility, making it a promising material for flexible and stretchable electrodes in soft bioelectronics. Liquid metal alloys are often used inside microfluidic channels, such as those composed of PDMS. Proper encapsulation methods are crucial to hold the liquid metal within microfluidic channels. Despite the complex fabrication processes and potential leakage risks, the devices fabricated using liquid metal exhibit both high stretchability and conductivity. For instance, Guo and Liu [339] developed an electrode using a gallium–indium (GaIn) alloy encapsulated in a PDMS elastomer (250 μm) that demonstrated soft mechanical properties (elastic modulus 1.05 MPa) and stable conductance even after repeated stretching cycles (~4% change in resistance after 7000 cycles of stretching at 15 applied strain). The electrode was able to stimulate the sciatic nerve of a bullfrog to make its lower limb move, without any observable cytotoxicity. Similarly, a microfluidic device featuring a eutectic gallium–indium (EGaIn) conductor (melting temperature 15.5 °C and low resistivity of $\sim 30 \times 10^{-6}$ Ω cm) can undergo stretching of up to 100%, with a resistance change of 0.24 Ω, and can be bent 180°, with a resistance change of 0.02 Ω [340–343]. In another approach, EGaIn wires positioned within a hollow-core poly(styrene-b-(ethylene-*co*-butylene)-b-styrene) elastomer retain their conductivity even when subjected to strains of up to 700%. This is attributed to the combination of the high conductivity of the liquid metal and the excellent elasticity of the elastomer. Since the metal is in a liquid state, it doesn't exhibit the stress limitations typical of solid conductors [342, 344]. Besides exhibiting minimal resistance when subjected to repetitive strain, EGaIn is also nontoxic, suggesting its potential compatibility in physiological environments [345, 346]. Traditionally, liquid metal–based devices have utilized a single liquid component to avoid mixing at their junction. Ota et al. [347] introduced a method to create a heterojunction between liquid metal and an ionic liquid within PDMS microchannels. This heterojunction served as a sensor for detecting temperature, humidity, and oxygen concentrations, enduring significant stretching (up to 90% strain) and twisting (up to 60°). Liquid metal also exhibits a phase-shifting property that can be exploited to fabricate intricate 3D structures. By dispensing liquid-phase EGaIn onto a cold plate and rapidly solidifying it, followed by encapsulation with an elastomer precursor, a solid-state 3D EGaIn structure is formed [348]. After the elastomer fully cross-links, the solid-state EGaIn structure is melted back into its liquid state and encapsulated by the elastomer again. This process results in highly conductive ($\sim$34,000 S cm^{-1}) soft 3D structures that can withstand deformations. Despite its softness and high conductivity, the application of liquid metal in soft bioelectronics is somewhat limited due to the necessity of securely entrapping it within elastomeric channels, leading to relatively thick device geometries.

5.2 Conductive Nanomaterials Coated on Soft Substrates

Another candidate for soft bioelectronics can be conductive nanomaterial network patterned on stretchable substrates [349, 350]. The percolated nanomaterial network creates pathways for electrical conduction, while the elastomeric substrate provides the necessary stretchability [351]. Materials like carbon nanotubes (CNTs) or metal nanowires can be applied onto the elastomer through surface treatments of both the nanomaterials and the elastomer substrates [352]. For instance, Lee et al. [353] demonstrated this by patterning a thin layer (80–330 nm) of CNTs on an O_2-treated elastomer (a mixture of Ecoflex and PDMS), which could withstand over 100% stretching.

Hydrogels, with their lower modulus and better tissue compatibility compared to elastomers, can also serve as soft substrates. For example, Gilshteyn et al. [354] created a conductive, stretchable, and transparent wearable electrode by depositing a single-walled CNT (SWCNT) film onto a tough hydrogel. The SWCNT–hydrogel film could serve as soft bioelectrodes owing to its resistance of $\sim$100 Ω cm^{-1}, transparency of 80% (at 550 nm), and stretchability of $\sim$100%. Woven fabrics provide another form of soft substrate, allowing for the creation of electrically conductive fabric that is lightweight, breathable, flexible, and stretchable. Various conductive materials, including metal films [355], carbon nanomaterials [356–358], metal nanomaterials [359], and conductive polymers [360, 361], can be coated onto these fabrics. For instance, silver nanowires (AgNWs) and polyurethane (PU) layers were coated on a carbon fiber fabric, resulting in a conductive fabric with high conductivity and durability [362]. This conductive fabric endured bending tests and was used to fabricate electromagnetic interference shielding devices. An alternative approach is to construct conductive fabric by weaving conductive fibers instead of depositing conductive films on insulative fabrics. Zhao et al. [363] coated polyacrylonitrile (PAN) fibers with copper (Cu) to create conductive fibers, which were then woven into a fabric (conductivity $\sim$0.15 Ω cm^{-1}). Additionally, parylene-coated insulative PAN fibers were used to fabricate the conductive fabric via conventional textile manufacturing processes. This conductive fabric showed customizable properties and was utilized in the development of a triboelectric wearable pressure sensor.

5.3 Conductive Polymer

Conjugated conductive polymers are composed of backbones with alternating double and single bonds that facilitate electron flow along the π bonds throughout the polymer chain [364]. When in contact with electrolytic biofluids, conductive polymers undergo ion/electron exchanges, making them suitable for interfacing with ion-rich tissues. Several organic polymers, such as polypyrrole [308], polyaniline (PANI) [365, 366], and poly(3,4-ethylenedioxythiophene) (PEDOT) [310], along with their modifications, have been investigated for their use as stretchable electrodes (Fig. 3) [311]. Their primary benefits include compatibility with biological systems and the enhancement of charge transfer through an increase in surface area [311]. Compared to traditional metals, conductive polymers offer superior electrochemical performance, including a high charge injection limit, high charge injection capacity, and low impedance, which are crucial for high-quality electrocorticography (ECoG) recording and stimulation. For instance, Ganji et al. [367, 368] conducted a study comparing the electrochemical properties of gold (Au), platinum (Pt), and PEDOT:PSS electrodes for ECoG recording. PEDOT:PSS electrode exhibited higher faradaic charge transfer and capacitive charge coupling than the metal electrodes, resulting in lower impedance and higher signal-to-noise ratio (SNR) [367]. Additionally, the charge injection capacity of PEDOT:PSS-coated electrodes was significantly higher (2.71 mC cm^{-2}, 3.2 times) compared to bare Pt electrodes, further enhancing their performance in neural recording and modulation in human subject [369]. The bulk resistivity of conducting polymers typically ranges between 10^{-1} and 10^4 Ω cm [307, 315] (as compared to metals ~10^6 Ω cm), which may be inconsequential for applications involving tissue data collection or actuation. Therefore, to address the inherent low conductivity and stretchability, conductive polymers are combined with nanomaterials, hydrogels, or ionic liquids [322, 370, 371]. In one approach, mixing PEDOT:PSS with an additive, i.e., a stretchable conductive enhancer (STEC enhancer that includes ionic compounds and liquids), reinforces both the stretchability and conductivity by softening polymer chains and forming a conductive network within the polymer matrix [371]. The resulting film exhibits high conductivity even under significant strain (>3100 > 4100 S cm^{-1} at 0% strain and a conductivity of >4100 S cm^{-1} at 100% strain) and maintains its conductivity after repeated stretching cycles (1000 cycles). Moreover, the material can be patterned to fabricate devices with complex geometries. Despite their remarkable electrochemical properties, conductive polymers still face challenges such as low conductivity and high Young's modulus. Stretchable electrodes can also be created by embedding metal-based or carbon-based nanoparticles into elastomers like polydimethylsiloxane (PDMS) or styrene-butadiene-styrene (SBS) [316]. Conductivity is achieved when the concentration of conductive particles within the elastomer exceeds the percolation threshold [372]. Stretching often leads to electrical failure due to material cracking, which isolates the conductive components. Materials with high aspect ratios, such as carbon nanotubes, can bridge these fractured regions [373]. However, their suitability for biomedical devices may be questioned due to concerns about biocompatibility [374–377].

5.4 Conductive Hydrogel Composites

Hydrogels are three-dimensional (3D) cross-linked porous polymer networks with high water content. With tissue-like mechanical features, biocompatibility, and unique electrical and chemical properties, hydrogels are highly appealing materials for soft bioelectronics by virtue of their bioinspired 3D structure, high surface area, high water content, tissue resemblance, stimuli-responsiveness, and tunable conductive channels to provide a seamless interface between the electronics and the biology [378–384]. However, conventional hydrogels usually do not possess electronic conductivity, and their ionic conductivity in physiological conditions is notably low, often ranging from 6 to 9 orders of magnitude lower than that of metals [385]. By incorporating conductive filler materials (electron conductors and/or ions) into hydrogel matrices, conductive hydrogel composites can be created [386, 387]. These composites benefit from the ion-rich nature of hydrogels and the percolation network formed by the conductive fillers, leading to properties such as high conductivity, low impedance, and high charge injection capacity [388]. Over the last few decades, several electron-conducting and/or ion-conducting hydrogels have been designed for potential applications in areas like tissue engineering, controlled drug delivery, diagnostics, e-skins, soft robotics, sensors, and actuators (summarized in Table 3).

Electron-conducting materials used in hydrogel fabrication are typically divided into three main categories [433]: (1) metallic nanoparticles/nanowires (such as Au, Ag, Cu), (2) carbon-based materials (including carbon grease, carbon nanoparticles/nanowires/CNTs, graphene, graphene oxide [GO], reduced GO [rGO] nanosheets), and (3) conducting polymers (like polyaniline, polypyrrole, poly(phenylene vinylene), polythiophene, poly(3,4-ethylenedioxythiophene): poly(styrene sulfonate) [PEDOT:PSS]). The conventional method for preparing electron-conducting hydrogels involves directly doping or blending these electron-conducting components into hydrogel matrices. However, this simple mixing approach often encounters challenges due to the high specific surface energy of nanomaterials, resulting in heterogeneous phase separation or filler aggregation, which can lead to diminished electrical conductivity and compromised mechanical properties [378, 387, 434]. To address this issue, rational design of polymer networks could be employed to optimize the distribution of the dispersed phase. Examples include PEDOT:PSS nanofibril hydrogels developed by mixing dimethyl sulfoxide (DMSO) with an aqueous PEDOT:PSS solution, followed by dry annealing and rehydration [322]. With an optimized DMSO content (13 vol.%), the hydrogel exhibited high electrical conductivity ($\sim$40 S cm^{-1} in deionized water), high stretchability (>35% strain), and a low Young's modulus ($\sim$2 MPa). In a separate study, Nam et al. [435] fabricated a neural interfacing device (injectable implant in the intracortical area for recording local field potential of epileptic events) using a hydrogel based on the supramolecular β-peptide (betaVhex) assembled with CNTs to form biocompatible and conductive electrical percolation networks (betaVhex/CNT complex) inside the hydrogel matrix. The betaVhex/CNT hydrogel nanocomposite showed a modulus of $\sim$1500 Pa and conductivity of 0.1345 S m^{-1}. Further, Lim et al. [436] developed a composite of silver nanowires (AgNWs) embedded in alginate hydrogel. The alginate hydrogel allowed for cross-linking in the presence of AgNWs without requiring covalent bond formation. The composite film, formed by drying the hydrogel–AgNW solution, was patterned using laser cutting. With Young's modulus of 15.5 MPa, the AgNW–alginate

Table 3 Typical examples of different classes of conducting hydrogels and their applications

Type	Conductive component	Materials	Conductivity ($S\ m^{-1}$)	Optical property	Applications
Electron-CHs	Metallic NPs/NWs	Ag [389]/Au [390]/Cu [391] NPs	$0.057–2.2 \times 10^7$	Transparent or semitransparent	Biosensor drug delivery, tissue engineering
	Carbon-based materials	CNTs [354, 380, 388]	0.072–8.2	Mostly black or semitransparent	Strain sensor, tissue engineering, biosensor, supercapacitor, drug delivery
		GO/rGO/graphene [392–394]	$1.56 \times 10^{-5}–0.66 \times 10^3$		
	Conducting polymers	PANI [395, 396]	$1 \times 10^{-3}–0.04 \times 10^3$	Mostly black and rarely transparent	Tissue engineering, biosensor, drug delivery, supercapacitor, bioelectrode, strain sensor
		PPy [397]	0.3–70		
		PEDOT:PSS [322, 398]	$1.5 \times 10^{-3}–4.0 \times 10^3$		
	Hybrid	Pt [399]/Ag [400]/(GO + rGO) [401]/SWCNTs [402] + PANI	13.64–21.0	Usually black or semitransparent	Biosensor, energy storage, supercapacitor, tissue engineering, pressure sensor, strain sensor
		Au [403]/(Fe + Co) [404]/Fe_3O_4 [405]/CNTs [406] + PPy	34.93		
		Graphene/GO + Ni/$Ni(OH)_2$ [407, 408]	–		
		GO+PEDOT:PSS [409]	1.25×10^3		
		GO + CuS + carbon dot [410]	–		

Ion-CHs	Acids	H_2SO_4 [411]/H_3PO_4 [412]	8.2	Transparent	Supercapacitor
	Metallic salts	LiCl [413]/Na^+ [385]/Ca^{2+} [414]/Al^{3+} [415]/Fe^{3+} [416]/Tb^{3+} [417]/K^+ [418]	1.25×10^{-3}–0.185×10^3	Transparent or semitransparent	Supercapacitor, nanogenerator, solar cell, actuator, electronic eel/fish, displays, touch panels, pressure sensor, strain sensor
		Na_2S + sulfur + NaOH [419]	11.53–31.35		
	Ionic liquids	1-Ethyl-3-methylimidazolium chloride [420]	0.04–2.32	Semitransparent	Supercapacitor
Electron and ion-CHs	Electron-conductive components and ions	Na^+ + Au NPs [421]	0.45–2.92	Usually black or opaque	Supercapacitor, biosensor, pressure sensor, battery, strain sensor, tissue engineering
		HCl [422]/$HClO_4$ [423]/(Na^+ + SWCNTs) [424] + PANI	–		
		H_2SO_4 [425]/Na^+ [426] + PEDOT	0.88×10^3		
		Na^+ + Ca^{2+} + SWCNTs [427]	–		
		Fe^{3+} + PPy [428]/rGO [429]	9.16×10^{-5}–7.0×10^3		
		Mg^{2+} + PEDOT:PSS + PPy [430]	–		
		Zn^{2+} [431]/CuPcTs [432] + PPy	0.012×10^3–0.78×10^3		

Abbreviations: *CHs* conductive hydrogels, *NPs* nanoparticles, *NW* nanowires, *GO* graphene oxide, *rGO* reduced graphene oxide, *CNTs* carbon nanotubes, *PEDOT:PSS* poly(3,4-ethylenedioxythiophene) polystyrene sulfonate, *SWCNTs* single-walled carbon nanotubes, *PPy* polypyrrole, *PANI* polyaniline hydrogel composite could stretch up to 30%. Notably, the nanocomposite exhibited extremely low impedance (around 8.5 Ω) across all frequency ranges (100–105 Hz). This versatile nanocomposite found applications in wearable devices like wearable antennas and supercapacitors.

Abundance of water within the porous structure of hydrogels facilitates the free movement of water molecules throughout the network. This high ion mobility in water enables the design of ion-conducting hydrogels, typically achieved by directly introducing free ions [92]. Materials capable of generating free ions in water fall into three categories [437]: (1) acids (such as HCl, H_2SO_4, H_3PO_4), (2) metallic salts (like $NaCl/Na_2SO_4$, KCl, LiCl, $LiClO_4$, $FeCl_3/FeNO_3$, $CaCO_3/CaCl_2$, $TbCl_3$, $AlCl_3$), and (3) ionic liquids (for instance, 1-ethyl-3-methylimidazolium chloride). Ion-conducting hydrogels offer several advantages compared to electron-conducting hydrogels, expanding their applications in flexible electronics. Firstly, unlike many electron-conducting hydrogels that tend to exhibit dark colors, ion-conducting hydrogels can be easily made transparent, making them suitable for applications requiring high transparency, such as see-through touch panels. Secondly, mobile ions in ion-conducting hydrogels can create electric double layers with surfaces rich or deficient in electrons, enabling them to serve as electrolytes in solid batteries and play a crucial role in actuators. Additionally, some ion-CHs possess water retention capabilities or antifreezing properties, enhancing the durability and adaptability of flexible electronics to various environments. Examples of ion-conducting hydrogels include a highly sensitive sensing array designed by Sun et al. [438] using CH/elastomer/CH as building blocks, which could detect both the location and the pressure (ranging from 0 to 40 kPa) of touch via recording capacitance variations of the dielectric layer. Liu et al. [439] devised a PAM hydrogel containing host–guest interactions between 1-benzyl-3-vinylimidazolium and cucurbit[8]uril as a pressure sensor, which has a stretchability of ~10,000% and can support objects with 2000 times weight of itself. Larson et al. [440] fabricated a stretchable multicolor display using photopatterning and transfer printing techniques. They achieved this by sandwiching an 8 × 8-pixel-patterned ZnS–silicone light-emitting layer between LiCl-PAM hydrogel electrode layers.

In broad sense, electron-conducting hydrogels are extensively used in bioengineering applications like biosensors, tissue engineering systems, and drug delivery devices due to their superior compatibility with biological tissues [437]. In contrast, ion-conducting hydrogels, characterized by ion gradients and transparency, are highly suitable for energy storage and conversion devices such as actuators and nanogenerators [437]. Additionally, their transparency is advantageous for see-through equipment like displays and touch panels.

5.5 *Elastomeric Nanocomposites*

Recently, elastomeric nanocomposites are developed by incorporating conductive nanomaterials into elastomers [441]. They offer both high stretchability and conductivity. The conductive nanomaterials create a 3D conductive percolation network within the elastomeric matrix, which remains intact and/or rearranged during mechanical deformation. Various types of conductive nanomaterials, such as graphene flakes [442], carbon nanotubes (CNTs) [443], metal nanoparticles [444, 445], and metal nanowires [446], have been explored as filler materials for these nanocomposites.

Carbon-based nanomaterials, including carbon blacks, CNTs, and graphene flakes, are favored as fillers due to their deformable mechanical properties, electrical conductivity, and ease of chemical functionalization. CNT–PDMS nanocomposites, for example, have been utilized in soft bioelectronics [447–450], with the conductivity of the composite increasing with the CNT ratio, for example, 3×10^{-5} S m^{-1} at 2 wt%, 0.015 S m^{-1} at 3 wt%, 0.38 S m^{-1} at 6 wt%, and 2.3 S m^{-1} at 10 wt% [447]. However, with increasing CNT ratio, Young's modulus also increased, from 3.4 MPa (pure PDMS) to 10 MPa (10 wt% CNTs in PDMS). Therefore, finding the right balance of CNT content is extremely important to maintain both conductivity and softness. The moderate conductivity of CNTs can further be enhanced by improving their dispersion. Ultrasonication is commonly used for this purpose [451]. Additionally, combining multiple types of carbon nanomaterials, such as graphene nanoflakes and CNTs, in PDMS has shown promising results, with enhanced conductivity (stretchability of 100% and a resistivity of ~100 Ω cm) compared to CNTs alone [103]. However, if the proportion of 2D carbon fillers to carbon nanotubes (CNTs) exceeds the optimized ratio, the synergistic effect diminishes.

Metal nanomaterials, including nanoparticles, nanoflakes, and nanowires, offer high conductivity among other fillers and are essential for achieving high-performance conductive nanocomposites. Strategies to maintain the percolation network of these fillers under mechanical deformation are crucial. Various approaches have been proposed to ensure the formation of dense percolation networks, including the use of two filler components, high aspect ratio fillers, and welding methods, for example, nanocomposites containing two kinds of conductive fillers [Ag nanoparticles (nm-scale) and Ag flakes (μm-scale)] in fluorine rubber with a surfactant [452]. These nanocomposites exhibit high conductivity (>4000 to 935 S m^{-1} at 400% strain) and elasticity. The nanocomposite could be printed on a textile, to be used as a wearable pressure and temperature sensor. Park et al. [316] developed a nanocomposite made of ultralong AgNWs and a poly(styrene-butadiene-styrene) elastomer. The nanocomposite showed a high conductivity of 11,210 S m^{-1}, and by making a mesh-type electrode array, it could be wrapped around the heart. As it shares mechanical characteristics similar to those of cardiac muscle, it can establish optimal contact with the pulsating heart without impeding the heart's natural systolic/diastolic motion. A recent strategy for incorporating metal atoms into elastomers involves thermal radiation–assisted metal deposition method [453]. This method enhances the electrode's surface area, resulting in low impedance and maintaining high conductivity even under stretching. Despite their high performance, elastomeric nanocomposites face challenges such as poor electrochemical properties due to the dense elastomer structure (that blocks penetration of ions and fluids) and low cellular affinity (due to hydrophobic nature) compared to hydrogels. Choosing the appropriate material depends on the specific application requirements.

6 Summary and Future Aspects

Advancements in soft wearable bioelectronics have revolutionized high-fidelity biosignal monitoring and real-time feedback therapy. These innovations hinge primarily on (1) designing of novel devices, (2) the integration of soft materials as constituents, and (3) advanced surface treatment methodologies. Flexible device designs facilitate seamless integration with target tissues. Particularly, the integration of soft materials enhances deformability of a device, which in turn promotes better interaction with biological tissues. Surface treatments further enhance device–tissue interfacing. Some soft bioelectronics even incorporate therapeutic functions like controlled drug release, ensuring long-term performance without adverse biological effects.

Despite significant progress, challenges persist. First, advanced fabrication techniques tailored for soft materials are needed, as conventional methods (e.g., microelectromechanical systems (MEMS) processes) often damage them. For example, majority of elastomers and hydrogels swell or degrade in the organic solvents and etchants that are typically used in MEMS processes, and high temperature methods (e.g., thermal evaporation for metal films) often lead to structural damage to soft materials. Second, composite materials with functional fillers face uniformity issues at small scales, necessitating further development of nanoscale materials and optimization for their uniform dispersion in the polymer matrix. Third, ensuring medical-grade biocompatibility is crucial, requiring extensive clinical trials to confirm safety. Additionally, soft encapsulation materials must resist degradation in the body. Forth, system optimization and device integration are also necessary for complex biomedical applications, requiring miniaturization and advanced packaging techniques.

The evolution of soft materials and innovative device design has propelled soft bioelectronics forward. Future advancements will require customization for specific applications and diseases, emphasizing collaboration between academia, industry, and clinicians. These efforts hold promise for transformative breakthroughs in clinical protocols and healthcare industries.

References

1. Rienzo, M.D., Meriggi, P., Rizzo, F., Castiglioni, P., Lombardi, C., Ferratini, M., Parati, G.: Textile technology for the vital signs monitoring in telemedicine and extreme environments. IEEE Trans. Inf. Technol. Biomed. **14**(3), 711–717 (2010)

2. Shuvo, I.I., Shah, A., Dagdeviren, C.: Electronic textile sensors for decoding vital body signals: state-of-the-art review on characterizations and recommendations. Adv. Intell. Syst. **4**(4), 2100223 (2022)
3. Yu, C., Yue, Z., Shi, M., Jiang, L., Chen, S., Yao, M., Yu, Q., Wu, X., Zhang, H., Yao, F., et al.: An intrapericardial injectable hydrogel patch for mechanical–electrical coupling with infarcted myocardium. ACS Nano. **16**(10), 16234–16248 (2022)
4. Hong, Y.J., Jeong, H., Cho, K.W., Lu, N., Kim, D.-H.: Wearable and implantable devices for cardiovascular healthcare: from monitoring to therapy based on flexible and stretchable electronics. Adv. Funct. Mater. **29**(19), 1808247 (2019)
5. Wang, L., Liu, Y., Ye, G., He, Y., Li, B., Guan, Y., Gong, B., Mequanint, K., Xing, M.M.Q., Qiu, X.: Injectable and conductive cardiac patches repair infarcted myocardium in rats and minipigs. Nat. Biomed. Eng. **5**(10), 1157–1173 (2021)
6. Roy, A., Samanta, S., Singha, K., Maity, P., Kumari, N., Ghosh, A., Dhara, S., Pal, S.: Development of a thermoresponsive polymeric composite film using cross-linked β-cyclodextrin embedded with carbon quantum dots as a transdermal drug carrier. ACS Appl. Bio Mater. **3**(5), 3285–3293 (2020)
7. Molina-Molina, A., Ruiz-Malagón, E.J., Carrillo-Pérez, F., Roche-Seruendo, L.E., Damas, M., Banos, O., García-Pinillos, F.: Validation of mDurance, a wearable surface electromyography system for muscle activity assessment. Front. Physiol. **11**, 606287 (2020)
8. Khokhlova, L., Belcastro, M., Torchia, P., O'Flynn, B., Tedesco, S.: Wearable textile-based device for human lower-limbs kinematics and muscle activity sensing. In: Wearables in Healthcare, pp. 70–81. Springer International Publishing, Cham (2021)
9. Ashcroft, K., Robinson, T., Condell, J., Penpraze, V., White, A., Bird, S.P.: An investigation of surface EMG shorts-derived training load during treadmill running. Sensors. **23**(15), 6998 (2023)
10. Lim, H.-R., Kim, H.S., Qazi, R., Kwon, Y.-T., Jeong, J.-W., Yeo, W.-H.: Advanced soft materials, sensor integrations, and applications of wearable flexible hybrid electronics in healthcare, energy, and environment. Adv. Mater. **32**(15), 1901924 (2020)
11. White, M.S., Kaltenbrunner, M., Głowacki, E.D., Gutnichenko, K., Kettlgruber, G., Graz, I., Aazou, S., Ulbricht, C., Egbe, D.A.M., Miron, M.C., et al.: Ultrathin, highly flexible and stretchable PLEDs. Nat. Photonics. **7**(10), 811–816 (2013)
12. Sekitani, T., Nakajima, H., Maeda, H., Fukushima, T., Aida, T., Hata, K., Someya, T.: Stretchable active-matrix organic light-emitting diode display using printable elastic conductors. Nat. Mater. **8**(6), 494–499 (2009)
13. Martinez, R.V., Branch, J.L., Fish, C.R., Jin, L., Shepherd, R.F., Nunes, R.M.D., Suo, Z., Whitesides, G.M.: Robotic tentacles with three-dimensional mobility based on flexible elastomers. Adv. Mater. **25**(2), 205–212 (2013)
14. Shepherd, R.F., Stokes, A.A., Freake, J., Barber, J., Snyder, P.W., Mazzeo, A.D., Cademartiri, L., Morin, S.A., Whitesides, G.M.: Using explosions to power a soft robot. Angew. Chem. Int. Ed. **52**(10), 2892–2896 (2013)
15. Roh, Y., Kim, M., Won, S.M., Lim, D., Hong, I., Lee, S., Kim, T., Kim, C., Lee, D., Im, S., et al.: Vital signal sensing and manipulation of a microscale organ with a multifunctional soft gripper. Sci. Robot. **6**(59), eabi6774 (2021)
16. Nicotra, M., Moncalero, M., Messori, M., Fabbri, E., Fiorini, M., Colonna, M.: Thermo-mechanical and impact properties of polymeric foams used for snow sports protective equipment. Proc. Eng. **72**, 678–683 (2014)
17. Wang, C., Hwang, D., Yu, Z., Takei, K., Park, J., Chen, T., Ma, B., Javey, A.: User-interactive electronic skin for instantaneous pressure visualization. Nat. Mater. **12**(10), 899–904 (2013)
18. Kim, D.-H., Lu, N., Ma, R., Kim, Y.-S., Kim, R.-H., Wang, S., Wu, J., Won, S.M., Tao, H., Islam, A., et al.: Epidermal electronics. Science. **333**(6044), 838–843 (2011)
19. Krupenkin, T., Taylor, J.A.: Reverse electrowetting as a new approach to high-power energy harvesting. Nat. Commun. **2**(1), 448 (2011)
20. Sue, C.-Y., Tsai, N.-C.: Human powered MEMS-based energy harvest devices. Appl. Energy. **93**, 390–403 (2012)

21. Viventi, J., Kim, D.-H., Vigeland, L., Frechette, E.S., Blanco, J.A., Kim, Y.-S., Avrin, A.E., Tiruvadi, V.R., Hwang, S.-W., Vanleer, A.C., et al.: Flexible, foldable, actively multiplexed, high-density electrode array for mapping brain activity in vivo. Nat. Neurosci. **14**(12), 1599–1605 (2011)
22. Minev, I.R., Musienko, P., Hirsch, A., Barraud, Q., Wenger, N., Moraud, E.M., Gandar, J., Capogrosso, M., Milekovic, T., Asboth, L., et al.: Electronic dura mater for long-term multimodal neural interfaces. Science. **347**(6218), 159–163 (2015)
23. Gong, S., Schwalb, W., Wang, Y., Chen, Y., Tang, Y., Si, J., Shirinzadeh, B., Cheng, W.: A wearable and highly sensitive pressure sensor with ultrathin gold nanowires. Nat. Commun. **5**(1), 3132 (2014)
24. Schwartz, G., Tee, B.C.K., Mei, J., Appleton, A.L., Kim, D.H., Wang, H., Bao, Z.: Flexible polymer transistors with high pressure sensitivity for application in electronic skin and health monitoring. Nat. Commun. **4**(1), 1859 (2013)
25. Lin, J., Fu, R., Zhong, X., Yu, P., Tan, G., Li, W., Zhang, H., Li, Y., Zhou, L., Ning, C.: Wearable sensors and devices for real-time cardiovascular disease monitoring. Cell Rep. Phys. Sci. **2**(8), 100541 (2021)
26. Stauffer, F., Thielen, M., Sauter, C., Chardonnens, S., Bachmann, S., Tybrandt, K., Peters, C., Hierold, C., Vörös, J.: Skin conformal polymer electrodes for clinical ECG and EEG recordings. Adv. Healthc. Mater. **7**(7), 1700994 (2018)
27. Kim, S.J., Cho, K.W., Cho, H.R., Wang, L., Park, S.Y., Lee, S.E., Hyeon, T., Lu, N., Choi, S.H., Kim, D.-H.: Stretchable and transparent biointerface using cell-sheet–graphene hybrid for electrophysiology and therapy of skeletal muscle. Adv. Funct. Mater. **26**(19), 3207–3217 (2016)
28. Wu, W., Li, L., Li, Z., Sun, J., Wang, L.: Extensible integrated system for real-time monitoring of cardiovascular physiological signals and limb health. Adv. Mater. **35**(51), 2304596 (2023)
29. Zhao, H., O'Brien, K., Li, S., Shepherd, R.F.: Optoelectronically innervated soft prosthetic hand via stretchable optical waveguides. Sci. Robot. **1**(1), eaai7529 (2016)
30. Dagdeviren, C., Shi, Y., Joe, P., Ghaffari, R., Balooch, G., Usgaonkar, K., Gur, O., Tran, P.L., Crosby, J.R., Meyer, M., et al.: Conformal piezoelectric systems for clinical and experimental characterization of soft tissue biomechanics. Nat. Mater. **14**(7), 728–736 (2015)
31. Arab Hassani, F., Jin, H., Yokota, T., Someya, T., Thakor, N.V.: Soft sensors for a sensing-actuation system with high bladder voiding efficiency. Sci Adv. **6**(18), eaba0412 (2020)
32. Yamada, T., Hayamizu, Y., Yamamoto, Y., Yomogida, Y., Izadi-Najafabadi, A., Futaba, D.N., Hata, K.: A stretchable carbon nanotube strain sensor for human-motion detection. Nat. Nanotechnol. **6**(5), 296–301 (2011)
33. Li, Q., Li, J., Tran, D., Luo, C., Gao, Y., Yu, C., Xuan, F.: Engineering of carbon nanotube/polydimethylsiloxane nanocomposites with enhanced sensitivity for wearable motion sensors. J. Mater. Chem. C. **5**(42), 11092–11099 (2017)
34. Kim, S., Oh, J., Jeong, D., Bae, J.: Direct wiring of eutectic gallium–indium to a metal electrode for soft sensor systems. ACS Appl. Mater. Interfaces. **11**(22), 20557–20565 (2019)
35. Gao, W., Emaminejad, S., Nyein, H.Y.Y., Challa, S., Chen, K., Peck, A., Fahad, H.M., Ota, H., Shiraki, H., Kiriya, D., et al.: Fully integrated wearable sensor arrays for multiplexed in situ perspiration analysis. Nature. **529**(7587), 509–514 (2016)
36. Li, R., Qi, H., Ma, Y., Deng, Y., Liu, S., Jie, Y., Jing, J., He, J., Zhang, X., Wheatley, L., et al.: A flexible and physically transient electrochemical sensor for real-time wireless nitric oxide monitoring. Nat. Commun. **11**(1), 3207 (2020)
37. Babu, M., Lautman, Z., Lin, X., Sobota, M.H.B., Snyder, M.P.: Wearable devices: implications for precision medicine and the future of health care. Annu. Rev. Med. **75**(1), 401–415 (2024)
38. Jia, M., Rolandi, M.: Soft and ion-conducting materials in bioelectronics: from conducting polymers to hydrogels. Adv. Healthc. Mater. **9**(5), 1901372 (2020)
39. Tham, J.C.K.: Wearable writing: enriching student peer review with point-of-view video feedback using Google glass. J. Tech. Writ. Commun. **47**(1), 22–55 (2016)

40. Shimura, T., Sato, S., Zalar, P., Matsuhisa, N.: Engineering the comfort-of-wear for next generation wearables. Adv. Electron. Mater. **9**(9), 2200512 (2023)
41. Wang, C., Xia, K., Wang, H., Liang, X., Yin, Z., Zhang, Y.: Advanced carbon for flexible and wearable electronics. Adv. Mater. **31**(9), 1801072 (2019)
42. Koydemir, H.C., Ozcan, A.: Wearable and implantable sensors for biomedical applications. Annu. Rev. Anal. Chem. **11**(1), 127–146 (2018)
43. Breteler, M.J.M., KleinJan, E.J., Dohmen, D.A.J., Leenen, L.P.H., van Hillegersberg, R., Ruurda, J.P., van Loon, K., Blokhuis, T.J., Kalkman, C.J.: Vital signs monitoring with wearable sensors in high-risk surgical patients: a clinical validation study. Anesthesiology. **132**(3), 424–439 (2020)
44. Sarah, E.S., Hyun-Sung, A., Danae, M.D., John, M.N., Jung-Min, L.: How accurate are the wrist-based heart rate monitors during walking and running activities? Are they accurate enough? BMJ Open Sport Exerc. Med. **2**(1), e000106 (2016)
45. Yamamoto, Y., Yamamoto, D., Takada, M., Naito, H., Arie, T., Akita, S., Takei, K.: Efficient skin temperature sensor and stable gel-less sticky ECG sensor for a wearable flexible healthcare patch. Adv. Healthc. Mater. **6**(17), 1700495 (2017)
46. Nemati, E., Deen, M.J., Mondal, T.: A wireless wearable ECG sensor for long-term applications. IEEE Commun. Mag. **50**(1), 36–43 (2012)
47. Ishihara, K., Shi, X., Fukazawa, K., Yamaoka, T., Yao, G., Wu, J.Y.: Biomimetic-engineered silicone hydrogel contact lens materials. ACS Appl. Bio Mater. **6**(9), 3600–3616 (2023)
48. Waltz, E.: Drugs go wireless. Nat. Biotechnol. **34**(1), 15–18 (2016)
49. Glikson, M., Friedman, P.A.: The implantable cardioverter defibrillator. Lancet. **357**(9262), 1107–1117 (2001)
50. Sarica, C., Iorio-Morin, C., Aguirre-Padilla, D.H., Najjar, A., Paff, M., Fomenko, A., Yamamoto, K., Zemmar, A., Lipsman, N., Ibrahim, G.M., et al.: Implantable pulse generators for deep brain stimulation: challenges, complications, and strategies for practicality and longevity. Front. Hum. Neurosci. **15**, 708481 (2021)
51. Steiger, C., Abramson, A., Nadeau, P., Chandrakasan, A.P., Langer, R., Traverso, G.: Ingestible electronics for diagnostics and therapy. Nat. Rev. Mater. **4**(2), 83–98 (2019)
52. Boutry, C.M., Beker, L., Kaizawa, Y., Vassos, C., Tran, H., Hinckley, A.C., Pfattner, R., Niu, S., Li, J., Claverie, J., et al.: Biodegradable and flexible arterial-pulse sensor for the wireless monitoring of blood flow. Nat. Biomed. Eng. **3**(1), 47–57 (2019)
53. Zhou, Z., Zhang, H., Liu, J., Huang, W.: Flexible electronics from intrinsically soft materials. Giant. **6**, 100051 (2021)
54. Liu, Y., Pharr, M., Salvatore, G.A.: Lab-on-skin: a review of flexible and stretchable electronics for wearable health monitoring. ACS Nano. **11**(10), 9614–9635 (2017)
55. Wang, C., Wang, C., Huang, Z., Xu, S.: Materials and structures toward soft electronics. Adv. Mater. **30**(50), 1801368 (2018)
56. Hussain, A.M., Hussain, M.M.: CMOS-technology-enabled flexible and stretchable electronics for internet of everything applications. Adv. Mater. **28**(22), 4219–4249 (2016)
57. Khan, Y., Ostfeld, A.E., Lochner, C.M., Pierre, A., Arias, A.C.: Monitoring of vital signs with flexible and wearable medical devices. Adv. Mater. **28**(22), 4373–4395 (2016)
58. Avila, R.: Curvature-matching mechanics in skin-based bioelectronics to minimize interfacial stresses. J. Appl. Mech. **91**(4), 045001 (2023)
59. Kim, D.W., Song, K.-I., Seong, D., Lee, Y.S., Baik, S., Song, J.H., Lee, H.J., Son, D., Pang, C.: Electrostatic–mechanical synergistic in situ multiscale tissue adhesion for sustainable residue-free bioelectronics interfaces. Adv. Mater. **34**(5), 2105338 (2022)
60. Jeong, J.-W., Shin, G., Park Sung, I., Yu Ki, J., Xu, L., Rogers John, A.: Soft materials in neuroengineering for hard problems in neuroscience. Neuron. **86**(1), 175–186 (2015)
61. Head, S.J., Mokhles, M.M., Osnabrugge, R.L.J., Pibarot, P., Mack, M.J., Takkenberg, J.J.M., Bogers, A.J.J.C., Kappetein, A.P.: The impact of prosthesis–patient mismatch on long-term survival after aortic valve replacement: a systematic review and meta-analysis of 34 observational studies comprising 27 186 patients with 133 141 patient-years. Eur. Heart J. **33**(12), 1518–1529 (2012)

62. Grill, W.M., Norman, S.E., Bellamkonda, R.V.: Implanted neural interfaces: biochallenges and engineered solutions. Annu. Rev. Biomed. Eng. **11**(1), 1–24 (2009)
63. Polikov, V.S., Tresco, P.A., Reichert, W.M.: Response of brain tissue to chronically implanted neural electrodes. J. Neurosci. Methods. **148**(1), 1–18 (2005)
64. Jia, M., Kim, J., Nguyen, T., Duong, T., Rolandi, M.: Natural biopolymers as proton conductors in bioelectronics. Biopolymers. **112**(7), e23433 (2021)
65. Levin, M., Pezzulo, G., Finkelstein, J.M.: Endogenous bioelectric signaling networks: exploiting voltage gradients for control of growth and form. Annu. Rev. Biomed. Eng. **19**(1), 353–387 (2017)
66. Aqrawe, Z., Montgomery, J., Travas-Sejdic, J., Svirskis, D.: Conducting polymers as electrode coatings for neuronal multi-electrode arrays. Trends Biotechnol. **35**(2), 93–95 (2017)
67. Long, R., Hui, C.-Y., Gong, J.P., Bouchbinder, E.: The fracture of highly deformable soft materials: a tale of two length scales. Annu. Rev. Condensed Matter Phys. **12**(1), 71–94 (2021)
68. Hassan, P.A., Verma, G., Ganguly, R.: 1—Soft materials—properties and applications. In: Banerjee, S., Tyagi, A.K. (eds.) Functional Materials, pp. 1–59. Elsevier, London (2012)
69. Hu, C., Pané, S., Nelson, B.J.: Soft micro- and nanorobotics. Annu. Rev. Control Robot. Auton. Syst. **1**(1), 53–75 (2018)
70. Choi, S., Lee, H., Ghaffari, R., Hyeon, T., Kim, D.-H.: Recent advances in flexible and stretchable bio-electronic devices integrated with nanomaterials. Adv. Mater. **28**(22), 4203–4218 (2016)
71. Chen, Y., Lu, S., Zhang, S., Li, Y., Qu, Z., Chen, Y., Lu, B., Wang, X., Feng, X.: Skin-like biosensor system via electrochemical channels for noninvasive blood glucose monitoring. Sci. Adv. **3**(12), e1701629 (2017)
72. Chortos, A., Bao, Z.: Skin-inspired electronic devices. Mater. Today. **17**(7), 321–331 (2014)
73. Ma, Y.Z., Sobernheim, D., Garzon, J.R.: Chapter 19—Glossary for unconventional oil and gas resource evaluation and development. In: Ma, Y.Z., Holditch, S.A. (eds.) Unconventional Oil and Gas Resources Handbook, pp. 513–526. Gulf Professional Publishing, Boston (2016)
74. Kimbell, G., Azad, M.A.: Chapter 15—3D printing: bioinspired materials for drug delivery. In: Nurunnabi, M. (ed.) Bioinspired and Biomimetic Materials for Drug Delivery, pp. 295–318. Woodhead Publishing (2021)
75. Gupta, S., Navaraj, W.T., Lorenzelli, L., Dahiya, R.: Ultra-thin chips for high-performance flexible electronics. npj Flexible Electron. **2**(1), 8 (2018)
76. Someya, T., Bao, Z., Malliaras, G.G.: The rise of plastic bioelectronics. Nature. **540**(7633), 379–385 (2016)
77. Park, S.I., Brenner, D.S., Shin, G., Morgan, C.D., Copits, B.A., Chung, H.U., Pullen, M.Y., Noh, K.N., Davidson, S., Oh, S.J., et al.: Soft, stretchable, fully implantable miniaturized optoelectronic systems for wireless optogenetics. Nat. Biotechnol. **33**(12), 1280–1286 (2015)
78. Gong, S., Cheng, W.: One-dimensional nanomaterials for soft electronics. Adv. Electron. Mater. **3**(3), 1600314 (2017)
79. Koo, J.H., Kim, D.C., Shim, H.J., Kim, T.-H., Kim, D.-H.: Flexible and stretchable smart display: materials, fabrication, device design, and system integration. Adv. Funct. Mater. **28**(35), 1801834 (2018)
80. Trung, T.Q., Lee, N.-E.: Recent progress on stretchable electronic devices with intrinsically stretchable components. Adv. Mater. **29**(3), 1603167 (2017)
81. Herbert, R., Kim, J.-H., Kim, Y.S., Lee, H.M., Yeo, W.-H.: Soft material-enabled, flexible hybrid electronics for medicine, healthcare, and human-machine interfaces. Materials. **11**(2), 187 (2018)
82. Oh, S.Y., Hong, S.Y., Jeong, Y.R., Yun, J., Park, H., Jin, S.W., Lee, G., Oh, J.H., Lee, H., Lee, S.-S., et al.: Skin-attachable, stretchable electrochemical sweat sensor for glucose and pH detection. ACS Appl. Mater. Interfaces. **10**(16), 13729–13740 (2018)
83. Yao, S., Swetha, P., Zhu, Y.: Nanomaterial-enabled wearable sensors for healthcare. Adv. Healthc. Mater. **7**(1), 1700889 (2018)

84. Wen, Z., Yang, Y., Sun, N., Li, G., Liu, Y., Chen, C., Shi, J., Xie, L., Jiang, H., Bao, D., et al.: A wrinkled PEDOT:PSS film based stretchable and transparent triboelectric nanogenerator for wearable energy harvesters and active motion sensors. Adv. Funct. Mater. **28**(37), 1803684 (2018)
85. Rivnay, J., Inal, S., Salleo, A., Owens, R.M., Berggren, M., Malliaras, G.G.: Organic electrochemical transistors. Nat. Rev. Mater. **3**(2), 17086 (2018)
86. Lee, Y., Howe, C., Mishra, S., Lee, D.S., Mahmood, M., Piper, M., Kim, Y., Tieu, K., Byun, H.-S., Coffey, J.P., et al.: Wireless, intraoral hybrid electronics for real-time quantification of sodium intake toward hypertension management. Proc. Natl. Acad. Sci. **115**(21), 5377–5382 (2018)
87. Kaltenbrunner, M., Sekitani, T., Reeder, J., Yokota, T., Kuribara, K., Tokuhara, T., Drack, M., Schwödiauer, R., Graz, I., Bauer-Gogonea, S., et al.: An ultra-lightweight design for imperceptible plastic electronics. Nature. **499**(7459), 458–463 (2013)
88. An, B.W., Heo, S., Ji, S., Bien, F., Park, J.-U.: Transparent and flexible fingerprint sensor array with multiplexed detection of tactile pressure and skin temperature. Nat. Commun. **9**(1), 2458 (2018)
89. Lipomi, D.J., Vosgueritchian, M., Tee, B.C.K., Hellstrom, S.L., Lee, J.A., Fox, C.H., Bao, Z.: Skin-like pressure and strain sensors based on transparent elastic films of carbon nanotubes. Nat. Nanotechnol. **6**(12), 788–792 (2011)
90. Miyamoto, A., Lee, S., Cooray, N.F., Lee, S., Mori, M., Matsuhisa, N., Jin, H., Yoda, L., Yokota, T., Itoh, A., et al.: Inflammation-free, gas-permeable, lightweight, stretchable on-skin electronics with nanomeshes. Nat. Nanotechnol. **12**(9), 907–913 (2017)
91. Huang, J., Zhu, H., Chen, Y., Preston, C., Rohrbach, K., Cumings, J., Hu, L.: Highly transparent and flexible nanopaper transistors. ACS Nano. **7**(3), 2106–2113 (2013)
92. Yang, C., Suo, Z.: Hydrogel ionotronics. Nat. Rev. Mater. **3**(6), 125–142 (2018)
93. Jang, K.-I., Han, S.Y., Xu, S., Mathewson, K.E., Zhang, Y., Jeong, J.-W., Kim, G.-T., Webb, R.C., Lee, J.W., Dawidczyk, T.J., et al.: Rugged and breathable forms of stretchable electronics with adherent composite substrates for transcutaneous monitoring. Nat. Commun. **5**(1), 4779 (2014)
94. Karp, J.M., Langer, R.: Dry solution to a sticky problem. Nature. **477**(7362), 42–43 (2011)
95. Drotlef, D.-M., Amjadi, M., Yunusa, M., Sitti, M.: Bioinspired composite microfibers for skin adhesion and signal amplification of wearable sensors. Adv. Mater. **29**(28), 1701353 (2017)
96. Hwang, D.-G., Trent, K., Bartlett, M.D.: Kirigami-inspired structures for smart adhesion. ACS Appl. Mater. Interfaces. **10**(7), 6747–6754 (2018)
97. Webb, R.C., Bonifas, A.P., Behnaz, A., Zhang, Y., Yu, K.J., Cheng, H., Shi, M., Bian, Z., Liu, Z., Kim, Y.-S., et al.: Ultrathin conformal devices for precise and continuous thermal characterization of human skin. Nat. Mater. **12**(10), 938–944 (2013)
98. Lewis, J.: Material challenge for flexible organic devices. Mater. Today. **9**(4), 38–45 (2006)
99. Fan, Y.J., Li, X., Kuang, S.Y., Zhang, L., Chen, Y.H., Liu, L., Zhang, K., Ma, S.W., Liang, F., Wu, T., et al.: Highly robust, transparent, and breathable epidermal electrode. ACS Nano. **12**(9), 9326–9332 (2018)
100. Zhang, Y., Zhang, L., Cui, K., Ge, S., Cheng, X., Yan, M., Yu, J., Liu, H.: Flexible electronics based on micro/nanostructured paper. Adv. Mater. **30**(51), 1801588 (2018)
101. Sun, B., McCay, R.N., Goswami, S., Xu, Y., Zhang, C., Ling, Y., Lin, J., Yan, Z.: Gas-permeable, multifunctional on-skin electronics based on laser-induced porous graphene and sugar-templated elastomer sponges. Adv. Mater. **30**(50), 1804327 (2018)
102. Reeder, J.T., Choi, J., Xue, Y., Gutruf, P., Hanson, J., Liu, M., Ray, T., Bandodkar, A.J., Avila, R., Xia, W., et al.: Waterproof, electronics-enabled, epidermal microfluidic devices for sweat collection, biomarker analysis, and thermography in aquatic settings. Sci. Adv. **5**(1), eaau6356 (2019)
103. Kim, T., Park, J., Sohn, J., Cho, D., Jeon, S.: Bioinspired, highly stretchable, and conductive dry adhesives based on 1D–2D hybrid carbon nanocomposites for all-in-one ECG electrodes. ACS Nano. **10**(4), 4770–4778 (2016)

104. Adly, N., Weidlich, S., Seyock, S., Brings, F., Yakushenko, A., Offenhäusser, A., Wolfrum, B.: Printed microelectrode arrays on soft materials: from PDMS to hydrogels. npj Flexible Electron. **2**(1), 15 (2018)
105. Roy, A., Maity, P.P., Dhara, S., Pal, S.: Biocompatible, stimuli-responsive hydrogel of chemically crosslinked β-cyclodextrin as amoxicillin carrier. J. Appl. Polym. Sci. **135**(10), 45939 (2018)
106. Chiappini, C., De Rosa, E., Martinez, J.O., Liu, X., Steele, J., Stevens, M.M., Tasciotti, E.: Biodegradable silicon nanoneedles delivering nucleic acids intracellularly induce localized in vivo neovascularization. Nat. Mater. **14**(5), 532–539 (2015)
107. Wang, C., Li, X., Hu, H., Zhang, L., Huang, Z., Lin, M., Zhang, Z., Yin, Z., Huang, B., Gong, H., et al.: Monitoring of the central blood pressure waveform via a conformal ultrasonic device. Nat. Biomed. Eng. **2**(9), 687–695 (2018)
108. Xiong, R., Grant, A.M., Ma, R., Zhang, S., Tsukruk, V.V.: Naturally-derived biopolymer nanocomposites: interfacial design, properties and emerging applications. Mater. Sci. Eng. R Rep. **125**, 1–41 (2018)
109. Manna, K., Patra, P., Roy, A., Roy, R.K., Chaitanya Sunka, K., Dhara, S., Patra, N., Pal, S.: Amino acid inspired alginate-based pH sensitive polymeric micelles via reversible addition–fragmentation chain transfer polymerization. ACS Appl. Polym. Mater. **4**(6), 4432–4444 (2022)
110. Kim, D.-H., Ghaffari, R., Lu, N., Wang, S., Lee, S.P., Keum, H., D'Angelo, R., Klinker, L., Su, Y., Lu, C., et al.: Electronic sensor and actuator webs for large-area complex geometry cardiac mapping and therapy. Proc. Natl. Acad. Sci. **109**(49), 19910–19915 (2012)
111. Kang, S.-K., Murphy, R.K.J., Hwang, S.-W., Lee, S.M., Harburg, D.V., Krueger, N.A., Shin, J., Gamble, P., Cheng, H., Yu, S., et al.: Bioresorbable silicon electronic sensors for the brain. Nature. **530**(7588), 71–76 (2016)
112. Wu, X., Zhou, J., Huang, J.: Integration of biomaterials into sensors based on organic thin-film transistors. Macromol. Rapid Commun. **39**(15), 1800084 (2018)
113. Rim, Y.S., Bae, S.-H., Chen, H., De Marco, N., Yang, Y.: Recent progress in materials and devices toward printable and flexible sensors. Adv. Mater. **28**(22), 4415–4440 (2016)
114. Park, B., Kim, J., Kang, D., Jeong, C., Kim, K.S., Kim, J.U., Yoo, P.J., Kim, T.-i.: Dramatically enhanced mechanosensitivity and signal-to-noise ratio of nanoscale crack-based sensors: effect of crack depth. Adv. Mater. **28**(37), 8130–8137 (2016)
115. Nur, R., Matsuhisa, N., Jiang, Z., Nayeem, M.O.G., Yokota, T., Someya, T.: A highly sensitive capacitive-type strain sensor using wrinkled ultrathin gold films. Nano Lett. **18**(9), 5610–5617 (2018)
116. Shi, X., Wang, H., Xie, X., Xue, Q., Zhang, J., Kang, S., Wang, C., Liang, J., Chen, Y.: Bioinspired ultrasensitive and stretchable MXene-based strain sensor via nacre-mimetic microscale "brick-and-mortar" architecture. ACS Nano. **13**(1), 649–659 (2019)
117. Shi, X., Liu, S., Sun, Y., Liang, J., Chen, Y.: Lowering internal friction of 0D–1D–2D ternary nanocomposite-based strain sensor by fullerene to boost the sensing performance. Adv. Funct. Mater. **28**(22), 1800850 (2018)
118. Kim, Y.-S., Lu, J., Shih, B., Gharibans, A., Zou, Z., Matsuno, K., Aguilera, R., Han, Y., Meek, A., Xiao, J., et al.: Scalable manufacturing of solderable and stretchable physiologic sensing systems. Adv. Mater. **29**(39), 1701312 (2017)
119. Bariya, M., Shahpar, Z., Park, H., Sun, J., Jung, Y., Gao, W., Nyein, H.Y.Y., Liaw, T.S., Tai, L.-C., Ngo, Q.P., et al.: Roll-to-roll gravure printed electrochemical sensors for wearable and medical devices. ACS Nano. **12**(7), 6978–6987 (2018)
120. Yin, X.-Y., Zhang, Y., Cai, X., Guo, Q., Yang, J., Wang, Z.L.: 3D printing of ionic conductors for high-sensitivity wearable sensors. Mater. Horizons. **6**(4), 767–780 (2019)
121. Rogers, J.A., Someya, T., Huang, Y.: Materials and mechanics for stretchable electronics. Science. **327**(5973), 1603–1607 (2010)
122. Li, Y., Li, N., Liu, W., Prominski, A., Kang, S., Dai, Y., Liu, Y., Hu, H., Wai, S., Dai, S., et al.: Achieving tissue-level softness on stretchable electronics through a generalizable soft interlayer design. Nat. Commun. **14**(1), 4488 (2023)

123. Feiner, R., Dvir, T.: Tissue–electronics interfaces: from implantable devices to engineered tissues. Nat. Rev. Mater. **3**(1), 17076 (2017)
124. Lacour, S.P., Courtine, G., Guck, J.: Materials and technologies for soft implantable neuroprostheses. Nat. Rev. Mater. **1**(10), 16063 (2016)
125. Hong, G., Lieber, C.M.: Novel electrode technologies for neural recordings. Nat. Rev. Neurosci. **20**(6), 330–345 (2019)
126. Roy, A., Manna, K., Ray, P.G., Dhara, S., Pal, S.: β-Cyclodextrin-based ultrahigh stretchable, flexible, electro- and pressure-responsive, adhesive, transparent hydrogel as motion sensor. ACS Appl. Mater. Interfaces. **14**(15), 17065–17080 (2022)
127. Dai, Y., Hu, H., Wang, M., Xu, J., Wang, S.: Stretchable transistors and functional circuits for human-integrated electronics. Nat. Electron. **4**(1), 17–29 (2021)
128. Jiang, F., Zhou, X., Lv, J., Chen, J., Chen, J., Kongcharoen, H., Zhang, Y., Lee, P.S.: Stretchable, breathable, and stable lead-free perovskite/polymer nanofiber composite for hybrid triboelectric and piezoelectric energy harvesting. Adv. Mater. **34**(17), 2200042 (2022)
129. Wang, W., Wang, S., Rastak, R., Ochiai, Y., Niu, S., Jiang, Y., Arunachala, P.K., Zheng, Y., Xu, J., Matsuhisa, N., et al.: Strain-insensitive intrinsically stretchable transistors and circuits. Nat. Electron. **4**(2), 143–150 (2021)
130. Decataldo, F., Cramer, T., Martelli, D., Gualandi, I., Korim, W.S., Yao, S.T., Tessarolo, M., Murgia, M., Scavetta, E., Amici, R., et al.: Stretchable low impedance electrodes for bioelectronic recording from small peripheral nerves. Sci. Rep. **9**(1), 10598 (2019)
131. Zheng, Y.-Q., Liu, Y., Zhong, D., Nikzad, S., Liu, S., Yu, Z., Liu, D., Wu, H.-C., Zhu, C., Li, J., et al.: Monolithic optical microlithography of high-density elastic circuits. Science. **373**(6550), 88–94 (2021)
132. Guan, Y.-S., Ershad, F., Rao, Z., Ke, Z., da Costa, E.C., Xiang, Q., Lu, Y., Wang, X., Mei, J., Vanderslice, P., et al.: Elastic electronics based on micromesh-structured rubbery semiconductor films. Nat. Electron. **5**(12), 881–892 (2022)
133. Gong, S., Yap, L.W., Zhu, B., Cheng, W.: Multiscale soft–hard interface design for flexible hybrid electronics. Adv. Mater. **32**(15), 1902278 (2020)
134. Jankowski, A.F., Tsakalakos, T.: The effect of strain on the elastic constants of noble metals. J. Phys. F Metal Phys. **15**(6), 1279 (1985)
135. Roberts, R.J., Rowe, R.C.: The Young's modulus of pharmaceutical materials. Int. J. Pharm. **37**(1), 15–18 (1987)
136. Fedorchenko, A.I., Wang, A.-B., Cheng, H.H.: Thickness dependence of nanofilm elastic modulus. Appl. Phys. Lett. **94**(15), 152111 (2009)
137. Wang, L., Lu, N.: Conformability of a thin elastic membrane laminated on a soft substrate with slightly wavy surface. J. Appl. Mech. **83**(4), 041007 (2016)
138. Wang, S., Li, M., Wu, J., Kim, D.-H., Lu, N., Su, Y., Kang, Z., Huang, Y., Rogers, J.A.: Mechanics of epidermal electronics. J. Appl. Mech. **79**(3), 031022 (2012)
139. Jeong, J.-W., Yeo, W.-H., Akhtar, A., Norton, J.J.S., Kwack, Y.-J., Li, S., Jung, S.-Y., Su, Y., Lee, W., Xia, J., et al.: Materials and optimized designs for human-machine interfaces via epidermal electronics. Adv. Mater. **25**(47), 6839–6846 (2013)
140. Atcha, H., Jairaman, A., Holt, J.R., Meli, V.S., Nagalla, R.R., Veerasubramanian, P.K., Brumm, K.T., Lim, H.E., Othy, S., Cahalan, M.D., et al.: Mechanically activated ion channel Piezo1 modulates macrophage polarization and stiffness sensing. Nat. Commun. **12**(1), 3256 (2021)
141. Konai, M.M., Bhattacharjee, B., Ghosh, S., Haldar, J.: Recent progress in polymer research to tackle infections and antimicrobial resistance. Biomacromolecules. **19**(6), 1888–1917 (2018)
142. Singh, G., Chanda, A.: Mechanical properties of whole-body soft human tissues: a review. Biomed. Mater. **16**(6), 062004 (2021)
143. Taylor, Z., Miller, K.: Reassessment of brain elasticity for analysis of biomechanisms of hydrocephalus. J. Biomech. **37**(8), 1263–1269 (2004)
144. Leipzig, N.D., Shoichet, M.S.: The effect of substrate stiffness on adult neural stem cell behavior. Biomaterials. **30**(36), 6867–6878 (2009)

145. Balguid, A., Rubbens, M.P., Mol, A., Bank, R.A., Bogers, A.J.J.C., van Kats, J.P., de Mol, B.A.J.M., Baaijens, F.P.T., Bouten, C.V.C.: The role of collagen cross-links in biomechanical behavior of human aortic heart valve leaflets—relevance for tissue engineering. Tissue Eng. **13**(7), 1501–1511 (2007)
146. Hasan, A., Ragaert, K., Swieszkowski, W., Selimović, Š., Paul, A., Camci-Unal, G., Mofrad, M.R.K., Khademhosseini, A.: Biomechanical properties of native and tissue engineered heart valve constructs. J. Biomech. **47**(9), 1949–1963 (2014)
147. Liang, X., Boppart, S.A.: Biomechanical properties of in vivo human skin from dynamic optical coherence elastography. IEEE Trans. Biomed. Eng. **57**(4), 953–959 (2010)
148. Jacquet, E., Chambert, J., Pauchot, J., Sandoz, P.: Intra- and inter-individual variability in the mechanical properties of the human skin from in vivo measurements on 20 volunteers. Skin Res. Technol. **23**(4), 491–499 (2017)
149. Hopcroft, M.A., Nix, W.D., Kenny, T.W.: What is the Young's modulus of silicon? J. Microelectromech. Syst. **19**(2), 229–238 (2010)
150. Petrova, H., Perez-Juste, J., Zhang, Z., Zhang, J., Kosel, T., Hartland, G.V.: Crystal structure dependence of the elastic constants of gold nanorods. J. Mater. Chem. **16**(40), 3957–3963 (2006)
151. Nawrocki, R.A., Jin, H., Lee, S., Yokota, T., Sekino, M., Someya, T.: Self-adhesive and ultra-conformable, sub-300 nm dry thin-film electrodes for surface monitoring of biopotentials. Adv. Funct. Mater. **28**(36), 1803279 (2018)
152. Li, T., Huang, Z., Suo, Z., Lacour, S.P., Wagner, S.: Stretchability of thin metal films on elastomer substrates. Appl. Phys. Lett. **85**(16), 3435–3437 (2004)
153. Lacour, S.P., Chan, D., Wagner, S., Li, T., Suo, Z.: Mechanisms of reversible stretchability of thin metal films on elastomeric substrates. Appl. Phys. Lett. **88**(20), 204103 (2006)
154. Mai, Y.-W., Cotterell, B.: On the essential work of ductile fracture in polymers. Int. J. Fract. **32**(2), 105–125 (1986)
155. Alkhadra, M.A., Root, S.E., Hilby, K.M., Rodriquez, D., Sugiyama, F., Lipomi, D.J.: Quantifying the fracture behavior of brittle and ductile thin films of semiconducting polymers. Chem. Mater. **29**(23), 10139–10149 (2017)
156. Xia, Z.C., Hutchinson, J.W.: Crack patterns in thin films. J. Mech. Phys. Solids. **48**(6), 1107–1131 (2000)
157. Lu, N., Suo, Z., Vlassak, J.J.: The effect of film thickness on the failure strain of polymer-supported metal films. Acta Mater. **58**(5), 1679–1687 (2010)
158. Kim, D.-H., Ahn, J.-H., Choi, W.M., Kim, H.-S., Kim, T.-H., Song, J., Huang, Y.Y., Liu, Z., Lu, C., Rogers, J.A.: Stretchable and foldable silicon integrated circuits. Science. **320**(5875), 507–511 (2008)
159. Zhang, Y., Wang, S., Li, X., Fan, J.A., Xu, S., Song, Y.M., Choi, K.-J., Yeo, W.-H., Lee, W., Nazaar, S.N., et al.: Experimental and theoretical studies of serpentine microstructures bonded to prestrained elastomers for stretchable electronics. Adv. Funct. Mater. **24**(14), 2028–2037 (2014)
160. Frank, I.W., Tanenbaum, D.M., van der Zande, A.M., McEuen, P.L.: Mechanical properties of suspended graphene sheets. J. Vacuum Sci. Technol. B. **25**(6), 2558–2561 (2007)
161. Lee, C., Wei, X., Kysar, J.W., Hone, J.: Measurement of the elastic properties and intrinsic strength of monolayer graphene. Science. **321**(5887), 385–388 (2008)
162. Georgakilas, V., Perman, J.A., Tucek, J., Zboril, R.: Broad family of carbon nanoallotropes: classification, chemistry, and applications of fullerenes, carbon dots, nanotubes, graphene, nanodiamonds, and combined superstructures. Chem. Rev. **115**(11), 4744–4822 (2015)
163. Zhu, X., Qian, Z., Chen, X., Liu, L., Sheng, C., Gu, W.: Electrohydrodynamics-printed silver nanoparticle flexible pressure sensors with improved gauge factor. IEEE Sensors J. **21**(5), 5836–5844 (2021)
164. Ha, M., Lim, S., Ko, H.: Wearable and flexible sensors for user-interactive health-monitoring devices. J. Mater. Chem. B. **6**(24), 4043–4064 (2018)
165. Huang, S., Liu, Y., Zhao, Y., Ren, Z., Guo, C.F.: Flexible electronics: stretchable electrodes and their future. Adv. Funct. Mater. **29**(6), 1805924 (2019)

166. Bae, W.G., Kim, D., Kwak, M.K., Ha, L., Kang, S.M., Suh, K.Y.: Enhanced skin adhesive patch with modulus-tunable composite micropillars. Adv. Healthc. Mater. **2**(1), 109–113 (2013)
167. Chen, J., Zheng, J., Gao, Q., Zhang, J., Zhang, J., Omisore, O.M., Wang, L., Li, H.: Polydimethylsiloxane (PDMS)-based flexible resistive strain sensors for wearable applications. Appl. Sci. **8**(3), 345 (2018)
168. Lee, H.-R., Kim, C.-C., Sun, J.-Y.: Stretchable ionics—a promising candidate for upcoming wearable devices. Adv. Mater. **30**(42), 1704403 (2018)
169. Kang, J., Son, D., Wang, G.-J.N., Liu, Y., Lopez, J., Kim, Y., Oh, J.Y., Katsumata, T., Mun, J., Lee, Y., et al.: Tough and water-insensitive self-healing elastomer for robust electronic skin. Adv. Mater. **30**(13), 1706846 (2018)
170. Li, R., Zhang, K., Cai, L., Chen, G., He, M.: Highly stretchable ionic conducting hydrogels for strain/tactile sensors. Polymer. **167**, 154–158 (2019)
171. Qiao, H., Qi, P., Zhang, X., Wang, L., Tan, Y., Luan, Z., Xia, Y., Li, Y., Sui, K.: Multiple weak H-bonds lead to highly sensitive, stretchable, self-adhesive, and self-healing ionic sensors. ACS Appl. Mater. Interfaces. **11**(8), 7755–7763 (2019)
172. Du, Y., Ge, J., Li, Y., Ma, P.X., Lei, B.: Biomimetic elastomeric, conductive and biodegradable polycitrate-based nanocomposites for guiding myogenic differentiation and skeletal muscle regeneration. Biomaterials. **157**, 40–50 (2018)
173. Ling, S., Kaplan, D.L., Buehler, M.J.: Nanofibrils in nature and materials engineering. Nat. Rev. Mater. **3**(4), 18016 (2018)
174. Beri, P., Matte, B.F., Fattet, L., Kim, D., Yang, J., Engler, A.J.: Biomaterials to model and measure epithelial cancers. Nat. Rev. Mater. **3**(11), 418–430 (2018)
175. Chen, G., Matsuhisa, N., Liu, Z., Qi, D., Cai, P., Jiang, Y., Wan, C., Cui, Y., Leow, W.R., Liu, Z., et al.: Plasticizing silk protein for on-skin stretchable electrodes. Adv. Mater. **30**(21), 1800129 (2018)
176. Hou, C., Xu, Z., Qiu, W., Wu, R., Wang, Y., Xu, Q., Liu, X.Y., Guo, W.: A biodegradable and stretchable protein-based sensor as artificial electronic skin for human motion detection. Small. **15**(11), 1805084 (2019)
177. Liu, R., Kuang, X., Deng, J., Wang, Y.-C., Wang, A.C., Ding, W., Lai, Y.-C., Chen, J., Wang, P., Lin, Z., et al.: Shape memory polymers for body motion energy harvesting and self-powered mechanosensing. Adv. Mater. **30**(8), 1705195 (2018)
178. Ico, G., Showalter, A., Bosze, W., Gott, S.C., Kim, B.S., Rao, M.P., Myung, N.V., Nam, J.: Size-dependent piezoelectric and mechanical properties of electrospun P(VDF-TrFE) nanofibers for enhanced energy harvesting. J. Mater. Chem. A. **4**(6), 2293–2304 (2016)
179. Ghosh, S.K., Mandal, D.: Synergistically enhanced piezoelectric output in highly aligned 1D polymer nanofibers integrated all-fiber nanogenerator for wearable nano-tactile sensor. Nano Energy. **53**, 245–257 (2018)
180. Jang, H., Dai, Z., Ha, K.-H., Ameri, S.K., Lu, N.: Stretchability of PMMA-supported CVD graphene and of its electrical contacts. 2D Mater. **7**(1), 014003 (2020)
181. Li, T., Suo, Z., Lacour, S.P., Wagner, S.: Compliant thin film patterns of stiff materials as platforms for stretchable electronics. J. Mater. Res. **20**(12), 3274–3277 (2005)
182. Gray, D.S., Tien, J., Chen, C.S.: High-conductivity elastomeric electronics. Adv. Mater. **16**(5), 393–397 (2004)
183. Xu, S., Zhang, Y., Cho, J., Lee, J., Huang, X., Jia, L., Fan, J.A., Su, Y., Su, J., Zhang, H., et al.: Stretchable batteries with self-similar serpentine interconnects and integrated wireless recharging systems. Nat. Commun. **4**(1), 1543 (2013)
184. Jeong, J.-W., Kim, M.K., Cheng, H., Yeo, W.-H., Huang, X., Liu, Y., Zhang, Y., Huang, Y., Rogers, J.A.: Capacitive epidermal electronics for electrically safe, long-term electrophysiological measurements. Adv. Healthc. Mater. **3**(5), 642–648 (2014)
185. Jeong, H., Wang, L., Ha, T., Mitbander, R., Yang, X., Dai, Z., Qiao, S., Shen, L., Sun, N., Lu, N.: Modular and reconfigurable wireless E-tattoos for personalized sensing. Adv. Mater. Technol. **4**(8), 1900117 (2019)

186. Jang, K.-I., Chung, H.U., Xu, S., Lee, C.H., Luan, H., Jeong, J., Cheng, H., Kim, G.-T., Han, S.Y., Lee, J.W., et al.: Soft network composite materials with deterministic and bio-inspired designs. Nat. Commun. **6**(1), 6566 (2015)
187. Liu, Z., Yu, M., Lv, J., Li, Y., Yu, Z.: Dispersed, porous nanoislands landing on stretchable nanocrack gold films: maintenance of stretchability and controllable impedance. ACS Appl. Mater. Interfaces. **6**(16), 13487–13495 (2014)
188. Jang, N.-S., Kim, K.-H., Ha, S.-H., Jung, S.-H., Lee, H.M., Kim, J.-M.: Simple approach to high-performance stretchable heaters based on Kirigami patterning of conductive paper for wearable thermotherapy applications. ACS Appl. Mater. Interfaces. **9**(23), 19612–19621 (2017)
189. Qi, K., He, J., Wang, H., Zhou, Y., You, X., Nan, N., Shao, W., Wang, L., Ding, B., Cui, S.: A highly stretchable nanofiber-based electronic skin with pressure-, strain-, and flexion-sensitive properties for health and motion monitoring. ACS Appl. Mater. Interfaces. **9**(49), 42951–42960 (2017)
190. Choi, S., Han, S.I., Jung, D., Hwang, H.J., Lim, C., Bae, S., Park, O.K., Tschabrunn, C.M., Lee, M., Bae, S.Y., et al.: Highly conductive, stretchable and biocompatible Ag–Au core–sheath nanowire composite for wearable and implantable bioelectronics. Nat. Nanotechnol. **13**(11), 1048–1056 (2018)
191. Zhu, C., Chortos, A., Wang, Y., Pfattner, R., Lei, T., Hinckley, A.C., Pochorovski, I., Yan, X., To, J.W.F., Oh, J.Y., et al.: Stretchable temperature-sensing circuits with strain suppression based on carbon nanotube transistors. Nat. Electron. **1**(3), 183–190 (2018)
192. Guo, W., Zheng, P., Huang, X., Zhuo, H., Wu, Y., Yin, Z., Li, Z., Wu, H.: Matrix-independent highly conductive composites for electrodes and interconnects in stretchable electronics. ACS Appl. Mater. Interfaces. **11**(8), 8567–8575 (2019)
193. Choi, C., Lee, Y., Cho, K.W., Koo, J.H., Kim, D.-H.: Wearable and implantable soft bioelectronics using two-dimensional materials. Acc. Chem. Res. **52**(1), 73–81 (2019)
194. Zheng, W., Huang, W., Gao, F., Yang, H., Dai, M., Liu, G., Yang, B., Zhang, J., Fu, Y.Q., Chen, X., et al.: Kirigami-inspired highly stretchable nanoscale devices using multidimensional deformation of monolayer MoS_2. Chem. Mater. **30**(17), 6063–6070 (2018)
195. Ma, Y., Liu, N., Li, L., Hu, X., Zou, Z., Wang, J., Luo, S., Gao, Y.: A highly flexible and sensitive piezoresistive sensor based on MXene with greatly changed interlayer distances. Nat. Commun. **8**(1), 1207 (2017)
196. Yang, Y., Sun, N., Wen, Z., Cheng, P., Zheng, H., Shao, H., Xia, Y., Chen, C., Lan, H., Xie, X., et al.: Liquid-metal-based super-stretchable and structure-designable triboelectric nanogenerator for wearable electronics. ACS Nano. **12**(2), 2027–2034 (2018)
197. Yoon, Y., Kim, S., Kim, D., Kauh, S.K., Lee, J.: Four degrees-of-freedom direct writing of liquid metal patterns on uneven surfaces. Adv. Mater. Technol. **4**(2), 1800379 (2019)
198. Pan, C., Kumar, K., Li, J., Markvicka, E.J., Herman, P.R., Majidi, C.: Visually imperceptible liquid-metal circuits for transparent, stretchable electronics with direct laser writing. Adv. Mater. **30**(12), 1706937 (2018)
199. Zhang, Q., Liu, X., Duan, L., Gao, G.: Ultra-stretchable wearable strain sensors based on skin-inspired adhesive, tough and conductive hydrogels. Chem. Eng. J. **365**, 10–19 (2019)
200. Shao, C., Wang, M., Meng, L., Chang, H., Wang, B., Xu, F., Yang, J., Wan, P.: Mussel-inspired cellulose nanocomposite tough hydrogels with synergistic self-healing, adhesive, and strain-sensitive properties. Chem. Mater. **30**(9), 3110–3121 (2018)
201. Wang, S., Xu, J., Wang, W., Wang, G.-J.N., Rastak, R., Molina-Lopez, F., Chung, J.W., Niu, S., Feig, V.R., Lopez, J., et al.: Skin electronics from scalable fabrication of an intrinsically stretchable transistor array. Nature. **555**(7694), 83–88 (2018)
202. Tinka, A., Wu, Q., Weekly, K., Oroza, C.A., Beard, J., Bayen, A.M.: Heterogeneous fleets of active and passive floating sensors for river studies. J. Field Robot. **33**(5), 618–638 (2016)
203. Yu, Y., Yu, D., Sadigh, B., Orme, C.A.: Space- and time-resolved small angle X-ray scattering to probe assembly of silver nanocrystal superlattices. Nat. Commun. **9**(1), 4211 (2018)
204. Kaltenbrunner, M., White, M.S., Głowacki, E.D., Sekitani, T., Someya, T., Sariciftci, N.S., Bauer, S.: Ultrathin and lightweight organic solar cells with high flexibility. Nat. Commun. **3**(1), 770 (2012)

205. Kaltenbrunner, M., Adam, G., Głowacki, E.D., Drack, M., Schwödiauer, R., Leonat, L., Apaydin, D.H., Groiss, H., Scharber, M.C., White, M.S., et al.: Flexible high power-per-weight perovskite solar cells with chromium oxide–metal contacts for improved stability in air. Nat. Mater. **14**(10), 1032–1039 (2015)
206. Ren, H., Cui, N., Tang, Q., Tong, Y., Zhao, X., Liu, Y.: High-performance, ultrathin, ultraflexible organic thin-film transistor array via solution process. Small. **14**(33), 1801020 (2018)
207. Liang, Y., Xiao, P., Wang, S., Shi, J., He, J., Zhang, J., Huang, Y., Chen, T.: Scalable fabrication of free-standing, stretchable CNT/TPE ultrathin composite films for skin adhesive epidermal electronics. J. Mater. Chem. C. **6**(25), 6666–6671 (2018)
208. Yan, C., Deng, W., Jin, L., Yang, T., Wang, Z., Chu, X., Su, H., Chen, J., Yang, W.: Epidermis-inspired ultrathin 3D cellular sensor array for self-powered biomedical monitoring. ACS Appl. Mater. Interfaces. **10**(48), 41070–41075 (2018)
209. Ding, Y., Xu, T., Onyilagha, O., Fong, H., Zhu, Z.: Recent advances in flexible and wearable pressure sensors based on piezoresistive 3D monolithic conductive sponges. ACS Appl. Mater. Interfaces. **11**(7), 6685–6704 (2019)
210. Kim, J., Sempionatto, J.R., Imani, S., Hartel, M.C., Barfidokht, A., Tang, G., Campbell, A.S., Mercier, P.P., Wang, J.: Simultaneous monitoring of sweat and interstitial fluid using a single wearable biosensor platform. Adv. Sci. **5**(10), 1800880 (2018)
211. Kabiri Ameri, S., Ho, R., Jang, H., Tao, L., Wang, Y., Wang, L., Schnyer, D.M., Akinwande, D., Lu, N.: Graphene electronic tattoo sensors. ACS Nano. **11**(8), 7634–7641 (2017)
212. Lopes, P.A., Paisana, H., De Almeida, A.T., Majidi, C., Tavakoli, M.: Hydroprinted electronics: ultrathin stretchable Ag–In–Ga E-skin for bioelectronics and human–machine interaction. ACS Appl. Mater. Interfaces. **10**(45), 38760–38768 (2018)
213. Wang, R., Zhai, Q., An, T., Gong, S., Cheng, W.: Stretchable gold fiber-based wearable textile electrochemical biosensor for lactate monitoring in sweat. Talanta. **222**, 121484 (2021)
214. An, T., Ling, Y., Gong, S., Zhu, B., Zhao, Y., Dong, D., Yap, L.W., Wang, Y., Cheng, W.: A wearable second skin-like multifunctional supercapacitor with vertical gold nanowires and electrochromic polyaniline. Adv. Mater. Technol. **4**(3), 1800473 (2019)
215. Fox, M.: Optical Properties of Solids, 2nd edn. Oxford University Press, New York (2010)
216. Wang, D., Zhang, Y., Lu, X., Ma, Z., Xie, C., Zheng, Z.: Chemical formation of soft metal electrodes for flexible and wearable electronics. Chem. Soc. Rev. **47**(12), 4611–4641 (2018)
217. Zhang, B., Korolj, A., Lai, B.F.L., Radisic, M.: Advances in organ-on-a-chip engineering. Nat. Rev. Mater. **3**(8), 257–278 (2018)
218. Wongkaew, N., Simsek, M., Griesche, C., Baeumner, A.J.: Functional nanomaterials and nanostructures enhancing electrochemical biosensors and lab-on-a-chip performances: recent progress, applications, and future perspective. Chem. Rev. **119**(1), 120–194 (2019)
219. Wang, X., Zhang, Y., Zhang, X., Huo, Z., Li, X., Que, M., Peng, Z., Wang, H., Pan, C.: A highly stretchable transparent self-powered triboelectric tactile sensor with metallized nanofibers for wearable electronics. Adv. Mater. **30**(12), 1706738 (2018)
220. Lee, D., Kim, J., Kim, H., Heo, H., Park, K., Lee, Y.: High-performance transparent pressure sensors based on sea-urchin shaped metal nanoparticles and polyurethane microdome arrays for real-time monitoring. Nanoscale. **10**(39), 18812–18820 (2018)
221. Feng, P., Ji, H., Zhang, L., Luo, X., Leng, X., He, P., Feng, H., Zhang, J., Ma, X., Zhao, W.: Highly stretchable patternable conductive circuits and wearable strain sensors based on polydimethylsiloxane and silver nanoparticles. Nanotechnology. **30**(18), 185501 (2019)
222. Duan, S., Wang, Z., Zhang, L., Liu, J., Li, C.: A highly stretchable, sensitive, and transparent strain sensor based on binary hybrid network consisting of hierarchical multiscale metal nanowires. Adv. Mater. Technol. **3**(6), 1800020 (2018)
223. Wang, Z., Zhang, L., Liu, J., Li, C.: Highly stretchable, sensitive, and transparent strain sensors with a controllable in-plane mesh structure. ACS Appl. Mater. Interfaces. **11**(5), 5316–5324 (2019)
224. Neilson, J., Cataldi, P., Derby, B.: Graphene-based transparent flexible strain gauges with tunable sensitivity and strain range. ACS Appl. Nano Mater. **6**(23), 21763–21774 (2023)

225. Wang, P., Peng, Z., Li, M., Wang, Y.: Stretchable transparent conductive films from long carbon nanotube metals. Small. **14**(38), 1802625 (2018)
226. Hong, M.-S., Choi, G.-M., Kim, J., Jang, J., Choi, B., Kim, J.-K., Jeong, S., Leem, S., Kwon, H.-Y., Hwang, H.-B., et al.: Biomimetic chitin–silk hybrids: an optically transparent structural platform for wearable devices and advanced electronics. Adv. Funct. Mater. **28**(24), 1705480 (2018)
227. Koh, L.-D., Yeo, J., Lee, Y.Y., Ong, Q., Han, M., Tee, B.C.K.: Advancing the frontiers of silk fibroin protein-based materials for futuristic electronics and clinical wound-healing (Invited review). Mater. Sci. Eng. C. **86**, 151–172 (2018)
228. Mestach, L., Huygens, S., Goossens, A., Gilissen, L.: Allergic contact dermatitis caused by acrylic-based medical dressings and adhesives. Contact Derm. **79**(2), 81–84 (2018)
229. Zulkowski, K.: Understanding moisture-associated skin damage, medical adhesive-related skin injuries, and skin tears. Adv. Skin Wound Care. **30**(8), 372 (2017)
230. Lichterfeld-Kottner, A., El Genedy, M., Lahmann, N., Blume-Peytavi, U., Büscher, A., Kottner, J.: Maintaining skin integrity in the aged: a systematic review. Int. J. Nurs. Stud. **103**, 103509 (2020)
231. Hansen, W.R., Autumn, K.: Evidence for self-cleaning in gecko setae. Proc. Natl. Acad. Sci. **102**(2), 385–389 (2005)
232. Kim, S., Cheung, E., Sitti, M.: Wet self-cleaning of biologically inspired elastomer mushroom shaped microfibrillar adhesives. Langmuir. **25**(13), 7196–7199 (2009)
233. Feng, L., Li, S., Li, Y., Li, H., Zhang, L., Zhai, J., Song, Y., Liu, B., Jiang, L., Zhu, D.: Super-hydrophobic surfaces: from natural to artificial. Adv. Mater. **14**(24), 1857–1860 (2002)
234. Autumn, K., Sitti, M., Liang, Y.A., Peattie, A.M., Hansen, W.R., Sponberg, S., Kenny, T.W., Fearing, R., Israelachvili, J.N., Full, R.J.: Evidence for van der Waals adhesion in gecko setae. Proc. Natl. Acad. Sci. **99**(19), 12252–12256 (2002)
235. Kwak, M.K., Jeong, H.-E., Suh, K.Y.: Rational design and enhanced biocompatibility of a dry adhesive medical skin patch. Adv. Mater. **23**(34), 3949–3953 (2011)
236. Carbone, G., Pierro, E., Gorb, S.N.: Origin of the superior adhesive performance of mushroom-shaped microstructured surfaces. Soft Matter. **7**(12), 5545–5552 (2011)
237. Chun, S., Kim, D.W., Baik, S., Lee, H.J., Lee, J.H., Bhang, S.H., Pang, C.: Conductive and stretchable adhesive electronics with miniaturized octopus-like suckers against dry/wet skin for biosignal monitoring. Adv. Funct. Mater. **28**(52), 1805224 (2018)
238. Kim, J.-H., Kim, S.-R., Kil, H.-J., Kim, Y.-C., Park, J.-W.: Highly conformable, transparent electrodes for epidermal electronics. Nano Lett. **18**(7), 4531–4540 (2018)
239. Di, X., Kang, Y., Li, F., Yao, R., Chen, Q., Hang, C., Xu, Y., Wang, Y., Sun, P., Wu, G.: Poly(N-isopropylacrylamide)/polydopamine/clay nanocomposite hydrogels with stretchability, conductivity, and dual light- and thermo-responsive bending and adhesive properties. Colloids Surf. B: Biointerfaces. **177**, 149–159 (2019)
240. Lu, Q., Oh, D.X., Lee, Y., Jho, Y., Hwang, D.S., Zeng, H.: Nanomechanics of cation–π interactions in aqueous solution. Angew. Chem. Int. Ed. **52**(14), 3944–3948 (2013)
241. Han, L., Yan, L., Wang, M., Wang, K., Fang, L., Zhou, J., Fang, J., Ren, F., Lu, X.: Transparent, adhesive, and conductive hydrogel for soft bioelectronics based on light-transmitting polydopamine-doped polypyrrole nanofibrils. Chem. Mater. **30**(16), 5561–5572 (2018)
242. Oh, J.H., Hong, S.Y., Park, H., Jin, S.W., Jeong, Y.R., Oh, S.Y., Yun, J., Lee, H., Kim, J.W., Ha, J.S.: Fabrication of high-sensitivity skin-attachable temperature sensors with bioinspired microstructured adhesive. ACS Appl. Mater. Interfaces. **10**(8), 7263–7270 (2018)
243. Wolf, M.P., Salieb-Beugelaar, G.B., Hunziker, P.: PDMS with designer functionalities—properties, modifications strategies, and applications. Prog. Polym. Sci. **83**, 97–134 (2018)
244. Young Shik, S., Keunchang, C., Sun Hee, L., Seok, C., Sung-Jin, P., Chanil, C., Dong-Chul, H., Jun Keun, C.: PDMS-based micro PCR chip with Parylene coating. J. Micromech. Microeng. **13**(5), 768 (2003)
245. Song, E., Fang, H., Jin, X., Zhao, J., Jiang, C., Yu, K.J., Zhong, Y., Xu, D., Li, J., Fang, G., et al.: Thin, transferred layers of silicon dioxide and silicon nitride as water and ion barriers for implantable flexible electronic systems. Adv. Electron. Mater. **3**(8), 1700077 (2017)

246. Khan, M.R., Trlica, C., So, J.-H., Valeri, M., Dickey, M.D.: Influence of water on the interfacial behavior of gallium liquid metal alloys. ACS Appl. Mater. Interfaces. **6**(24), 22467–22473 (2014)
247. Hirsch, A., Dejace, L., Michaud, H.O., Lacour, S.P.: Harnessing the rheological properties of liquid metals to shape soft electronic conductors for wearable applications. Acc. Chem. Res. **52**(3), 534–544 (2019)
248. Wang, X., Yao, W., Guo, R., Yang, X., Tang, J., Zhang, J., Gao, W., Timchenko, V., Liu, J.: Soft and moldable Mg-doped liquid metal for conformable skin tumor photothermal therapy. Adv. Healthc. Mater. **7**(14), 1800318 (2018)
249. Liu, H., Qing, H., Li, Z., Han, Y.L., Lin, M., Yang, H., Li, A., Lu, T.J., Li, F., Xu, F.: Paper: A promising material for human-friendly functional wearable electronics. Mater. Sci. Eng. R Rep. **112**, 1–22 (2017)
250. Wang, M., Gorham, J.M., Killgore, J.P., Omidvar, M., Lin, H., DelRio, F.W., Cox, L.M., Zhang, Z., Ding, Y.: Formation of a crack-free, hybrid skin layer with tunable surface topography and improved gas permeation selectivity on elastomers using gel–liquid infiltration polymerization. ACS Appl. Mater. Interfaces. **9**(33), 28100–28106 (2017)
251. Zhang, J., Cheng, Y., Tebyetekerwa, M., Meng, S., Zhu, M., Lu, Y.: "Stiff–soft" binary synergistic aerogels with superflexibility and high thermal insulation performance. Adv. Funct. Mater. **29**(15), 1806407 (2019)
252. Chang, H., Sun, Z., Yuan, Q., Ding, F., Tao, X., Yan, F., Zheng, Z.: Thin film field-effect phototransistors from bandgap-tunable, solution-processed, few-layer reduced graphene oxide films. Adv. Mater. **22**(43), 4872–4876 (2010)
253. Du, W., Nie, J., Ren, Z., Jiang, T., Xu, L., Dong, S., Zheng, L., Chen, X., Li, H.: Inflammation-free and gas-permeable on-skin triboelectric nanogenerator using soluble nanofibers. Nano Energy. **51**, 260–269 (2018)
254. Liu, W., Song, M.-S., Kong, B., Cui, Y.: Flexible and stretchable energy storage: recent advances and future perspectives. Adv. Mater. **29**(1), 1603436 (2017)
255. Kang, J., Tok, J.B.H., Bao, Z.: Self-healing soft electronics. Nat. Electron. **2**(4), 144–150 (2019)
256. Thakur, V.K., Kessler, M.R.: Self-healing polymer nanocomposite materials: a review. Polymer. **69**, 369–383 (2015)
257. Blaiszik, B.J., Kramer, S.L.B., Olugebefola, S.C., Moore, J.S., Sottos, N.R., White, S.R.: Self-healing polymers and composites. Annu. Rev. Mater. Res. **40**(1), 179–211 (2010)
258. Yang, Y., Ding, X., Urban, M.W.: Chemical and physical aspects of self-healing materials. Prog. Polym. Sci. **49–50**, 34–59 (2015)
259. Son, D., Kang, J., Vardoulis, O., Kim, Y., Matsuhisa, N., Oh, J.Y., To, J.W.F., Mun, J., Katsumata, T., Liu, Y., et al.: An integrated self-healable electronic skin system fabricated via dynamic reconstruction of a nanostructured conducting network. Nat. Nanotechnol. **13**(11), 1057–1065 (2018)
260. Cordier, P., Tournilhac, F., Soulié-Ziakovic, C., Leibler, L.: Self-healing and thermoreversible rubber from supramolecular assembly. Nature. **451**(7181), 977–980 (2008)
261. Roy, A., Manna, K., Dey, S., Pal, S.: Chemical modification of β-cyclodextrin towards hydrogel formation. Carbohydr. Polym. **306**, 120576 (2023)
262. Jing, X., Mi, H.-Y., Lin, Y.-J., Enriquez, E., Peng, X.-F., Turng, L.-S.: Highly stretchable and biocompatible strain sensors based on mussel-inspired super-adhesive self-healing hydrogels for human motion monitoring. ACS Appl. Mater. Interfaces. **10**(24), 20897–20909 (2018)
263. Wang, T., Zhang, Y., Liu, Q., Cheng, W., Wang, X., Pan, L., Xu, B., Xu, H.: A self-healable, highly stretchable, and solution processable conductive polymer composite for ultrasensitive strain and pressure sensing. Adv. Funct. Mater. **28**(7), 1705551 (2018)
264. Wang, Y., Huang, F., Chen, X., Wang, X.-W., Zhang, W.-B., Peng, J., Li, J., Zhai, M.: Stretchable, conductive, and self-healing hydrogel with super metal adhesion. Chem. Mater. **30**(13), 4289–4297 (2018)
265. Yang, Y., Guan, L., Li, X., Gao, Z., Ren, X., Gao, G.: Conductive organohydrogels with ultrastretchability, antifreezing, self-healing, and adhesive properties for motion detection and signal transmission. ACS Appl. Mater. Interfaces. **11**(3), 3428–3437 (2019)

266. Khatib, M., Zohar, O., Haick, H.: Self-healing soft sensors: from material design to implementation. Adv. Mater. **33**(11), 2004190 (2021)
267. Black, J.: Biological Performance of Materials: Fundamentals of Biocompatibility, 4th edn. CRC Press (2005)
268. Wang, N., Yang, A., Fu, Y., Li, Y., Yan, F.: Functionalized organic thin film transistors for biosensing. Acc. Chem. Res. **52**(2), 277–287 (2019)
269. Wang, Z., Chen, J., Cong, Y., Zhang, H., Xu, T., Nie, L., Fu, J.: Ultrastretchable strain sensors and arrays with high sensitivity and linearity based on super tough conductive hydrogels. Chem. Mater. **30**(21), 8062–8069 (2018)
270. Wang, C., Wang, S., Chen, G., Kong, W., Ping, W., Dai, J., Pastel, G., Xie, H., He, S., Das, S., et al.: Flexible, bio-compatible nanofluidic ion conductor. Chem. Mater. **30**(21), 7707–7713 (2018)
271. Chen, J., Oh, S.K., Nabulsi, N., Johnson, H., Wang, W., Ryou, J.-H.: Biocompatible and sustainable power supply for self-powered wearable and implantable electronics using III-nitride thin-film-based flexible piezoelectric generator. Nano Energy. **57**, 670–679 (2019)
272. Kamath, S., Bhattacharyya, D., Padukudru, C., Timmons, R.B., Tang, L.: Surface chemistry influences implant-mediated host tissue responses. J. Biomed. Mater. Res. A. **86A**(3), 617–626 (2008)
273. Cattaneo, G., Bräuner, C., Siekmeyer, G., Ding, A., Bauer, S., Wohlschlögel, M., Lang, L., Hierlemann, T., Akimov, M., Schlensak, C., et al.: In vitro investigation of chemical properties and biocompatibility of neurovascular braided implants. J. Mater. Sci. Mater. Med. **30**(6), 67 (2019)
274. Tang, L., Mou, L., Zhang, W., Jiang, X.: Large-scale fabrication of highly elastic conductors on a broad range of surfaces. ACS Appl. Mater. Interfaces. **11**(7), 7138–7147 (2019)
275. Xu, G., Cheng, C., Liu, Z., Yuan, W., Wu, X., Lu, Y., Low, S.S., Liu, J., Zhu, L., Ji, D., et al.: Battery-free and wireless epidermal electrochemical system with all-printed stretchable electrode array for multiplexed in situ sweat analysis. Adv. Mater. Technol. **4**(7), 1800658 (2019)
276. Yu, X., Shou, W., Mahajan, B.K., Huang, X., Pan, H.: Materials, processes, and facile manufacturing for bioresorbable electronics: a review. Adv. Mater. **30**(28), 1707624 (2018)
277. Hwang, S.-W., Park, G., Cheng, H., Song, J.-K., Kang, S.-K., Yin, L., Kim, J.-H., Omenetto, F.G., Huang, Y., Lee, K.-M., et al.: 25th Anniversary Article: Materials for high-performance biodegradable semiconductor devices. Adv. Mater. **26**(13), 1992–2000 (2014)
278. Tan, M.J., Owh, C., Chee, P.L., Kyaw, A.K.K., Kai, D., Loh, X.J.: Biodegradable electronics: cornerstone for sustainable electronics and transient applications. J. Mater. Chem. C. **4**(24), 5531–5558 (2016)
279. Costerton, J.W., Montanaro, L., Arciola, C.R.: Biofilm in implant infections: its production and regulation. Int. J. Artif. Organs. **28**(11), 1062–1068 (2005)
280. Lee, J., Teh, S.-Y., Lee, A., Kim, H.H., Lee, C., Shung, K.K.: Single beam acoustic trapping. Appl. Phys. Lett. **95**(7), 073701 (2009)
281. Hwang, S.-W., Tao, H., Kim, D.-H., Cheng, H., Song, J.-K., Rill, E., Brenckle, M.A., Panilaitis, B., Won, S.M., Kim, Y.-S., et al.: A physically transient form of silicon electronics. Science. **337**(6102), 1640–1644 (2012)
282. Hwang, S.-W., Song, J.-K., Huang, X., Cheng, H., Kang, S.-K., Kim, B.H., Kim, J.-H., Yu, S., Huang, Y., Rogers, J.A.: High-performance biodegradable/transient electronics on biodegradable polymers. Adv. Mater. **26**(23), 3905–3911 (2014)
283. Zheng, Y.F., Gu, X.N., Witte, F.: Biodegradable metals. Mater. Sci. Eng. R Rep. **77**, 1–34 (2014)
284. Yin, L., Cheng, H., Mao, S., Haasch, R., Liu, Y., Xie, X., Hwang, S.-W., Jain, H., Kang, S.-K., Su, Y., et al.: Dissolvable metals for transient electronics. Adv. Funct. Mater. **24**(5), 645–658 (2014)
285. Yin, L., Farimani, A.B., Min, K., Vishal, N., Lam, J., Lee, Y.K., Aluru, N.R., Rogers, J.A.: Mechanisms for hydrolysis of silicon nanomembranes as used in bioresorbable electronics. Adv. Mater. **27**(11), 1857–1864 (2015)

286. Hwang, S.-W., Park, G., Edwards, C., Corbin, E.A., Kang, S.-K., Cheng, H., Song, J.-K., Kim, J.-H., Yu, S., Ng, J., et al.: Dissolution chemistry and biocompatibility of single-crystalline silicon nanomembranes and associated materials for transient electronics. ACS Nano. **8**(6), 5843–5851 (2014)
287. Kang, S.-K., Park, G., Kim, K., Hwang, S.-W., Cheng, H., Shin, J., Chung, S., Kim, M., Yin, L., Lee, J.C., et al.: Dissolution chemistry and biocompatibility of silicon- and germanium-based semiconductors for transient electronics. ACS Appl. Mater. Interfaces. **7**(17), 9297–9305 (2015)
288. Kang, S.-K., Hwang, S.-W., Cheng, H., Yu, S., Kim, B.H., Kim, J.-H., Huang, Y., Rogers, J.A.: Dissolution behaviors and applications of silicon oxides and nitrides in transient electronics. Adv. Funct. Mater. **24**(28), 4427–4434 (2014)
289. Li, R., Cheng, H., Su, Y., Hwang, S.-W., Yin, L., Tao, H., Brenckle, M.A., Kim, D.-H., Omenetto, F.G., Rogers, J.A., et al.: An analytical model of reactive diffusion for transient electronics. Adv. Funct. Mater. **23**(24), 3106–3114 (2013)
290. Kang, S.-K., Koo, J., Lee, Y.K., Rogers, J.A.: Advanced materials and devices for bioresorbable electronics. Acc. Chem. Res. **51**(5), 988–998 (2018)
291. Lendlein, A., Langer, R.: Biodegradable, elastic shape-memory polymers for potential biomedical applications. Science. **296**(5573), 1673–1676 (2002)
292. Soppimath, K.S., Aminabhavi, T.M., Kulkarni, A.R., Rudzinski, W.E.: Biodegradable polymeric nanoparticles as drug delivery devices. J. Control. Release. **70**(1), 1–20 (2001)
293. Miyamoto, H., Ogura, Y., Hashizoe, M., Kunou, N., Honda, Y., Ikada, Y.: Biodegradable scleral implant for intravitreal controlled release of fluconazole. Curr. Eye Res. **16**(9), 930–935 (1997)
294. Venkatraman, S.S., Tan, L.P., Joso, J.F.D., Boey, Y.C.F., Wang, X.: Biodegradable stents with elastic memory. Biomaterials. **27**(8), 1573–1578 (2006)
295. Hermawan, H., Dubé, D., Mantovani, D.: Developments in metallic biodegradable stents. Acta Biomater. **6**(5), 1693–1697 (2010)
296. Eppley, B.L.: Use of resorbable plates and screws in pediatric facial fractures. J. Oral Maxillofac. Surg. **63**(3), 385–391 (2005)
297. Lei, T., Guan, M., Liu, J., Lin, H.-C., Pfattner, R., Shaw, L., McGuire, A.F., Huang, T.-C., Shao, L., Cheng, K.-T., et al.: Biocompatible and totally disintegrable semiconducting polymer for ultrathin and ultralightweight transient electronics. Proc. Natl. Acad. Sci. **114**(20), 5107–5112 (2017)
298. Yu, K.J., Kuzum, D., Hwang, S.-W., Kim, B.H., Juul, H., Kim, N.H., Won, S.M., Chiang, K., Trumpis, M., Richardson, A.G., et al.: Bioresorbable silicon electronics for transient spatiotemporal mapping of electrical activity from the cerebral cortex. Nat. Mater. **15**(7), 782–791 (2016)
299. Gao, Z., Agarwal, A., Trigg, A.D., Singh, N., Fang, C., Tung, C.-H., Fan, Y., Buddharaju, K.D., Kong, J.: Silicon nanowire arrays for label-free detection of DNA. Anal. Chem. **79**(9), 3291–3297 (2007)
300. Patolsky, F., Zheng, G., Lieber, C.M.: Fabrication of silicon nanowire devices for ultrasensitive, label-free, real-time detection of biological and chemical species. Nat. Protoc. **1**(4), 1711–1724 (2006)
301. Zheng, G., Patolsky, F., Cui, Y., Wang, W.U., Lieber, C.M.: Multiplexed electrical detection of cancer markers with nanowire sensor arrays. Nat. Biotechnol. **23**(10), 1294–1301 (2005)
302. Salvatore, G.A., Sülzle, J., Dalla Valle, F., Cantarella, G., Robotti, F., Jokic, P., Knobelspies, S., Daus, A., Büthe, L., Petti, L., et al.: Biodegradable and highly deformable temperature sensors for the internet of things. Adv. Funct. Mater. **27**(35), 1702390 (2017)
303. Guo, Y., Zhong, M., Fang, Z., Wan, P., Yu, G.: A wearable transient pressure sensor made with MXene nanosheets for sensitive broad-range human–machine interfacing. Nano Lett. **19**(2), 1143–1150 (2019)
304. Kim, D.-H., Kim, Y.-S., Amsden, J., Panilaitis, B., Kaplan, D.L., Omenetto, F.G., Zakin, M.R., Rogers, J.A.: Silicon electronics on silk as a path to bioresorbable, implantable devices. Appl. Phys. Lett. **95**(13), 133701 (2009)

305. Tao, H., Hwang, S.-W., Marelli, B., An, B., Moreau, J.E., Yang, M., Brenckle, M.A., Kim, S., Kaplan, D.L., Rogers, J.A., et al.: Silk-based resorbable electronic devices for remotely controlled therapy and in vivo infection abatement. Proc. Natl. Acad. Sci. **111**(49), 17385–17389 (2014)
306. Makadia, H.K., Siegel, S.J.: Poly lactic-co-glycolic acid (PLGA) as biodegradable controlled drug delivery carrier. Polymers. **3**(3), 1377–1397 (2011)
307. McCoul, D., Hu, W., Gao, M., Mehta, V., Pei, Q.: Recent advances in stretchable and transparent electronic materials. Adv. Electron. Mater. **2**(5), 1500407 (2016)
308. Wang, C., Zheng, W., Yue, Z., Too, C.O., Wallace, G.G.: Buckled, stretchable polypyrrole electrodes for battery applications. Adv. Mater. **23**(31), 3580–3584 (2011)
309. Li, L., Zhang, Y., Lu, H., Wang, Y., Xu, J., Zhu, J., Zhang, C., Liu, T.: Cryopolymerization enables anisotropic polyaniline hybrid hydrogels with superelasticity and highly deformation-tolerant electrochemical energy storage. Nat. Commun. **11**(1), 62 (2020)
310. Hansen, T.S., West, K., Hassager, O., Larsen, N.B.: Highly stretchable and conductive polymer material made from poly(3,4-ethylenedioxythiophene) and polyurethane elastomers. Adv. Funct. Mater. **17**(16), 3069–3073 (2007)
311. Samba, R., Herrmann, T., Zeck, G.: PEDOT–CNT coated electrodes stimulate retinal neurons at low voltage amplitudes and low charge densities. J. Neural Eng. **12**(1), 016014 (2015)
312. Gozen, B.A., Tabatabai, A., Ozdoganlar, O.B., Majidi, C.: High-density soft-matter electronics with micron-scale line width. Adv. Mater. **26**(30), 5211–5216 (2014)
313. Xu, S., Yan, Z., Jang, K.-I., Huang, W., Fu, H., Kim, J., Wei, Z., Flavin, M., McCracken, J., Wang, R., et al.: Assembly of micro/nanomaterials into complex, three-dimensional architectures by compressive buckling. Science. **347**(6218), 154–159 (2015)
314. Yin, R., Xu, Z., Mei, M., Chen, Z., Wang, K., Liu, Y., Tang, T., Priydarshi, M.K., Meng, X., Zhao, S., et al.: Soft transparent graphene contact lens electrodes for conformal full-cornea recording of electroretinogram. Nat. Commun. **9**(1), 2334 (2018)
315. Owens, R.M., Malliaras, G.G.: Organic electronics at the interface with biology. MRS Bull. **35**(6), 449–456 (2010)
316. Park, J., Choi, S., Janardhan, A.H., Lee, S.-Y., Raut, S., Soares, J., Shin, K., Yang, S., Lee, C., Kang, K.-W., et al.: Electromechanical cardioplasty using a wrapped elasto-conductive epicardial mesh. Sci. Transl. Med. **8**(344), 344ra386 (2016)
317. Xu, J.-L., Liu, Y.-H., Gao, X., Shen, S., Wang, S.-D.: Toward wearable electronics: a lightweight all-solid-state supercapacitor with outstanding transparency, foldability and breathability. Energy Storage Mater. **22**, 402–409 (2019)
318. Wu, H., Liu, Q., Du, W., Li, C., Shi, G.: Transparent polymeric strain sensors for monitoring vital signs and beyond. ACS Appl. Mater. Interfaces. **10**(4), 3895–3901 (2018)
319. Wang, L., Daoud, W.A.: Hybrid conductive hydrogels for washable human motion energy harvester and self-powered temperature-stress dual sensor. Nano Energy. **66**, 104080 (2019)
320. Duan, J., Wen, H., Zong, S., Li, T., Lv, H., Liu, L.: Soft/hard controllable conversion galactomannan ionic conductive hydrogel as a flexible sensor. ACS Appl. Electron. Mater. **3**(11), 5000–5014 (2021)
321. Liu, C., Li, Q., Wang, S., Liu, W., Fang, N.X., Feng, S.-P.: Ion regulation in double-network hydrogel module with ultrahigh thermopower for low-grade heat harvesting. Nano Energy. **92**, 106738 (2022)
322. Lu, B., Yuk, H., Lin, S., Jian, N., Qu, K., Xu, J., Zhao, X.: Pure PEDOT:PSS hydrogels. Nat Commun. **10**(1), 1043 (2019)
323. Gu, H., Zhang, H., Ma, C., Sun, H., Liu, C., Dai, K., Zhang, J., Wei, R., Ding, T., Guo, Z.: Smart strain sensing organic–inorganic hybrid hydrogels with nano barium ferrite as the cross-linker. J. Mater. Chem. C. **7**(8), 2353–2360 (2019)
324. Sun, Z., Dong, C., Chen, B., Li, W., Hu, H., Zhou, J., Li, C., Huang, Z.: Strong, tough, and anti-swelling supramolecular conductive hydrogels for amphibious motion sensors. Small. **19**(44), 2303612 (2023)
325. Chen, X., Hao, W., Lu, T., Wang, T., Shi, C., Zhao, Y., Liu, Y.: Stretchable zwitterionic conductive hydrogels with semi-interpenetrating network based on polyaniline for flexible strain sensors. Macromol. Chem. Phys. **222**(24), 2100165 (2021)

326. Su, G., Zhang, Y., Zhang, X., Feng, J., Cao, J., Zhang, X., Zhou, T.: Soft yet tough: a mechanically and functionally tissue-like organohydrogel for sensitive soft electronics. Chem. Mater. **34**(3), 1392–1402 (2022)
327. Gao, Y., Jia, F., Gao, G.: Transparent and conductive amino acid-tackified hydrogels as wearable strain sensors. Chem. Eng. J. **375**, 121915 (2019)
328. Zhou, Y., Wan, C., Yang, Y., Yang, H., Wang, S., Dai, Z., Ji, K., Jiang, H., Chen, X., Long, Y.: Highly stretchable, elastic, and ionic conductive hydrogel for artificial soft electronics. Adv. Funct. Mater. **29**(1), 1806220 (2019)
329. Hao, X.P., Li, C.Y., Zhang, C.W., Du, M., Ying, Z., Zheng, Q., Wu, Z.L.: Self-shaping soft electronics based on patterned hydrogel with stencil-printed liquid metal. Adv. Funct. Mater. **31**(47), 2105481 (2021)
330. Zhang, X., Liu, W., Cai, J., Huang, J., Qiu, X.: Equip the hydrogel with armor: strong and super tough biomass reinforced hydrogels with excellent conductivity and anti-bacterial performance. J. Mater. Chem. A. **7**(47), 26917–26926 (2019)
331. Nie, Z., Peng, K., Lin, L., Yang, J., Cheng, Z., Gan, Q., Chen, Y., Feng, C.: A conductive hydrogel based on nature polymer agar with self-healing ability and stretchability for flexible sensors. Chem. Eng. J. **454**, 139843 (2023)
332. Kadumudi, F.B., Hasany, M., Pierchala, M.K., Jahanshahi, M., Taebnia, N., Mehrali, M., Mitu, C.F., Shahbazi, M.-A., Zsurzsan, T.-G., Knott, A., et al.: The manufacture of unbreakable bionics via multifunctional and self-healing silk–graphene hydrogels. Adv. Mater. **33**(35), 2100047 (2021)
333. Chae, A., Murali, G., Lee, S.-Y., Gwak, J., Kim, S.J., Jeong, Y.J., Kang, H., Park, S., Lee, A.S., Koh, D.-Y., et al.: Highly oxidation-resistant and self-healable MXene-based hydrogels for wearable strain sensor. Adv. Funct. Mater. **33**(24), 2213382 (2023)
334. Zhou, Q., Lyu, J., Wang, G., Robertson, M., Qiang, Z., Sun, B., Ye, C., Zhu, M.: Mechanically strong and multifunctional hybrid hydrogels with ultrahigh electrical conductivity. Adv. Funct. Mater. **31**(40), 2104536 (2021)
335. Cai, Y., Shen, J., Ge, G., Zhang, Y., Jin, W., Huang, W., Shao, J., Yang, J., Dong, X.: Stretchable Ti3C2Tx MXene/carbon nanotube composite based strain sensor with ultrahigh sensitivity and tunable sensing range. ACS Nano. **12**(1), 56–62 (2018)
336. Yang, Y., Shi, L., Cao, Z., Wang, R., Sun, J.: Strain sensors with a high sensitivity and a wide sensing range based on a Ti3C2Tx (MXene) nanoparticle–nanosheet hybrid network. Adv. Funct. Mater. **29**(14), 1807882 (2019)
337. Pang, Y., Zhang, K., Yang, Z., Jiang, S., Ju, Z., Li, Y., Wang, X., Wang, D., Jian, M., Zhang, Y., et al.: Epidermis microstructure inspired graphene pressure sensor with random distributed spinosum for high sensitivity and large linearity. ACS Nano. **12**(3), 2346–2354 (2018)
338. Yin, B., Liu, X., Gao, H., Fu, T., Yao, J.: Bioinspired and bristled microparticles for ultrasensitive pressure and strain sensors. Nat. Commun. **9**(1), 5161 (2018)
339. Guo, R., Liu, J.: Implantable liquid metal-based flexible neural microelectrode array and its application in recovering animal locomotion functions. J. Micromech. Microeng. **27**(10), 104002 (2017)
340. Kim, H.-J., Son, C., Ziaie, B.: A multiaxial stretchable interconnect using liquid-alloy-filled elastomeric microchannels. Appl. Phys. Lett. **92**(1), 011904 (2008)
341. Siegel, A.C., Bruzewicz, D.A., Weibel, D.B., Whitesides, G.M.: Microsolidics: fabrication of three-dimensional metallic microstructures in poly(dimethylsiloxane). Adv. Mater. **19**(5), 727–733 (2007)
342. Dickey, M.D., Chiechi, R.C., Larsen, R.J., Weiss, E.A., Weitz, D.A., Whitesides, G.M.: Eutectic gallium-indium (EGaIn): a liquid metal alloy for the formation of stable structures in microchannels at room temperature. Adv. Funct. Mater. **18**(7), 1097–1104 (2008)
343. Khondoker, M.A.H., Sameoto, D.: Fabrication methods and applications of microstructured gallium based liquid metal alloys. Smart Mater. Struct. **25**(9), 093001 (2016)
344. So, J.-H., Thelen, J., Qusba, A., Hayes, G.J., Lazzi, G., Dickey, M.D.: Reversibly deformable and mechanically tunable fluidic antennas. Adv. Funct. Mater. **19**(22), 3632–3637 (2009)

345. Chiechi, R.C., Weiss, E.A., Dickey, M.D., Whitesides, G.M.: Eutectic gallium–indium (EGaIn): a moldable liquid metal for electrical characterization of self-assembled monolayers. Angew. Chem. Int. Ed. **47**(1), 142–144 (2008)
346. Lide, D.R.: CRC Handbook of Chemistry and Physics, vol. 85, 85th edn. CRC Press (2004)
347. Ota, H., Chen, K., Lin, Y., Kiriya, D., Shiraki, H., Yu, Z., Ha, T.-J., Javey, A.: Highly deformable liquid-state heterojunction sensors. Nat. Commun. **5**(1), 5032 (2014)
348. Gannarapu, A., Gozen, B.A.: Freeze-printing of liquid metal alloys for manufacturing of 3D, conductive, and flexible networks. Adv. Mater. Technol. **1**(4), 1600047 (2016)
349. Choi, M.K., Park, I., Kim, D.C., Joh, E., Park, O.K., Kim, J., Kim, M., Choi, C., Yang, J., Cho, K.W., et al.: Thermally controlled, patterned graphene transfer printing for transparent and wearable electronic/optoelectronic system. Adv. Funct. Mater. **25**(46), 7109–7118 (2015)
350. Son, D., Koo, J.H., Song, J.-K., Kim, J., Lee, M., Shim, H.J., Park, M., Lee, M., Kim, J.H., Kim, D.-H.: Stretchable carbon nanotube charge-trap floating-gate memory and logic devices for wearable electronics. ACS Nano. **9**(5), 5585–5593 (2015)
351. Kim, J., Ghaffari, R., Kim, D.-H.: The quest for miniaturized soft bioelectronic devices. Nat. Biomed. Eng. **1**(3), 0049 (2017)
352. Koo, J.H., Song, J.-K., Kim, D.-H.: Solution-processed thin films of semiconducting carbon nanotubes and their application to soft electronics. Nanotechnology. **30**(13), 132001 (2019)
353. Lee, E., Kim, H.J., Park, Y., Lee, S., Lee, S.Y., Ha, T., Shin, H.-J., Kim, Y., Kim, J.: Direct patterning of a carbon nanotube thin layer on a stretchable substrate. Micromachines. **10**(8), 530 (2019)
354. Gilshteyn, E.P., Lin, S., Kondrashov, V.A., Kopylova, D.S., Tsapenko, A.P., Anisimov, A.S., Hart, A.J., Zhao, X., Nasibulin, A.G.: A one-step method of hydrogel modification by single-walled carbon nanotubes for highly stretchable and transparent electronics. ACS Appl. Mater. Interfaces. **10**(33), 28069–28075 (2018)
355. Li, X., Li, Y., Guan, T., Xu, F., Sun, J.: Durable, highly electrically conductive cotton fabrics with healable superamphiphobicity. ACS Appl. Mater. Interfaces. **10**(14), 12042–12050 (2018)
356. Yang, M., Pan, J., Xu, A., Luo, L., Cheng, D., Cai, G., Wang, J., Tang, B., Wang, X.: Conductive cotton fabrics for motion sensing and heating applications. Polymers. **10**(6), 568 (2018)
357. Lin, Z.-I., Lou, C.-W., Pan, Y.-J., Hsieh, C.-T., Huang, C.-H., Huang, C.-L., Chen, Y.-S., Lin, J.-H.: Conductive fabrics made of polypropylene/multi-walled carbon nanotube coated polyester yarns: mechanical properties and electromagnetic interference shielding effectiveness. Compos. Sci. Technol. **141**, 74–82 (2017)
358. Hu, L., Pasta, M., La Mantia, F., Cui, L., Jeong, S., Deshazer, H.D., Choi, J.W., Han, S.M., Cui, Y.: Stretchable, porous, and conductive energy textiles. Nano Lett. **10**(2), 708–714 (2010)
359. Jia, L.-C., Xu, L., Ren, F., Ren, P.-G., Yan, D.-X., Li, Z.-M.: Stretchable and durable conductive fabric for ultrahigh performance electromagnetic interference shielding. Carbon. **144**, 101–108 (2019)
360. Tseghai, G.B., Mengistie, D.A., Malengier, B., Fante, K.A., Van Langenhove, L.: PEDOT:PSS-based conductive textiles and their applications. Sensors. **20**(7), 1881 (2020)
361. Tadesse, M.G., Mengistie, D.A., Chen, Y., Wang, L., Loghin, C., Nierstrasz, V.: Electrically conductive highly elastic polyamide/lycra fabric treated with PEDOT:PSS and polyurethane. J. Mater. Sci. **54**(13), 9591–9602 (2019)
362. Song, W.-L., Gong, C., Li, H., Cheng, X.-D., Chen, M., Yuan, X., Chen, H., Yang, Y., Fang, D.: Graphene-based sandwich structures for frequency selectable electromagnetic shielding. ACS Appl. Mater. Interfaces. **9**(41), 36119–36129 (2017)
363. Zhao, Z., Huang, Q., Yan, C., Liu, Y., Zeng, X., Wei, X., Hu, Y., Zheng, Z.: Machine-washable and breathable pressure sensors based on triboelectric nanogenerators enabled by textile technologies. Nano Energy. **70**, 104528 (2020)
364. Sunwoo, S.-H., Ha, K.-H., Lee, S., Lu, N., Kim, D.-H.: Wearable and implantable soft bioelectronics: device designs and material strategies. Annu. Rev. Chem. Biomol. Eng. **12**(1), 359–391 (2021)

365. Liu, T., Finn, L., Yu, M., Wang, H., Zhai, T., Lu, X., Tong, Y., Li, Y.: Polyaniline and polypyrrole pseudocapacitor electrodes with excellent cycling stability. Nano Lett. **14**(5), 2522–2527 (2014)
366. Dubarry, M., Liaw, B.Y.: Identify capacity fading mechanism in a commercial $LiFePO_4$ cell. J. Power Sources. **194**(1), 541–549 (2009)
367. Ganji, M., Elthakeb, A.T., Tanaka, A., Gilja, V., Halgren, E., Dayeh, S.A.: Scaling effects on the electrochemical performance of poly(3,4-ethylenedioxythiophene) (PEDOT), Au, and Pt for electrocorticography recording. Adv. Funct. Mater. **27**(42), 1703018 (2017)
368. Ganji, M., Tanaka, A., Gilja, V., Halgren, E., Dayeh, S.A.: Scaling effects on the electrochemical stimulation performance of Au, Pt, and PEDOT:PSS electrocorticography arrays. Adv. Funct. Mater. **27**(42), 1703019 (2017)
369. Ganji, M., Kaestner, E., Hermiz, J., Rogers, N., Tanaka, A., Cleary, D., Lee, S.H., Snider, J., Halgren, M., Cosgrove, G.R., et al.: Development and translation of PEDOT:PSS microelectrodes for intraoperative monitoring. Adv. Funct. Mater. **28**(12), 1700232 (2018)
370. Fang, Y., Li, Y., Li, Y., Ding, M., Xie, J., Hu, B.: Solution-processed submicron free-standing, conformal, transparent, breathable epidermal electrodes. ACS Appl. Mater. Interfaces. **12**(21), 23689–23696 (2020)
371. Wang, Y., Zhu, C., Pfattner, R., Yan, H., Jin, L., Chen, S., Molina-Lopez, F., Lissel, F., Liu, J., Rabiah, N.I., et al.: A highly stretchable, transparent, and conductive polymer. Sci. Adv. **3**(3), e1602076 (2017)
372. Carpi, F., De Rossi, D., Kornbluh, R., Pelrine, R., Sommer-Larsen, P. (eds.): Dielectric Elastomers as Electromechanical Transducers. Elsevier, Amsterdam (2008)
373. Hu, L., Yuan, W., Brochu, P., Gruner, G., Pei, Q.: Highly stretchable, conductive, and transparent nanotube thin films. Appl. Phys. Lett. **94**(16), 161108 (2009)
374. Smart, S.K., Cassady, A.I., Lu, G.Q., Martin, D.J.: The biocompatibility of carbon nanotubes. Carbon. **44**(6), 1034–1047 (2006)
375. Tian, F., Cui, D., Schwarz, H., Estrada, G.G., Kobayashi, H.: Cytotoxicity of single-wall carbon nanotubes on human fibroblasts. Toxicol. In Vitro. **20**(7), 1202–1212 (2006)
376. Alarifi, S., Ali, D., Verma, A., Almajhdi, F.N., Al-Qahtani, A.A.: Single-walled carbon nanotubes induce cytotoxicity and DNA damage via reactive oxygen species in human hepatocarcinoma cells. In Vitro Cell. Dev. Biol. Anim. **50**(8), 714–722 (2014)
377. Dong, J., Ma, Q.: Advances in mechanisms and signaling pathways of carbon nanotube toxicity. Nanotoxicology. **9**(5), 658–676 (2015)
378. Yuk, H., Lu, B., Zhao, X.: Hydrogel bioelectronics. Chem. Soc. Rev. **48**(6), 1642–1667 (2019)
379. Zhang, Y., Gong, M., Wan, P.: MXene hydrogel for wearable electronics. Matter. **4**(8), 2655–2658 (2021)
380. Aregueta-Robles, U.A., Woolley, A.J., Poole-Warren, L.A., Lovell, N.H., Green, R.A.: Organic electrode coatings for next-generation neural interfaces. Front Neuroeng. **7**, 15 (2014)
381. Roy, A., Guha Ray, P., Manna, K., Banerjee, C., Dhara, S., Pal, S.: Poly(N-vinyl imidazole) cross-linked β-cyclodextrin hydrogel for rapid hemostasis in severe renal arterial hemorrhagic model. Biomacromolecules. **22**(12), 5256–5269 (2021)
382. Roy, A., Manna, K., Pal, S.: Recent advances in various stimuli-responsive hydrogels: from synthetic designs to emerging healthcare applications. Mater. Chem. Front. **6**(17), 2338–2385 (2022)
383. Roy, A., Maity, P.P., Bose, A., Dhara, S., Pal, S.: β-Cyclodextrin based pH and thermoresponsive biopolymeric hydrogel as a dual drug carrier. Mater. Chem. Front. **3**(3), 385–393 (2019)
384. Roy, A., Guha Ray, P., Bose, A., Dhara, S., Pal, S.: pH-responsive copolymeric network gel using methacrylated β-cyclodextrin for controlled codelivery of hydrophilic and hydrophobic drugs. ACS Appl. Bio Mater. **5**(7), 3530–3543 (2022)
385. Keplinger, C., Sun, J.-Y., Foo, C.C., Rothemund, P., Whitesides, G.M., Suo, Z.: Stretchable, transparent, ionic conductors. Science. **341**(6149), 984–987 (2013)

386. Cha, G.D., Lee, W.H., Lim, C., Choi, M.K., Kim, D.-H.: Materials engineering, processing, and device application of hydrogel nanocomposites. Nanoscale. **12**(19), 10456–10473 (2020)
387. Choi, S., Han, S.I., Kim, D., Hyeon, T., Kim, D.-H.: High-performance stretchable conductive nanocomposites: materials, processes, and device applications. Chem. Soc. Rev. **48**(6), 1566–1595 (2019)
388. Wang, H., Biswas, S.K., Zhu, S., Lu, Y., Yue, Y., Han, J., Xu, X., Wu, Q., Xiao, H.: Self-healable electro-conductive hydrogels based on core-shell structured nanocellulose/carbon nanotubes hybrids for use as flexible supercapacitors. Nanomaterials. **10**(1), 112 (2020)
389. Fantino, E., Chiappone, A., Roppolo, I., Manfredi, D., Bongiovanni, R., Pirri, C.F., Calignano, F.: 3D printing of conductive complex structures with in situ generation of silver nanoparticles. Adv. Mater. **28**(19), 3712–3717 (2016)
390. Navaei, A., Saini, H., Christenson, W., Sullivan, R.T., Ros, R., Nikkhah, M.: Gold nanorod-incorporated gelatin-based conductive hydrogels for engineering cardiac tissue constructs. Acta Biomater. **41**, 133–146 (2016)
391. Wei, Q.-B., Luo, Y.-L., Zhang, C.-H., Fan, L.-H., Chen, Y.-S.: Assembly of Cu nanoparticles in a polyacrylamide grafted poly(vinyl alcohol) copolymer matrix and vapor-induced response. Sensors Actuators B Chem. **134**(1), 49–56 (2008)
392. Liu, Q., Zhang, M., Huang, L., Li, Y., Chen, J., Li, C., Shi, G.: High-quality graphene ribbons prepared from graphene oxide hydrogels and their application for strain sensors. ACS Nano. **9**(12), 12320–12326 (2015)
393. Xiao, X., Wu, G., Zhou, H., Qian, K., Hu, J.: Preparation and property evaluation of conductive hydrogel using poly(vinyl alcohol)/polyethylene glycol/graphene oxide for human electrocardiogram acquisition. Polymers. **9**(7), 259 (2017)
394. Liang, Y., Zhao, X., Hu, T., Chen, B., Yin, Z., Ma, P.X., Guo, B.: Adhesive hemostatic conducting injectable composite hydrogels with sustained drug release and photothermal antibacterial activity to promote full-thickness skin regeneration during wound healing. Small. **15**(12), 1900046 (2019)
395. Mano, N., Yoo, J.E., Tarver, J., Loo, Y.-L., Heller, A.: An electron-conducting cross-linked polyaniline-based redox hydrogel, formed in one step at pH 7.2, wires glucose oxidase. J. Am. Chem. Soc. **129**(22), 7006–7007 (2007)
396. Guarino, V., Alvarez-Perez, M.A., Borriello, A., Napolitano, T., Ambrosio, L.: Conductive PANi/PEGDA macroporous hydrogels for nerve regeneration. Adv. Healthc. Mater. **2**(1), 218–227 (2013)
397. Hur, J., Im, K., Kim, S.W., Kim, J., Chung, D.-Y., Kim, T.-H., Jo, K.H., Hahn, J.H., Bao, Z., Hwang, S., et al.: Polypyrrole/agarose-based electronically conductive and reversibly restorable hydrogel. ACS Nano. **8**(10), 10066–10076 (2014)
398. Rong, Q., Lei, W., Chen, L., Yin, Y., Zhou, J., Liu, M.: Anti-freezing, conductive self-healing organohydrogels with stable strain-sensitivity at subzero temperatures. Angew. Chem. Int. Ed. **56**(45), 14159–14163 (2017)
399. Li, L., Wang, Y., Pan, L., Shi, Y., Cheng, W., Shi, Y., Yu, G.: A nanostructured conductive hydrogels-based biosensor platform for human metabolite detection. Nano Lett. **15**(2), 1146–1151 (2015)
400. Das, S., Chakraborty, P., Mondal, S., Shit, A., Nandi, A.K.: Enhancement of energy storage and photoresponse properties of folic acid–polyaniline hybrid hydrogel by in situ growth of Ag nanoparticles. ACS Appl. Mater. Interfaces. **8**(41), 28055–28067 (2016)
401. Li, P., Jin, Z., Peng, L., Zhao, F., Xiao, D., Jin, Y., Yu, G.: Stretchable all-gel-state fiber-shaped supercapacitors enabled by macromolecularly interconnected 3D graphene/nanostructured conductive polymer hydrogels. Adv. Mater. **30**(18), 1800124 (2018)
402. Guo, Y., Zheng, K., Wan, P.: A flexible stretchable hydrogel electrolyte for healable all-in-one configured supercapacitors. Small. **14**(14), 1704497 (2018)
403. Rong, Q., Han, H., Feng, F., Ma, Z.: Network nanostructured polypyrrole hydrogel/Au composites as enhanced electrochemical biosensing platform. Sci. Rep. **5**(1), 11440 (2015)
404. Yang, J., Wang, X., Li, B., Ma, L., Shi, L., Xiong, Y., Xu, H.: Novel iron/cobalt-containing polypyrrole hydrogel-derived trifunctional electrocatalyst for self-powered overall water splitting. Adv. Funct. Mater. **27**(17), 1606497 (2017)

405. Shi, Y., Zhang, J., Bruck, A.M., Zhang, Y., Li, J., Stach, E.A., Takeuchi, K.J., Marschilok, A.C., Takeuchi, E.S., Yu, G.: A tunable 3D nanostructured conductive gel framework electrode for high-performance lithium ion batteries. Adv. Mater. **29**(22), 1603922 (2017)
406. Deng, Z., Guo, Y., Zhao, X., Ma, P.X., Guo, B.: Multifunctional stimuli-responsive hydrogels with self-healing, high conductivity, and rapid recovery through host–guest interactions. Chem. Mater. **30**(5), 1729–1742 (2018)
407. Mao, L., Guan, C., Huang, X., Ke, Q., Zhang, Y., Wang, J.: 3D graphene-nickel hydroxide hydrogel electrode for high-performance supercapacitor. Electrochim. Acta. **196**, 653–660 (2016)
408. Wang, R., Jayakumar, A., Xu, C., Lee, J.-M.: $Ni(OH)_2$ nanoflowers/graphene hydrogels: a new assembly for supercapacitors. ACS Sustain. Chem. Eng. **4**(7), 3736–3742 (2016)
409. Javadi, M., Gu, Q., Naficy, S., Farajikhah, S., Crook, J.M., Wallace, G.G., Beirne, S., Moulton, S.E.: Conductive tough hydrogel for bioapplications. Macromol. Biosci. **18**(2), 1700270 (2018)
410. De, B., Kuila, T., Kim, N.H., Lee, J.H.: Carbon dot stabilized copper sulphide nanoparticles decorated graphene oxide hydrogel for high performance asymmetric supercapacitor. Carbon. **122**, 247–257 (2017)
411. Wang, H., Zhu, B., Jiang, W., Yang, Y., Leow, W.R., Wang, H., Chen, X.: A mechanically and electrically self-healing supercapacitor. Adv. Mater. **26**(22), 3638–3643 (2014)
412. Huang, Y., Zhong, M., Shi, F., Liu, X., Tang, Z., Wang, Y., Huang, Y., Hou, H., Xie, X., Zhi, C.: An intrinsically stretchable and compressible supercapacitor containing a polyacrylamide hydrogel electrolyte. Angew. Chem. Int. Ed. **56**(31), 9141–9145 (2017)
413. Pu, X., Liu, M., Chen, X., Sun, J., Du, C., Zhang, Y., Zhai, J., Hu, W., Wang, Z.L.: Ultrastretchable, transparent triboelectric nanogenerator as electronic skin for biomechanical energy harvesting and tactile sensing. Sci. Adv. **3**(5), e1700015 (2017)
414. Lei, Z., Wang, Q., Sun, S., Zhu, W., Wu, P.: A bioinspired mineral hydrogel as a self-healable, mechanically adaptable ionic skin for highly sensitive pressure sensing. Adv. Mater. **29**(22), 1700321 (2017)
415. Si, Y., Wang, L., Wang, X., Tang, N., Yu, J., Ding, B.: Ultrahigh-water-content, superelastic, and shape-memory nanofiber-assembled hydrogels exhibiting pressure-responsive conductivity. Adv. Mater. **29**(24), 1700339 (2017)
416. Liu, S., Li, K., Hussain, I., Oderinde, O., Yao, F., Zhang, J., Fu, G.: A conductive self-healing double network hydrogel with toughness and force sensitivity. Chemistry. **24**(25), 6632–6638 (2018)
417. Lai, J., Zhou, H., Wang, M., Chen, Y., Jin, Z., Li, S., Yang, J., Jin, X., Liu, H., Zhao, W.: Recyclable, stretchable and conductive double network hydrogels towards flexible strain sensors. J. Mater. Chem. C. **6**(48), 13316–13324 (2018)
418. Schroeder, T.B.H., Guha, A., Lamoureux, A., VanRenterghem, G., Sept, D., Shtein, M., Yang, J., Mayer, M.: An electric-eel-inspired soft power source from stacked hydrogels. Nature. **552**(7684), 214–218 (2017)
419. Huo, Z., Tao, L., Wang, S., Wei, J., Zhu, J., Dong, W., Liu, F., Chen, S., Zhang, B., Dai, S.: A novel polysulfide hydrogel electrolyte based on low molecular mass organogelator for quasi-solid-state quantum dot-sensitized solar cells. J. Power Sources. **284**, 582–587 (2015)
420. Liu, X., Wu, D., Wang, H., Wang, Q.: Self-recovering tough gel electrolyte with adjustable supercapacitor performance. Adv. Mater. **26**(25), 4370–4375 (2014)
421. He, X., Zhang, C., Wang, M., Zhang, Y., Liu, L., Yang, W.: An electrically and mechanically autonomic self-healing hybrid hydrogel with tough and thermoplastic properties. ACS Appl. Mater. Interfaces. **9**(12), 11134–11143 (2017)
422. Hao, G.-P., Hippauf, F., Oschatz, M., Wisser, F.M., Leifert, A., Nickel, W., Mohamed-Noriega, N., Zheng, Z., Kaskel, S.: Stretchable and semitransparent conductive hybrid hydrogels for flexible supercapacitors. ACS Nano. **8**(7), 7138–7146 (2014)
423. Wang, K., Zhang, X., Li, C., Sun, X., Meng, Q., Ma, Y., Wei, Z.: Chemically crosslinked hydrogel film leads to integrated flexible supercapacitors with superior performance. Adv. Mater. **27**(45), 7451–7457 (2015)

424. Liao, M., Wan, P., Wen, J., Gong, M., Wu, X., Wang, Y., Shi, R., Zhang, L.: Wearable, healable, and adhesive epidermal sensors assembled from mussel-inspired conductive hybrid hydrogel framework. Adv. Funct. Mater. **27**(48), 1703852 (2017)
425. Yao, B., Wang, H., Zhou, Q., Wu, M., Zhang, M., Li, C., Shi, G.: Ultrahigh-conductivity polymer hydrogels with arbitrary structures. Adv. Mater. **29**(28), 1700974 (2017)
426. Kleber, C., Bruns, M., Lienkamp, K., Rühe, J., Asplund, M.: An interpenetrating, microstructurable and covalently attached conducting polymer hydrogel for neural interfaces. Acta Biomater. **58**, 365–375 (2017)
427. Tai, Y., Mulle, M., Aguilar Ventura, I., Lubineau, G.: A highly sensitive, low-cost, wearable pressure sensor based on conductive hydrogel spheres. Nanoscale. **7**(35), 14766–14773 (2015)
428. Darabi, M.A., Khosrozadeh, A., Mbeleck, R., Liu, Y., Chang, Q., Jiang, J., Cai, J., Wang, Q., Luo, G., Xing, M.: Skin-inspired multifunctional autonomic-intrinsic conductive self-healing hydrogels with pressure sensitivity, stretchability, and 3D printability. Adv. Mater. **29**(31), 1700533 (2017)
429. Jing, X., Mi, H.-Y., Peng, X.-F., Turng, L.-S.: Biocompatible, self-healing, highly stretchable polyacrylic acid/reduced graphene oxide nanocomposite hydrogel sensors via mussel-inspired chemistry. Carbon. **136**, 63–72 (2018)
430. Yu, C., Wang, C., Liu, X., Jia, X., Naficy, S., Shu, K., Forsyth, M., Wallace, G.G.: A cytocompatible robust hybrid conducting polymer hydrogel for use in a magnesium battery. Adv. Mater. **28**(42), 9349–9355 (2016)
431. Shi, Y., Wang, M., Ma, C., Wang, Y., Li, X., Yu, G.: A conductive self-healing hybrid gel enabled by metal ligand supramolecule and nanostructured conductive polymer. Nano Lett. **15**(9), 6276–6281 (2015)
432. Wang, Y., Shi, Y., Pan, L., Ding, Y., Zhao, Y., Li, Y., Shi, Y., Yu, G.: Dopant-enabled supramolecular approach for controlled synthesis of nanostructured conductive polymer hydrogels. Nano Lett. **15**(11), 7736–7741 (2015)
433. Dong, R., Ma, P.X., Guo, B.: Conductive biomaterials for muscle tissue engineering. Biomaterials. **229**, 119584 (2020)
434. Gaharwar, A.K., Peppas, N.A., Khademhosseini, A.: Nanocomposite hydrogels for biomedical applications. Biotechnol. Bioeng. **111**(3), 441–453 (2014)
435. Nam, J., Lim, H.-K., Kim, N.H., Park, J.K., Kang, E.S., Kim, Y.-T., Heo, C., Lee, O.-S., Kim, S.-G., Yun, W.S., et al.: Supramolecular peptide hydrogel-based soft neural interface augments brain signals through a three-dimensional electrical network. ACS Nano. **14**(1), 664–675 (2020)
436. Lim, C., Shin, Y., Jung, J., Kim, J.H., Lee, S., Kim, D.-H.: Stretchable conductive nanocomposite based on alginate hydrogel and silver nanowires for wearable electronics. APL Mater. **7**(3), 031502 (2018)
437. Peng, Q., Chen, J., Wang, T., Peng, X., Liu, J., Wang, X., Wang, J., Zeng, H.: Recent advances in designing conductive hydrogels for flexible electronics. InfoMat. **2**(5), 843–865 (2020)
438. Sun, J.-Y., Keplinger, C., Whitesides, G.M., Suo, Z.: Ionic skin. Adv. Mater. **26**(45), 7608–7614 (2014)
439. Liu, J., Tan, C.S.Y., Yu, Z., Li, N., Abell, C., Scherman, O.A.: Tough supramolecular polymer networks with extreme stretchability and fast room-temperature self-healing. Adv. Mater. **29**(22), 1605325 (2017)
440. Larson, C., Peele, B., Li, S., Robinson, S., Totaro, M., Beccai, L., Mazzolai, B., Shepherd, R.: Highly stretchable electroluminescent skin for optical signaling and tactile sensing. Science. **351**(6277), 1071–1074 (2016)
441. Kim, D.C., Shim, H.J., Lee, W., Koo, J.H., Kim, D.-H.: Material-based approaches for the fabrication of stretchable electronics. Adv. Mater. **32**(15), 1902743 (2020)
442. Ke, K., Sang, Z., Chen, X., Rohm, K., Manas-Zloczower, I.: Tuning mechanical and electrical properties of elastomer composites with hybrid filler network containing graphene for stretchable strain sensors. Adv. Eng. Mater. **24**(2), 2100703 (2022)

443. Hong, S., Lee, J., Do, K., Lee, M., Kim, J.H., Lee, S., Kim, D.-H.: Stretchable electrode based on laterally combed carbon nanotubes for wearable energy harvesting and storage devices. Adv. Funct. Mater. **27**(48), 1704353 (2017)
444. Park, M., Im, J., Shin, M., Min, Y., Park, J., Cho, H., Park, S., Shim, M.-B., Jeon, S., Chung, D.-Y., et al.: Highly stretchable electric circuits from a composite material of silver nanoparticles and elastomeric fibres. Nat. Nanotechnol. **7**(12), 803–809 (2012)
445. Kim, Y., Zhu, J., Yeom, B., Di Prima, M., Su, X., Kim, J.-G., Yoo, S.J., Uher, C., Kotov, N.A.: Stretchable nanoparticle conductors with self-organized conductive pathways. Nature. **500**(7460), 59–63 (2013)
446. Joo, H., Jung, D., Sunwoo, S.-H., Koo, J.H., Kim, D.-H.: Material design and fabrication strategies for stretchable metallic nanocomposites. Small. **16**(11), 1906270 (2020)
447. Du, J., Wang, L., Shi, Y., Zhang, F., Hu, S., Liu, P., Li, A., Chen, J.: Optimized CNT-PDMS flexible composite for attachable health-care device. Sensors. **20**(16), 4523 (2020)
448. Lind, J.U., Busbee, T.A., Valentine, A.D., Pasqualini, F.S., Yuan, H., Yadid, M., Park, S.-J., Kotikian, A., Nesmith, A.P., Campbell, P.H., et al.: Instrumented cardiac microphysiological devices via multimaterial three-dimensional printing. Nat. Mater. **16**(3), 303–308 (2017)
449. Chu, M., Nguyen, T., Pandey, V., Zhou, Y., Pham, H.N., Bar-Yoseph, R., Radom-Aizik, S., Jain, R., Cooper, D.M., Khine, M.: Respiration rate and volume measurements using wearable strain sensors. npj Dig. Med. **2**(1), 8 (2019)
450. Zheng, Y., Li, Y., Dai, K., Wang, Y., Zheng, G., Liu, C., Shen, C.: A highly stretchable and stable strain sensor based on hybrid carbon nanofillers/polydimethylsiloxane conductive composites for large human motions monitoring. Compos. Sci. Technol. **156**, 276–286 (2018)
451. Dang, Z.-M., Yuan, J.-K., Zha, J.-W., Zhou, T., Li, S.-T., Hu, G.-H.: Fundamentals, processes and applications of high-permittivity polymer–matrix composites. Prog. Mater. Sci. **57**(4), 660–723 (2012)
452. Matsuhisa, N., Inoue, D., Zalar, P., Jin, H., Matsuba, Y., Itoh, A., Yokota, T., Hashizume, D., Someya, T.: Printable elastic conductors by in situ formation of silver nanoparticles from silver flakes. Nat. Mater. **16**(8), 834–840 (2017)
453. Liu, Z., Wang, H., Huang, P., Huang, J., Zhang, Y., Wang, Y., Yu, M., Chen, S., Qi, D., Wang, T., et al.: Highly stable and stretchable conductive films through thermal-radiation-assisted metal encapsulation. Adv. Mater. **31**(35), 1901360 (2019)

Functional Polymers for Impact Absorption

Pedro Lance, Laureano Ortellado, and Leopoldo R. Gómez

Abstract This chapter provides an in-depth review of the behavior of functional polymers under high-strain-rate deformations, focusing on their response to shock and blast waves in both bulk systems and thin film states. Emphasizing the critical role polymers play under extreme conditions, this study underscores their significance in testing the structural integrity and behavioral response of materials subjected to severe pressure. This research is particularly pertinent given the escalating interest in high-strain-rate phenomena, with implications for a broad range of industries, including aerospace, defense, and automotive sectors.

Here, we focus on the latest advancements in understanding the shock response of diverse polymeric materials. This includes a detailed examination of thermoplastic polymers, thermoplastic elastomers, and vitrimers, delving into their distinctive properties and behaviors when faced with extreme conditions. A key aspect of our review is the exploration of the shock puncture phenomena in membranes, offering crucial insights into the resilience and reliability of these materials in real-world applications. Methodologically, the chapter blends experimental data with theoretical analysis to provide a comprehensive understanding of the subject. It also methodically organizes the presentation of results into specific polymer categories, ensuring thorough coverage of the topic. Ultimately, this chapter aims to present an in-depth revision of the current understanding of the intricate physical processes behind polymer shock absorption, shedding light on aspects of polymer behavior under high-strain-rate conditions that have remained partially unexplored until recent years.

Keywords Functional polymers · Shock waves · Adiabatic compression · Impact absorption

P. Lance · L. Ortellado · L. R. Gómez (✉)
Departmento de Física—IFISUR, Universidad Nacional del Sur—CONICET, Bahía Blanca, Buenos Aires, Argentina
e-mail: lgomez@uns.edu.ar

V. Kumar, Y. K. Mishra (eds.), *Soft Materials for Functional Applications*, Advanced Structured Materials 225, https://doi.org/10.1007/978-981-97-9468-3_12

Table 1 Different strain rates $\dot{\epsilon}$ and compression velocities u_p lead to different types of deformation mechanisms and physicochemical processes [6]

Strain rate $\dot{\epsilon}$	Velocity u_p	Deformation mechanism
~10^0 s^{-1}	~10^1 m/s	Predominantly elastic deformation/localized plasticity
~10^2–10^3 s^{-1}	~10^2 m/s	Increased plasticity and dissipation, recrystallization
~10^4–10^5 s^{-1}	~10^0 km/s	Plastic deformation, adiabatic shear banding, fragmentation
~10^6–10^8 s^{-1}	~10^1 km/s	Shock waves, phase transformations, very high-temperature rise
~10^8 s^{-1}	$\gtrsim 10^1$ km/s	Ultra-high temperature and pressure conditions

1 Introduction

Shock waves and blast waves are intense phenomena that rigorously test the structural integrity and behavioral response of various materials under extreme pressure conditions [1–3]. These abrupt events can lead to a significant pressure increase, prompting a pronounced nonlinear response in the material. The results can vary from significant deformation to total material failure or fracture, which may induce phase transitions in the microstructure [4, 5].

High-strain-rate responses of materials have garnered increasing attention due to their significant implications in both fundamental research and technological applications. This interest spans a wide variety of materials, including metals, amorphous solids, ceramics, and notably, polymers. The strain rate, represented by $\dot{\epsilon}$, indicates the speed at which a material deforms under external forces. The deformation mechanism triggered can differ based on the strain rate, and understanding these nuances is crucial for a range of applications, from materials testing to analyzing shocks and explosions. Table 1 provides an overview of strain rates and the related deformation mechanisms observed in different materials. However, it's pivotal to consider these values as general references, as they might differ considerably based on the specific system.

Recent research suggests that polymeric materials hold promise in offering robust resistance to these intense events. Compared to traditional shock-absorbing materials like metals, polymers offer several distinct advantages. Their lower density, rate-dependent behavior, adjustable transparency, and intrinsic self-healing capabilities mark them as promising candidates for a variety of applications. The range of these applications includes aerospace, defense, automotive industries, and infrastructure, highlighting a promising direction for polymers in shock absorption and resistance [7–9].

Polymers stand out due to their diverse components, phases, and interactions, making them prime candidates for shock absorption. Modern synthesis techniques now enable the manipulation of crucial polymer features, such as glass transitions, crystallinity, entanglements, crosslinking, and chain architecture, among others. This allows for the production of materials like crystalline and semicrystalline polymers, block copolymers with precisely controlled compositions [10], elastomers

[11], polyurethanes [12], polycarbonates, and more complex systems like ionomers [13] and vitrimers [14].

In this chapter, we offer a comprehensive review of both experimental and theoretical advancements in understanding the shock response of various polymeric materials, in their bulk and thin film states. Notably, although research into the shock response of polymers dates back to the mid-twentieth century, the precise physical processes that result in shock absorption in polymers are still not fully understood.

This chapter is structured as follows: in Sect. 2, we will revisit the core and fundamental properties of shock waves propagating through condensed matter media. In Sect. 3, we review the main experimental and computational techniques used to study shocks in polymers over the past decades. Section 4 presents a review of experimental and theoretical results on the shock response of various polymer systems. The presentation of results is organized into four subsections: thermoplastic polymers (Sect. 4.1), thermoplastic elastomers (Sect. 4.2), vitrimers (Sect. 4.3), and membranes (Sect. 4.4). The final conclusions and outlooks are presented in Sect. 5.

2 Some General Properties of Shocks in Polymers

Shock waves are abrupt disturbances that can travel through gases, liquids, or solids, leading to rapid changes in properties like pressure and density. In general, studying the propagation of shock waves can help understand how materials behave under extreme conditions, potentially leading to the development of stronger and more resilient systems for impact and blast absorption.

Key parameters gauge the power of a shock wave and how materials respond to it. Therefore, polymer chains near the surface exhibit increased mobility, which enhances local toughness and improves overall impact absorption. The Mach number (M), or the ratio of the shock wave speed u_s to the local sound speed c_0, is another measure. A shock with $M \gg 1$ is a "strong" shock, whereas one with $M \gtrsim 1$ is considered a "weak" shock, roughly traveling at the speed of sound of the media. Shock thickness or width represents the distance over which physical properties alter and is determined by the medium and shock strength. The shock width is related to microscopic mechanisms of molecular relaxation and local energy dissipation within the material. For instance, in monatomic gases, the shock width is about four mean free paths, indicating roughly four molecular collisions to shift from one state to another. In polyatomic gases, shocks are wider due to more molecular interactions required for equilibrium. Thus, the molecular complexity in both architectures and interactions can enhance shock absorption. As we will discuss throughout the chapter, the richness of polymers in architectures and interactions makes them ideal materials for the design of materials with shock-absorbing properties in a desired range.

In a controlled shock compression experiment, a material is uniformly compressed at a high speed by a piston or plate moving with velocity u_p. After an

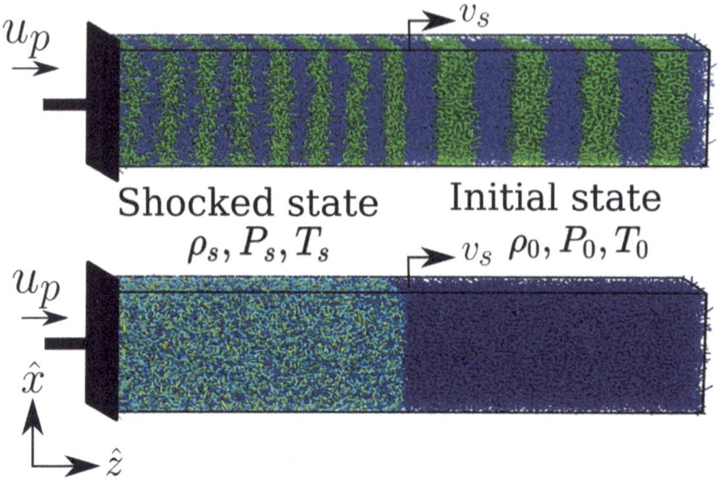

Fig. 1 Piston compression of a lamellae-forming block copolymer (simulations). The compression of the systems at velocity u_p leads to the formation of a shock front traveling through the system at velocity v_s. The shock is easily observed when plotting the local particle velocity in the polymer (below) and can also be visualized through the local density and pressure. The passage of the shock leads to a transition in the material from an initial state characterized by macroscopic variables ρ_0, P_0, and T_0, to a final state characterized by ρ_s, P_s, and T_s

initial transient, where a wave develops in the material, this compression produces a uniform shock wave traveling through the media at a higher velocity v_s; see Fig. 1. In general, the passage of the shock induces a rapid transition in the material's state. Since the transition between the initial and shocked states occurs quickly, the system lacks time for heat exchange. Thus, from a thermodynamic perspective, the passage of the shock can be considered an adiabatic transition.

Being an adiabatic process, high-strain-rate deformations and shock compression lead to the local heating of the material, which can further enhance yielding and induce phase transitions. Frost and Ashby derived an expression for the critical strain rate beyond which adiabatic behavior dominates [15, 16]. Their expression hinges on the critical strain initiating instability under strictly adiabatic conditions:

$$\epsilon_\text{c} = \frac{-m\, C_p}{\partial\sigma/\partial T} \tag{1}$$

where the stress–strain relationship of the material is assumed to be $\sigma = \sigma_0(\epsilon/\epsilon_0)^m$ and C_p denotes the specific heat at constant pressure.

$$\dot{\epsilon}_\text{c} = \frac{\epsilon_\text{c}}{\tau} = \frac{2m\alpha K}{R^2 \partial\sigma/\partial T} \tag{2}$$

In this equation, $\tau = C_p R^2/2\alpha K$ represents the characteristic time for thermal diffusion. R signifies the sample's characteristic size, K stands for the thermal conductivity, and α is a constant typically around 2. For materials like polycarbonate, the strain rate might be as low as $\dot{\epsilon} = 0.2\ \text{s}^{-1}$, given standard values like $R = 3$ mm, $K = 0.2$ W/m °C, $\frac{\partial\sigma}{\partial T} = 0.39$ MPa/°C, and $n = 1$ [17].

Shock wave behavior is governed by conservation laws, which outline changes in physical attributes across the shock's boundary [1–3]. Stemming from core principles of mass, momentum, and energy conservation, these laws define the system by its initial and shocked states (Fig. 1). Uniform shocks undergo changes only across this boundary. These principles are captured in the Rankine–Hugoniot jump conditions:

$$\begin{cases} \rho_0 v_s = \rho_s \left(v_s - u_p\right) \\ P_s - P_0 = \rho_0 v_s u_p \\ P_s u_p = \rho_0 v_s \left(1/2 u_p^2 + e_s - e_0\right) \end{cases} \tag{3}$$

Indices 0 and s represent pre-shock and post-shock values of density ρ, pressure P, and internal energy e, with the frame of reference being the shock front.

The Hugoniot curve (or shock adiabat) is essential for understanding the traits of regions impacted by shocks. Located in the pressure–volume (P–V) plane, it signifies the adiabatic nature of the shock compression process. It's instrumental for deriving a material's equation of state. Hence, shock compression tests have been widely used to explore the high-pressure equation of state for various substances [2, 3].

The shock Hugoniot can be derived from simulations or the Rankine–Hugoniot jump conditions:

$$\begin{cases} V/V_0 = 1 - u_p/v_s \\ P - P_0 = \rho_0 v_s u_p \end{cases} \tag{4}$$

Given the function $v_s = v_s(u_p)$, these equations provide the Hugoniot curve implicitly.

The shock impedance relationship is a widely observed association in a multitude of materials. It presents a linear approximation between the shock speed (v_s) and particle velocity (u_p) as demonstrated by the following equation:

$$v_s = c_0 + S u_p \tag{5}$$

Within this formula, c_0 symbolizes the bulk sound speed in the undisturbed material, while S is a dimensionless constant referred to as the shock impedance or the slope of the Hugoniot curve.

The linear approximation is typically valid for weak shocks, but its accuracy diminishes for stronger shocks due to nonlinear influences. In polymers, a linear relationship between compression and shock velocity is observed for various systems such as polyethylene, epoxy, polycarbonate, and several others at low to moderate compression velocities [18–25]. This relationship is captured by Eq. (5). However, most systems exhibit deviations from this linearity at higher compression velocities. Such deviations can result from several factors, including pressure-induced phase transitions and changes in covalent bonding due to pressure [18].

3 Shocks in Polymers: Techniques for Advanced Characterization

Currently, there are several experimental and computational techniques used to study the shock response of polymeric materials across a wide range of strain rate deformations. Here, we focus on those techniques that have been most commonly used in recent decades to unveil the correlation between shock response and structure in various polymeric materials.

3.1 Experimental Techniques

The Hopkinson pressure bar, one of the pioneering techniques to analyze material shock responses, was originally devised by Hopkinson [26] to measure pressures generated by bullet impacts or gun cotton detonations. Hopkinson employed a cylindrical steel bar, shooting one end with a rifle bullet to produce a pressure pulse that traveled through the bar. By affixing a smaller rod of identical material and cross-section to the opposite end, he was able to gauge the momentum of the direct wave using a ballistic pendulum.

Kolsky [27] adapted the original experiment to measure the properties of various materials, leading to the development of what is now known as the Kolsky bar or split-Hopkinson pressure bar (see Fig. 2b). The design of this device has been further refined to conduct high-strain-rate experiments in tension [28] and torsion [29]. In a typical setup, the split-Hopkinson pressure bar features a specimen sandwiched between two comparatively long bars, termed the incident (or input) and transmitted (or output) bars. The free end of the incident bar is struck by another relatively short bar, sending a pulse through the system. By recording the pulse in both the incident and transmitter bars, properties of the specimen, such as stress and strain, can be determined. Greater strain can be achieved if the specimen is directly struck (without the incident bar); however, this approach can cause problematic displacements at the impact face. Typical strain rates for this method range from 100 to 10^4 s^{-1} [30, 31].

Various experimental techniques utilize the acceleration of a projectile to investigate materials under high strains. Depending on the specific focus of the experiment, the material under scrutiny can either be the projectile itself or the target it impacts. The design choices for these projectiles span a broad spectrum, including spherical

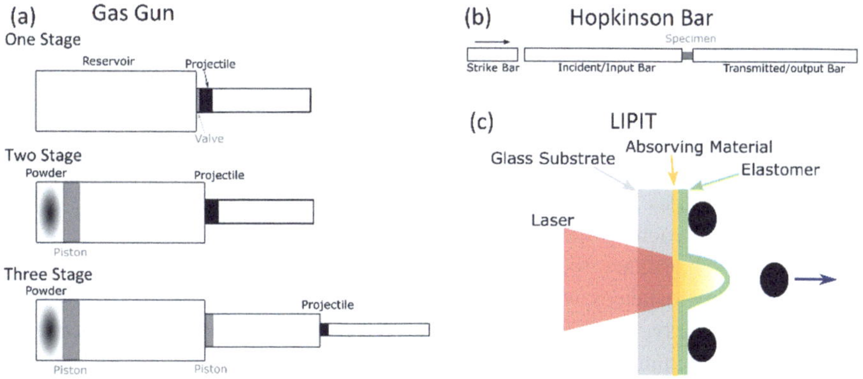

Fig. 2 Different experimental setups for high strain materials behavior. (**a**) Light gas gun of one-, two-, and three-stage schemes. (**b**) Representation of a split-Hopkinson pressure bar experiment. (**c**) Scheme of a laser-induced projectile impact test (LIPIT) system for particle acceleration

shapes, employing either a single projectile [32–34] or multiple projectiles in a single shot [35–38]; cylindrical designs [39]; plate configurations [40–42]; and bullet-inspired models [43, 44].

To set these projectiles into motion, various acceleration methodologies are adopted, such as expansion of a light gas [32, 45], laser-driven propulsion [36, 40–42], and electrical acceleration [39]. The selection of a specific projectile design and its corresponding acceleration technique is informed by the goals of the experiment. Parameters like the projectile's mass, velocity, and shape play pivotal roles in ensuring the accuracy and relevance of the outcomes.

The most straightforward method for accelerating a projectile to significant velocities is through gas expansion. Various apparatuses utilizing this principle are available, with the one-stage, two-stage, and three-stage gas guns being the most prevalent (Fig. 2a). The single-stage gas gun essentially comprises a reservoir containing high-pressure gas, isolated from a barrel by a valve. Upon opening the valve, the gas expands and propels the projectile [35].

For even higher velocities, the two-stage gas gun is employed. This system is partitioned into three segments: a powder chamber, a pump tube, and a launch tube. The pump tube, filled with a light gas, is positioned between the powder chamber and launch tube, separated by a piston and a diaphragm, respectively. Combustion within the powder chamber compresses the gas, elevating its pressure to a point where the diaphragm ruptures, consequently releasing the gas that drives the projectile forward [32, 45].

The three-stage gas gun extends this concept, now incorporating a powder chamber, two successive pump tubes, and a launch tube [38, 46]. Although augmenting the number of stages enhances the projectile's velocity, it also confines the allowable projectile mass to reduced values and demands a larger installation footprint for the apparatus.

Over recent decades, a novel projectile acceleration technique has emerged, known as the laser-induced projectile impact test (LIPIT) as depicted in Fig. 2c or laser-driven flyer (LDF). This method draws on principles akin to the gas expansion utilized in gas guns. In this setup, a laser pulse targets a specific absorbing material, causing it to heat up and undergo a phase transition. This induces a rapid material expansion which propels one or multiple projectiles. Commonly employed projectiles include spheres [33, 36, 37, 47, 48] and plates [40–42], with microprojectiles reaching speeds up to approximately 4 km/s [36, 41].

Although the primary focus in projectile acceleration techniques is often the target, there are instances where the projectile itself is the subject of study [47, 49, 50]. Specifically, when the projectile is cylindrical and meets certain criteria, the experimental approach is termed the Taylor impact test, a method extensively employed to assess material properties under high strain rates.

In the realm of space exploration, testing materials against high-velocity projectile impacts is crucial, especially given the threat posed by high-speed space dust particles. The Van der Graaff accelerator serves as a quintessential laboratory instrument to mimic space particle conditions [51–54]. This device can accelerate micron and submicron particles to velocities around 80 km/s [51, 52]. However, one limitation is the necessity for particles to sustain a surface charge, thereby typically restricting the use to smaller particles.

3.2 Simulation Techniques

With advancements in computational capabilities and methodologies, simulations have gained prominence in recent years. Compared to experimental setups, they offer cost efficiency, allowing one system to analyze numerous material properties. While simulations grant insights into scales unreachable by experiments, such as the atomic scale (on the order of nanometers or less), they also face the limitation of system size due to computational time constraints.

Polymers are typically studied using molecular dynamics (MD), which tracks the evolution of an interacting particle system by solving Newton's equations of motion numerically. Interactions between particles are defined by interatomic potentials or force fields tailored to the desired system. Often, polymers are represented using a bead–spring model, where beads symbolize individual or groups of atoms. In this model, consecutive beads are linked either via a harmonic potential:

$$E_{\text{harmonic}} = k\,(r - r_0)^2$$

or a finite extensible nonlinear elastic (FENE) potential:

$$E_{\text{FENE}} = -k\,r_0^2\,\ln\left[1 - \left(\frac{r}{r_0}\right)^2\right]$$

Here, k is the spring constant, r_0 is the equilibrium position, and r is the distance between particles. Nonbonded particles generally interact via a Lennard–Jones 12-6 potential:

$$E_{\mathrm{LJ}} = 4\epsilon\left[\left(\frac{\sigma}{r}\right)^{12} - \left(\frac{\sigma}{r}\right)^{6}\right]$$

where ϵ and σ are specific constants. Systems employing these potentials are often referred to as the Kremer–Grest model [55]. In some instances, other nonbonded potentials, like the Lennard–Jones 9-6 [56–58], are utilized. Additionally, multi-particle potentials, such as angle or torsional potentials, are employed to introduce attributes like chain stiffness or diverse phases like crystalline or glassy [56, 59, 60].

Modifying these potentials, adding new ones, or altering chains enables the simulation of varied polymer types like copolymers [61–63], ionomers [64–70], vitrimers [71], branch polymers [72, 73], and ring polymers [74], among others.

MD simulations are categorized into atomistic and coarse-grained models. In atomistic models, each particle corresponds to a single atom, whereas in coarse-grained models, a particle represents a group of atoms. While atomistic models afford detailed representations, their computational demands can be high. Conversely, coarse-grained models provide efficient simulations of larger systems at the expense of detail. The choice between them hinges on the researcher's objectives.

Figure 3 presents snapshots of systems simulated using diverse coarse-grained models. Figure 3a showcases a system with an angle potential enhancing chain rigidity, evident from the chain's straightness. Figure 3b depicts a semicrystalline system with crystalline (blue) and amorphous (red) domains. In Fig. 3c, a diblock copolymer is displayed, highlighting varied potentials for different particle types. Figure 3d portrays an ionomer system where charged particles (blue and yellow) cluster within an amorphous polymer (red) matrix.

To investigate the energy absorption of polymer systems under high strain rates, piston compression simulations are commonly employed [61, 75–77]. Numerous such simulations have mirrored the phase transitions observed experimentally. Parameters like shock wave properties, order metrics, and chain attributes are analyzed to emulate real material behaviors.

Quantum effects arising from the shock compression of polymers can be explored using the density functional theory (DFT) [78, 79]. While quantum calculations provide deeper insights into molecular configurations, structural adjustments under confinement, and electronic structures, they further constrain the system size and computational time. Consequently, capturing certain phase transitions in materials through DFT becomes challenging, given the extended time scales associated with notable reorganization.

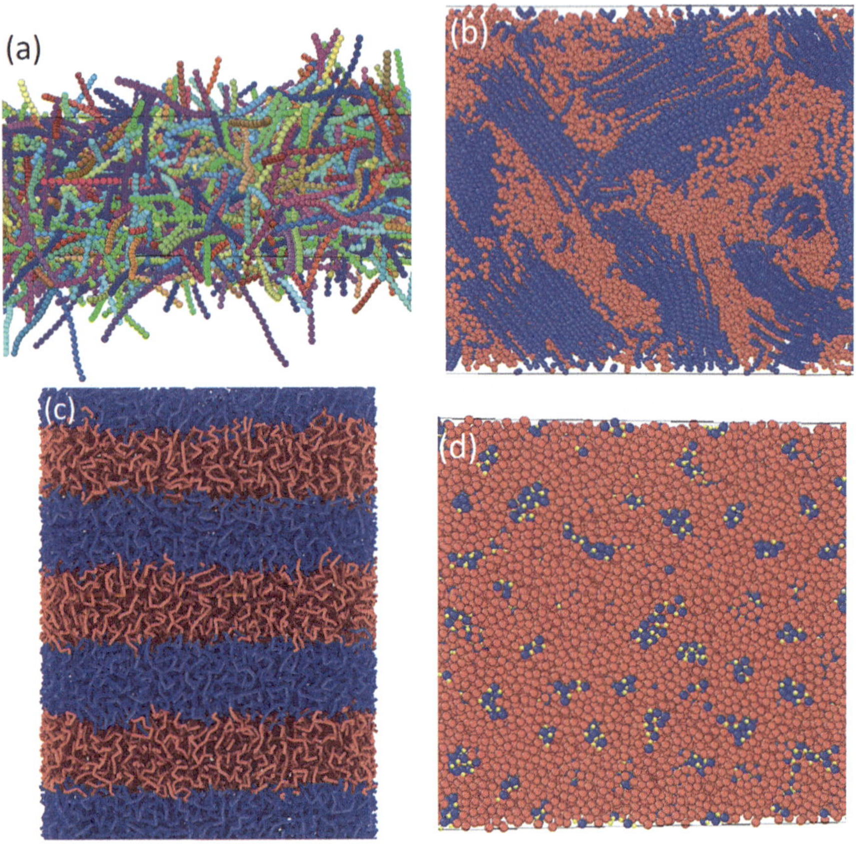

Fig. 3 MD simulations of various polymeric systems. (**a**) By introducing an angular potential between bonded beads, both semiflexible and stiff polymers can be effectively modeled. (**b**) MD representation of a semicrystalline polymer with crystalline monomers in blue and amorphous ones in red. (**c**) A self-assembled diblock copolymer with different colors indicating the two immiscible polymer blocks. (**d**) Ionomers and charged polymers. Particle colors denote uncharged (red), negatively charged (blue), and positively charged (yellow) entities. In ionomers, charged particles undergo self-assembly into multipolar clusters, which can disintegrate and reconfigure during deformation

4 Shock Response of Polymers

Polymers, with their varied microstructures shaped by diverse components and synthesis methods, exhibit a broad spectrum of mechanical behaviors under varying conditions [80]. When subjected to mechanical loading, their responses can be classified into three primary regimes: quasistatic, intermediate rate, and shock. These responses are not only temperature-dependent but also closely tied to the duration under load.

In the quasistatic domain, a polymer's strength varies as its microstructure evolves with temperature. In its glassy state, a polymer possesses high stiffness, which considerably reduces during the glass transition phase, turning the material more rubbery. As temperatures further escalate, the polymer exhibits a more viscous behavior, flowing until it ultimately decomposes. It's crucial to note that the speed of deformation—high strain rates in particular—coupled with low temperatures can significantly constrain the deformation modes of the polymer chains within.

In a seminal work [19], Walley and Field obtained a comprehensive dataset that highlights the compressive stress–strain responses across a wide spectrum of strain rates for an assortment of polymers. Using a Hopkinson bar, they mapped the strain rate sensitivity of the uniaxial stress–strain compressive behavior for 17 polymers under ambient conditions. Their findings spanned strain rates ranging from 10^{-2} to $10^4\ s^{-1}$.

Their study presented a notable observation: a linear relationship between the compressive stress at a specific strain and the logarithm of the strain rate. However, exceptions to this trend were identified in polymers like acetal, high-density polyethylene, and polytetrafluoroethylene. More interestingly, the bulk of the polymers manifested a strength increase in tandem with rising strain rates, especially noticeable around $10^3 s^{-1}$. This behavior allowed polymers to be categorized into three main groups based on their strain rate sensitivity [3, 19, 24]:

1. Polymers showcasing heightened strain rate sensitivity (Fig. 4a).
2. Those with reduced strain rate sensitivity (Fig. 4b).
3. A set that exhibited negligible changes in strain rate sensitivity within this range (Fig. 4c)

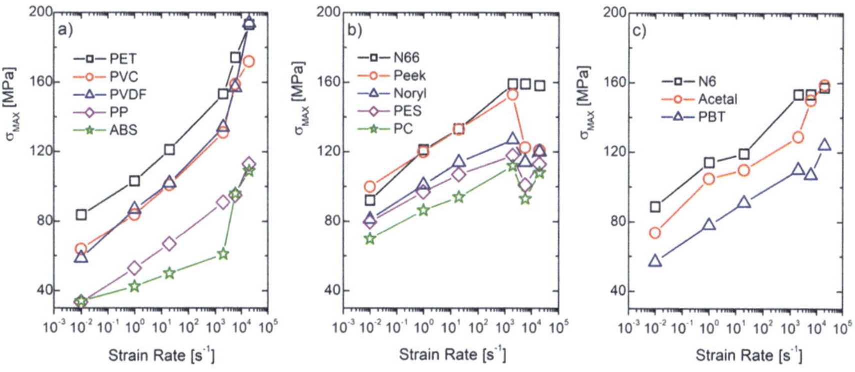

Fig. 4 Response of polymers at low to intermediate deformation rates. (**a**) Some polymers exhibit enhanced strain rate sensitivity at high strain rates, while others demonstrate reduced sensitivity (**b**). Certain polymers, however, display negligible changes across varying strain rates within the studied deformation velocity ranges (**c**). (Experimental data sourced from the research by Walley and Field [19])

While it's evident that intermediate strain rate responses are potentially governed by the inherent microstructure of the material, the exact relationship between a polymer's strain rate behavior and its microstructural properties is yet to be conclusively determined.

There is also a remarkable experimental work by Carter and Marsh who proved the response of a variety of polymers in a range of strong shock compression [18]. Using explosive lenses, this work reached compression velocities of the order of 1–5 km/s. The most salient feature of this work was the observance of a high-pressure phase transition (~20–30 GPa) in several polymers studied. The transition is observed as a change in the slope in the Hugoniot plot $v_s - u_p$ (see Fig. 5a) and is in general also associated with an abrupt change in volume. Between other systems this transition has been observed for polyurethane (PU), cellulose acetate (CA), polystyrene (PS), polysulfone (PSU), phenoxy (PK), polycarbonate (PC), and polyimide (PI) [18]. As this transition was rather general, and observed in both thermoset and thermoplastic polymers, it was argued that the nature of this transition has been associated with a major reordering on the atomic level, where the high pressure compresses the system such that breaking and reformation of covalent bonds can occur. The most prevailing reorganization would occur in polymers with open ring structures made up of covalently bonded carbons. Unfortunately, the nature of this transition is still not completely clear and even quantum mechanics

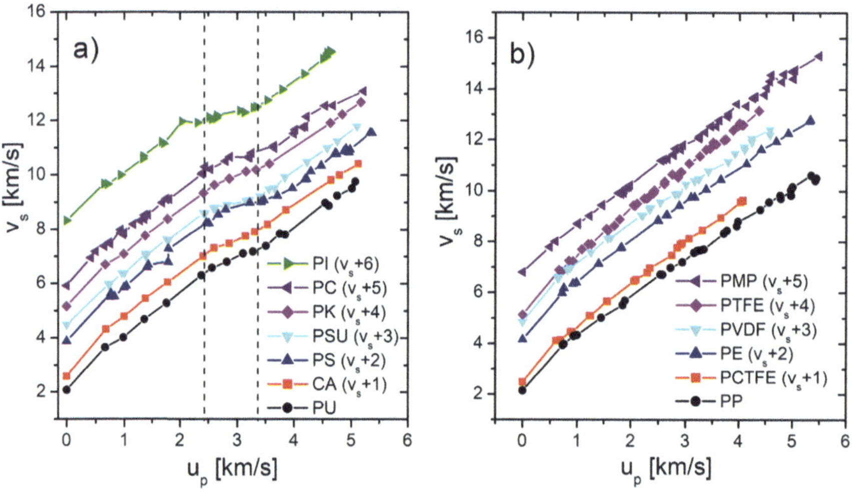

Fig. 5 Shock Hugoniot of polymers in the fast compression regime obtained by Carter and Marsh [18]. In both panels, the curves have been shifted as indicated in order to improve visualization. (**a**) For some polymers, there is a clear phase transition in the system around $u_p \sim 3$ km/s, where there is a change in the slope in the shock velocity v_s. (**b**) Some systems display more a continuous variation in the shock velocity with the compression velocity, such that no transition is clearly identified. (Data reported at [18]; see the same work for more details on the polymeric systems and more shock data on other polymers)

calculations could not be used to capture the transition [78]. On the contrary, there are some polymeric systems studied by Carter and Marsh which do not show a so clear phase transition in the strain rate regime u_p ~ 0–5 km/s (Fig. 5b). Between other systems are polypropylene (PP), polychlorotrifluoroethylene (PCFTE), polyethylene (PE), polyvinylidene fluoride (PVDF), polytetrafluoroethylene (PFFE), and poly-4-methyl-1-pentene (PMP).

In what follows, we present a revision of more recent experimental and simulation results in polymers having different morphologies and properties.

4.1 *Thermoplastic Polymers*

Polymers can be broadly categorized into thermoplastics and thermosets. Thermoplastics are composed of polymer chains of finite lengths, whereas thermosets are chemically crosslinked, forming three-dimensional networks that represent infinite molecular weights. These structural distinctions lead to unique applications for each. Thermoplastics are commonly shaped by cooling their highly viscous melts, a process that solidifies them quickly, ensuring efficient processing. On the other hand, thermosets are produced by reacting small molecular precursors with low viscosity. Even though their formation can be slower due to reaction kinetics, their low viscosity and fluidity make them perfect for crafting structural composites. This section will delve into the high-strain-rate behavior of thermoplastic polymers as demonstrated in both experimental and simulation settings.

Glassy Amorphous Polymers

The mechanical response of various glassy polymers, including polymethylmethacrylate (PMMA), polycarbonate (PC), polyvinylchloride (PVC), and epoxies has been examined in relation to strain rate variations. Generally, the compressive stress–strain curves for these materials exhibit similar trends. Figure 6 depicts a schematic of the typical observed response: an initial linear phase precedes a nonlinear segment culminating in a stress peak. In general, glassy shows an increased strain rate sensitivity at higher strain rates, such that the peak in the stress–strain curve increases with the strain rate. Subsequently, a phase of strain softening emerges, which then gives way to a phase of strain hardening.

Hasan and Boyce [81], striving to shed light on the mechanics underpinning this behavior, have proposed a compelling theory. They interpret the stress–strain response in terms of the evolution of shear transformation sites. To break it down, the material initially contains sites poised for transformation during the testing phase (Fig. 6A). As stress mounts, these transformation sites, especially those possessing high local free volume, start to yield and undergo metamorphosis (Fig. 6B). This process accumulates energy in the segments of the material that remain untouched,

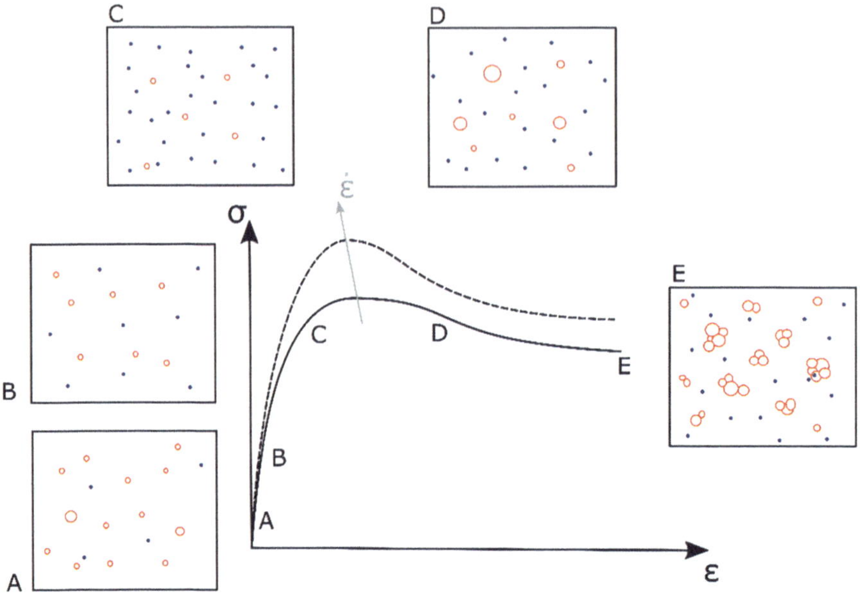

Fig. 6 Diagram illustrating the typical compressive stress–strain behavior of glassy polymers. The curve starts with a linear response, transitioning into a nonlinear pattern as strain intensifies, culminating in a stress peak and subsequent softening. Generally, the stress peak amplifies with an increasing strain rate, represented by $\dot{\epsilon}$. The different stages A–E correspond to the Hasan and Boyce model [81]

termed the "matrix." This accumulated energy generates a counteracting force, or back stress, which initially impedes further transformation.

Yet, when the applied stress escalates, transformation sites with elevated resistance or activation energy become active (Fig. 6C). This activity renders the stress–strain curve increasingly nonlinear. As the magnitude of stress continues to climb, there comes a juncture where the surrounding material can no longer bear additional transformation strain energy. Consequently, new defect sites are spawned (Fig. 6D). These defects usher in strain softening, a phase where the freshly minted sites, due to their reduced activation energy, become readily transformable. In due course, a dynamic equilibrium ensues, marked by the widespread presence of active zones throughout the material, facilitating a continuous plastic flow (Fig. 6E) [81].

Under conditions of exceptionally high strain, a distinct phenomenon comes into play. The polymer chains within the material resist alignment, resulting in strain hardening. However, elevating the strain rate introduces another factor: adiabatic heating, as elaborated in Sect. 2. This heat generation tends to offset the strain hardening effect. In certain instances, it can even dominate the hardening effect altogether [82–86]. Notably, PMMA experiences total failure at high strain rates. This failure arises from the molecular chains' inability to execute specific lateral movements quickly.

Crystalline Polymers

Numerous studies have delved into the behavior of semicrystalline polymers under high strain rates. Typically, these polymers consist of a "rigid" crystalline phase interspersed within a "soft" amorphous matrix. Among the polymers analyzed are various forms of polyethylene (PE), poly(*p*-phenylene terephthalamide) (PPTA), polytetrafluoroethylene (PTFE), and polypropylene (PP). These semicrystalline materials generally exhibit increased sensitivity to strain rates. However, research from Okereke et al. [87] showed that PP presents a distinct stress peak followed by softening. This phenomenon is ascribed to changes in its noncrystalline section, which makes up approximately 35%, and the polymer's adiabatic heating which leads to a softer response.

Of these, PTFE has received the most comprehensive research attention [18–20, 88–90]. Its composition includes a solid crystalline phase set within a pliable amorphous matrix. PTFE's behavior is crystallinity-dependent; increasing its crystallinity enhances its strength, reflecting patterns seen in particulate composites. Notably, PTFE exhibits phase transitions: transitioning from Phase II to Phase IV at 19 °C and from Phase IV to Phase I at 30 °C under standard atmospheric conditions [91]. While both Phases II and IV are highly crystalline, Phase I is sometimes termed a mesophase. Shocking material within Phase II (<0.7 GPa) results in a decrease in crystallinity, decrease in Young's modulus, decrease in yield stress, and little change in microstructure. Conversely, transition from Phase II into Phase III (>0.7 GPa) results in increased crystallinity, an apparently independence of pressure, increase in Young's modulus, increase in yield stress, and significant microstructural refinement [92]. The polymer's likelihood to fracture, either being brittle (in Phase II) or ductile (in Phases I and IV), is also determined by its crystalline phase [91].

Polyethylene (PE) is another focal point of study, especially given the diversity brought by its molecular structure and conformations. Brown et al.'s research [93] across HDPE, UHMWPE, and PEX variants revealed that UHMWPE and PEX both display post-yield strain hardening, whereas HDPE has a pronounced yield stress followed by almost purely plastic flow. For all PE varieties, the stress at specific strains linearly correlates with the strain rate, signifying the lack of phase transitions impacting their reactions. Recent findings also suggest the existence of crystalline configurations of shock-compressed polyethylene up to a pressure of 200 GPa [94]. The lattice structure aligns with the *A2/m* phase, which had been detected at pressures of only up to 40 GPa under static compression.

Additionally, molecular dynamics simulations were employed to study the effects of interfaces in semicrystalline polyethylene under shock loads [77]. Results indicated that energy reflection was more proficient in mitigating weak shocks than strong ones. For polyethylene samples with two interfaces, reflection was more diminished than from a singular interface, especially in slender amorphous zones. This surprising result stems from two nanoscale mechanisms. First, the early onset of an elastic precursor wave compresses smaller amorphous sections, thereby reducing the impedance difference. Second, the rigidity of thin amorphous domains, which are greater than bulk amorphous PE due to confinement by rigid crystalline PE regions, diminishes as their width expands. In summary, for ideal energy reflection and shock dampening, amorphous sections should surpass 50 nm in thickness.

4.2 Thermoplastic Elastomers

Thermoplastic elastomers (TPEs) consist of polymer chains, interplaying segments of strong interactions with those of weak ones. This unique composition results in microphase-separated materials, defined by their distinct soft and hard block morphologies [95, 96]. These are frequently referred to as segmented elastomers.

The emergence of these microphase-separated structures primarily stems from the chemical disparity between the hard and soft segments. While mere blending could induce macrophase separation due to this disparity, copolymerization ensures the formation of covalent bonds between the segments. This bond formation wards off macroscopic phase separation, paving the way for a microphase-separated texture.

Broadly speaking, thermoplastic elastomers can be categorized into the following six types [95, 96]:

- Styrenic thermoplastic elastomers
- Multiblock copolymers
- Hard polymer–elastomer blends
- Graft copolymers
- Ionomers
- Core–shell structures

A requisite for these materials is the presence of at least one elastomeric phase and one hard phase. Notably, the hard phase(s) must transform into a soft and fluid state at elevated temperatures, ensuring the entire material behaves thermoplastically. The subsequent subsections delve into the performance of specific thermoplastic elastomers in impact absorption applications.

Polyurethanes, Polyureas, and Polyurethane Ureas

With hard blocks containing urethane or urea groups, strong hydrogen-bonding interactions confer exceptional properties, such as a high modulus [11, 12]. Modern polyureas and polyurethanes are celebrated for their superior impact absorption, mechanical resilience, and resistance to wear and corrosion. These polymers find broad application in industrial coatings, ranging from transportation pipelines and chemical tanks to vehicle hulls including ships and cars [97, 98].

Polyurethanes result from the exothermic reaction between di- or poly-isocyanates and hydroxyl compounds, notably polyols. Conversely, polyureas emerge when dior poly-isocyanates react with active hydrogen groups, such as amines. At its core, polyurea involves a reaction between a difunctional amine (H_2N–R–NH_2) and a difunctional isocyanate (OCN–R′–NCO), where R represents a linear hydrocarbon and R′ an aromatic unit. This gives a polymer featuring alternating soft (R) and hard (R′) segments. The formulation can be enriched with additives, diluents, and dyes to tailor its properties [97, 98]. Remarkably, the isocyanate-amine reaction, often void of catalysts, culminates swiftly, permitting polyurea synthesis under diverse conditions. The vast versatility of polyurethanes and polyureas stems from the myriad of isocyanates, polyols, and amines available, facilitating the creation of custom materials.

Polyurethane ureas, combining both urethane and urea linkages, present a dynamic chemistry. On a molecular scale, urethane linkages emerge from isocyanates interfacing with alcohols, while urea linkages originate from isocyanates interacting with amines. This unique synthesis yields polymers endowed with a spectrum of structural and mechanical attributes.

Polyurethanes and polyureas undergo self-assembly, resulting in nanoscale segregated structures. These structures are made of hard segments, identifiable by their high glass transition temperatures (T_g), embedded within a matrix of linear hydrocarbon chains, forming the soft phase with a lower T_g. In polyureas, the segregation is mainly due to hydrogen bond formation between urea linkages across hard segments. These hard domains, varying in size from about 10 to 100 nm based on chain length, act as physical crosslinks, categorizing the material as a thermoplastic elastomer. This complex structure bestows polyureas with their distinct mechanical properties, striking a balance between rigidity and elasticity. At room temperature, while the soft domains retain their elastomeric nature by staying above their T_g, the hard domains contribute to the material's toughness and stiffness by remaining below their respective T_g.

Microphase separation in polyurethanes and polyureas emerges from the incompatibility between hard and soft segments and can be thermodynamically depicted using the free energy of mixing, ΔF [12]:

$$\Delta F = \Delta U - T\Delta S \tag{6}$$

Here, ΔU and ΔS denote enthalpy and entropy of mixing. Thermodynamic incompatibility usually yields $\Delta F > 0$, resulting in phase separation. However, in these elastomers, covalent bonds between segments allow only nanoscale self-separation. The morphologies of the phase separation depend on factors like volume fraction ϕ, polymerization degree N, and Flory–Huggins interaction parameter χ [99]:

$$\chi_{AB} = Z\left[\epsilon_{AB} - \left(\epsilon_{AA} + \epsilon_{BB}\right)/2\right]/k_{\mathrm{B}}T \tag{7}$$

A positive χ suggests a net repulsion between components A and B. When χ is small, the system undergoes microphase separation. As temperature increases, an order-to-disorder transition (ODT) occurs due to χ's inverse temperature dependence.

Hansen proposed a theory to account for hydrogen bonding in thermodynamics, introducing the solubility parameter, δ, comprising components δ_h (hydrogen bonding), δ_d (dispersive forces), and δ_p (dipole forces) [100]:

$$\delta^2 = \delta_h^2 + \delta_d^2 + \delta_p^2 \tag{8}$$

This is directly linked to the Flory–Huggins interaction parameter [101]:

$$\chi_{hs} = \chi_{\Delta S} + \frac{\nu(\delta_{hs} - \delta_{ss})^2}{RT} \tag{9}$$

In thermoplastic polyurethane (TPU), factors ϕ and N are influenced by polymerization stoichiometry, while χ relates to the choice of segments. For instance, polyether-based polyurethanes show more phase separation than polyester-based ones.

Research methodologies such as split-Hopkinson bars, drop-weight tests, and simulations have extensively explored the behavior of polyurea under varied strain rates. One standout characteristic of this material is its exceptional wave absorption capacity. This ability possibly originates from several mechanisms, including wave scattering between its hard and soft components, resonance within hard regions at elevated frequencies, and the breaking of hydrogen bonds during impact events.

There's a wealth of research focusing on polyurea's mechanical attributes. However, fully grasping the science underpinning its distinct abilities remains elusive. A salient point is that polyurea's behavior markedly shifts with changing strain rates, displaying notable energy dissipation. Sarva's research underscores this phenomenon, identifying a shift from rubbery characteristics to more leatherlike responses across varying strain rates [102]. Remarkably, its shear resistance at high strain rates is on par with high-strength steels.

Polyurea's resistance to impacts and shocks is governed by different mechanisms. Its prowess in impact resistance is credited to energy absorption during its transition from a rubbery to a glassy phase [103]. Bogoslovov's team found that a slower relaxation rate in polyurea supports this transition during ballistic impacts, a conclusion further backed by Ransom's work [104, 105]. Additionally, Fragiadakis pointed out that variations in the ratio of hard to soft segments within polyurea molecules don't significantly influence its transitional behavior or impact resistance [106].

Simulation studies present a nuanced view of polyurea's shock response. It appears that its exemplary shock absorption isn't merely due to the glassification of rubbery regions but is attributed to its hard regions, interfaces between soft and hard domains, and hydrogen bond disruptions [76, 107–111]. There's a prevailing theory suggesting that alterations within these hard regions, including atomic movements and hydrogen bond dynamics, play a role in dispersing shock energy. Research by Qiao posits that during deformation, hard domains remain interconnected through soft segments, enhancing energy absorption during shocks [109]. Molecular dynamic simulations by Manav and Ortiz indicated that under shock conditions, polyurea's hard domains experience less compression compared to soft ones, accompanied by significant hydrogen bond disruptions [111]. Their work also spotlighted perturbations in phenyl ring configurations under intense shock pressures.

Block Copolymers

Copolymers result from the bonding of distinct monomers within a polymer chain through covalent bonds. This category encompasses various types, including branched, random, and alternating structures, among others. An important subgroup, block copolymers, emerges when distinct polymer segments are interconnected into a single chain using covalent bonds [10, 112]. When the interaction between these segments is unfavorable, a unique phenomenon occurs: under a specific temperature threshold, they self-assemble into complex morphologies. These morphologies are dictated by the relative chain lengths of each block, temperature, and chemical repulsion between the segments. This behavior results in a complex phase diagram [113]. Interestingly, these distinct blocks within the copolymers can exhibit varying mechanical responses. As a result, the overall material response can be finely tuned, contingent upon the specific morphology that emerges from the self-assembly process.

One of the pioneering studies exploring the behavior of nanolayered materials was conducted by Lee et al. [36]. They conducted LIPIT experiments on thick plates of polystyrene and polydimethylsiloxane (PS-b-PDMS) diblock copolymer. At room temperature, the PS block is below its glass transition temperature (T_g), which is approximately 103 °C. The PDMS block is above its T_g ($T_g \sim -125$ °C), rendering it rubbery in nature. Moreover, the elastic modulus of the PS block is almost 5000 times higher than that of the PDMS block. This vast difference in stiffness effectively results in a composite of hard–soft nanolayers.

The nanolayers can be aligned such that the normals to the layers are either parallel or perpendicular to the impact direction, imparting distinct physical responses to the shock dynamics. Scanning electron microscopy (SEM) images near the bullet reveal the effects of this compression, including fragmentation induced by the elevated temperature, ultimately culminating in a melt phase. Additionally, this compression induces the formation of kink bands due to the inherent mechanical instability of the layers during compression. When the nanolayer's normal is parallel to the impact direction, this compression effect becomes more pronounced, intensifying the generation of the melt phase. Conversely, when the layer normals are oriented perpendicular to the impact direction, the region in front of the bullet undergoes robust reorientation events through kinking without experiencing significant lamellar compression.

Notably, the parallel orientation arrests the bullet before the perpendicular orientation, enhancing the ballistic absorption by 30% although the perpendicular orientation has a static modulus three times higher. Impacts in materials oriented parallel to the layers showed that the reflection of shock waves, when passing through an interface with a mechanical impedance mismatch, can aid in transferring momentum in the opposite direction [114]. It is believed that the same mechanisms are responsible for enhancing absorption at the microscale.

Coarse-grained molecular dynamics simulations were employed to study the transition to the melt phase via adiabatic shock compression [61]. When subjected to high strain rates, the injection of kinetic energy causes the transformation of the strongly segregated lamellar phase into a less-segregated state, or even melt. Interestingly, the morphologies observed after the passage of the shock resemble those observed at equilibrium, at a temperature dictated by the compression velocity.

Polydimethylsiloxane (PDMS), exhibiting both glassy and rubbery phases, has shown the ability to induce a nonlinear tribological response. In a study conducted by Kim et al., the tribological behavior of PS homopolymer (PS83) and PS-PDMS block copolymers (PS87-PDMS40 and PS83-PDMS83) was investigated using microbullets (diameters of 10–30 μm). These materials were subjected to impact a plane solid target at a 45° angle using a LIPIT setup [115]. To create the microbullets, a solution of the polymer and solvent was atomized by an ultrasonic device into a hot air column. Rapid solvent evaporation prevented long-range phase separation.

Ultrafast stroboscopic imaging was employed to capture the impact and subsequent rebound. Remarkably, minimal adhesion to the target substrate was observed, with the majority of energy dissipated through the plastic deformation of the bullets. Two velocity components—normal to and tangent to the target—were recorded. The normal residual velocity (v_r) of PS demonstrated a plateau for impact velocities above 300 m/s. In contrast, PS-PDMS exhibited nonlinear behavior at high impact velocities. This nonlinear behavior was also evident when quantifying the coefficient of friction (μ_k^*). For PS83, μ_k^* remained nearly constant up to 300 m/s, followed by an increase peaking at 530 m/s. The authors proposed a hypothesis that the contact surface temperature surpasses the dynamic glass transition temperature of PS. This leads to the creation of a visco-plastic layer with increased chain mobility. This results in a nonlinear relationship between the real contact area and the apparent contact pressure during high-velocity impacts. The presence of a rubbery matrix in PS-PDMS further accentuates this effect, resulting in a nonlinear friction coefficient with a more pronounced peak at slower velocities compared to PS.

A similar study was conducted by Shan et al., in which they manufactured layered microspheres of PS-b-PDMS (2–3 μm) [116]. The process produced layered microspheres with 20–30 shells. Upon subjecting these microspheres to impacts onto a solid target surface at varying incident angles, the researchers explored the microstructural response under high-rate strains. In cases of normal impacts at low velocities, the particles adhere and deform into an axisimetrically bell-like shape. Subsequent scanning electron microscopy (SEM) imaging showcased that the lamellar period in the lower portion was compressed, while the upper section remained undeformed. As velocities increased, the particles further flattened, adopting an overall lens-like configuration. Interestingly, at high-velocity tilted-angled impacts, the deformation is no longer axisymmetric generating stepped layer deposit on the substrate via friction. This process is possible since the PS-b-PDMS of 42–32 kDa limits the entanglements in the mid-layer region, effectively creating a slip plane. This facilitates the deposition of a structured polymer coating.

Ionomers

Ionomers are polymeric materials comprised of hydrophobic organic chains interspersed with small proportions of ionic groups (up to approximately ~15 mol%) [13]. These charged particles cluster together to form aggregates known as multiplets. These multiplets create regions of restricted mobility, acting similar to inter- or intra-chain physical crosslinks. The number, shape, and size of these regions are influenced by multiple system properties, including the type of ionic molecule, degree of neutralization, chain flexibility, and the dielectric constant of the polymer backbone, among others. When the ionic concentration exceeds a certain threshold, these regions of restricted mobility begin to overlap. Different structural formations lead to variations in system properties, impacting toughness, transparency, flexibility, viscosity, conductivity, and more.

Ethylene–methacrylic acid (EMAA) stands out as a widely used and prevalent ionomer:

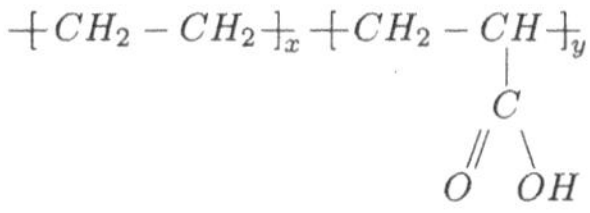

Various experimental techniques corroborate this model. X-ray scattering of these materials reveals the so-called ionomer peak, which corresponds to the clusters formed by charged particles [68, 117]. Register et al. observed that this peak remains even at elevated temperatures, reinforcing the idea that ionic clusters persist after the melting of the semicrystalline matrix.

Differential scanning calorimetry (DSC) experiments [43, 118–120] have confirmed the order–disorder transition of ionic clusters upon heating. In an initial heating of the system, two distinct peaks appear on the heat flow vs. temperature graph. The first peak represents the phase transition of the ionic cluster, while the second indicates the melting of the semicrystalline polymer matrix. As the temperature subsequently decreases, a peak corresponding to the crystallization of the polymer matrix is evident, but the ionomer peak is not immediately apparent. Only after the system undergoes aging does the ionomer phase transition become noticeable [118, 119]. This is consistent with the understanding that ionic clusters have a longer relaxation time compared to the polymer matrix.

Dynamic mechanical analysis has revealed changes in the viscoelastic relaxation of EMAA with varying degrees of neutralization. Unneutralized EMAA displays a distinct β' peak in the loss modulus, which is attributed to the short-range motion of chains in the amorphous phase crosslinked via hydrogen-bonded acids. As neutralization progresses, this peak bifurcates into two: a β peak tied to amorphous polyethylene domains and an α peak reflecting the ionic order–disorder transition [119, 121, 122].

Tensile tests indicate that an increase in system neutralization enhances the Young's modulus while diminishing material extension [119, 121]. Kalista et al. found that beyond a specific degree of EMAA neutralization with Na, the material's maximum strain significantly decreases under relatively swift strain velocities. The reduced mobility of the more neutralized system prevents it from adjusting rapidly enough to the applied strain.

A key motivation behind the extensive research on these materials is their potential self-healing capability after a rupture under certain conditions. The synergy between the semicrystalline polymer matrix and ionic clusters positions them as promising self-healing materials, and they don't require external consumable substances, unlike materials equipped with healing-agent-filled capsules or tubules. Ideally, ionomers could self-heal repeatedly. Numerous studies have delved into the self-healing capacity of these materials, primarily focusing on the effects of neutralization degree, neutralizing cation, and temperature [43, 119–121]. The primary test method is projectile impact testing, employing varying impact velocities and projectile shapes. Kalista and colleagues have also explored other rupture methods [121].

Regarding the projectile impact test, many researchers propose a two-stage self-healing mechanism. In the first stage, the projectile's energy heats the polymer film locally to a molten state, after which the elastic melt contracts to close the puncture. In the subsequent stage, the still-fluid polymer seals through autohesion and chain interdiffusion. After a relatively extended duration, the physical crosslinks created by the ionic groups are expected to regenerate, restoring the original structural integrity. It's crucial to note that this description requires the polymer's molten state. Ionomers can undergo this process since they can achieve a molten state post typical projectile impact, maintaining partial structure due to the ionic clusters, thereby enhancing the initial elastic rebound. The re-crystallization of the polymer matrix drives the second stage, which is swift, while strength restoration via ionic clusters necessitates more extended periods.

The elevation in temperature beyond the melt state of ionomers during a projectile impact test has been identified using image analysis [43] and thermal IR [123]. Varley et al. delineated three distinct regions in the material post-impact: an outer domain with a ductile elastic network aiding in the polymer rebound, a mid-zone characterized by a molten viscous polymeric flow, and a central area exhibiting not only viscous flow but also pronounced deformation and polymer fibril failures. Using DSC measurements, Kalista et al. [121] confirmed the disappearance of the ionic order–disorder transition after impact, a phenomenon present even in healed materials.

Though many EMAA ionomers feature the interplay of ionic clusters and a semicrystalline polymer matrix, self-healing capacities can markedly differ. These differences primarily hinge on the neutralization degree and the specific cation used. Kalista et al. [121] examined the healing dependency on the neutralization degree and temperature. Their investigation encompassed three neutralization levels: 0%, 30%, and 60% with Na, spanning a temperature spectrum from −50 to 140 °C. All systems manifested healing between T_i(~50 °C) and T_m(~90 °C). Remarkably, even unneutralized systems demonstrated healing, attributable to hydrogen bonding. Systems with 30% neutralization showcased healing throughout the assessed temperature range, making them the most adaptable self-healing candidates.

El Choufi et al. [119] explored the self-healing prowess of EMAA neutralized with various cations (Na, Li, Zn, and Mg) across different degrees. They deduced the pivotal role of ample ionic clusters in these systems, underscoring the requisite for a significant, yet not overly high, neutralization degree. Excessive neutralization often culminated in brittleness and compromised healing. Intermediate degrees of neutralization, particularly around ~50% with Li neutralization, proved optimal for healing.

The marriage of ionic clusters with semicrystalline phases in judiciously neutralized ionomers endows them with self-healing capabilities. This union facilitates both elastic and thermoplastic behaviors, requisite for the biphasic healing mechanism. While insufficient ionic aggregates at low neutralization levels can hinder the healing onset, an overly high degree can impede the full realization of the two-stage healing. However, an intermediate neutralization level, contingent upon the cation and polymer characteristics, appears to strike the right balance, offering robust ionic interactions without sacrificing essential elasticity.

4.3 Vitrimers

Vitrimers represent a cutting-edge class of polymers characterized by their exchangeable covalent bonds, which can be reshuffled upon cleavage [124, 125]. These bonds can be activated by various stimuli such as thermal or chemical changes, offering a customizable balance between rigidity and plasticity.

Essentially, vitrimers are polymer networks that encompass dynamic covalent bonds, partaking in topology-conserving exchange reactions. When exposed to high temperatures, vitrimers flow and mimic the characteristics of viscoelastic liquids. Conversely, at low temperatures, they exhibit thermoset-like properties due to the slowed exchange reactions. The unique ability of vitrimers to undergo network rearrangements grants them exceptional mechanical recyclability and thermal healing capabilities, along with rapid stress relaxations—traits not observed in traditional crosslinked thermoset polymers. Furthermore, they can manifest shape memory attributes, facilitating autonomous assembly into intricate shapes by utilizing dynamic crosslinks and exchangeable chemical bonds. Over recent years, vitrimers have garnered significant interest as a strong contender for creating recyclable materials, underlined by their advanced self-healing properties.

An intriguing application of vitrimers lies in their potential as shock-absorbing materials. Their bond exchange mechanism is thermally activated, enabling vitrimers to absorb energy from shocks through localized reshuffling of bonds during the shock's passage. A recent exploration of shock wave energy dissipation in poly(dimethylsiloxane) vitrimers, conducted through experiments and simulations [126], revealed that intensifying the density of dynamic bonds systematically augmented energy dissipation, as evidenced by the reduction in peak pressure of a laser-induced shock wave. Notably, the shock absorption capacity of vitrimers surpasses that of permanent non-dynamic networks and even polyurea.

In a more novel application, vitrimers have been utilized in the development of cartilage-inspired gels, showcasing good anti-impact performance. These innovative gels offer exceptional protection for brittle objects when compared with conventional materials such as foam, rubber, wood, plastic, and steel [127].

One of the defining attributes of vitrimers is their energy absorption capability through the bond exchange mechanism, allowing them to absorb significant energy while preserving structural integrity. This contrasts sharply with behaviors observed in other materials, such as semicrystalline polymers, where a shock may induce a local phase transition (like melting), thus jeopardizing the integrity and subsequent thermomechanical response. Vitrimers' unique properties and adaptability make them a promising frontier in the field of advanced materials, with applications and potentials still unfolding.

4.4 *Shock Puncture of Polymer Membranes*

The lightweight transparency of polymer plates has made them indispensable for protective applications, including safety goggles, aircraft windscreens, and police riot shields. Notably, polycarbonate exhibit superior ballistic performance normalized by weight density compared to structured metals, mainly due to its high yield strain and ductility [128, 129].

The advent of LIPIT setups has revolutionized the testing of mechanical behavior in thinner confined polymer films at high-rate-strain conditions [36]. To measure the impact absorption of thin films, high-speed cameras are typically used to record the projectile's speed before and after impact. By neglecting air drag, the energy of puncture (E_{p}) is calculated as the difference in kinetic energy:

$$E_{\mathrm{p}} = \frac{1}{2} m_{\mathrm{p}} \left(v_{\mathrm{i}}^2 - v_{\mathrm{r}}^2 \right) \tag{10}$$

where v_{i} is the projectile's velocity before impact, v_{r} is the velocity after impact, and m_{p} is the projectile's mass. Puncture energy (E_{p}) increases with the density (ρ) and thickness (h) of the thin film. To compare the mitigation efficiency of different materials and film thicknesses, E_{p} is normalized by the mass of the target within the projectile strike area (A). This mass-insensitive quantity is called specific puncture energy (E_{p}^*), defined as $E_{\mathrm{p}}^* = \frac{E_{\mathrm{p}}}{\rho A h}$.

When polymers are confined, the chain packing becomes more pronounced, increasing the influence of entropic effects and enhancing interfacial interactions with decreasing thickness [130]. Consequently, the confinement results in significant changes to the mechanical properties of polymer thin films, although the quantitative variations depend heavily on the technique employed for measurement [131].

Among the properties influenced by confinement, chain mobility increases due to the reduction of entanglements near the surface of the polymer thin film [132]. Therefore, polymer chains near the surface exhibit increased mobility, which enhances local toughness and improves overall impact absorption. This conclusion has been supported by Hyon et al., where they impacted free-standing glassy polystyrene thin films (75–500 nm thickness) by silica microprojectiles (3.7 μm diameter) [133]. As a result, employing layers of optimal thicknesses of polymer thin films, instead of a single monolithic film of the same overall mass, could serve as a promising strategy for effectively arresting microparticle impacts at high velocity [134]. They also identified the main mitigation mechanism as the initiation of crazes that grow as they heat up adiabatically, leading to a high-rate stretched plastic flow that eventually ruptures. Polymer thin films exhibit superior absorption capabilities compared to other materials. For instance, glassy polystyrene thin films have approximately twice the specific puncture energy of multilayer graphene with similar thickness and five times greater than macroscopic steel [133].

Zhu et al. conducted coarse-grained molecular dynamic simulations to explore nanoimpacts on thin films of methyl methacrylate (PMMA) [135]. Their focus was on investigating the relationship between ballistic absorption and film thickness at the nanoscale. The researchers proposed that the specific puncture energy (E_{p}^{*}) scales inversely with the square root of the film thickness. Moreover, the penetration process during impacts was divided into two stages. The initial stage involved local compression, while the second stage corresponded to global film deformation. In high-velocity impacts on thin films, the latter stage dominated, providing an explanation for the enhanced absorption observed in thinner films.

The capacity of polymers to form large chains that entangle allows them to propagate stress far from the puncture zone. Chan et al. investigated the role of entanglements in glassy polycarbonate (PC) thin films with varying chain lengths and degrees of plasticizers [136]. Comparing the specific puncture energy with the estimated entanglement density reveals that entanglements enhance energy absorption. Furthermore, since the work of fracture is proportional to the area beneath the stress–strain curve, the specific puncture energy increases with the size of the hole generated by the impact.

Additionally, Xie and Lee investigated the ballistic impact on atactic polystyrene (PS) using a LIPIT setup [33]. The choice of atactic PS ensured a narrow polydispersity, and the selection of a wide range of molecular weights allowed them to experimentally study the role of entanglements in the dissipation process. Their findings indicated that mitigation is enhanced for higher molecular masses and, consequently, higher entanglement densities. This gap between lighter and heavier polymers becomes more pronounced in low-velocity impacts, as the delocalized energy dissipation process contributes less at higher projectile velocities (v_{i}).

To study the failure mechanisms associated with the nanoscale ballistic impacts, Bowman et al. have developed a large-scale atomistic molecular dynamics simulation [137]. They subjected ultrathin films of amorphous rubbery polyethylene and amorphous glassy polystyrene to impacts. During low-velocity impacts, the absorption of energy was attributed to chain disentanglement, with the rubbery polyethylene exhibiting greater energy absorption compared to the glassy polystyrene due to its higher chain mobility in the rubbery state. In contrast, at high-velocity impacts, the glassy polystyrene displayed superior energy mitigation compared to rubbery polyethylene. This effect arose from an increase in chain scission and chain pullout generated by elevated temperatures of the shock surpassing the glass transition temperature (T_g) of the material. Surprisingly, at high-velocity impacts, both glassy polystyrene and rubbery polyethylene exhibited significant specific puncture energy values, rivaling those of multilayer graphene.

Ballistic impact on thin films is a rich physical process involving momentum transfer, adiabatic heating, propagation of multiples mechanical waves, and crack propagation, among others. To gain a deeper understanding of this process, Chen et al. developed a comprehensive analytical model of the energy dissipation process accounting kinetic energy transfer, the work of fracture, the elastic bending, and the thermal work [48]. Using this model, they derived an expression for the ballistic limit (v_0), which represents the maximum projectile velocity a material can withstand without perforation. The analytical predictions were then compared with experimental results obtained through LIPIT experiments conducted on polycarbonate (PC) thin films. Their findings revealed that for an impact velocity of 55 m/s, the critical film thickness of polycarbonate is 190 nm, beyond which perforation occurs.

Polymer-grafted nanoparticles (PGNs) have emerged as promising materials for impact mitigation. These materials consist of long polymeric chains grafted to hard nanoparticles, which enables better quasistatic mechanical performance by synergistically coupling the mechanical properties of both components. In a study conducted by Hyon et al., LIPIT experiments were performed on thin films of SiO_2–PS and Fe_3O_4–PS PGNs, with variations in the molecular weight of the polymer part, and compared to PS homopolymer thin films [138]. The results showed that SiO_2–PS231 exhibited superior impact absorption, particularly at higher velocities and for thinner films. Moreover, PS267 outperformed Fe_3O_4PS250, indicating a complex interplay between nanoparticle shape, size, volume fraction, polymer graft density, and polymer entanglement length, which may enhance or limit the impact mitigation process. Similarly, Chen et al. conducted research on silica-polymethylacrylate (PMA) PGNs thin films with varying polymer lengths and examined the impact using LIPIT [139]. Interestingly, they observed a peak in specific puncture energy just before the system reached full entanglement. Through quasi-elastic neutron scattering and X-ray photon correlation spectroscopy, they found that larger chains, while enhancing entanglements, also hindered nanoparticle diffusion. This intricate balance between the two effects resulted in an optimal performance situated in an intermediate regime. Further research and understanding of these complex interactions will undoubtedly contribute to the development of improved impact-resistant materials for various applications.

Semicrystalline polymers like polyethylene (PE), poly(*p*-phenylene terephthalamide) (PPTA), and polytetrafluoroethylene (PTFE) have displayed exceptional ballistic performance [24, 90, 91]. Notably, Cai and Thevamaran synthesized poly(vinylidene fluoride-co-trifluoroethylene) (P(VDF-TrFE)) thin films, which demonstrated impressive ballistic absorption in LIPIT experiments [140]. They successfully controlled the crystalline structure and polymer chain conformation by varying the film's thickness and using thermal annealing. Reducing the crystalline domains through film thinning or non-annealing increased the segmental mobility of polymer chains in the amorphous phase, resulting in enhanced specific puncture energy. Additionally, the presence of polar C–F bonds created dipole–dipole interactions, leading to significantly increased intermolecular attractive forces. This allowed substantial energy dissipation during interactions between the chains, particularly when they were oriented along the stretching direction. Measurements of E_{p}^{*} for P(VDF-TrFE) revealed that its ballistic performance is comparable to that of PS (polystyrene) at high impacts, approximately five times that of steel. What's more, P(VDF-TrFE) exhibited outstanding performance across a broad range of velocities. For example, at low-velocity impacts (400 m/s), the E_{p}^{*} was approximately 5.4 times higher compared to PS.

5 Conclusions

In this chapter, we embarked on a detailed examination of polymeric materials and their proficiency in mitigating shocks and impacts. While we didn't canvass every conceivable system, our focus was firmly on contemporary polymeric systems that had seen significant exploration in recent times.

Through both experimental and simulated studies, it was underscored that certain molecular traits—like the relaxation time spectrum—are paramount in the domain of impact absorption. Polymers, in their adaptability, emerged prominently. Their inherent characteristics can be finetuned, making them suitable for a wide range of applications, with shock absorption being a primary focus.

We discussed several molecular attributes that can be altered using refined synthesis methods, such as glass transitions, crystallinity levels, entanglements, chain patterns, crosslinking techniques, a spectrum between semiflexibility and rigidity, the interplay of soft with tough regions, formation of self-assembling structures (seen in copolymers), electrical charge zones, multipolar domains, and dynamic covalent linkages.

For example, we elucidated the role of transitions from rubbery states to more glass-like states in bolstering the absorption potential of polyureas. A system's crystallinity can function as a barrier to shock waves and handle impacts by transitioning into a melting phase. Aspects like the intricacies of entanglements and the architecture of chains proved crucial in the absorption spectrum from low to moderate strain rates. Semiflexibility, through its chain buckling mechanism, was highlighted as beneficial for absorbing impacts. The balance between soft and rigid molecular regions played a pivotal role in the shock-absorbing capabilities of materials like polyurethanes, polyureas, and their kin. Furthermore, the study of ionomers, with their charge-defined domains, and vitrimers, equipped with dynamic bonds, suggested a promising future in shock absorption.

In conclusion, the attributes we spotlighted are instrumental in understanding how a material responds to impacts across varying strain rates. By integrating these features, we can better comprehend and potentially design systems optimized for wide-ranging impact absorption.

Acknowledgments We acknowledge support from Universidad Nacional del Sur, the National Research Council of Argentina (CONICET), and the Fondo para la Investigación Científica y Tecnológica (FONCYT, Grant PICT-2021-1272). LRG acknowledges support from the Alexander von Humboldt Foundation through the Georg Forster and return fellowships.

References

1. Zel'Dovich, Y.B., Raizer, Y.P.: Physics of Shock Waves and High-Temperature Hydrodynamic Phenomena. Courier Corporation (2002)
2. Meyers, M.A.: Dynamic Behavior of Materials. John Wiley & Sons (1994)
3. Bourne, N.: Materials in Mechanical Extremes: Fundamentals and Applications. Cambridge University Press (2013)
4. Duvall, G., Graham, R.: Phase transitions under shock-wave loading. Rev. Mod. Phys. **49**(3), 523 (1977)
5. Kadau, K., Germann, T.C., Lomdahl, P.S., Holian, B.L.: Microscopic view of structural phase transitions induced by shock waves. Science. **296**(5573), 1681–1684 (2002)
6. Jonas, G., Zukas, J.A.: Mechanics of penetration: analysis and experiment. Int. J. Eng. Sci. **16**(11), 879–903 (1978)
7. National Research Council, et al.: Opportunities in Protection Materials Science and Technology for Future Army Applications. National Academies Press (2011)
8. Shojaei, B., Najafi, M., Yazdanbakhsh, A., Abtahi, M., Zhang, C.: A review on the applications of polyurea in the construction industry. Polym. Adv. Technol. **32**(8), 2797–2812 (2021)
9. Mills, N., Fitzgerald, C., Gilchrist, A., Verdejo, R.: Polymer foams for personal protection: cushions, shoes and helmets. Compos. Sci. Technol. **63**(16), 2389–2400 (2003)
10. Hadjichristidis, N., Pispas, S., Floudas, G.: Block Copolymers: Synthetic Strategies, Physical Properties, and Applications. John Wiley & Sons (2003)
11. Barsoum, R.G.S.: Elastomeric Polymers with High Rate Sensitivity: Applications in Blast, Shockwave, and Penetration Mechanics. William Andrew (2015)
12. He, Y., Xie, D., Zhang, X.: The structure, microphase-separated morphology, and property of polyurethanes and polyureas. J. Mater. Sci. **49**, 7339–7352 (2014)
13. Tant, M.R., Mauritz, K.A., Wilkes, G.L.: Ionomers: Synthesis, Structure, Properties and Applications. Springer (2012)
14. Denissen, W., Winne, J.M., Du Prez, F.E.: Vitrimers: permanent organic networks with glass-like fluidity. Chem. Sci. **7**(1), 30–38 (2016)
15. Frost, H.J., Ashby, M.F.: Deformation-Mechanism Maps: The Plasticity and Creep of Metals and Ceramics. Pergamon Press, Oxford (1982) 175 p
16. Sargent, P., Ashby, M.: The Presentation of High Strain-Rates on Deformation Mechanism Maps. Engineering Department, Cambridge University (1983)

17. Dawson, P., Swallowe, G., Xinwu, Z.: Temperature rises during high-rate deformation of polymers. J. Phys. IV. **1**(C3), 3–701 (1991)
18. Carter, W.J., Marsh, S.P.: Hugoniot Equation of State of Polymers. Technical Report. Los Alamos National Lab. (LANL), Los Alamos, NM (1995)
19. Walley, S., Field, J.: Strain rate sensitivity of polymers in compression from low to high rates. DYMAT J. **1**(3), 211–227 (1994)
20. Bourne, N.K.: On the shock response of polymers to extreme loading. J. Dyn. Behav. Mater. **2**, 33–42 (2016)
21. Millett, J., Bourne, N.: The shock induced equation of state of three simple polymers. J. Phys. D. Appl. Phys. **37**(20), 2901 (2004)
22. Millett, J., Bourne, N., Akhavan, J.: The response of hydroxy-terminated polybutadiene to one-dimensional shock loading. J. Appl. Phys. **95**(9), 4722–4727 (2004)
23. Millett, J., Brown, E., Gray, G., Bourne, N.K., Wood, D., Appleby-Thomas, G.: The effects of changing chemistry on the shock response of basic polymers. J. Dyn. Behav. Mater. **2**, 326–336 (2016)
24. Siviour, C.R., Jordan, J.L.: High strain rate mechanics of polymers: a review. J. Dyn. Behav. Mater. **2**, 15–32 (2016)
25. Jordan, J.L., Casem, D.T., Robinette, J.: Hugoniot and dynamic strength in polyurea. J. Appl. Phys. **131**(16), 165903 (2022)
26. Hopkinson, B.: X. A method of measuring the pressure produced in the detonation of high, explosives or by the impact of bullets. Philos. Trans. R. Soc. Lond. A Pap. Math. Phys. Charact. **213**(497–508), 437–456 (1914)
27. Kolsky, H.: An investigation of the mechanical properties of materials at very high rates of loading. Proc. Phys. Soc. Sect. B. **62**(11), 676 (1949)
28. Harding, J., Wood, E., Campbell, J.: Tensile testing of materials at impact rates of strain. J. Mech. Eng. Sci. **2**(2), 88–96 (1960)
29. Duffy, J., Campbell, J., Hawley, R.: On the use of a torsional split Hopkinson bar to study rate effects in 1100-0 aluminum. J. Appl. Mech. **38**, 83 (1971)
30. Field, J.E., Walley, T.M., Proud, W., Goldrein, H., Siviour, C.: Review of experimental techniques for high-rate deformation and shock studies. Int. J. Impact Eng. **30**(7), 725–775 (2004)
31. Gama, B.A., Lopatnikov, S.L., Gillespie Jr., J.W.: Hopkinson bar experimental technique: a critical review. Appl. Mech. Rev. **57**(4), 223–250 (2004)
32. Kawai, N., Tsurui, K., Hasegawa, S., Sato, E.: Single microparticle launching method using two-stage light-gas gun for simulating hypervelocity impacts of micrometeoroids and space debris. Rev. Sci. Instrum. **81**(11), 115105 (2010)
33. Xie, W., Lee, J.-H.: Dynamics of entangled networks in ultrafast perforation of polystyrene nanomembranes. Macromolecules. **53**(5), 1701–1705 (2020)
34. Lee, J.-H., Loya, P.E., Lou, J., Thomas, E.L.: Dynamic mechanical behavior of multilayer graphene via supersonic projectile penetration. Science. **346**(6213), 1092–1096 (2014)
35. Barrios, S., Lance, P., Abate, A.A., Prieto, G., García, N.A., Piqueras, C.M., Vega, D.A., Satti, A., Gómez, L.R.: Velocity distributions in a gas-gun microparticle accelerator. Rev. Sci. Instrum. **93**(10), 105101 (2022)
36. Lee, J.-H., Veysset, D., Singer, J.P., Retsch, M., Saini, G., Pezeril, T., Nelson, K.A., Thomas, E.L.: High strain rate deformation of layered nanocomposites. Nat. Commun. **3**(1), 1164 (2012)
37. Veysset, D., Hsieh, A.J., Kooi, S., Maznev, A.A., Masser, K.A., Nelson, K.A.: Dynamics of supersonic microparticle impact on elastomers revealed by real– time multi–frame imaging. Sci. Rep. **6**(1), 25577 (2016)
38. Thornhill, T., Chhabildas, L., Reinhart, W., Davidson, D.: Particle launch to 19 km/s for micro-meteoroid simulation using enhanced three-stage light gas gun hypervelocity launcher techniques. Int. J. Impact Eng. **33**(1–12), 799–811 (2006)
39. Uhlig, W.C., Berning, P.R., Bartkowski, P.T., Coppinger, M.J.: Electrically-launched mm-sized hypervelocity projectiles. Int. J. Impact Eng. **137**, 103441 (2020)

40. Swift, D.C., Niemczura, J.G., Paisley, D.L., Johnson, R.P., Luo, S.-N., Tierney, T.E.: Laser-launched flyer plates for shock physics experiments. Rev. Sci. Instrum. **76**(9), 093907 (2005)
41. Brown, K.E., Shaw, W.L., Zheng, X., Dlott, D.D.: Simplified laser-driven flyer plates for shock compression science. Rev. Sci. Instrum. **83**(10), 103901 (2012)
42. Zhou, X., Miao, Y.-R., Shaw, W.L., Suslick, K.S., Dlott, D.D.: Shock wave energy absorption in metal–organic framework. J. Am. Chem. Soc. **141**(6), 2220–2223 (2019)
43. Varley, R.J., Zwaag, S.: Towards an understanding of thermally activated self-healing of an ionomer system during ballistic penetration. Acta Mater. **56**(19), 5737–5750 (2008)
44. Sundaresan, V.B., Morgan, A., Castellucci, M.: Self-healing of ionomeric polymers with carbon fibers from medium-velocity impact and resistive heating. Smart Mater. Res. **2013**, 1 (2013)
45. Moritoh, T., Kawai, N., Nakamura, K.G., Kondo, K.-I.: Optimization of a compact two-stage light-gas gun aiming at a velocity of 9 km/s. Rev. Sci. Instrum. **72**(11), 4270–4272 (2001)
46. Piekutowski, A., Poormon, K.: Development of a three-stage, light-gas gun at the University of Dayton Research Institute. Int. J. Impact Eng. **33**(1–12), 615–624 (2006)
47. Xie, W., Alizadeh-Dehkharghani, A., Chen, Q., Champagne, V.K., Wang, X., Nardi, A.T., Kooi, S., Müftü, S., Lee, J.-H.: Dynamics and extreme plasticity of metallic microparticles in supersonic collisions. Sci. Rep. **7**(1), 5073 (2017)
48. Chen, S.H., Souna, A.J., Soles, C.L., Stranick, S.J., Chan, E.P.: Using microprojectiles to study the ballistic limit of polymer thin films. Soft Matter. **16**(16), 3886–3890 (2020)
49. Willmott, G., Radford, D.: Taylor impact of glass rods. J. Appl. Phys. **97**(9), 093522 (2005)
50. Gao, C., Iwamoto, T.: Instrumented Taylor impact test for measuring stress-strain curve through single trial. Int. J. Impact Eng. **157**, 103980 (2021)
51. Mocker, A., Bugiel, S., Auer, S., Baust, G., Colette, A., Drake, K., Fiege, K., Grün, E., Heckmann, F., Helfert, S., et al.: A 2 mV Van de Graaff accelerator as a tool for planetary and impact physics research. Rev. Sci. Instrum. **82**(9), 095111 (2011)
52. Burchell, M.J., Cole, M.J., McDonnell, J., Zarnecki, J.C.: Hypervelocity impact studies using the 2 mV Van de Graaff accelerator and two-stage light gas gun of the University of Kent at Canterbury. Meas. Sci. Technol. **10**(1), 41 (1999)
53. Hasegawa, S., Hamabe, Y., Fujiwara, A., Yano, H., Sasaki, S., Ohashi, H., Kawamura, T., Nogami, K.-I., Kobayashi, K., Iwai, T., et al.: Microparticle acceleration for hypervelocity experiments by a 3.75 mV Van de Graaff accelerator and a 100kv electrostatic accelerator in Japan. Int. J. Impact Eng. **26**(1–10), 299–308 (2001)
54. Shibata, H., Kobayashi, K., Iwai, T., Hamabe, Y., Sasaki, S., Hasegawa, S., Yano, H., Fujiwara, A., Ohashi, H., Kawamura, T., et al.: Microparticle acceleration by a Van de Graaff accelerator and application to space and material sciences. Radiat. Phys. Chem. **60**(4–5), 277–282 (2001)
55. Kremer, K., Grest, G.S.: Dynamics of entangled linear polymer melts: a molecular-dynamics simulation. J. Chem. Phys. **92**(8), 5057–5086 (1990)
56. Meyer, H., Müller-Plathe, F.: Formation of chain-folded structures in supercooled polymer melts. J. Chem. Phys. **115**(17), 7807–7810 (2001)
57. Reith, D., Meyer, H., Müller-Plathe, F.: Mapping atomistic to coarse-grained polymer models using automatic simplex optimization to fit structural properties. Macromolecules. **34**(7), 2335–2345 (2001)
58. Sommer, J.-U., Luo, C.: Molecular dynamics simulations of semicrystalline polymers: crystallization, melting, and reorganization. J. Polym. Sci. B Polym. Phys. **48**(21), 2222–2232 (2010)
59. Waheed, N., Ko, M., Rutledge, G.: Molecular simulation of crystal growth in long alkanes. Polymer. **46**(20), 8689–8702 (2005)
60. Gee, R.H., Lacevic, N., Fried, L.E.: Atomistic simulations of spinodal phase separation preceding polymer crystallization. Nat. Mater. **5**(1), 39–43 (2006)
61. Ortellado, L., Vega, D.A., Gómez, L.R.: Shock melting of lamellae-forming block copolymers. Phys. Rev. E. **106**(4), 044502 (2022)

62. Seo, Y., Brown, J.R., Hall, L.M.: Effect of tapering on morphology and interfacial behavior of diblock copolymers from molecular dynamics simulations. Macromolecules. **48**(14), 4974–4982 (2015)
63. Murat, M., Grest, G.S., Kremer, K.: Statics and dynamics of symmetric diblock copolymers: a molecular dynamics study. Macromolecules. **32**(3), 595–609 (1999)
64. Hall, L.M., Seitz, M.E., Winey, K.I., Opper, K.L., Wagener, K.B., Stevens, M.J., Frischknecht, A.L.: Ionic aggregate structure in ionomer melts: effect of molecular architecture on aggregates and the ionomer peak. J. Am. Chem. Soc. **134**(1), 574–587 (2012)
65. Hall, L.M., Stevens, M.J., Frischknecht, A.L.: Dynamics of model ionomer melts of various architectures. Macromolecules. **45**(19), 8097–8108 (2012)
66. Sampath, J., Hall, L.M.: Impact of ionic aggregate structure on ionomer mechanical properties from coarse-grained molecular dynamics simulations. J. Chem. Phys. **147**(13), 134901 (2017)
67. Sampath, J., Hall, L.M.: Effect of neutralization on the structure and dynamics of model ionomer melts. Macromolecules. **51**(2), 626–637 (2018)
68. Buitrago, C.F., Bolintineanu, D.S., Seitz, M.E., Opper, K.L., Wagener, K.B., Stevens, M.J., Frischknecht, A.L., Winey, K.I.: Direct comparisons of x-ray scattering and atomistic molecular dynamics simulations for precise acid copolymers and ionomers. Macromolecules. **48**(4), 1210–1220 (2015)
69. Bolintineanu, D.S., Stevens, M.J., Frischknecht, A.L.: Atomistic simulations predict a surprising variety of morphologies in precise ionomers. ACS Macro Lett. **2**(3), 206–210 (2013)
70. Bolintineanu, D.S., Stevens, M.J., Frischknecht, A.L.: Influence of cation type on ionic aggregates in precise ionomers. Macromolecules. **46**(13), 5381–5392 (2013)
71. Zhao, H., Wei, X., Fang, Y., Gao, K., Yue, T., Zhang, L., Ganesan, V., Meng, F., Liu, J.: Molecular dynamics simulation of the structural, mechanical, and reprocessing properties of vitrimers based on a dynamic covalent polymer network. Macromolecules. **55**(4), 1091–1103 (2022)
72. Jabbarzadeh, A., Atkinson, J., Tanner, R.: Effect of molecular shape on rheological properties in molecular dynamics simulation of star, h, comb, and linear polymer melts. Macromolecules. **36**(13), 5020–5031 (2003)
73. Wessels, M.G., Jayaraman, A.: Molecular dynamics simulation study of linear, bottlebrush, and star-like amphiphilic block polymer assembly in solution. Soft Matter. **15**(19), 3987–3998 (2019)
74. Halverson, J.D., Lee, W.B., Grest, G.S., Grosberg, A.Y., Kremer, K.: Molecular dynamics simulation study of nonconcatenated ring polymers in a melt. I. Statics. J. Chem. Phys. **134**(20), 204905 (2011)
75. Vega, D.A., Lance, P., Zorzi, E., Register, R.A., Gómez, L.R.: Shock compression of semiflexible polymers. Soft Matter. **19**(32), 6131–6139 (2023)
76. Arman, B., Reddy, A.S., Arya, G.: Viscoelastic properties and shock response of coarse-grained models of multiblock versus diblock copolymers: insights into dissipative properties of polyurea. Macromolecules. **45**(7), 3247–3255 (2012)
77. Elder, R.M., O'Connor, T.C., Chantawansri, T.L., Sliozberg, Y.R., Sirk, T.W., Yeh, I.-C., Robbins, M.O., Andzelm, J.W.: Shock-wave propagation and reflection in semicrystalline polyethylene: a molecular-level investigation. Phys. Rev. Mater. **1**(4), 043606 (2017)
78. Chantawansri, T.L., Sirk, T.W., Byrd, E.F., Andzelm, J.W., Rice, B.M.: Shock Hugoniot calculations of polymers using quantum mechanics and molecular dynamics. J. Chem. Phys. **137**(20), 204901 (2012)
79. Dewapriya, M., Miller, R.: Quantum and classical molecular dynamics simulations of shocked polyurea and polyurethane. Comput. Mater. Sci. **203**, 111166 (2022)
80. Ferry, J.D.: Viscoelastic Properties of Polymers. John Wiley & Sons (1980)
81. Hasan, O., Boyce, M.: A constitutive model for the nonlinear viscoelastic viscoplastic behavior of glassy polymers. Polym. Eng. Sci. **35**(4), 331–344 (1995)

82. Kendall, M.J., Siviour, C.R.: Experimentally simulating adiabatic conditions: approximating high-rate polymer behavior using low rate experiments with temperature profiles. Polymer. **54**(18), 5058–5063 (2013)
83. Arruda, E.M., Boyce, M.C., Jayachandran, R.: Effects of strain rate, temperature and thermomechanical coupling on the finite strain deformation of glassy polymers. Mech. Mater. **19**(2–3), 193–212 (1995)
84. Kendall, M., Siviour, C.: Experimentally simulating high-rate behaviour: rate and temperature effects in polycarbonate and PMMA. Philos. Trans. R. Soc. A Math. Phys. Eng. Sci. **372**(2015), 20130202 (2014)
85. Richeton, J., Ahzi, S., Vecchio, K.S., Jiang, F.C., Adharapurapu, R.R.: Influence of temperature and strain rate on the mechanical behavior of three amorphous polymers: characterization and modeling of the compressive yield stress. Int. J. Solids Struct. **43**(7–8), 2318–2335 (2006)
86. Kendall, M.J., Froud, R.F., Siviour, C.R.: Novel temperature measurement method & thermodynamic investigations of amorphous polymers during high-rate deformation. Polymer. **55**(10), 2514–2522 (2014)
87. Okereke, M., Buckley, C., Siviour, C.: Compression of polypropylene across a wide range of strain rates. Mech. Time Depend. Mater. **16**, 361–379 (2012)
88. Rae, P., Dattelbaum, D.: The properties of poly (tetrafluoroethylene)(ptfe) in compression. Polymer. **45**(22), 7615–7625 (2004)
89. Brown, E., Dattelbaum, D., Brown, D., Rae, P., Clausen, B.: A new strain path to inducing phase transitions in semi-crystalline polymers. Polymer. **48**(9), 2531–2536 (2007)
90. Jordan, J.L., Siviour, C.R., Foley, J.R., Brown, E.N.: Compressive properties of extruded polytetrafluoroethylene. Polymer. **48**(14), 4184–4195 (2007)
91. Brown, E.N., Dattelbaum, D.M.: The role of crystalline phase on fracture and microstructure evolution of polytetrafluoroethylene (PTFE). Polymer. **46**(9), 3056–3068 (2005)
92. Brown, E., Trujillo, C., Gray, G., Rae, P., Bourne, N.: Soft recovery of polytetrafluoroethylene shocked through the crystalline phase II-III transition. J. Appl. Phys. **101**(2), 024916 (2007)
93. Brown, E., Willms, R., Gray, G., Rae, P., Cady, C., Vecchio, K., Flowers, J., Martinez, M.: Influence of molecular conformation on the constitutive response of polyethylene: a comparison of HDPE, UHMWPE, and PEX. Exp. Mech. **47**, 381–393 (2007)
94. Hartley, N., Brown, S., Cowan, T., Cunningham, E., Döppner, T., Falcone, R., Fletcher, L., Frydrych, S., Galtier, E., Gamboa, E., et al.: Evidence for crystalline structure in dynamically-compressed polyethylene up to 200 GPa. Sci. Rep. **9**(1), 4196 (2019)
95. Holden, G., Bishop, E., Legge, N.R.: Thermoplastic elastomers. J. Polym. Sci. C Polym. Symp. **26**, 37–57 (1969)
96. Holden, G.: Thermoplastic elastomers. In: Applied Plastics Engineering Handbook, pp. 77–91. Elsevier (2011)
97. Iqbal, N., Tripathi, M., Parthasarathy, S., Kumar, D., Roy, P.: Polyurea coatings for enhanced blast-mitigation: a review. RSC Adv. **6**(111), 109706–109717 (2016)
98. Zhang, R., Huang, W., Lyu, P., Yan, S., Wang, X., Ju, J.: Polyurea for blast and impact protection: a review. Polymers. **14**(13), 2670 (2022)
99. Doi, M.: Soft Matter Physics. Oxford University Press USA, New York (2013)
100. Hansen, C.M.: The universality of the solubility parameter. Ind. Eng. Chem. Prod. Res. Dev. **8**(1), 2–11 (1969)
101. Camberlin, Y., Pascault, J.P.: Phase segregation kinetics in segmented linear polyurethanes: relations between equilibrium time and chain mobility and between equilibrium degree of segregation and interaction parameter. J. Polym. Sci. Polym. Phys. Ed. **22**(10), 1835–1844 (1984)
102. Sarva, S.S., Deschanel, S., Boyce, M.C., Chen, W.: Stress–strain behavior of a polyurea and a polyurethane from low to high strain rates. Polymer. **48**(8), 2208–2213 (2007)
103. Bogoslovov, R., Roland, C., Gamache, R.: Impact-induced glass transition in elastomeric coatings. Appl. Phys. Lett. **90**(22), 221910 (2007)
104. Ransom, T.C., Ahart, M., Hemley, R.J., Roland, C.M.: Vitrification and density scaling of polyurea at pressures up to 6 GPa. Macromolecules. **50**(20), 8274–8278 (2017)

105. Ransom, T.C., Ahart, M., Hemley, R.J., Roland, C.M.: Acoustic properties and density of polyurea at pressure up to 13.5 GPa through Brillouin scattering spectroscopy. J. Appl. Phys. **123**(19), 195102 (2018)
106. Fragiadakis, D., Gamache, R., Bogoslovov, R., Roland, C.: Segmental dynamics of polyurea: effect of stoichiometry. Polymer. **51**(1), 178–184 (2010)
107. Grujicic, M., Yavari, R., Snipes, J., Ramaswami, S., Runt, J., Tarter, J., Dillon, G.: Molecular-level computational investigation of shock-wave mitigation capability of polyurea. J. Mater. Sci. **47**, 8197–8215 (2012)
108. Grujicic, M., Snipes, J., Ramaswami, S., Yavari, R., Ramasubramanian, M.: Meso-scale computational investigation of shock-wave attenuation by trailing release wave in different grades of polyurea. J. Mater. Eng. Perform. **23**, 49–64 (2014)
109. Qiao, J., Amirkhizi, A.V., Schaaf, K., Nemat-Nasser, S., Wu, G.: Dynamic mechanical and ultrasonic properties of polyurea. Mech. Mater. **43**(10), 598–607 (2011)
110. Grujicic, M., Yavari, R., Snipes, J., Ramaswami, S., Jiao, T., Clifton, R.: Experimental and computational study of the shearing resistance of polyurea at high pressures and high strain rates. J. Mater. Eng. Perform. **24**, 778–798 (2015)
111. Manav, M., Ortiz, M.: Molecular dynamics study of the shock response of polyurea. Polymer. **212**, 123109 (2021)
112. Hamley, I.W., et al.: Developments in Block Copolymer Science and Technology. Wiley Online Library (2004)
113. Leibler, L.: Theory of microphase separation in block copolymers. Macromolecules. **13**(6), 1602–1617 (1980)
114. Zhuang, S., Ravichandran, G., Grady, D.E.: An experimental investigation of shock wave propagation in periodically layered composites. J. Mech. Phys. Solids. **51**(2), 245–265 (2003)
115. Kim, A., Muftu, S., Thomas, E.L., Lee, J.-H.: Extreme tribological characteristics of copolymers induced by dynamic rheological instability. ACS Appl. Polym. Mater. **3**(9), 4413–4418 (2021)
116. Shan, W., Weisbord, I., Feng, X., Hyon, J., Manesi, G.-M., Avgeropoulos, A., Segal-Peretz, T., Thomas, E.L.: Layered thin film deposition via extreme interbrush slip in a lamellar block copolymer. Macromolecules. **55**(20), 9022–9029 (2022)
117. Register, R.A., Yu, X.-H., Cooper, S.L.: Effects of matrix polarity and ambient aging on the morphology of sulfonated polyurethane ionomers. Polym. Bull. **22**, 565–571 (1989)
118. Tadano, K., Hirasawa, E., Yamamoto, H., Yano, S.: Order-disorder transition of ionic clusters in ionomers. Macromolecules. **22**(1), 226–233 (1989)
119. El Choufi, N., Mustapha, S., Tehrani-Bagha, A.R., Grady, B.P.: Self-healability of poly (ethylene-co-methacrylic acid): effect of ionic content and neutralization. Polymers. **14**(17), 3575 (2022)
120. Kalista Jr., S.J., Ward, T.C.: Thermal characteristics of the self-healing response in poly (ethylene-co-methacrylic acid) copolymers. J. R. Soc. Interface. **4**(13), 405–411 (2007)
121. Kalista, S.J., Pflug, J.R., Varley, R.J.: Effect of ionic content on ballistic self-healing in EMAA copolymers and ionomers. Polym. Chem. **4**(18), 4910–4926 (2013)
122. Tachino, H., Hara, H., Hirasawa, E., Kutsumizu, S., Tadano, K., Yano, S.: Dynamic mechanical relaxations of ethylene ionomers. Macromolecules. **26**(4), 752–757 (1993)
123. Fall, R.: Puncture reversal of ethylene ionomer-mechanistic studies. http://scholar.lib.vt.edu/theses/availale/etd-08312001-084412/unrestricted/etd.PDF (2001)
124. Montarnal, D., Capelot, M., Tournilhac, F., Leibler, L.: Silica-like malleable materials from permanent organic networks. Science. **334**(6058), 965–968 (2011)
125. Zheng, N., Xu, Y., Zhao, Q., Xie, T.: Dynamic covalent polymer networks: a molecular platform for designing functions beyond chemical recycling and self-healing. Chem. Rev. **121**(3), 1716–1745 (2021)
126. Lee, J., Jing, B.B., Porath, L.E., Sottos, N.R., Evans, C.M.: Shock wave energy dissipation in catalyst-free poly (dimethylsiloxane) vitrimers. Macromolecules. **53**(12), 4741–4747 (2020)
127. Cheng, J., Zhang, Z., Liu, K., Ma, C., Zhang, G.: Cartilage-inspired smart anti-impact gel with highly stable and tailored properties. Cell Rep. Phys. Sci. **4**(3), 101289 (2023)

128. Radin, J., Goldsmith, W.: Normal projectile penetration and perforation of layered targets. Int. J. Impact Eng. **7**(2), 229–259 (1988)
129. Wright, S., Fleck, N., Stronge, W.: Ballistic impact of polycarbonate-an experimental investigation. Int. J. Impact Eng. **13**(1), 1–20 (1993)
130. Hashim, A.A.: Polymer Thin Films. InTech, Vienna (2010)
131. Vogt, B.D.: Mechanical and viscoelastic properties of confined amorphous polymers. J. Polym. Sci. B Polym. Phys. **56**(1), 9–30 (2018)
132. García, N.A., Barrat, J.-L.: Entanglement reduction induced by geometrical confinement in polymer thin films. Macromolecules. **51**(23), 9850–9860 (2018)
133. Hyon, J., Lawal, O., Fried, O., Thevamaran, R., Yazdi, S., Zhou, M., Veysset, D., Kooi, S.E., Jiao, Y., Hsiao, M.-S., et al.: Extreme energy absorption in glassy polymer thin films by supersonic micro-projectile impact. Mater. Today. **21**(8), 817–824 (2018)
134. Liu, P., Strano, M.S.: Toward ambient armor: can new materials change longstanding concepts of projectile protection? Adv. Funct. Mater. **26**(6), 943–954 (2016)
135. Zhu, Y., Giuntoli, A., Hansoge, N., Lin, Z., Keten, S.: Scaling for the inverse thickness dependence of specific penetration energy in polymer thin film impact tests. J. Mech. Phys. Solids. **161**, 104808 (2022)
136. Chan, E.P., Xie, W., Orski, S.V., Lee, J.-H., Soles, C.L.: Entanglement density-dependent energy absorption of polycarbonate films via supersonic fracture. ACS Macro Lett. **8**(7), 806–811 (2019)
137. Bowman, A.L., Chan, E.P., Lawrimore, W.B., Newman, J.K.: Supersonic impact response of polymer thin films via large-scale atomistic simulations. Nano Lett. **21**(14), 5991–5997 (2021)
138. Hyon, J., Gonzales, M., Streit, J.K., Fried, O., Lawal, O., Jiao, Y., Drummy, L.F., Thomas, E.L., Vaia, R.A.: Projectile impact shock-induced deformation of one-component polymer nanocomposite thin films. ACS Nano. **15**(2), 2439–2446 (2021)
139. Chen, S.H., Souna, A.J., Stranick, S.J., Jhalaria, M., Kumar, S.K., Soles, C.L., Chan, E.P.: Controlling toughness of polymer-grafted nanoparticle composites for impact mitigation. Soft Matter. **18**(2), 256–261 (2022)
140. Cai, J., Thevamaran, R.: Superior energy dissipation by ultrathin semicrystalline polymer films under supersonic microprojectile impacts. Nano Lett. **20**(8), 5632–5638 (2020)

Optical Micromanipulation of Soft Materials: Applications in Devices and Technologies

Sanatan Halder, Debojit Chanda, Dibyendu Mondal, Sandip Kundu, and Manas Khan

Abstract Since its invention by Arthur Ashkin and colleagues at Bell Labs in the 1970s, optical micromanipulation, also known as optical tweezers or laser tweezers, has evolved remarkably to become one of the most convenient and versatile tools for studying soft materials, including biological systems. Arthur Ashkin received the Nobel Prize in Physics in 2018 for enabling these extraordinary scientific advancements. Essentially, a focused laser beam is used to apply and measure minuscule forces from a few piconewtons to femtonewtons by utilizing light–matter interaction at mesoscopic length scales. Combined with advanced microscopy and position-sensing techniques, optical micromanipulations enable us to investigate diverse aspects of functional soft materials. These include studying mechanical responses through force–elongation measurements, examining the structural properties of complex fluids employing microrheology, analyzing chemical compositions using spectroscopy, and sorting cells through single-cell analysis. Furthermore, it is utilized in various soft matter–based devices, such as laser scissors and optical motors in microfluidic channels. This chapter presents an overview of optical micromanipulation techniques by describing fundamental theories and explaining the design considerations of conventional single-trap and dual-trap setups as well as recent improvisations. We further discuss their capabilities and applications in probing exotic soft matter systems and in developing widely utilized devices and technologies based on functional soft materials.

Keywords Optical tweezers · Laser scissors · Microrheology · Force spectroscopy · Biomedical devices

S. Halder · D. Chanda · D. Mondal · S. Kundu · M. Khan (✉)
Department of Physics, Indian Institute of Technology Kanpur, Kanpur, India
e-mail: mkhan@iitk.ac.in

V. Kumar, Y. K. Mishra (eds.), *Soft Materials for Functional Applications*,
Advanced Structured Materials 225, https://doi.org/10.1007/978-981-97-9468-3_13

1 Introduction

The ability of light to exert force on matter was first recognized by Kepler in the early seventeenth century, as he described that the tails of comets are deflected in response to the light pressure produced by the sun's light [1]. However, it was in the late nineteenth century that the radiation pressure that acts along the direction of light propagation was understood, describing light propagation through Maxwell's theory of electromagnetism [2]. In initial experiments, Nichols and Hull [3] and Lebedew [4] were able to detect the effect of radiation pressure of light on macroscopic objects using arc lamps as the light source and a torsion balance for detection. Intense and coherent light from lasers [5] makes it possible to generate sufficiently large forces that are relevant to many microscopic phenomena in physics and biology. Moreover, advancements in various microscopy and imaging techniques, such as confocal imaging and two-photon excitation, have made it possible to visualize and track the dynamics of smaller objects, even up to a few nanometers. In a seminal contribution, Arthur Ashkin and co-workers first demonstrated that a laser beam can be used to hold, that is, to trap and manipulate particles [6, 7]. In a series of subsequent experiments, they demonstrated the optical micromanipulation of atoms [8–10] and biological matter [11, 12]. Within a few years, Block et al. used an optical trap to apply forces and thus manipulate *Escherichia coli* flagella [13] and single kinesin motors [14]. It was then accepted that optical tweezers provide the most flexible tool that uses light to noninvasively trap and manipulate microscopic objects for ultrafine positioning, measurement, and control. Over time, it has become the most popular and convenient tool for studying soft materials and biological systems, as it provides more powerful yet more convenient ways to probe and manipulate these systems. Arthur Ashkin was later awarded the *2018 Nobel Prize in Physics* (half of the prize) for his invention of optical tweezers [15].

In their simplest configuration, optical tweezers are set up with a single laser beam tightly focused down to the diffraction limit, thereby creating the steep light intensity gradients required to hold dielectric particles in three dimensions at the point of focus. Over the years, this conventional form has evolved with many improvisations and modifications to provide sophisticated and versatile tools for holding, manipulating, probing, and sorting particles, starting from single molecules to entire mammalian cells and colloidal particles (Fig. 1) in diverse environments, such as liquids, complex fluids, and even air. Currently, there are multiple available optical micromanipulation techniques, including optical tweezers, optical scissors, and optical spanners. The domain of their applicability has widened substantially. Employing these tools, forces of up to $\sim$200 pN can be applied with sub-piconewton resolution. The emerging applications being quite diverse and extensive, optical tweezers have been a focal point of interdisciplinary science, with revolutionary applications in soft materials, including biological systems [14, 16–20]. It has also proven useful in various other areas of physics [21–31], nanotechnology [32–35], and chemistry [36]. Many interesting review articles and books have been written

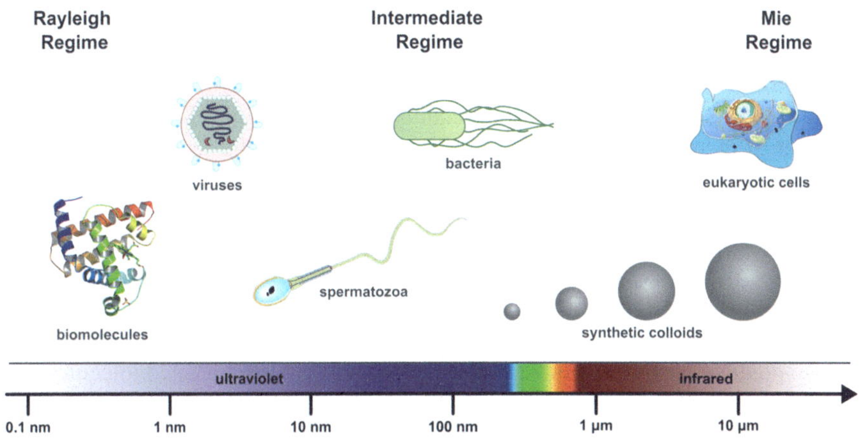

Fig. 1 Typical examples of systems trapped and manipulated using optical tweezers in different regimes based on the size of the object with respect to the wavelength of the light beam used. If the object size is smaller or greater than the wavelength, it falls within the Rayleigh or Mie regime. In the intermediate regime, the size of the object is comparable to the wavelength of the laser. Trapping lasers are mostly used in the visible or near-infrared spectral regions

on optical trapping and manipulation, their design, and their applications [21–23, 27–31, 37–48].

This chapter cannot cover the full extent or diversity of the applications of optical micromanipulation techniques in soft matter–based systems and devices but hopes to provide an introduction to optical micromanipulation, including the fundamental phenomena of its operations, some of the latest advancements, its principal capabilities, and finally a few important applications. It is organized as follows: in the next section, we discuss optical micromanipulation techniques including the theory of optical trapping at two different regimes and considerations and design of the conventional and various contemporary realizations of optical tweezers. Next, we explain the most utilized capabilities of optical tweezers. Finally, some of the important applications of optical micromanipulation in soft material–based systems, devices, and technologies are discussed.

2 Optical Micromanipulation Techniques

Optical micromanipulation techniques enable us to hold and precisely control the movement of micron- and submicron-sized particles using a laser beam. The basic principle that this relatively contemporary tool relies on is the interaction between the electromagnetic field of the light beam and a dielectric particle. A focused laser beam, with higher energy density, is required to enhance the interaction and thus enables successful optical micromanipulation. For simplicity, let us first consider the change in momentum of the photons when a light beam passes through an interface between two materials with different refractive indices. The photons change their speed and direction of propagation, and hence the momentum, upon crossing the interface. This difference in the momentum generates a finite force, which is of the order of piconewton and exerted onto the interface. If the second medium is a micron-sized particle suspended in the first medium, the force exerted by this means can become significant to manipulate the particle.

In 1969, Arthur Ashkin first showed that it is possible to accelerate a micron-sized dielectric particle, suspended in liquid, using a laser beam [49]. A year later, in 1970, Ashkin reported that two counter-propagating laser beams can apply precisely (un)balancing forces to hold and manipulate such particles and water droplets in air. He also predicted the extension of optical manipulation to atoms and molecules [6]. Finally, in 1986, Ashkin and colleagues could successfully trap dielectric particles with size ranging from ~25 nm to 10 μm with a tightly focused single laser beam [7]. This technique became popular as *optical tweezers* and has been proven to be very convenient and useful in various fields of research and applications, particularly with soft materials. Arthur Ashkin was awarded the Nobel prize in Physics in 2018 for inventing the *optical tweezers* and their application to biological systems.

2.1 Theory of Optical Micromanipulation

When a dielectric particle enters the electromagnetic field of a laser beam and interacts with it, both electric field strength and polarization play important roles. However, for simplicity, we consider a linearly polarized laser beam and nonbirefringent particles. The force exerted by the laser field on the dielectric particle can be decomposed into two principal components of different origins: gradient force and scattering force. The scattering of the laser beam from the particle gives rise to a scattering force, which acts along the direction of propagation of the beam if it has a symmetric intensity profile. Thus, the scattering force pushes the particle along, inducing a destabilizing effect. In contrast, the gradient force is conservative in nature and stabilizes the particle by pushing it toward the point of focus. As the name suggests, the gradient force field is proportional to the spatial gradient of intensity. Therefore, a beam with a Gaussian lateral intensity profile (TEM_{00} mode) creates a restoring force field centered at the point of maximum intensity, that is, the point of focus. The origin of the gradient force can be explained in two different ways depending on the size of the particle. For particles that are larger than the wavelength of the trapping laser, simple ray optics principles, such as the refraction of the light beam through the particle, can explain the gradient force in the Mie regime. In this regime, the transparency of the particles at the trapping wavelength plays an important role. Particles with higher transmittance experience a dominant gradient force, whereas the scattering force is stronger when the reflectance is greater. In the Rayleigh regime, where the particle size is smaller than the wavelength of the trapping laser, we must consider the particle as a Rayleigh scatterer and use Maxwell's laws of electromagnetism to understand the origin of the gradient force.

Mie Regime

In the Mie regime, where the particle diameter is larger than the laser wavelength, diffractive effects can be neglected, and the geometrical optics approach can be safely employed [40, 43, 46, 50–52]. The laser beam is decomposed into rays with appropriate intensities and directions that propagate along straight lines in media with a uniform refractive index. The rays have the characteristics of a plane wave that changes direction upon reflection and refraction, following Fresnel's equations. When a transparent microparticle is within the spatial intensity gradient of the laser, the refraction of rays with varied intensity through the particle causes a change in the total momentum of the light beam. Figure 2 pictorially shows the origin of the gradient force in Mie regime. When an incident ray with momentum $\overrightarrow{p_1}$ is refracted through the particle and emerges with momentum $\overrightarrow{p_2}$, the particle experiences a

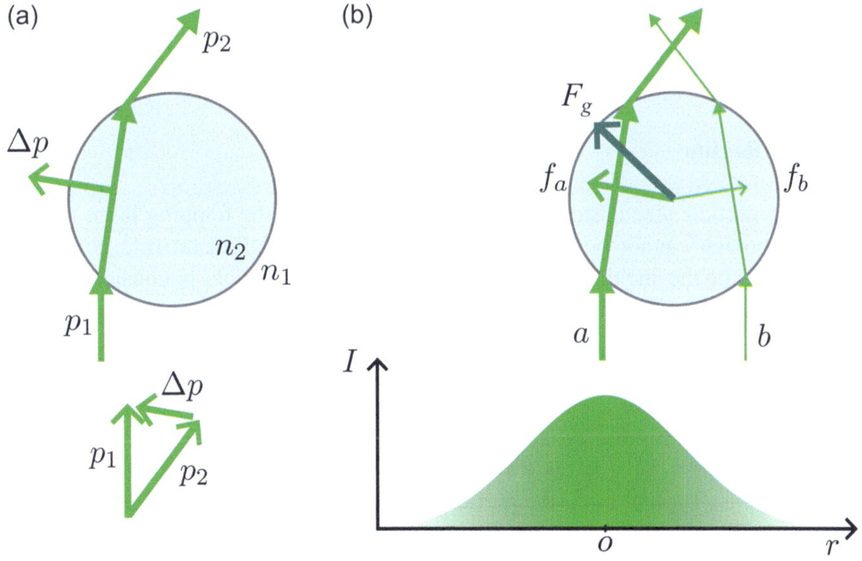

Fig. 2 Origin of the gradient force in Mie regime. (**a**) A ray with momentum p_1 gets transmitted through a microparticle and emerges with momentum p_2. The difference in momentum, Δp, causes a motion of the particle to conserve the linear momentum. (**b**) When the rays (two typical ones are marked as a and b) transmit through the particle, the change in their momentum gives rise to forces (e.g., f_a and f_b, respectively). Depending on the intensity field of the laser at the position of the particle, different rays that pass through the particle have varied momentum. A ray from the central region of the laser profile (e.g., a) will have higher momentum and consequently applies stronger force (f_a) to the particle as compared to a ray from the periphery (e.g., b and corresponding force f_b). Thus, the net force always acts along the direction of the highest intensity

change in momentum of $\Delta \vec{p}$, where $\vec{p_1} = \vec{p_2} + \Delta \vec{p}$, according to the conservation of linear momentum. Based on the position of the particle in the laser intensity field, rays with varied intensities, and thus different momenta, pass through the particle, exerting forces accordingly. Therefore, the sum of the forces acting on the particle depends on its position. However, the net force is always directed toward the point of highest intensity and pushes the particle toward the point of focus, irrespective of its initial position. This net force, which is proportional to the spatial gradient of light intensity, is called the gradient force.

Some parts of the rays are reflected at the surface of the particle depending on the reflectance of the particle at the wavelength of the trapping laser. The change in momentum upon reflection results in a new force. For the symmetric intensity profile of the laser beam, these forces add up to cancel all the lateral components and generate a force that pushes the particle along the direction of propagation. This phenomenon is known as the scattering force. Figure 3 shows the origin and direction of the scattering force. While the gradient force acts along the direction of maximum intensity and hence the direction changes depending on the position of the particle in the laser intensity field, the scattering force always applies along the propagation direction, irrespective of the position of the particle. The particle in Fig. 3a is positioned at an off-axis position, whereas in Fig. 3b, it is positioned on axis. In both cases, the scattering force acts vertically upward, which is the propagation direction, although the gradient force changes its direction.

Rayleigh Regime

When the particle size is smaller than the wavelength of the trapping laser, the ray optics approach cannot be used as diffraction plays a significant role. Since the perturbation of the incident wavefront is minimal, the particle is considered as a Rayleigh scatterer. The trapping force is then derived using electromagnetic laws [40, 43, 53, 54]. When a dielectric particle experiences the electric field associated with the light beam, it becomes polarized because of its finite polarizability. For a particle with positive polarizability, relative refractive index being greater than one, the energy is minimized as it moves to the highest field intensity. This gives rise to the gradient force in the Rayleigh regime and draws the particle toward the point of focus. The net force the particle experiences in this process is the sum of gradient and scattering force. The Lorentz force acting on the induced dipole in the electromagnetic field gives rise to the gradient force. A dielectric particle with polarizability α collects an induced polarization $\vec{p}\left(\vec{r}, t\right) = \alpha \vec{E}\left(\vec{r}, t\right)$, where $\vec{E}\left(\vec{r}, t\right)$ is the electromagnetic field of the laser beam. Therefore, the Lorentz force can be given by

$$\vec{F}\left(\vec{r}, t\right) = \left[\vec{p}\left(\vec{r}, t\right) \cdot \nabla\right] \vec{E} + \frac{d\vec{p}\left(\vec{r}, t\right)}{dt} \times \vec{B}\left(\vec{r}, t\right) \tag{1}$$

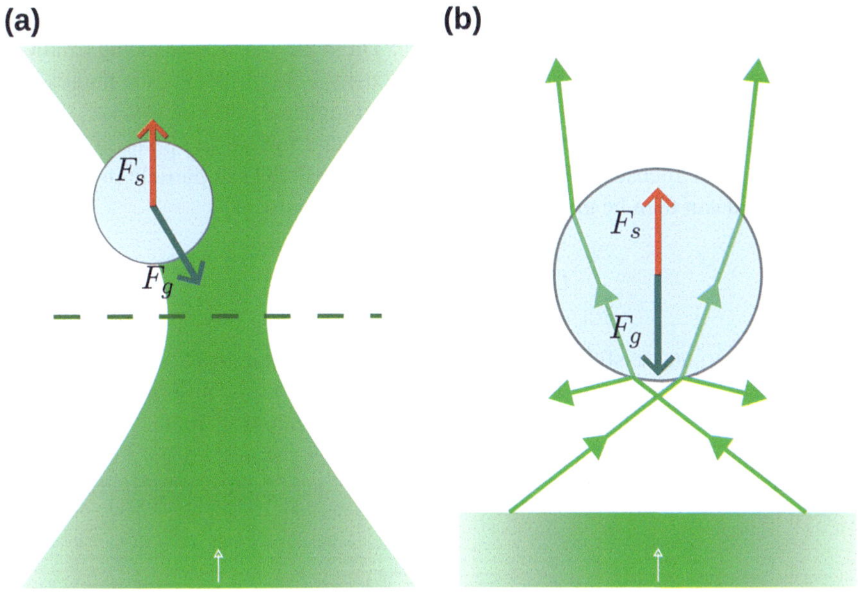

Fig. 3 Origin and direction of the scattering force in Mie regime. (**a**) The directions of the gradient force (F_g) and the scattering force (F_s) are shown for an off-axis particle. While the gradient force is directed toward the point of maximum intensity, the direction of scattering force remains along the propagation direction of the laser beam. (**b**) A pair of rays from a focused laser beam is shown to fall on the particle which is situated on the axis, just above the point of focus (where the rays intersect). Since the lateral components balance each other, the net scattering force always acts along the direction of propagation of the laser beam, directly opposing the gradient force

$$= \alpha\left[\left(\vec{E}(\vec{r},t)\cdot\nabla\right)\vec{E}(\vec{r},t) + \frac{d\vec{E}(\vec{r},t)}{dt}\times\vec{B}(\vec{r},t)\right] \tag{2}$$

At frequencies much smaller than that of the laser light (10^{14} Hz), the particle experiences a steady force which is time average of $\vec{F}(\vec{r},t)$. This gives the simplified expression for the gradient force, which can be written as

$$\vec{F}_g(\vec{r}) = \left\langle \vec{F}(\vec{r},t)\right\rangle_t = \frac{1}{2}\alpha\nabla\left\langle \vec{E}(\vec{r},t)^2\right\rangle_t = \frac{1}{4}\alpha\nabla\left|E(\vec{r})\right|^2 \tag{3}$$

Relating the intensity of the laser beam $I(\vec{r})$ with $\left|E(\vec{r})\right|^2$, using the equation $I(\vec{r}) = (n_1\epsilon_0 c/2)\left|E(\vec{r})\right|^2$, the gradient force simplifies to

$$\vec{F}_g(\vec{r}) = \frac{1}{2n_1\varepsilon_0 c}\alpha\nabla I(\vec{r}) \tag{4}$$

Here n_1, ϵ_0, and c are the refractive index of the medium, permittivity of the free space, and the velocity of light, respectively. Thus, the magnitude of the gradient force is proportional to the gradient of the intensity field, justifying the name. The direction of the force is always toward the point of highest light intensity, i.e., toward the laser beam axis for a Gaussian beam profile, and toward the point of focus if the laser beam is focused. The potential energy of a dielectric particle in the laser intensity field can then be given by

$$U = \frac{-\alpha}{n_1 \epsilon_0 c} \int I\left(\vec{r}\right) d^3 r \tag{5}$$

Therefore, to successfully trap a small dielectric particle, it needs to have high polarizability. Another force that arises in this scenario is because of the radiation pressure on the particle and, hence, is called the scattering force. The particle absorbs a part of the incident radiation and scatters it partially. A net force results toward incident photon flux, i.e., along the propagation of light. This scattering force is given by

$$\vec{F}_s = n_1 \frac{\sigma \left\langle \vec{S} \right\rangle_t}{c} \tag{6}$$

where n_1 is the refractive index of the medium, c is the velocity of light, $\left\langle \vec{S} \right\rangle_t$ is the time-averaged Poynting vector, and σ is the scattering cross-section of the particle. The scattering cross-section, σ, for a spherical particle is well known and given by

$$\sigma = \frac{8}{3} \pi (kr)^4 r^2 \left(\frac{n^2 - 1}{n^2 + 2} \right)^2 \tag{7}$$

k, r, and n are the wave vector of the incident light, radius, and the refractive index of the particle, respectively. Thus, the scattering force has a strong particle size dependence, through the scattering cross-section.

A particle situated away from the beam axis experiences the gradient force as a restoring force toward the beam waist. For a low curvature beam, i.e., a loosely focused laser beam, the component of the gradient force along a direction opposite to that of light propagation becomes too weak to balance a strong scattering force. Hence, the scattering force pushes the particle along the beam. To achieve a stable trap, one requires a counter-propagating beam to balance the respective scattering forces [6]. However, in a tightly focused laser beam, the gradient force is directed along the point of focus, and it possesses a component against the Poynting vector, or the scattering force. That component prevents the particle from being pushed away by the scattering force, resulting in a stable confinement of the particle in three dimensions, as is illustrated in Fig. 4.

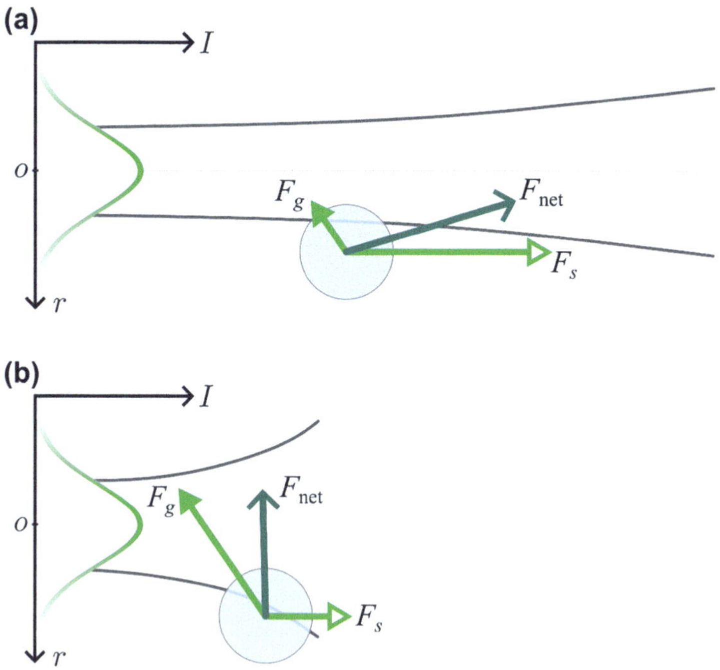

Fig. 4 Comparison of relative strength of the gradient force, F_g, and the scattering force (F_s), for two cases. (**a**) Low curvature beam, where the scattering force is stronger than the gradient force and the net force (F_{net}), pushes the particle away, and hence, the trap becomes unstable. (**b**) Tightly focused beam, where the component of F_g along the propagation direction of the laser balances the scattering force, thereby stabilizes the particle in the trap

Although the fundamental physics to hold and manipulate a microscopic particle remains the same, it can be realized in various ways, depending on the requirements. Designing a pair of optical tweezers with a Gaussian laser beam is the traditional method, as was shown by Arthur Ashkin in 1986 [7], and endless customization has been adopted since then. Different beam shapes, such as a Laguerre–Gaussian (LG) beam, which has a dark center resulting in reduced radiation pressure, or a helical beam, have been employed to transfer angular momentum to the particle alongside trapping [37, 55–59]. New trapping mechanisms such as trapping near a surface with an evanescent wave or near-field enhancement using metallic nanostructures have also been reported [60–72]. In the following section, we discuss the conventional method and some of the more relevant contemporary optical micromanipulation techniques.

2.2 Conventional Techniques

Single Optical Trap

Starting with a two counter-propagating laser beam trapping setup, where the scattering force of the opposing beams is balanced at the middle where a micron-sized particle gets trapped [6], it was soon realized that the use of a single laser beam is more convenient [7]. As previously discussed, a single laser beam can trap a dielectric particle when the gradient force is strengthened by tightly focusing the laser beam. A TEM_{00} laser beam, which has a nearly Gaussian lateral intensity profile and does not change as the laser propagates, is the most popular choice for setting up an optical trap because it can be focused tighter and has a lower divergence. However, there are a few design considerations that are important to follow to set up an optical trap that is fully steerable in three dimensions [40, 43]. The optical layout of such an optical trap is shown in Fig. 5, and the importance of each optical element is discussed in the following paragraphs.

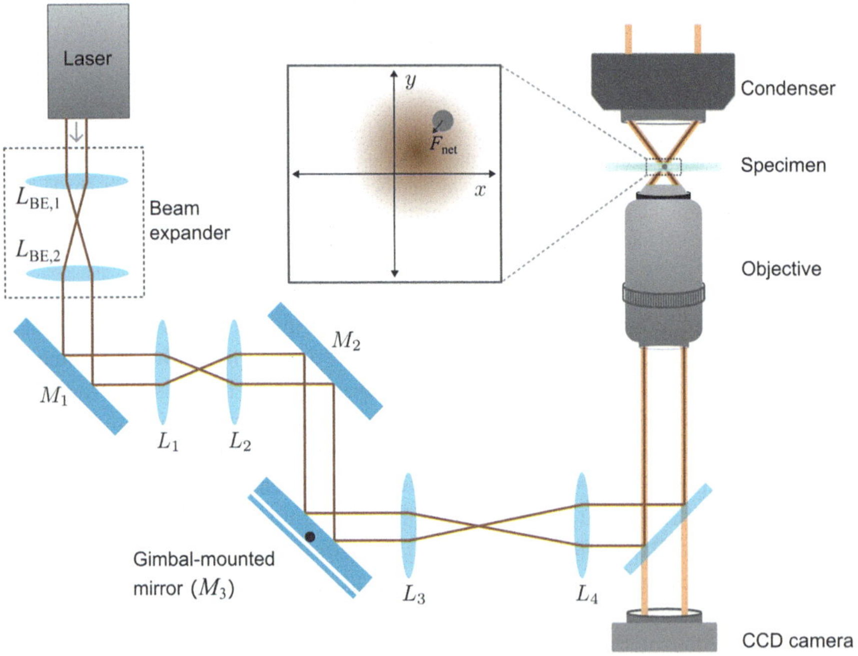

Fig. 5 Optical diagram of a single-trap experimental setup. The component names have been abbreviated as $L_{BE,1}$, $L_{BE,2}$, L_1, L_2, L_3, and L_4 (lenses); M_1 and M_2 (plane mirrors); and M_3 (gimbal-mounted mirror). Brown lines depict the ray path of the laser. The enlarged view of the specimen plane shows a microsphere in an optical trap formed by a laser beam with TEM_{00} mode; the particle is slightly displaced from the center of the trap; thus, a net force (F_{net}) acts on the particle pulling it toward the trap center

First, the laser beam is collimated and appropriately expanded using a beam expander. A beam expander (BE) is essentially a telescopic arrangement composed of a pair of biconvex or plano-convex lenses of focal lengths $f_{BE,1}$ and $f_{BE,2}$ placed at a separation of ($f_{BE,1} + f_{BE,2}$). The telescopic arrangement only changes the beam diameter by a ratio of $f_{BE,2}/f_{BE,1}$ without affecting the convergence of the beam. However, the inherent divergence of the laser beam can be corrected to obtain a perfectly collimated beam by appropriately increasing the separation between lenses. In principle, the expanded beam can be tightly focused through a high numerical aperture (N.A.) objective to form a functional optical trap. N.A. is defined as the half-angle θ of the widest cone of light that can enter or exit the lens system. For a microscopic objective, N.A. is defined as N.A. $= n_1 \sin\theta$, where n_1 is the refractive index of the medium. A more convergent beam (larger θ) produced by a high N.A. objective is used for more efficient trapping in advanced applications. This is because the gradient force is stronger when the light rays are bent to a greater degree; in other words, the peripheral rays contribute more effectively to the trapping efficiency compared to rays at the central part of the beam. Hence, oil immersion objectives with N.A. as high as 1.4 or 1.45 are used for advanced and most effective optical tweezer setups. To utilize a high N.A. objective to its fullest potential, the entrance aperture (the back aperture, through which the laser beam enters) of the objective should be completely filled up by the laser beam. For this reason, the laser beam was first expanded. However, in practical applications, the trap must be steered in three dimensions, that is, both in the focal plane ($X - Y$) and along the optic axis (Z). We must use other optical elements to steer the trap freely in three dimensions without affecting its strength. To create a movable trap that provides the same trapping strength irrespective of its position (X and Y) in the field of view, two important conditions must be fulfilled.

1. The laser beam pivots around the center of the entrance aperture of the microscope objective.
2. The laser beam must retain the same degree of filling of the entrance aperture when it pivots.

The second condition is conveniently satisfied if the laser beam is expanded slightly wider such that it overfills the entrance aperture of the microscope objective. We will later see how this will also take care of the Z-movement of the trap with unaffected trapping strength.

To satisfy the first condition, a gimbal-mounted mirror must be placed in the conjugate plane of the back aperture of the objective. This is achieved by imaging the entrance aperture of the objective on a gimbal-mounted mirror using a system of lenses. The gimbal-mounted mirror can then be tilted to achieve the desired rotation of the laser beam around the center of the mirror. Because the mirror is placed at the conjugate plane of the entrance aperture, a small tilting of the mirror around its nominal position, which is defined as the position at which the laser beam enters the objective parallel to its optic axis, will cause the laser beam to pivot around the center of the entrance aperture without any lateral movement. This, in turn, results in the movement of the trap in the focal plane ($X - Y$), as shown in Fig. 6. A slight

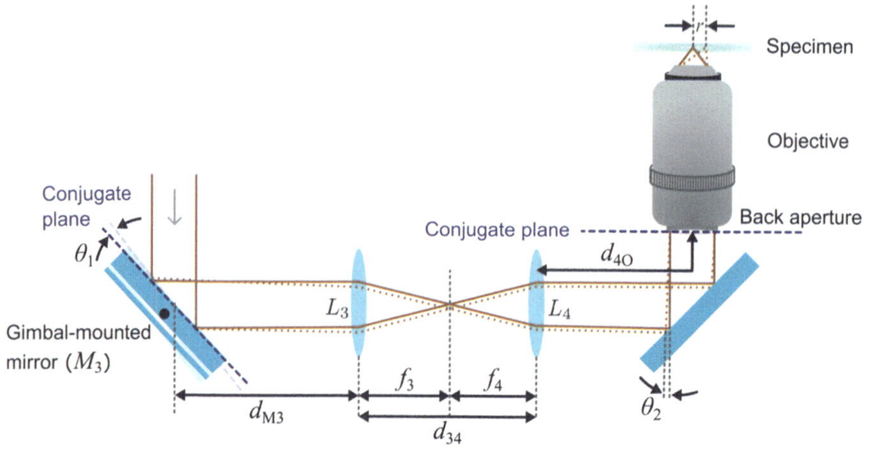

Fig. 6 The schematic depicts how an optical trap is steered in the focal plane ($X - Y$) by tilting a gimbal-mounted mirror (M_3). This is achieved by placing two lenses L_3 and L_4 in such a way that M_3 is at the conjugate plane of the objective back aperture. In this configuration, if M_3 is rotated by an angle θ_1, the laser beam will enter the objective at an angle θ_2, shifting the laser focus by a distance r in the focal plane. Solid and dotted brown lines show the laser beam at two different orientations of the gimbal-mounted mirror

overfilling of the back aperture by the laser beam ensures that it always completely fills the aperture when it pivots around the center.

Typically, a telescopic arrangement of two lenses is placed between the objective and the gimbal-mounted mirror for this purpose, maintaining the convergence of the beam. Thus, positioning the optical trap at a desired point in the focal plane by tilting the gimbal-mounted mirror ensures that the trap will not shift along the optic axis, that is, to a different Z plane. The trapping strength also remains the same irrespective of its movement in the $X - Y$ plane, as the beam fully fills the entrance aperture of the objective at all smaller angles (with respect to the optic axis) of incidence.

To image the objective entrance aperture at the center of the gimbal-mounted mirror, that is, to make them conjugate planes, the two lenses between L_3 and L_4, with corresponding focal lengths f_3 and f_4, should be placed at specific positions. If we denote the distances between the gimbal-mounted mirror and L_3, L_3 and L_4, and the L_4 and the microscope objective as d_{M3}, d_{34}, and d_{4O}, respectively, using the thin lens formalism, we can derive the distance d_{M3} in terms of the other parameters, as

$$d_{M3} = f_3 \frac{f_4 d_{34} - d_{4O}(d_{34} - f_4)}{f_4(d_{34} - f_3) - d_{4O}(d_{34} - f_3 - f_4)} \tag{8}$$

In general, the lens system with L_3 and L_4 is afocal to maintain convergence of the beam; therefore, we can set $d_{34} = f_3 + f_4$. Under these conditions, the expression for d_{M3} becomes

$$d_{M3} = \frac{f_3}{f_4}\left(f_3 + f_4 - \frac{f_3}{f_4}d_{4O}\right) \tag{9}$$

In the most trivial case, d_{4O} is set as f_4 to make d_{M3} equal to f_3. With this arrangement, if the tilt angle of the gimbal-mounted mirror is θ_1, the corresponding angular deviation of the laser beam at the entrance aperture of the objective, θ_2, can be given by

$$\theta_2 = -2\frac{f_3}{f_4}\theta_1 \tag{10}$$

The shift of the optical trap in the focal plane, that is, in the $X - Y$ plane, r, is thus connected to θ_2 through the effective focal length of the objective, f_O, as

$$r = f_O\theta_2 = -2f_O\frac{f_3}{f_4}\theta_1 \tag{11}$$

Therefore, the efficiency of steering the trap position by tilting a gimbal-mounted mirror is proportional to the ratio f_3/f_4 and f_O, which becomes a constant with the choice of the objective. The ratio f_3/f_4 is chosen as a compromise between the efficiency of the ability to move the trap in the focal plane and the expansion of the beam, given by the reciprocal, f_4/f_3, to overfill the objective back aperture.

In an alternative realization, the entrance aperture of the objective is imaged onto the first lens of the two-lens afocal system, that is, L_3. With this arrangement, lens L_3 can be moved laterally (in a plane perpendicular to the optic axis) to achieve a corresponding movement of the beam at the back aperture of the objective and, hence, a shift of the optical trap in the $X - Y$ plane. This realization also ensures a constant degree of filling of the objective entrance aperture, as the beam pivots around its center for small movements of the lens. However, there are some constraints on the focal lengths f_3 and f_4 required for this arrangement to work faithfully.

To move the optical trap freely along the Z direction with unchanged trapping strength, a second afocal optical system, formed by a pair of lenses L_1 and L_2 with corresponding focal lengths f_1 and f_2, is placed between the beam expander and the gimbal-mounted mirror. This can be realized with either two positive (biconvex or plano-convex) lenses or one positive and one negative (concave) lens. To create an afocal optical system, the separation between the lenses, d_{12}, is nominally fixed as the sum of the focal lengths of the two lenses, $f_1 + f_2$. By moving one of the lenses, say L_1, slightly along the optic axis, the otherwise collimated laser beam can be made slightly divergent or convergent, as depicted in Fig. 7. This results in

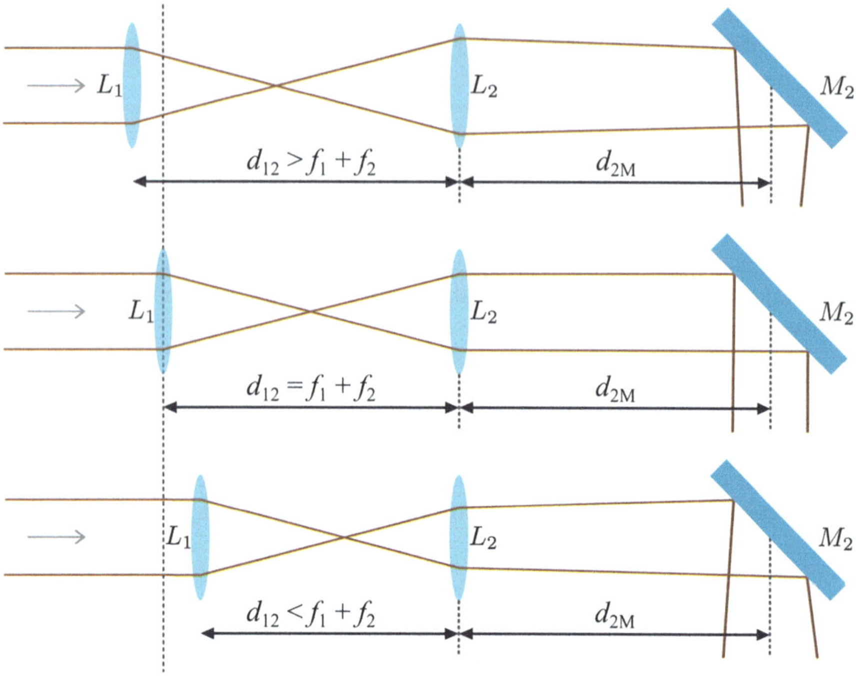

Fig. 7 The schematic demonstrates shifting the optical trap along the optic axis (Z direction) by changing the convergence of the trapping beam. Here, the beam becomes convergent to collimated to divergent, as the lens L_1 is shifted toward the right making the separation between the lenses L_1 and L_2 smaller in three steps, from top to bottom

the beam focusing at a different Z plane as it emerges from the objective, thereby moving the trap at a different plane along the optic axis.

However, it is crucial to maintain the same degree of filling of the objective back aperture, as the beam is made convergent or divergent to move the trap along the Z direction. The size of the laser beam around the center of the gimbal-mounted mirror reflects the degree of overfilling at the objective entrance aperture because the mirror is at the conjugate plane. Using thin lens formalism, the size of the laser beam at the gimbal-mounted mirror, denoted as the beam diameter D_M, can be expressed in terms of the size of the incoming beam diameter, D_{in}, and the distances between the optical elements, that is, d_{2M}, the distance between the gimbal-mounted mirror and L_2, and d_{12}, and the distance between lenses L_1 and L_2:

$$D_M = -\left[\frac{d_{2M}}{f_1} - \frac{(d_{12} - f_1)(d_{2M} - f_2)}{f_1 f_2}\right] D_{in} \tag{12}$$

From the above equation, it is evident that if d_{2M} is set to f_2, the size of the laser beam at the gimbal-mounted mirror and therefore at the entrance aperture of the objective will be independent of d_{12}, that is, the separation between lenses L_1 and L_2. Therefore, d_{12} can be varied appropriately to realize the desired convergence of the laser beam and, hence, the placement of the trap at the required Z plane without affecting the degree of overfilling at the objective entrance aperture. The size of the beam at the gimbal-mounted mirror, in this realization, is given by

$$D_{\mathrm{M}} = \frac{-f_2}{f_1} D_{\mathrm{in}} \tag{13}$$

This equation is valid regardless of whether the beam is collimated, convergent, or divergent. Note that a convergent beam (for $d_{12} > f_1 + f_2$) will focus at the lower Z plane, whereas a divergent beam (for $d_{12} < f_1 + f_2$) will form a trap at a higher Z plane. For a perfectly collimated beam (for $d_{12} = f_1 + f_2$), the focal plane of the trap coincides with the imaging plane of the objective. Therefore, with two sets of afocal arrangements, one before and one after the gimbal-mounted mirror, the resultant expansion of the beam can be estimated as

$$D_{\mathrm{O}} = \frac{f_2}{f_1} \frac{f_4}{f_3} D_{\mathrm{in}} \tag{14}$$

where D_{O} denotes the beam diameter at the entrance aperture of the objective. Considering the expansion of the laser beam that is coming out of the laser head with diameter D_{L} provided by the beam expander, the total expansion factor of the laser beam is calculated as

$$D_{\mathrm{O}} = \frac{f_{\mathrm{BE},2}}{f_{\mathrm{BE},1}} \frac{f_2}{f_1} \frac{f_4}{f_3} D_{\mathrm{L}} \tag{15}$$

Dual Optical Traps

In most practical applications, having two independently steerable traps side-by-side enhances the experimental capabilities by many folds [26]. Using a polarizing beam-splitter cube, a laser beam can be conveniently split into two orthogonally polarized beams (Fig. 8 inset), each creating an independent trap at the specimen plane [43]. This is particularly convenient because the two laser beams can be merged into a common optical path entering the objective using a second polarizing beam-splitter cube (which works as a beam-merging cube) with negligible loss in laser power. Most of the high-quality lasers that are commonly used for optical micromanipulation applications are linearly polarized, and the plane of polarization can be rotated in the desired direction using a half-wave ($\lambda/2$) plate. Thus, by rotating the $\lambda/2$ plate, which is placed just prior to the beam-splitter cube, the plane of polarization of the incoming beam, and hence the relative intensities of the parallel and perpendicular polarizations that govern the splitting ratio, can

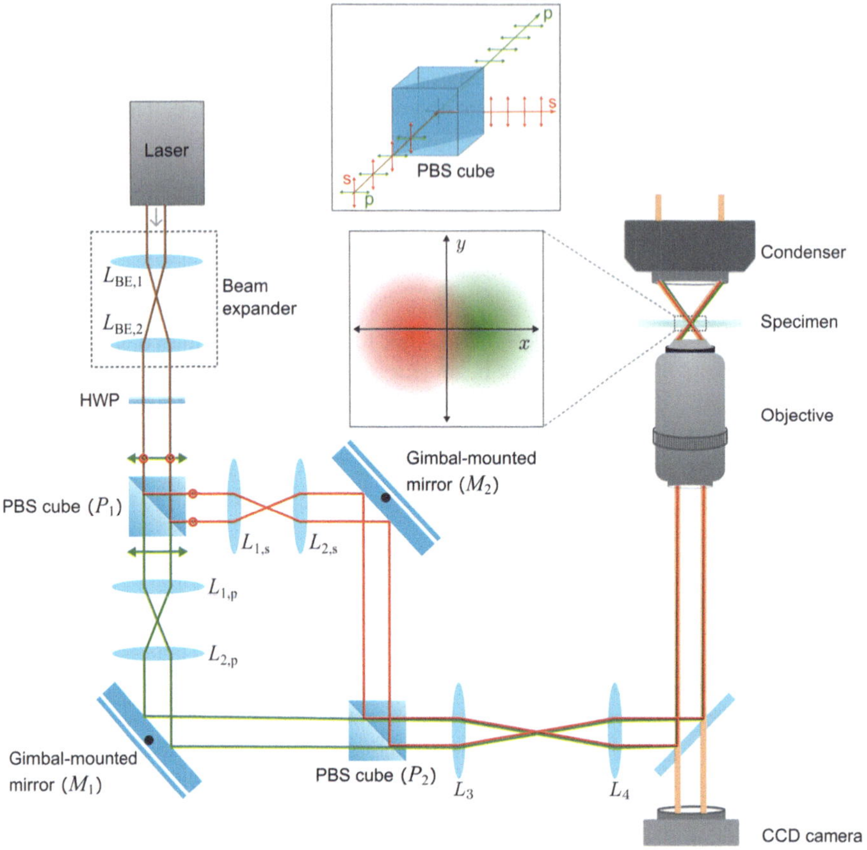

Fig. 8 Optical layout of a dual-trap experimental setup. The component names have been abbreviated as HWP, half-wave ($\lambda/2$) plate; $L_{BE,1}$, $L_{BE,2}$, $L_{1,s}$, $L_{2,s}$, $L_{1,p}$, $L_{2,p}$, L_3, and L_4 (lenses), M_1 and M_2 (gimbal-mounted mirrors), and P_1 and P_2 (polarizing beam-splitter cubes). The red and green lines show the ray path of the s- and p-polarized components of the laser beam. The enlarged view of the specimen plane shows two optical traps in red and green originating from s- and p-polarized components, respectively. The traps are separated by some distance as the gimbal-mounted mirrors, M_1 and M_2, are not perfectly parallel, making a slight angle between them. *Inset*: Working principle of a polarizing beam-splitter (PBS) cube. It splits an incident light into two orthogonally polarized beams by allowing one polarization to pass undeflected (p) and reflecting the orthogonal polarization (s)

be conveniently controlled. Moreover, the two trapping beams with orthogonal polarizations do not interfere with each other and hence can be manipulated independently, as if they are from two different lasers. Forming two traps from orthogonal polarizations will be beneficial only if the two traps can be steered independently in three dimensions. Hence, the optical elements that allow steering of the laser beam in three dimensions must be independent for these two beams with orthogonal polarizations, as shown in Fig. 8. Because the polarizing beam-splitter cube allows one polarization to pass undeflected and reflects the orthogonal polarization, the two beams move along perpendicular directions after being split by the cube. Two gimbal-mounted mirrors are used to bring the two beams back to the second polarizing beam-splitter cube at a right angle. The second cube merges two beams to subsequently follow a common path. Moreover, the two gimbal-mounted mirrors, one for each beam, steer the beams independently, resulting in decoupled lateral movements (in $X - Y$) of the corresponding traps in their respective focal planes. It is important to note that the two beams of orthogonal polarizations create a small angle between them as they enter the objective; hence, they are focused at different points in the focal plane. The angle between the beams is tuned using the two gimbal-mounted mirrors to achieve the desired separation between the traps or the specific positioning of each of the traps in the $X - Y$ plane. The afocal lens arrangement formed by L_1 and L_2, which allows Z-movement of the trap, is also placed separately in the beam paths of the two beams, allowing independent vertical shifting of the traps. After the second beam-splitter cube, the two beams pass through the same set of optics before entering the objective at a small relative angle.

Any optical micromanipulation technique remains incomplete, unless positional fluctuations of the trapped microsphere or other specimens can be detected. In a dual-trap setup, sensing the positions of both trapped objects in real time with a high spatial resolution is required. In a later section, we discuss the position detection techniques and their importance in practical applications.

2.3 *Contemporary Techniques*

There have been many improvisations on traditional single Gaussian beam optical tweezers, and other completely new tweezer configurations have been developed in recent years. In this section, we discuss the improvements and developments. Each of these techniques has specific applications and comes with its own limitations. However, the working principles remain the same—utilizing the photon momentum of a laser beam for manipulation—that is, to confine, push, or drag micron- and submicron-sized dielectric materials in a suspension.

Tweezers Based on Novel Light Modes

Non-Gaussian modes other than the TEM_{00} mode, which is commonly used in conventional optical tweezers setups, are also employed to trap and manipulate dielectric particles [21, 37, 55–59, 73–79]. Higher-order laser beams, such as the Hermite–Gaussian beam (TEM_{xy}), Laguerre–Gaussian (LG) beams (TEM_{pl}), and Bessel beams are typical examples. Optical tweezers based on LG beams offer unique possibilities and advantages over the conventional Gaussian beams. The most important advantages include better trapping efficiency along the axial (Z)

direction and the capability to trap particles with lower refractive index contrast or with higher reflectance and absorbance. An LG beam also possesses a well-defined orbital angular momentum that can apply torque and thus rotate a trapped particle without any external mechanical or electrical steering. Bessel beams are advantageous because they satisfy the Helmholtz equation and thus are propagation invariant, as they retain their cross-sectional intensity profile upon propagation, unlike a Gaussian beam. Bessel beams have concentric rings, where the zeroth order has a bright spot at the center and the higher orders possess a dark region at the center. Thus, a long array of traps with lateral translation, which acts as an optical conveyer belt transporting submicron particles over longer distances, can be formed by interference patterns from two Bessel beams.

Time-Shared Multiple Traps

Multiple optical traps or customized optical potentials are often required for many advanced applications such as molecular motor studies. The most convenient way to create multiple traps is by time-sharing a single beam with an acousto-optic deflector (AOD) or using a piezo-driven scanning mirror that can deflect the beam in the desired direction [46, 80]. These elements usually replace the gimbal-mounted mirror and thus sit at the conjugate plane of the objective back-entrance aperture. The AOD or piezo-driven mirror time shares the laser beam between each site with an appropriate residence time or scans over a desired path to create a customized potential pattern. A slower diffusing micron-sized object feels a time-averaged potential averaged over its self-diffusion timescale. Therefore, high-frequency switching of the beam from one site to another, or continuous scanning over a pattern, effectively creates a static trap pattern. One-dimensional line traps, circular traps, and arrays of optical traps are commonly created using this technique. Here, the limitation comes from the scanning or switching rate of the beam-deflecting device.

Holographic Optical Tweezers

Another convenient way of creating intricate trap patterns, even in three dimensions, is by using specially designed diffractive optical elements, specifically holographic techniques or phase-contrast methods [81–84]. The phase and intensity profiles of the laser beam are modulated by a reflecting or transmitting spatial light modulator (SLM). When the modulated beam is focused through the objective, the phase profile creates an intensity profile, which is the Fourier transform of the phase profile. An SLM is an effective dynamic hologram made from an array of liquid crystals that can induce a phase change from 0 to 2π at the trapping wavelength. Therefore, the SLM can be readily programmed for appropriate phase and intensity modulations to create a trap pattern with the desired intensity and phase profile. Dynamic holograms created on SLM are employed with various algorithms, and can create a large number of traps, where even the mode structure of each trap can be specified individually, arrays of optical vortices, or holographic line traps that can be placed at any desired position in three dimensions. Nucleation seeds with trap patterns matching the lattice structure for larger colloidal crystals or templates for advanced studies of cell growth can also be formed.

Optical Fiber–Based Traps

If the end of an optical fiber tip molded into a lens-like facet, it focuses the light to form a steep-gradient optical trap at the desired precise location. Thus, a single-mode optical fiber can create an optical trap in 3D without using any external optics [85–90]. However, if the shape of the tip of the fiber is not adequate to tightly focus the light beam, a counter-propagating beam from another fiber at the opposite end is required to balance the scattering force. This technique of two counter-propagating beams from two fibers has also been used to stretch microparticles, where the stretching force can be optimized by appropriately tuning the beam intensities. In an optical fiber–based setup, multiple traps with different mode structures can be formed using multimode optical fibers. Intricate excitation of different optical modes of the fiber is required to realize different trapping geometries. Cells and clusters of cells have been oriented and rotated under a microscope using optical fiber–based traps [91]. The main advantage of this technique over conventional optical tweezers is access to the precise trapping location and thus less distortion and loss of laser power and decoupling of trapping from imaging optics.

Near-Field Optical Trapping

Although conventional optical trapping is performed in the far-field region, operating in the near-field offers some advantages. The evanescent field, which is a residue optical field that "leaks" during total internal reflection and fades off at an exponential rate, is used to manipulate a microbead. Hence, this technique has a significantly reduced trapping region within approximately 100 nm from the surface [68, 69, 72, 92, 93]. A continuous evanescent field with directional sense is produced owing to multiple internal reflections when a light beam propagates through an optical waveguide. Thus, microparticles can be guided along the propagation direction of the light beam. A more recent version of evanescent field optical tweezers employs extended optical landscapes to propel a large number of particles along a preferred direction without the use of waveguides. This is termed "lensless optical trapping" (LOT).

Electromagnetically coupled pairs of metallic nanostructures illuminated by a laser beam produce a localized evanescent field; thus, an array of subwavelength plasmonic optical traps can be created to trap nanometer-sized particles suspended in water or air [94–97]. Another realization involves the use of an evanescent field near a nanoaperture or a nanostructured substrate. Although the working principle of these techniques is exciting and important, they are difficult to implement in practice. Moreover, the nanostructure-induced optical near fields reduce the trapping volume beyond the diffraction limit. However, if implemented, nanometric tweezers have the potential to provide nanoscale control of entities at significantly lower laser powers, allowing the nano-manipulation of fragile biological macromolecules.

3 Capabilities of Optical Micromanipulations

3.1 Application and Measurement of Force

The ability to manipulate micron- and submicron-sized particles by applying precise forces at the piconewton level makes optical tweezers a unique technique. Many new optical toolkits, such as optical spanners [37, 59] and holographic optical tweezers [63], have been developed over the years to widen its applicability. Here, we discuss how an optical trap can be employed to apply both scattering and gradient forces to a trapped particle and to a system to which the bead is attached. Moreover, optical tweezers are used to conveniently measure an external force with sub-piconewton to femtonewton resolution. However, proper calibration is required to either apply predetermined forces or measure externally applied forces.

Application of Optical Force

Two types of optical forces, the gradient force and scattering force, can be applied through a focused laser beam. Micron- and submicron-sized dielectric particles are manipulated using these forces in an appropriate fashion. Bulk media can also be perturbed on a micrometer length scale through a trapped bead. It is important to note that many design and system parameters must be tuned properly to maximize the effect of the applied force [40, 43].

The radiation pressure of the laser beam exerts a scattering force along the direction of beam propagation. Thus, it is nonconservative in nature and can propel micron-sized objects downstream [6]. To effectively push micron or submicron particles and transport them from one point to another using the scattering force, the gradient force that attracts them toward the focal spot must be reduced. This can be achieved in many ways, such as not focusing the beam tightly, removing the peripheral rays that contribute more to the gradient force, or slightly defocusing the beam as it enters the objective. The same radiation pressure can induce a torque on micro-objects that have finite chirality. A chiral object can transform a linear force to a torque by virtue of its geometry and thereby rotates about the optical axis, where the gradient force provides a hinge to the rotating particle [32, 34, 98–100]. A birefringent particle can also rotate and apply torque when trapped by circularly polarized light [101].

The gradient force is a restoring force; hence, it pulls a particle to a mechanical equilibrium at the point of focus. Thus, it can hold a micron- or submicron-sized object nonmechanically at the focal spot. Even moving micro-objects, such as motile bacteria or spermatozoa, can be held static by a gradient force. In the opposite scenario, a moving trap can pull a trapped bead and thereby apply an effective force on the system to which the bead is attached, for example, a cell membrane. Similarly, a flowing medium drags along a microbead until it becomes trapped in the gradient force field. In this case, the gradient force not only seizes the motion of the microbead but also applies shear to the flowing medium by creating a relative local displacement field. This constitutes a very effective method to apply shear in active microrheology experiments. However, this process changes significantly depending on the density, viscosity, or local microstructures in the medium and stiffness of the optical trap.

Measuring External Forces

The gradient force generated by a Gaussian (TEM_{00}) laser beam is a linear restoring force ($F_x = \kappa_x \Delta x$), where κ_x is the force constant along the x direction, for small displacements (Δx) from the center of the trap, similar to the force applied by a simple spring in a mechanical system [40, 43]. The gradient of the Gaussian intensity profile (Fig. 9a) is shown in Fig. 9b. The potential landscape corresponding to the linear region is parabolic ($\kappa_x x^2$) in shape (Fig. 9c) [26]. Thus, an optical trap not only applies force to the dielectric particle but can also measure the applied force on the trapped bead from its displacement from the center. The stiffness, that is, the force constant of an optical trap, must be determined prior to the measurement of external forces using different calibration techniques, which will be discussed later. In the case of a properly aligned optical trap with a symmetric beam profile, the lateral force constants κ_x and κ_y mostly have very close values. However, the axial

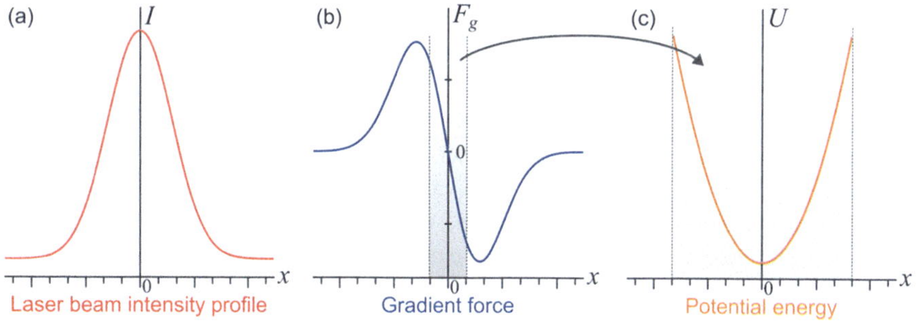

Fig. 9 The gradient force and associated potential corresponding to a Gaussian beam profile. (**a**) A tightly focused beam with Gaussian intensity distribution generates a gradient force that is linear near the center (**b**) and hence acts as a restoring force. (**c**) The potential landscape corresponding to the linear region is parabolic in shape

confinement is comparatively much weaker because of the lower intensity gradient along the axial direction, and the associated force constant (κ_z) is significantly smaller. When working with small particles, the effect of polarization becomes significant, and the κ_x and κ_y values may differ considerably.

Under the influence of an external force (other than the trapping forces), the bead is displaced from the center of the trap, which is otherwise the equilibrium position. At the displaced position, x_d, the gradient force balances the x-component of the external force, $F_{\mathrm{ext},x}$, and a new mechanical equilibrium is established, where

$$F_{\mathrm{ext},x} = \kappa_x x_\mathrm{d} \tag{16}$$

A similar equation can be written with y_d and κ_y if the external force has a nonzero y component. The displaced position of the trapped particle is usually tracked using a position-sensing diode or quadrant photodiode. Particle tracking from a recorded video can also provide the lateral displacement of the bead from the center of the trap, x_d. The external force is then precisely calculated using Eq. (16) and the value of the force constant, κ.

Using this technique, external forces, such as the drag force exerted by a flowing medium on a microbead or the force required to stretch a coiled molecule that is attached to the trapped bead and pulled from the other end, can be measured with high precision. Appropriate tuning of the force constant (by adjusting the laser beam power) to maximize the displacement of the trapped bead, which depends on the magnitude of the external force, enhances the sensitivity and reliability of the measurement.

Calibration of Force Constant

The stiffness or the force constant (κ) of an optical trap arises from the interaction between the dielectric particle in the trap and the electromagnetic field of the focused laser beam near the focal spot. Thus, its value depends on many parameters associated with the interaction, such as particle size, shape, refractive index, beam geometry, and laser power. However, in practice, the force constant is conveniently adjusted by changing the laser power if the properties of the trapped particles remain the same, which is often the case for a set of similar experiments. Therefore, trap stiffness is often calibrated against the laser power for a given experimental setting using the same particle and medium. The value of κ (κ_x and κ_y) can be determined using a number of methods, each with its own advantages. Here, we discuss the three most commonly used techniques: the equipartition, power spectrum, and drag force methods [40, 43].

To determine the trap stiffness using the equipartition method, the equilibrium positional fluctuations of a trapped particle due to spontaneous thermal excitations are recorded to calculate the average potential energy along X and Y. According to the equipartition theorem for a particle confined in a harmonic potential [40, 43],

$$\frac{1}{2}\kappa_i \left\langle q_i^2 \right\rangle = \frac{1}{2}k_B T, \quad i = x, y \tag{17}$$

$$\text{or,} \quad \kappa_i = \frac{k_B T}{\left\langle q_i^2 \right\rangle}, \quad i = x, y \tag{18}$$

where q_i denotes the generalized position measured along coordinates i that is x and y and k_B and T are the Boltzmann constant and temperature, respectively. Thus, the positional variance of a trapped particle provides trap stiffness using the equipartition theorem. In addition to its simplicity, the principal advantage of this method is that it does not explicitly depend on other parameters such as the size and shape of the particle or the viscosity of the surrounding medium. Nonetheless, in fairness, all the properties, such as the size and shape of the particle and the optical and viscous properties of the medium, will influence the trap stiffness through the positional fluctuations of the particle. The most crucial part of this method is to reliably track the position of the trapped particle. Because variance is an intrinsically biased estimator (it is derived from the square of a quantity and is therefore always positive), any added noise and drift in position measurements results in an increase in the overall variance, thereby decreasing the apparent stiffness estimate. However, the presence of these artifacts becomes apparent by the deviation of the positional probability distribution from a Gaussian distribution, which is expected for a harmonically bound Brownian particle. Figure 10a, b shows the typical position fluctuations of a trapped particle, which follows a Gaussian distribution. The trap stiffness is calculated from the variance of the probability distribution determined by the fitting parameter for a Gaussian fit. In Fig. 10d, the linear increase in the trap stiffness along x ($\kappa_{x,\text{equ}}$), measured using the equipartition method, is plotted against the laser power (P_L).

When a microbead of known (Stokes) radius (a) is trapped in a medium of known viscosity (η), the Langevin equation for a harmonically bound Brownian particle can be used to determine the force constant (κ) of the optical trap [40, 43]. The Langevin equation (in one dimension for simplicity) is given as

$$m\ddot{x}(t) + \gamma\dot{x}(t) + \kappa_x x(t) = f(t) \tag{19}$$

Here, $x(t)$ is the trajectory of the trapped Brownian particle of mass m and friction coefficient γ $(= 6\pi\eta a)$. The RHS, $f(t)$, represents the random Brownian forces. In the Fourier space, that is, as a function of the Fourier frequency, f, and defining $\kappa_x = 2\pi\gamma f_0$, this equation can be given by

$$|X(f)|^2 4\pi^2 \left(f_0^2 + f^2\right) = \frac{1}{\gamma^2}|Ff(t)|^2 \tag{20}$$

By definition, $|X(f)|^2 \equiv S_x$ and $|F\{f(t)\}|^2 \equiv S_f$ are the single-sided power spectral densities of $x(t)$, and $f(t)$, respectively. By applying the fluctuation dissipation

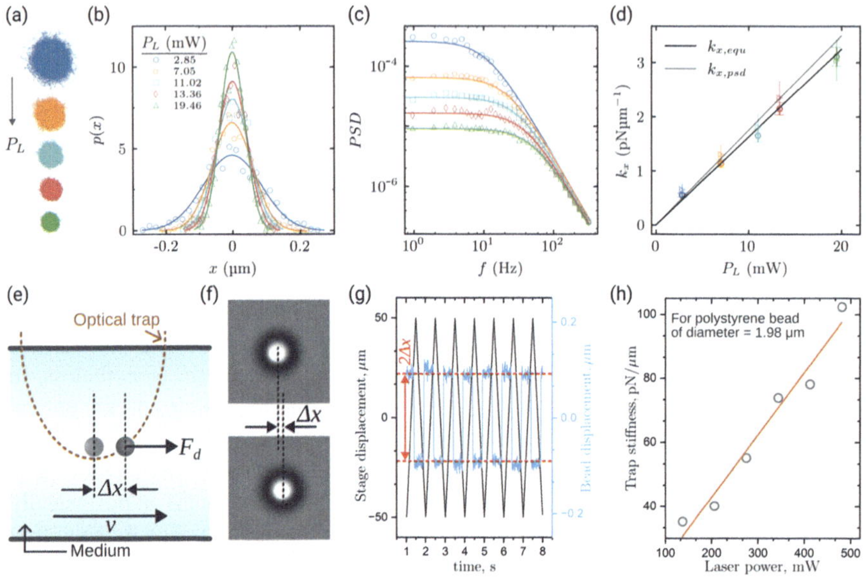

Fig. 10 Different methods for trap stiffness calibration. The confined trajectories of a microbead are captured at increasing trapping laser intensity (**a**) to obtain the positional probability distribution along X (**b**) or Y. The fitting of the positional probability distributions along X, $p(x)$, with Gaussian distributions provides the variance ($\langle x^2 \rangle$) for each $p(x)$. Moreover, good agreement with the Gaussian fits demonstrates the absence of artifacts. (**c**) The corresponding single-sided power spectral densities are fitted with Lorentzian and show good agreement. The linear relation of the force constant κ_x (along X) with laser power P_{L}, measured using equipartition and power spectral density methods, is shown in (**d**). Schematic (**e**) and microscope images (**f**) show the displacement (Δx) of a trapped microsphere in response to the drag force F_{d} applied by translating the sample stage at a constant speed v, inducing the same relative motion to the medium in reference to the trapped bead. The measured variation in the displacement of the bead and stage is shown in (**g**). The trap stiffness determined using the drag method too increases linearly with the laser power (**h**)

theorem, we obtain $S_f = 4\gamma k_{\mathrm{B}} T$. Therefore, Eq. (20) simplifies to

$$S_x = \frac{k_{\mathrm{B}} T}{\pi^2 \gamma} \frac{1}{f_0^2 + f^2} \tag{21}$$

The functional form of S_x describes a Lorentzian function with corner frequency f_0 and a plateau value S_0, occurring at $f \rightarrow 0$. Thus, fitting the single-sided power spectral density of the positional fluctuation, $x(t)$, of a trapped bead with a Lorentzian provides the value of f_0, which in turn determines the value of κ_x as $\kappa_x = 2\pi \gamma f_0$ [24]. Figure 10c shows a few typical variations in S_x at a monotonically varying laser power. The linear increase in the trap stiffness ($\kappa_{x,\mathrm{psd}}$) obtained from the Lorentzian fits to S_x with increasing laser power (P_{L}) is shown in Fig. 10d.

A trapped microbead can be displaced from its equilibrium position in the trap by applying a constant external force of known magnitude, and the trap stiffness can be determined simply by measuring the displacement of the bead in the most direct manner [40, 43]. A constant force of known magnitude can be conveniently applied by creating a relative streamlined flow of constant speed (v_x) along the X direction with respect to the trapped bead. Knowing the viscosity of the fluid (η) and Stokes radius of the bead (a), the drag coefficient (β) and the drag force (F_d) are calculated as $\beta = 6\pi\eta a$, and $F_d = \beta v_x$, respectively. The relative flow is usually generated by translating the sample stage at a constant speed (v_x) in a back-and-forth motion for repeated measurements, while the bead remains trapped at a displaced position (Δx), along the same direction, from the center of the trap. However, because the Stokes drag force that displaces the particle arises from the hydrodynamics around the trapped bead, the drag coefficient, including any surface proximity corrections, must be known. A triangular displacement of amplitude A_x and frequency f of the sample stage along the X direction results in a square-wave profile of the relative velocity and, hence, the force, and the corresponding displacement of the trapped microbead along the same X direction. For each period of motion, the bead trajectory can be given by

$$\Delta x(t) = \frac{\beta A_x f}{2\kappa_x}\left[1 - \exp\left(\frac{-\kappa_x}{\beta}t\right)\right] \tag{22}$$

where κ_x is the trap stiffness along the X direction. This equation also includes Faxen's correction because of the proximity of static surfaces. When the particle is trapped at a height ($\sim 50a$) where the Faxen's law correction is negligible, this equation reduces to the simple Stokes drag equation

$$\Delta x(t) = \frac{\beta}{\kappa_x}v_x(t) = \frac{6\pi\eta a v_x(t)}{\kappa_x} \tag{23}$$

Thus, following the drag method, the stiffness along X and Y can be determined as

$$\kappa_x = \frac{6\pi\eta a v_x(t)}{\Delta x(t)} \quad \text{and} \quad \kappa_y = \frac{6\pi\eta a v_y(t)}{\Delta y(t)} \tag{24}$$

Figure 10e shows the schematic of the experiment, and Fig. 10f, g exhibits a typical displacement of the position of the bead, as observed using video microscopy and measured using a quadrant photodiode for the corresponding triangular stage displacement profile, respectively. A linear relationship between trap stiffness and trapping laser power is shown in Fig. 10h.

For all trap stiffness calibration techniques, or in general, for all optical trap–based experiments and applications, the most crucial part is to record the instantaneous position of the trapped particle with finer resolution and at a high rate. The time series of the position coordinates of the trapped bead, $x(t)$ and $y(t)$, can be tracked in two broad ways, using video imaging or with a position-sensing module, such as a quadrant photodiode (QPD) or position-sensing diode (PSD). Although it is easier to integrate with the setup, the video imaging technique has two shortcomings: a relatively low bandwidth (typically at fps of 30–500) and critical dependence on the particle detection and tracking algorithm. In contrast, in the indirect method, the deviation of the forward or backscattered light from the trapped bead can be detected using a QPD or PSD to obtain the translational movement of the particle with high precision (sub-nanometer) and at a higher bandwidth ($\sim$20 kHz). This requires an additional set of optical elements to image the deviation of the scattered light onto the sensor, that is, the QPD. After passing through the trapped particle, the forward-scattered trapping laser is collected and collimated by the condenser. The translational movement of the trapped particle in the focal plane causes the collimated forward-scattered light to pivot around the center of the back aperture of the condenser. A lens arrangement is used to image the back aperture of the condenser on to the QPD, as shown in Fig. 11b [102]. Thus, even the slightest movement of the trapped particle in reference to the center of the trap results in an amplified deviation of the collimated beam on the QPD surface. The QPD signal is processed by a differential amplifier, which provides a linear variation in the output voltage as a measure of the translational movement of the bead. In an alternate arrangement, a separate collinear tracking laser is sent to the sample along with the trapping laser [43]. Because of the low power and small beam waist, the tracking laser does not generate a significant gradient force to add to the trapping efficiency. The tracking laser is collected and collimated by the objective itself after being backscattered by the trapped particle. In this case, the back aperture of the objective is imaged on the QPD using a lens arrangement (Fig. 11a). A similar deviation of the backscattered tracking laser on the surface of the QPD provides positional fluctuation of the trapped bead. While aligning the tracking laser, making it collinear with the trapping laser, is nontrivial, this method makes the tracking process and its efficiency independent of the trapping strength or the positioning of the trap. In contrast, position detection using the forward-scattered trapping laser provides the displacement of the bead in reference to the center of the trap; hence, it becomes a more convenient choice when working with a moving or scanning optical trap.

3.2 *Localized Heating*

When exposed to incident light, a scatterer in the form of a single atom, molecule, liquid, or solid radiates a part of the incident electromagnetic energy through its accelerated charge carriers. Apart from the reradiation of the energy, the charge carriers also absorb part of the incident energy to get thermally excited, raising the local temperature [40, 46]. Thus, the absorption of the trapping laser in the optical trap causes local heating. Although local heating due to absorption of the trapping laser light may cause damage to the sample that is in the trap or in the proximity of

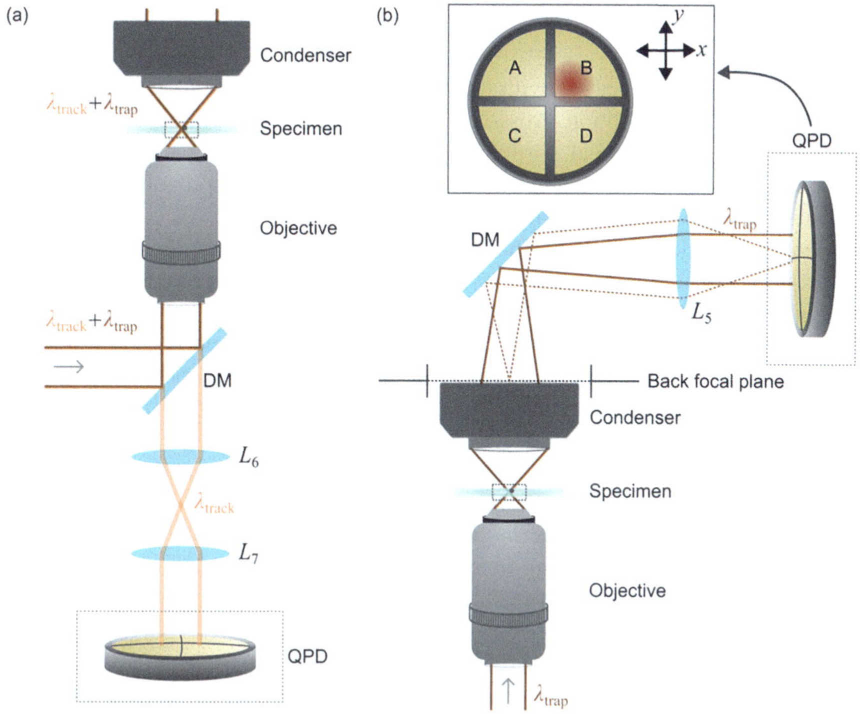

Fig. 11 Tracking the position of the trapped particle using QPD. (**a**) A separate tracking laser (λ_{track}), collinear to the trapping laser (λ_{trap}), is focused on the trapped particle, and the back scattered light is captured through the objective. The back aperture of the objective is imaged on the QPD to track the particle from the deviation of the back scattered light. (**b**) The trapping laser (λ_{trap}) becomes forward-scattered by the trapped particle and is collected by the condenser. The deviation of the forward-scattered light is captured by the QPD, which is placed at the conjugate plane of the back aperture of the condenser. The inset shows how the collimated light falls asymmetrically on the QPD and generates an output signal as the particle moves away from the center of the trap

the trap, it also opens up new applications. To minimize harm to biological samples, near-infrared lasers are mostly used in optical tweezers for biological applications because biological materials absorb less at longer wavelengths ($\lambda > 800$ nm), and water absorption, which peaks far-infrared (3 μm), is only moderate at near-IR wavelengths [103].

The absorption of light not only causes local heating but also induces a nonconservative force, F_A, that is similar and complementary to the scattering force (F_S) [104]. The resultant nonconservative force field (F_{NC}) can be given by

$$F_{NC} = F_S + F_A = \frac{n\left(\sigma_S + \sigma_A\right)}{c}\left\langle \overrightarrow{S_i} \right\rangle \tag{25}$$

where n is the refractive index of the medium; σ_S and σ_A are the scattering and absorption cross-sections, respectively; and $\langle \vec{S_i} \rangle$ is the time-averaged Poynting vector. This extra force, along with local heating, becomes significant for particles with asymmetric absorption cross-sections, such as Janus colloids, where one hemisphere of the spherical particles is metal coated, bringing in a wide range of experimental possibilities, from the thermophoretic active propulsion of Janus colloids to the measurement of local temperature in fluids.

Metal-coated Janus or metal nanoparticle-embedded colloids generate a local temperature gradient around them owing to absorption when placed in an optical trap. In the case of a half-coated Janus colloidal particle, the absorption is dominant at the metal-coated sides (for a 10 nm gold thin film, absorption $\sim$30% at λ = 492 nm) [105, 106], creating an asymmetric local temperature gradient around the particle. The Janus colloid experienced thermophoretic self-propulsion because of the asymmetric temperature gradient. While the self-propulsion direction is solely determined by the Soret coefficient (S_T), the speed (v) depends on the temperature gradient (∇T) as well and is given by $v = -DS_T\nabla T$, where D is the diffusivity of the particle [107, 108]. This thermophoretic self-propulsion provides an excellent system for conveniently studying free or confined active Brownian dynamics under a defocused or tightly focused optical trap, respectively. Local asymmetric heating of carbon-capped Janus particles under laser radiation also causes dissociation of a critical mixture on the hotter side, resulting in diffusiophoretic self-propulsion of the particles [109]. In contrast, metal nanoparticle–embedded colloids generate a circularly symmetric temperature gradient along the radial direction. The azimuthally symmetric radial thermophoretic force induces an effective confinement of particles, resulting in optothermal trapping [110]. The temperature field at the microscopic length scale can also be mapped using optical tweezer–based fluorescent thermometry. For this, a temperature-sensitive fluorescent dye (Rhodamine B, Rhodamine 6G, etc.) is excited by a focused laser beam, usually an optical trap, working at a wavelength within the excitation window, and its emission, whose intensity also depends on the local temperature, is imaged to map the temperature of the fluid [111–114]. Although the quantum efficiency, and hence the emission of these dyes, changes with temperature, their absorption depends on several factors, including the pH of the medium, which is also dependent on the local temperature.

Many biological processes are modulated by temperature, such as enzymatic reaction rates and opening/closing of membrane channels [115]. The heterogeneity in temperature might also have a regulatory effect on intracellular biomolecules and their functions [116]. Thus, these phenomena can be regulated by controlling the local temperature using an optical trap. The use of gold nanoparticles to enhance local heating provides an effective solution for many applications, including photothermal cancer therapy [117], contraction of cardiomyocytes [118], and vesicle transport [119]. This has been applied even at the single-molecule level, such as in the activation of a single myosin. The stepping rate of myosin V, which is attached to gold nanoparticles, was accelerated by a few folds, while the maximum force decreased to almost half of that of ones attached to polystyrene beads.

3.3 Cutting and Ablation

Different laser wavelengths (UV or IR) with varied energy per photon and the desired laser power of a focused laser microbeam are used to employ them as scalpels or scissors, allowing precise cutting with sub-micrometer resolution [120]. For the first time, in 1969, an argon ion laser was used to create half micron–diameter lesions in individual chromosomes [121, 122]. Subsequently, an increased energy density in the focused spot of the argon laser microbeam could alter chromosomes in live cells [123, 124]. Later, laser micromanipulation was performed independently of imaging, using dichroic mirrors [125, 126]. This simple and versatile laser microbeam was used in many similar studies over a decade until pulsed lasers and different dye lasers became available for application as laser scissors [127]. The ability of a focused laser beam to cut through or fuse a cell membrane or tissue, in its application as scissors, and for ablation of organelles has been utilized in many different forms based on the requirements. These wide range of applications include genetic surgery [121, 122], chromosome movements during mitosis or meiosis [128], and DNA damage recognition and repair [129–131]. These capabilities become enhanced manifolds when multiple microbeams are combined to work simultaneously as laser tweezers to hold, as scalpel and scissors to dissect, and for ablation.

A focused laser beam for trapping and ablation is also useful in other systems, such as ablative photodecomposition in solid materials for surface modification and fabrication of polymers [132, 133], as well as for microelectronic applications [134]. Both the high-energy state photochemical reaction and locally heated thermal reaction induced by high peak power and short pulse width lasers play crucial roles in the ablation of solid surfaces. A laser trapping–spectroscopy–ablation system was employed by coupling a CW and a pulsed Nd^{3+}:YAG laser with an optical microscope to achieve laser ablation along with simultaneous fluorescence spectroscopy with nanosecond time resolution on an optically trapped PMMA latex particle in water [135]. Laser ablation of an optically trapped microcapsule containing pyrene/toluene in water and optical micromanipulation of small pyrene/toluene oil droplets ejected from the capsule have also been demonstrated [136].

4 Optical Micromanipulations of Soft Materials for Devices and Technologies

4.1 Characterization of Viscoelastic Properties

The viscoelastic properties of soft materials, that is, their complex mechanical response to external perturbations, are arguably the most pertinent characteristics for determining their usability in devices and technologies. These properties have been conventionally studied in industries employing (bulk) rheology, where the entire sample is sheared and the stress response is measured. A more recent technique, invented in 1995, called microrheology, revealed that it is possible to probe the mechanical response of soft materials, even at microscopic length scales [137, 138]. Since then, it has been widely used to probe the local viscoelastic properties of materials, where heterogeneity plays an important role, and to study scale-dependent rheological properties. Although microrheology was invented to extract the viscoelastic properties of the environment from the trajectories of a probe microbead, optical trapping greatly enhances its capabilities. Optical micromanipulation offers a powerful platform for performing microrheological measurements from single molecules to near-macroscopic scales in both linear and nonlinear regimes. The principal advantage of optical tweezer–based microrheology measurements is that they allow measurements over a wider frequency range, and active shear can be applied to push the system to a nonlinear regime employing active microrheology techniques [24, 139–147].

The working principle of the more widely used passive microrheology is to extract the frequency-dependent elastic modulus ($G'(\omega)$) and viscous modulus ($G''(\omega)$) of soft materials from the thermal diffusion of an embedded microsphere through the generalized Stokes–Einstein relation [137, 138, 145, 148]. In an improvised optical tweezer–based version, the positional fluctuations of a trapped microbead conveniently provide the viscoelastic response ($G'(\omega)$ and $G''(\omega)$) of the surrounding fluid, relying on the fluctuation–dissipation theorem (FDT), over a wider frequency range. The FDT enables us to derive the viscous part ($\chi''(f)$) of the response function of the system, $\chi(f) = \chi'(f) + i\chi''(f)$, directly from the equilibrium positional fluctuation of the trapped bead, $x(t)$ (or $y(t)$), through the following equation [24, 139, 149]:

$$\chi''(f) = \frac{\pi}{2k_{\mathrm{B}}T} f S(f) \tag{26}$$

where k_{B}, T, and $S(f)$ are the Boltzmann constant, temperature, and single-sided power spectral density (PSD) of the positional time series $x(t)$, respectively. The real part, ($\chi'(f)$), representing the elastic response, is dependent on the viscous response ($\chi''(f)$) through the Kramers–Kronig relation and can be derived as

$$\chi''(f) = \frac{2}{\pi}\int_0^{\infty} dt \cos(ft) \int_0^{\infty} d\xi \sin(\xi t)\, \chi''(\xi) \tag{27}$$

The elastic response of the system is overestimated because the optical trap response is elastic in nature. To subtract the trap effect, the response function is corrected as

$$\alpha(f) = \frac{\chi(f)}{1 - \kappa\chi(f)} \tag{28}$$

where $\alpha(f)$ is the corrected system response function and κ is the trap stiffness. The complex shear modulus of the system, $G(f) = G'(f) + iG''(f)$, is related to the complex response function through

$$G(f) = \frac{1}{6\pi a \alpha(f)} \tag{29}$$

where a denotes the Stokes radius of the bead. Therefore, the storage and loss moduli can easily be calculated in terms of the response functions as

$$G'(f) = \frac{1}{6\pi a} \frac{\chi'}{\chi'^2 + \chi''^2} - \frac{\kappa}{6\pi a} \tag{30}$$

$$G''(f) = \frac{1}{6\pi a} \frac{\chi''}{\chi'^2 + \chi''^2} \tag{31}$$

Active microrheology is a more widely employed technique because it provides enhanced capabilities in terms of applying an active shear to the system. An optically trapped microsphere is often driven through a soft material, and the mechanical response to strain is measured from the displacement of the same trapped bead with respect to the center of the trap. Both the instantaneous velocity of the sample stage (v_x) and the displacement of the trapped bead (Δx) are measured at high spatial resolution and high bandwidth to obtain precise microrheological measurements. The strain rate ($\dot{\gamma}_x$) is measured from the velocity (v_x) as $\dot{\gamma}_x = 6v_x/a$, where the prefactor (=6 here) depends on the boundary condition [150], a is the Stokes radius of the trapped bead, and the stress response (σ_x) is obtained from $\sigma_x = \kappa \Delta x$, where κ is the stiffness of the trap. The strain rate–dependent viscosity ($\sigma(\dot{\gamma})$) has been conveniently measured for various soft materials that are more liquid-like using this simple method [24, 141, 146].

A trapped microsphere is also driven through a soft material with a high relative speed to apply a large shear rate, thereby pushing the system to a strongly nonlinear regime. When the relative movement suddenly stops, the system is left to relax. The stress relaxation of the system is recorded by the decreasing distance between the position of the trapped bead and the center of the trap, $\Delta x(t)$. The stress relaxation curve, $\sigma(t)$, reveals the relaxation times (τ) and respective relaxation mechanisms of the system [146, 147, 151] (Fig. 12). In an alternate measurement protocol, the optical trap is switched off with cessation of the drive. The microspheres exhibit recoil in response to the elastic recovery of the material as soon as it is set free by switching off the trap. These recoil trajectories also reveal the elastic stress recovery of soft materials [146, 152, 153].

The microrheological analogs of the oscillatory shear experiments are implemented by driving an optically trapped bead in a small-amplitude back-and-forth motion, following a sinusoidal strain rate profile ($\gamma(t)$) with a specific angular frequency (ω). The instantaneous displacement of the trapped bead from the center of the trap, $\Delta x(t)$, measures the time-varying stress profile, σ(t). Their relative

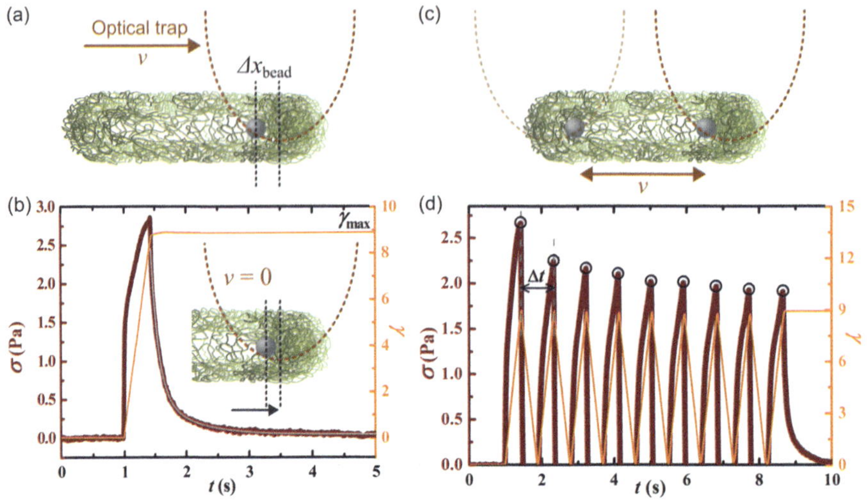

Fig. 12 Active microrheology of entangled biopolymer (DNA) network. (**a**) A microsphere was dragged through the network suspension by applying active shear to the system. The speed of the bead (v) provides the shear rate ($\dot{\gamma}$) or, in turn, the shear strain (γ), and the instantaneous displacement of the bead from the center of the trap (Δx_{bead}) measures the shear stress (σ). When the translational motion is stopped, the displaced bead returns to the center of the trap, exhibiting the stress relaxation behavior of the sample, as shown in the inset. (**b**) The stress buildup while the probe particle applies shear to the system and subsequent stress relaxation after cessation of shear are plotted (brown) using the left axis, while the strain profile (orange) is shown using the right axis. (**c**) In oscillatory shear measurement, the trapped bead is dragged to and fro with a sinusoidally varying or constant speed (v). (**d**) Typical time-varying stress (σ) and strain (γ) are plotted in brown and orange, respectively, using the left and right axes [147]

amplitude and phase difference ($\Delta\phi$) provide the storage modulus, $G'(\omega)$, and loss modulus, $G''(\omega)$, of the system at ω through

$$G'(\omega) = [\sigma_{\max}/\gamma_{\max}]\cos\Delta \qquad (32)$$

$$G''(\omega) = [\sigma_{\max}/\gamma_{\max}]\sin\Delta\phi \qquad (33)$$

$$\eta^*(\omega) = \left[\left(G'(\omega)\right)^2 + \left(G''(\omega)\right)^2\right]^{1/2}/\omega \qquad (34)$$

The measurement is repeated at varying frequencies (ω) to derive the frequency-dependent moduli, $G'(\omega)$, $G''(\omega)$ [144, 146, 154, 155]. Large-amplitude and high-strain-rate oscillatory shear measurements have also been performed using optical tweezer–based active microrheology [147].

Optical tweezer–based microrheology is a contemporary and evolving field. Many new measurement protocols are applied, and existing protocols are improvised to address the specific requirements of the study of viscoelasticity in soft materials. Some examples include dual oscillating trap active microrheology experiments [156] and simultaneous measurements of active and passive microrheology [24].

4.2 Dynamics Force Spectroscopy

Measuring and sensing dynamic changes in small forces, down to the order of a few piconewtons, in response to microscopic elongations or displacements, commonly called dynamic force spectroscopy, is a fundamental part of making devices for technologies based on biomaterials. Although other techniques are available for force spectroscopy, such as magnetic tweezers and atomic force microscopy (AFM), optical tweezers are arguably the most versatile and convenient single-molecule manipulation tool. It enables us to exert forces on the order of a few pN to an excess of 100 pN on particles ranging in size from nanometers to micrometers and simultaneously detect its displacements in three dimensions with sub-nanometer accuracy at sub-millisecond time resolution. Herein, we discuss the different applications of dynamic force spectroscopy using optical tweezers.

Optical tweezer–based dynamic force spectroscopy is widely used to pull out a membrane tether, which is a lipid nanotube from the cell surface, to measure the mechanical properties of the cell membrane and the underlying receptor–cytoskeleton linkages [157]. While a constant or linearly increasing force is applied by pulling a trapped bead attached to the cell membrane, a nonlinear transition in the force–distance plot signifies the formation of a membrane tether [158]. The tether force and membrane viscosity can be determined from the slope of the force–distance plot at the transition [159]. For example, it was revealed that membrane tension is a continuum property over the entire cell surface, and the force required to pull a tether is determined by membrane–cytoskeleton linkage [160]. More recent studies that combined fluorescence imaging with an optical trap demonstrated that actin exists in pulled membrane tethers [161] (Fig. 13). Free membrane-like behavior showing a movable tether–membrane junction has also been observed in neuronal axons [162].

Some cells, such as human erythrocytes (red blood cells or RBCs), become fully deformed when a membrane-attached microbead is pulled using optical tweezers [103, 163]. Mechanical properties, such as shear modulus and relaxation time, were studied by applying varied uniaxial deformations to RBCs. To this end, RBCs are elongated either by fixing one end of the cell to the cover glass and pulling it with a bead attached at the opposite end or by stretching it with two optical handles attached at diametrically opposite positions [163–168]. Employing an alternate approach, three time-shared optical traps were used to fold a trapped RBC, mimicking the deformation of an erythrocyte in narrow capillaries, to discover younger cells that have shorter relaxation times compared to the population [169].

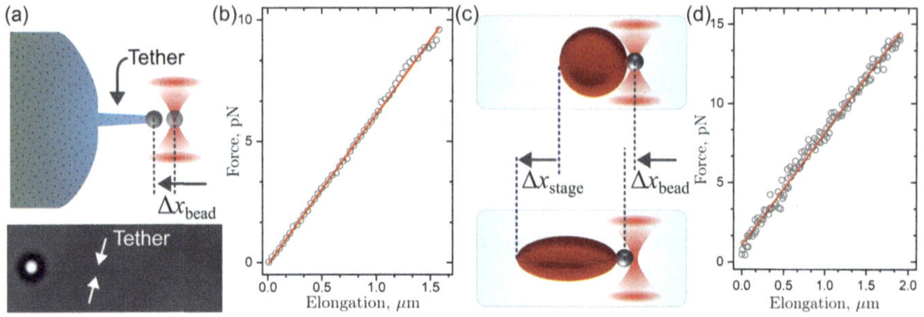

Fig. 13 Dynamic force spectroscopy on cells by measuring the force required for stretching at different elongation rates. (**a**) A tether is formed when a microbead attached to the cell membrane is pulled by a translating optical trap. The tether is a thin tubelike structure that remains invisible or poorly visible under bright-field microscopy. (**b**) The force–elongation plot provides the stiffness of the tether, which often varies with the pulling rate and elongation length. (**c**) In some cases, such as for human erythrocytes, the cell as a whole becomes elongated when a bead attached to its membrane is pulled by translating either the trap or the stage, that is, the substrate where one end of the cell is anchored. (**d**) A similar force–elongation plot is obtained, providing the elasticity of the cell, which contains elastic contributions from both the cell membrane and cytoskeleton

Dynamic force spectroscopy on malaria-infected red blood cells provides a great tool to distinguish healthy RBCs from pathologically infected RBCs by comparing their shear moduli. The Pf155/ring-infected erythrocyte surface antigen (RESA), introduced by *Plasmodium falciparum* merozoites, significantly reduced the deformability of infected RBCs and even more so at febrile temperatures [170]. The adhesive strength between merozoites and RBCs was also measured [171, 172], and Dantu RBCs, a blood group that protects against malaria, were found to have higher average tension, providing greater resistance against parasite invasion [173]. Weak optical traps have also been used to observe membrane fluctuations in red blood cells, allowing an indirect measurement of membrane bendability and stiffness [174]. Recent membrane fluctuation studies have exhibited a violation of the fluctuation–dissipation relation, demonstrating the nonequilibrium active nature of healthy RBC membrane flickering. Other notable studies on erythrocyte deformability using optical tweezers have been discussed in these review articles [175–180].

In another set of dynamic force spectroscopic applications, the stepping characteristics of highly dynamic motors in protein systems have been studied. This has helped us understand the fundamental aspects of how various motor proteins, such as myosin, kinesin, and dynein, move on the microtubules and actin filaments and, more importantly, how the cargo, such as molecules, droplets, and organelles, are transported by the motors along the cytoskeletal tracks over a long distance. Moreover, from a future application perspective, it is more important to discern how ATP hydrolysis fuels the mechanical work done by motors [181]. The pivotal roles and contributions of optical tweezer–based techniques in advancing this field over the last two decades have been well described in these review articles [182–

185]. The two-step working stroke of a myosin I head attached to an actin filament and the associated biochemical states of the actomyosin cycle were first discovered in 1999 by achieving a time resolution of 1 ms for the measurements [186]. Further improvements in the time resolution allowed the characterization of the two mechanical states of a fast myosin V motor [187]. It has been shown that working stroke for myosin decreases with the load, and the forward motion of kinesin-14 HSET can be stalled with a load of 1.1 pN [188]. Although the motility of motors is stalled if the load is increased above a certain value, a fluctuating force causes acceleration in kinesin motors [189].

Optical tweezers have also been used to study the force spectroscopy associated with the mechanical unfolding of proteins and nucleic acid structures [190, 191]. The force measurement capabilities of optical traps are ideally suited for nucleic acid folding, which corresponds to $\sim$15 pN force for $\sim$1 nm displacements. Using a birefringent particle in an optical trap made of circularly polarized light [61, 192, 193] or a chiral particle in a linearly polarized trap (see Sect. 4.5) can apply torque and induce rotation. The torque required to buckle DNA as a function of the applied load has been measured using this technique [193].

4.3 *Mechanosensitivity*

In many biological systems, the binding energies of molecular bonds and the opening of ion channels are governed by mechanical stress or other mechanical stimuli. Optical tweezers have become a convenient and popular tool to measure the mechanosensitivity, that is, the response of these systems to a mechanical stress or strain applied by the optical trap, and moreover to examine how extracellular mechanical stimuli such as pressure, stretching, and motion are transformed into intracellular biochemical signals via a number of mechanoreceptors on the cell membrane [194]. Mechanosensitive channels are transmembrane ion channels regulated by mechanical stimuli [195]. They are those that transfer external mechanical triggers inside the cells and convert them to biochemical signals and thus play crucial roles in heat and touch sensations. By addressing the complexity associated with these channels, nano-tweezing capabilities have been utilized to study how small forces affect important domains of mechanosensitive channels and their physiological behaviors. Piezo channels, which are large proteins with propeller-shaped structures [196], are another example of mechanosensitive channels. These review articles [20, 185, 197] summarize the applications of optical tweezers in various mechanosensitivity studies.

Fluorescence imaging coupled with dynamic force spectroscopy can directly image and correlate the external force acting on the mechanoreceptors to binding kinetics and biochemical signals [198]. For example, the optical tweezer–based technique revealed that the mechanosensor αβT CR's response is anisotropic, where its costimulatory activation is triggered by only a tangential external force but not by a normal force [199]. It was also demonstrated that in the absence of any mechanical stimuli, the initial TCR calcium triggering requires a much higher pMHC ligand density [200]. Calcium mobilization with increased membrane tension was observed and associated with the activation of the TRP-4 channel while pulling membrane tethers from *Caenorhabditis elegans* DAV neurons using a similar technique [201]. Nano-tweezers have been employed to control mechanosensitive channel behavior by applying membrane deformations in more complex systems, tapping or indenting cell membranes, and studying Ca^{2+} transients [202].

Mechanical force has also been shown to activate receptor proteins when the cells come in contact, including members of the Notch family. Single-molecule optical tweezers helped to probe the forces in Notch–ligand interactions [203–205], and finally to conclude that the proteolytic sensitivity increased when a Notch was stretched with a small force, even forces in the order of piconewton was sufficient to expose the cleavage site [206]. Another example of receptor–ligand binding probed using optical traps is the interaction between the von Willebrand factor (VWF) and platelet receptor glycoprotein Ibα(GPIbα), which mediates the attachment of platelets on vessel walls [207]. The capabilities of optical tweezers in resolving forces were utilized to analyze the single-molecule kinetics of VWF and GPIbα pairs [208, 209]. Other surface receptors, such as integrins, cadherins, selectins, and T-cell receptors, have also been probed using optical traps [194]. In similar mechanosensitivity studies employing optical tweezers, the binding probability and unbinding force between virus-coated colloids and erythrocytes were measured by first bringing them in contact and then applying a gradually increasing force until they detached [210]. Similarly, the binding strength and activation state of single fibrinogen–integrin pairs have been measured in living cells [211].

4.4 Biomedical Devices

Optical micromanipulation of cells and tissues using laser tweezers and scissors has opened up many possibilities in developmental research and biomedical applications. The potential of laser microsurgery for the isolation of single cells or groups of cells from a cell cluster was identified and employed early on [212, 213]. The dissection of cellular organelles and cytoskeletal filaments [214] or even chromosomes [38, 215] were also performed. An intricate technical setup named laser capture microdissection has also been developed (LCM) [216]. The radiation pressure of a high-intensity laser was used to blow a desired cell in order to isolate it from the tissue [217–219].

Focused laser beams provide a much better tool for cell fusion by transient permeabilization of cell membranes upon short-term laser irradiation and molecular injection (opto-injection). A method for successful cell fusion by bringing two cells in contact using optical traps and cutting the common wall with a few pulses of high-power UV laser microbeams was developed and demonstrated by Steubing et al. [220]. In the laser beam–assisted opto-injection technique, a temporary opening, that is, a small hole, is created in the cell membrane by focusing a single ultraviolet pulse on the phospholipid bilayer, and then, the foreign species or molecules are injected or allowed to pass through the hole in a passive manner. Improvements over the conventional glass capillary technique are in terms of success rates, less harm to the cells, and fine control over the hole size [41]. Laser-assisted genetic material transfer has been demonstrated in both animal and human cells [221–223]. Although the conventional technique is still applicable to animal cells, it is often not useful for plant cells because of their thicker cell membranes [224–226]. For example, while working with rapeseed cells, thin capillaries are fragile and cannot penetrate the cell membrane, and on the other hand, thicker capillaries damage the cells. However, photo-injection through laser holes in rapeseed cell membranes and tobacco [226, 227] have been successfully performed. This method is most commonly used to inject fluorescent biomolecules into the cytosol and nucleus of cells, such as Neuro-2A mouse neuroblastoma cells [228]. A fluorescent dye can be taken up through laser-created holes into plasmolyzed cells in less than 5 s.

Similar methods provide an effective way to bring two cells together, thus generating hybrid cells without losing cell functions. This method has been successfully applied to in vitro fertilization. Individual sperms are transported using an optical trap and brought into contact with egg cells [229]. Alternatively, a micrometer-sized hole is formed in the zona pellucida, and the viscous barrier is opened. Sperm cells then swim unassisted and slip through the hole into its interior, where the fertilization process continues. However, more recent techniques combine laser microbeam-assisted zona drilling with the optical trapping of sperm cells to enhance efficiency [230]. In vitro fertilization success rates have been compared between methods without laser support and with laser zona drilling and subzonal insemination assisted by optical tweezers in a model experiment with mouse gametes [231]. After in vitro fertilization, very early embryos are placed in the womb, where thinning of the zona facilitates easy implantation.

Optical tweezer–based techniques have also helped in various studies in biomedical research and development, such as the study of virus–cell adhesion, which is of considerable practical importance. A microsphere coated with virus particles was brought into contact, and its adhesion to the cells was studied using a double optical tweezers system [210]. An optical tweezer–based immunosensor was also constructed with these mechanisms in mind [232]. Optical micromanipulation-based techniques have also been applied to neuroscience. Optical tweezers have proven to be excellent tools for the manipulation of whole neurons [233, 234] or for indirectly probing synapses and receptors using attached particles. In other applications, tissue sections of early gastric tumors have been investigated [217, 218, 235]. The fight between human immune cells and cancer cells has also been observed using optical tweezers [236].

4.5 Optical Micromotors

A dielectric particle experiences a scattering force, or in other words, radiation pressure, in the light field of a laser beam. The scattering force pushes the particle along the propagation direction. While it can transport the particle from one point to another, in a more interesting scenario, when there is chirality either in the light field or in the particle geometry, it induces torque and thus rotational motion to the particle. Surface-micromachined structures were fabricated with chiral shapes. These specially designed chiral shapes, often made from transparent polymers using photolithography, exhibit spontaneous rotation, as the scattering force produces a torque owing to the finite chirality of the particles, while the gradient force holds them, providing a hinge. Another way to induce light-driven rotation is to use special beam modes with finite orbital angular momentum to form an optical trap. Using these special beam modes or dynamically changing beam shapes, non-chiral particles can also be rotated under the trap. In an alternative realization, optical tweezers with circularly polarized laser beams have been shown to induce torque on the birefringent particles. This light-induced rotation is of great importance because of its potential applications in optically driven micromachines, motors, actuators, and biological specimens. The first demonstration of an optically driven micromotor was realized by optically trapping 10–25 μm diameter silicon dioxide (SiO_2) structures made by ion beam etching [237]. Other examples include surface-micromachined SiO_2 rotors [238] and optically fabricated microstructures with various shapes (helices, sprinklers, and propellers) under optical traps [98]. Synthesized inorganic nanorods have been shown to rotate when trapped in optical tweezers, where their random surface roughness provides finite chirality [32, 34]. Even deformed human erythrocytes exhibit rotation under an optical trap because of their asymmetric chiral shapes upon crenation in a hypertonic buffer [99, 100]. Asymmetric non-chiral particles, such as rods and ellipsoids, often exhibit rotational motion when trapped in certain configurations where the asymmetric scattering force generates a torque [32, 34, 239, 240]. Symmetric particles with asymmetric scattering profiles, such as metal half-coated spherical Janus colloids, also experience torque owing to an imbalance in the optical forces acting on the two hemispheres [239].

Optical traps with circularly polarized light, where the photons carry spin angular momentum, are commonly used to induce spinning of trapped birefringent particles. This method has been employed for the rotation of elongated particles and a cluster of particles, in addition to birefringent particles [101]. However, the maximal torque experienced by the trapped object is much higher in the case of a birefringent particle than in the case of an elongated particle [241]. Rotation rates as high as 350 Hz and 10 MHz were achieved using a circularly polarized laser beam for an optically trapped micron-sized birefringent particle in water [54] and vacuum [242], respectively. Selective 3D trapping of chiral micro- and nanoparticles with circularly polarized light, where the focused beam can induce non-restoring or restoring forces on the chiral particle, has also been demonstrated [243, 244]. Deformed and crenated human erythrocytes also exhibit rotation under a circularly polarized optical trap, the rate of rotation (at constant laser power) being a measure of their deformation [245–247].

A structured laser beam carrying angular momentum also causes rotation of the trapped micro- and nano-objects because of the transfer of angular momentum. The concept of a light beam possessing orbital angular momentum, which is an azimuthally varying phase, was first introduced in 1992 [248]. However, the use of such a laser beam in an optical trap, which induces rotation to a trapped absorptive dielectric particle, was first demonstrated by the group of Rubinsztein–Dunlop [55, 56]. LG modes of circularly polarized light, carrying both orbital and spin angular momentum, have also been used to induce rotation on absorptive and birefringent particles about their own axis as well as the beam axis [58, 249, 250]. In this case, the rotation rate is proportional to the sum of the orbital and spin angular momenta [251]. This method has been used to study active microrheology, where the spinning of a wax disk applies torque to the surrounding viscoelastic medium and measures its response [252].

A laser beam with an externally modulated time-dependent intensity profile can form a dynamic trap, causing the trapped particles to move along the desired direction or rotate about a desired axis and rate. Following this concept, experimental studies have been carried out with the optical realization of a thermal ratchet where the laser beam intensity is modulated asymmetrically along a circle, causing circular motion of micron-sized particles trapped in the rim [253]. Similar experimental realizations with externally modulated optical traps include Feynman's ratchet with colloidal particles (1D) [254], reconfigurable two-dimensional (2D) rocking ratchet systems [255], and feedback-controlled flashing ratchets [256]. Transport of trapped microparticles has been achieved in a tilted line (one-dimensional) trap, thus, by applying an asymmetric gradient force [33, 257–259]. Particle transport can also be realized using bistable, circular, sinusoidally oscillating optical traps and pattern holographic tweezers [253, 257–263].

While most noise-driven micromachines, such as thermal ratchets and Brownian motors, rely on time-dependent forces, time-independent nonconservative force fields can also create stochastic toroidal motion of trapped particles, called Brownian vortices. When dielectric particles diffuse inside radially symmetric Gaussian profiled laser beams, they also experience scattering forces that are nonconservative in nature. The trapped particles thus experience circulatory bias and exhibit toroidal circulation inside the trap owing to the nonconservative solenoid component of the optical forces [264, 265]. Brownian vortices at the liquid–air interface have also been demonstrated by applying an external nonconservative drag force to a colloidal particle bound by optical tweezers [25].

4.6 Microfluidics Devices

In many studies and applications, microfluidic channels and devices have been proven to be immensely beneficial because they enable working with small sample volumes (microliters to nanoliters) and to fulfill special requirements, such as setting up customized flow patterns or maintaining desired ionic environments with the help of a buffer that needs to be in proximity yet separated by a semipermeable barrier. Another advantage of working with flows in a microfluidic channel is that in narrow channels, the flow is always in the low Reynolds number regime where viscosity dominates, and inertial effects may largely be ignored. In the absence of turbulence, mixing and other related phenomena are mostly governed by diffusion, which also slows down inside the narrow channels. This brings in interesting new applications, besides making it easier to observe flows under a microscope. Custom-designed micro-channels or microfluidic devices can easily be fabricated as per customized design with optically transparent material (such as PDMS film) using electron beam or laser-based lithography techniques [266].

Optical forces can be conveniently used to drive and actuate micro-components, such as micro-cogs [98] and micro-rotors (please see Sect. 4.5), which are rotated by exploiting the spin or angular momentum of the light beam or the chiral geometry of the rotor, to set up small pumps. Specific realizations include an optically driven microfluidic pump driven by two rotating birefringent vaterite particles in linearly polarized optical tweezers [267]. An optical trap was integrated with a microfluidic chamber by placing the trapping laser adjacent to the flow channel to make a fully portable and integrated lab-on-chip device for various applications [268].

Sorting cells, colloidal particles, or other microscopic objects is another important application. The easy integration of optical tweezers with microfluidic devices for single-cell manipulation has been demonstrated [42, 269–271]. Single-cell synthesis in microfluidic channels using optical manipulation results in homogeneous daughter cells, which is a significant enabler in cell therapy. Optical trap-integrated microfluidic devices are also used to study the response of a cell to its neighboring cells, other biomaterials or surfaces, and external chemical, mechanical, and electrical stimuli. High-throughput, automated, passive cell sorting in microfluidic channels has been proven to be more convenient than the conventionally used industry standard—fluorescence-activated cell sorter based on a flow cytometer. The new separation methods based on optical tweezers do not require fluorescent tagging because they distinguish cells based on their individual physical properties. Here, an optical force was employed to trap the cells of interest and for further recognition, sorting, transportation, and isolation from the batch [272–277]. Thus, this technique can achieve a very high throughput, which can be further enhanced by forming an extended 2D or 3D optical potential landscape. Further improvements are realized by using a scanning mirror [278], sharing the beam between multiple traps by acoustic optical deflectors (AODs) [279], and creating extended trap configurations with a spatial light modulator (SLM) [280]. The optical scattering force is also useful for sorting cells by pushing them along the laser propagation direction.

We have already discussed the process of characterizing erythrocyte deformability by their spinning speed under an optical trap [99, 100, 245–247], which can be performed more conveniently in a microfluidic channel.

Optical trap-assisted particle sorting and separation in a microfluidic chamber, that is, optical fractionation, has also been performed by tuning the competition between the Stokes drag that can be externally regulated by the flow speed, and the attractive potential provided by the trap, which depends on the particle size, refractive index, and laser power [281, 282]. Such sorting or separation offers new potential for the selection and enrichment of cell phenotypes. More recently, holographic techniques and evanescent waves have been employed to create an array of optical traps [283], and in another case, near-field optics have been used to enhance the plasmonic field [284] to simultaneously manipulate a large number of particles in microfluidic devices. Holographic-based tweezers, which provide the capability to generate dynamically programmable trap patterns, make it an easy tool for automated particle capture and sorting [280].

5 Conclusion

This chapter covers the theory of optical micromanipulations, design considerations of conventional single-trap and dual-trap systems, as well as recent improvements and adaptations. Furthermore, the diverse capabilities of this technique and some of its widely used applications have been discussed. Nevertheless, it is important to acknowledge that this chapter does not aim to provide an exhaustive exploration of this extensive field of research and application. There have been boundless adaptations and customizations of optical micromanipulations, which, as a whole, are beyond the scope of this chapter. Instead, this is an introductory overview of the advantages and possibilities of optical tweezers for investigating functional soft materials and designing application-oriented devices.

Optical micromanipulation has emerged as one of the most convenient and versatile tools for studying soft matter systems because of its ability to hold and manipulate micron and submicron objects in a noninvasive and nonmechanical manner. These capabilities make them particularly useful for working with biological samples such as living cells. Laser scissors and tweezers are used in various challenging biomedical applications, including laser-based surgeries. By exploiting the ability to measure calibrated forces at the pico- and femtonewton levels, optical micromanipulations are used for in vitro studies of mechanical responses and mechanosensitivity. The structural and viscoelastic properties of complex fluid systems are also commonly investigated using optical traps. In addition, complex and customized optical potential landscapes, such as holographic optical tweezers and time-shared optical trap patterns, are gaining popularity for investigating the statistical physics of soft, active, and biological systems.

The most crucial aspect of optical micromanipulation is choosing an appropriate laser beam depending on the study or application, as the fundamental physical phenomenon is light–matter interaction. The wavelength of the laser not only determines the trapping efficiency but also the extent of scattering, absorption, and photonic damage of the sample. Trapping energy, and hence the efficiency, increases with the photon energy and, thus, with the inverse of wavelength. However, the use of shorter wavelengths is not always recommended to enhance trapping efficiency. For particles that are predominantly optically transparent, it is more convenient to operate them in the Mie regime, that is, with longer wavelengths. Moreover, the scattering force pushes the trapped particle along the beam propagation direction, thereby destabilizing the trap. Hence, it is always intended to minimize the scattering and maximize the transmission of the trapping laser to achieve a stronger trap. The absorption of the laser by the trapped microparticle or the surrounding medium results in an increase in the local temperature. Raising the local temperature amplifies the spontaneous thermal fluctuations and induces convective fluid flow around the trapped particle, affecting the stability of the trap. In addition, increased temperatures often ruin biological samples beyond recovery. Another important consideration regarding the trapping wavelength is the photonic damage of the sample, such as the wavelength-dependent photo-induced breakdown of chemical bonds and the consequent degradation of the sample or medium being studied. Occasionally, pulsed laser beams are also utilized for optical micromanipulations to reduce heating, in contrast to continuous-wave laser beams, which are traditionally employed in optical tweezers. However, the required peak power of the laser pulses to achieve a trapping efficiency similar to that of a continuous-wave laser must be within the damage threshold of the sample or the optical setup.

Although optical micromanipulation-based diagnostics can detect diseases at an early stage of infection by analyzing the alteration of the physical or optical properties of the infected cells using a small sample volume, biochemistry-based tests continue to be the established benchmarks in the industry. Therefore, to design optical micromanipulation-based devices that are both convenient and economically viable for biomedical applications, the primary considerations are cost-effectiveness and portability. The key to building more compact and popular biomedical devices lies in the design and use of high-quality small-diode lasers and adaptive yet cost-effective optical elements constructed from responsive soft materials.

References

1. J. Kepler, De cometis libelli tres (augustae vindelicorum) (1619)
2. Poynting, J.H.: On the transfer of energy in the electromagnetic field. Philos. Trans. R. Soc. Lond. **175**, 343 (1884)
3. Nichols, E.F., Hull, G.F.: A preliminary communication on the pressure of heat and light radiation. Phys. Rev. Ser. I. **13**, 307 (1901)
4. Lebedew, P.: Untersuchungen über die druckkrafte des lichtes. Ann. Phys. **311**, 433 (1901)
5. Siegman, A.E.: Lasers. University Science Books, Mill Valley, CA (literaturangaben) (1986)
6. Ashkin, A.: Acceleration and trapping of particles by radiation pressure. Phys. Rev. Lett. **24**, 156 (1970)

7. Ashkin, A., Dziedzic, J., Bjorkholm, J., Chu, S.: Observation of a single-beam gradient force optical trap for dielectric particles. Opt. Lett. **11**, 288 (1986)
8. Ashkin, A.: Atomic-beam deflection by resonance-radiation pressure. Phys. Rev. Lett. **25**, 1321 (1970)
9. Chu, S., Hollberg, L., Bjorkholm, J.E., Cable, A., Ashkin, A.: Three-dimensional viscous confinement and cooling of atoms by resonance radiation pressure. Phys. Rev. Lett. **55**, 48 (1985)
10. Chu, S., Bjorkholm, J.E., Ashkin, A., Cable, A.: Experimental observation of optically trapped atoms. Phys. Rev. Lett. **57**, 314 (1986)
11. Ashkin, A., Dziedzic, J.M., Yamane, T.: Optical trapping and manipulation of single cells using infrared laser beams. Nature. **330**, 769 (1987)
12. Ashkin, A., Dziedzic, J.M.: Optical trapping and manipulation of viruses and bacteria. Science. **235**, 1517 (1987)
13. Block, S.M., Blair, D.F., Berg, H.C.: Compliance of bacterial flagella measured with optical tweezers. Nature. **338**, 514 (1989)
14. Block, S.M., Goldstein, L.S.B., Schnapp, B.J.: Bead movement by single kinesin molecules studied with optical tweezers. Nature. **348**, 348 (1990)
15. Arthur Ashkin Nobel Lecture entitled Optical tweezers and their application to biological systems (Stockholm). www.nobelprize.org/prizes/physics/2018/ashkin/lecture (2018)
16. Ashkin, A., Schütze, K., Dziedzic, J.M., Euteneuer, U., Schliwa, M.: Force generation of organelle transport measured in vivo by an infrared laser trap. Nature. **348**, 346 (1990)
17. Bustamante, C., Marko, J.F., Siggia, E.D., Smith, S.: Entropic elasticity of lambda-phage DNA. Science. **265**, 1599 (1994)
18. Finer, J.T., Simmons, R.M., Spudich, J.A.: Single myosin molecule mechanics: piconewton forces and nanometre steps. Nature. **368**, 113 (1994)
19. Bustamante, C., Bryant, Z., Smith, S.B.: Ten years of tension: single-molecule DNA mechanics. Nature. **421**, 423 (2003)
20. Neuman, K.C., Nagy, A.: Single-molecule force spectroscopy: optical tweezers, magnetic tweezers and atomic force microscopy. Nat. Methods. **5**, 491 (2008)
21. Grier, D.G.: A revolution in optical manipulation. Nature. **424**, 810 (2003)
22. Dholakia, K., Reece, P., Gu, M.: Optical micromanipulation. Chem. Soc. Rev. **37**, 42 (2008)
23. Jonas, A., Zemanek, P.: Light at work: the use of optical forces for particle manipulation, sorting, and analysis. Electrophoresis. **29**, 4813 (2008)
24. Khan, M., Sood, A.K.: Out-of-equilibrium microrheology using optical tweezers to probe directional viscoelastic properties under shear. Europhys. Lett. **92**, 48001 (2010)
25. Khan, M., Sood, A.K.: Tunable Brownian vortex at the interface. Phys. Rev. E. **83**, 041408 (2011)
26. Khan, M., Sood, A.K.: Irreversibility-to-reversibility crossover in transient response of an optically trapped particle. Europhys. Lett. **94**, 60003 (2011)
27. Padgett, M., Di Leonardo, R.: Holographic optical tweezers and their relevance to lab on chip devices. Lab Chip. **11**, 1196 (2011)
28. Padgett, M., Bowman, R.: Tweezers with a twist. Nat. Photonics. **5**, 343 (2011)
29. Dholakia, K., Cizmar, T.: Shaping the future of manipulation. Nat. Photonics. **5**, 335 (2011)
30. Juan, M.L., Righini, M., Quidant, R.: Plasmon nano-optical tweezers. Nat. Photonics. **5**, 349 (2011)
31. Polimeno, P., Magazzù, A., Iati, M.A., Patti, F., Saija, R., Boschi, C.D.E., Donato, M.G., Gucciardi, P.G., Jones, P.H., Volpe, G., Marago, O.M.: Optical tweezers and their applications. J. Quant. Spectrosc. Radiat. Transf. **218**, 131 (2018)
32. Khan, M., Sood, A.K., Deepak, F.L., Rao, C.N.R.: Nanorotors using asymmetric inorganic nanorods in an optical trap. Nanotechnology. **17**, S287 (2006)
33. Khan, M., Sood, A.K., Mohanty, S.K., Gupta, P.K., Arabale, G.V., Vijaymohanan, K., Rao, C.N.R.: Optical trapping and transportation of carbon nanotubes made easy by decorating with palladium. Opt. Express. **14**, 424 (2006)

34. Khan, M., Sood, A.K., Deepak, F.L., Rao, C.N.R.: Optically driven nanorotors: experiments and model calculations. J. Nanosci. Nanotechnol. **7**, 1800 (2007)
35. Marago, O.M., Jones, P.H., Gucciardi, P.G., Volpe, G., Ferrari, A.C.: Optical trapping and manipulation of nanostructures. Nat. Nanotechnol. **8**, 807 (2013)
36. Yi, C., Li, C.-W., Ji, S., Yang, M.: Microfluidics technology for manipulation and analysis of biological cells. Anal. Chim. Acta. **560**, 1 (2006)
37. Simpson, N.B., Allen, L., Padgett, M.J.: Optical tweezers and optical spanners with Laguerre–Gaussian modes. J. Mod. Opt. **43**, 2485 (1996)
38. Berns, M.W., Tadir, Y., Liang, H., Tromberg, B.: Chapter 5: Laser scissors and tweezers. In: Methods in Cell Biology, pp. 71–98. Elsevier (1997)
39. Grier, D.G.: Optical tweezers in colloid and interface science. Curr. Opin. Colloid Interface Sci. **2**, 264 (1997)
40. Sheetz, M.P., Sheetz, M. (eds.): Laser Tweezers in Cell Biology Methods in Cell Biology, vol. 55. Academic Press, San Diego, CA (1998)
41. Greulich, K.O.: Micromanipulation by Light in Biology and Medicine: The Laser Microbeam and Optical Tweezers. Birkhäuser, Boston (1999)
42. Zhang, H., Liu, K.-K.: Optical tweezers for single cells. J. R. Soc. Interface. **5**, 671 (2008)
43. Khan, M.: Optical Tweezers: Methods and Applications in Contemporary Research. Scholars' Press (2014)
44. Gennerich, A.: Optical Tweezers: Methods and Protocols. Humana Press, New York (2017)
45. Pesce, G., Jones, P.H., Marago, O.M., Volpe, G.: Optical tweezers: theory and practice. Eur. Phys. J. Plus. **135** (2020). https://doi.org/10.1140/epjp/s13360-020-00843-5
46. Bustamante, C., Chemla, Y., Liu, S., Wang, M.: Optical tweezers in single-molecule biophysics. Nat. Rev. Methods Primers. **1** (2021). https://doi.org/10.1038/s43586-021-00021-6
47. Yang, Y., Ren, Y.-X., Chen, M., Arita, Y., Rosales-Guzman, C.: Optical trapping with structured light: a review. Adv. Photonics. **3** (2021). https://doi.org/10.1117/1.ap.3.3.034001
48. Volpe, A.G., Marago, O.M., Rubinsztein-Dunlop, H., Pesce, G., Stilgoe, A.B., Volpe, G., Tkachenko, G., Truong, V.G., Chormaic, S.N., Kalantarifard, F., Elahi, P., Kall, M., Callegari, A., Marques, M.I., Neves, A.A.R., Moreira, W.L., Fontes, A., Cesar, C.L., Saija, R., Saidi, A., Beck, P., Eismann, J.S., Banzer, P., Fernandes, T.F.D., Pedaci, F., Bowen, W.P., Vaippully, R., Lokesh, M., Roy, B., Thalhammer-Thurner, G., Ritsch-Marte, M., Garcıa, L.P., Arzola, A.V., Castillo, I.P., Argun, A., Muenker, T.M., Vos, B.E., Betz, T., Cristiani, I., Minzioni, P., Reece, P.J., Wang, F., McGloin, D., Ndukaife, J.C., Quidant, R., Roberts, R.P., Laplane, C., Volz, T., Gordon, R., Hanstorp, D., Marmolejo, J.T., Bruce, G.D., Dholakia, K., Li, T., Brzobohatý, O., Simpson, S.H., Zemanek, P., Ritort, F., Roichman, Y., Bobkova, V., Wittkowski, R., Denz, C., Kumar, G.V.P., Foti, A., Donato, M.G., Gucciardi, P.G., Gardini, L., Bianchi, G., Kashchuk, A.V., Capitanio, M., Paterson, L., Jones, P.H., Berg-Sørensen, K., Barooji, Y.F., Oddershede, L.B., Pouladian, P., Preece, D., Adiels, C.B., De Luca, A.C., Magazzù, A., Bronte Ciriza, D., Iatı, M.A., Swartzlander, G.A.: Roadmap for optical tweezers. J. Phys. Photonics. **5**, 022501 (2023)
49. Ashkin, A.: Optical trapping and manipulation of neutral particles using lasers. Proc. Natl. Acad. Sci. **94**, 4853 (1997)
50. Ashkin, A.: Forces of a single-beam gradient laser trap on a dielectric sphere in the ray optics regime. Biophys. J. **61**, 569 (1992)
51. Visscher, K., Brakenhoff, G.J.: Theoretical study of optically induced forces on spherical particles in a single beam trap. II: Mie scatterers. Optik. **90**, 57 (1992)
52. Ren, K., Greha, G., Gouesbet, G.: Radiation pressure forces exerted on a particle arbitrarily located in a Gaussian beam by using the generalized Lorenz-Mie theory, and associated resonance effects. Opt. Commun. **108**, 343 (1994)
53. Visscher, K., Brakenhoff, G.J.: Theoretical study of optically induced forces on spherical particles in a single beam trap. I: Rayleigh scatterers. Optik. **89**, 174 (1992)
54. Harada, Y., Asakura, T.: Radiation forces on a dielectric sphere in the Rayleigh scattering regime. Opt. Commun. **124**, 529 (1996)

55. He, H., Friese, M.E.J., Heckenberg, N.R., Rubinsztein-Dunlop, H.: Direct observation of transfer of angular momentum to absorptive particles from a laser beam with a phase singularity. Phys. Rev. Lett. **75**, 826 (1995)
56. He, H., Heckenberg, N., Rubinsztein-Dunlop, H.: Optical particle trapping with higher-order doughnut beams produced using high efficiency computer generated holograms. J. Mod. Opt. **42**, 217 (1995)
57. Bains, S.: Helical beams give particles a whirl. Science. **273**, 36 (1996)
58. Friese, M.E.J., Enger, J., Rubinsztein-Dunlop, H., Heckenberg, N.R.: Optical angular-momentum transfer to trapped absorbing particles. Phys. Rev. A. **54**, 1593 (1996)
59. Simpson, N.B., Dholakia, K., Allen, L., Padgett, M.J.: Mechanical equivalence of spin and orbital angular momentum of light: an optical spanner. Opt. Lett. **22**, 52 (1997)
60. Almaas, E., Brevik, I.: Radiation forces on a micrometer-sized sphere in an evanescent field. J. Opt. Soc. Am. B. **12**, 2429 (1995)
61. Friese, M.E.J., Nieminen, T.A., Heckenberg, N.R., Rubinsztein-Dunlop, H.: Optical alignment and spinning of laser-trapped microscopic particles. Nature. **394**, 348 (1998)
62. Friese, M.E.J., Nieminen, T.A., Heckenberg, N.R., Rubinsztein-Dunlop, H.: Optical torque controlled by elliptical polarization. Opt. Lett. **23**, 1 (1998)
63. Zemanek, P., Jonas, A., Šramek, L., Liska, M.: Optical trapping of Rayleigh particles using a Gaussian standing wave. Opt. Commun. **151**, 273 (1998)
64. Fuhr, G., Schnelle, T., Müller, T., Hitzler, H., Monajembashi, S., Greulich, K.-O.: Force measurements of optical tweezers in electro-optical cages. Appl. Phys. Mater. Sci. Process. **67**, 385 (1998)
65. Heckenberg, N.R., Friese, M.E.J., Nieminen, T.A., Rubinsztein-Dunlop, H.: Optical vortices. In: Vasnecov, M.V. (ed.) Horizons in World Physics, vol. 228. Nova Science Publishers, Commack, NY (1999)
66. Reicherter, M., Haist, T., Wagemann, E.U., Tiziani, H.J.: Optical particle trapping with computer-generated holograms written on a liquid-crystal display. Opt. Lett. **24**, 608 (1999)
67. Zemanek, P., Jonas, A., Šramek, L., Liska, M.: Optical trapping of nanoparticles and microparticles by a Gaussian standing wave. Opt. Lett. **24**, 1448 (1999)
68. Clapp, A.R., Ruta, A.G., Dickinson, R.B.: Three-dimensional optical trapping and evanescent wave light scattering for direct measurement of long range forces between a colloidal particle and a surface. Rev. Sci. Instrum. **70**, 2627 (1999)
69. Lester, M., Nieto-Vesperinas, M.: Optical forces on microparticles in an evanescent laser field. Opt. Lett. **24**, 936 (1999)
70. Mogensen, P.C., Glückstad, J.: Dynamic array generation and pattern formation for optical tweezers. Opt. Commun. **175**, 75 (2000)
71. Schnelle, T., Müller, T., Reichle, C., Fuhr, G.: Combined dielectrophoretic field cages and laser tweezers for electrorotation. Appl. Phys. B Lasers Opt. **70**, 267 (2000)
72. Clapp, A.R., Dickinson, R.B.: Direct measurement of static and dynamic forces between a colloidal particle and a flat surface using a single-beam gradient optical trap and evanescent wave light scattering. Langmuir. **17**, 2182 (2001)
73. O'Neil, A.T., Padgett, M.J.: Axial and lateral trapping efficiency of Laguerre–Gaussian modes in inverted optical tweezers. Opt. Commun. **193**, 45 (2001)
74. Meyrath, T.P., Schreck, F., Hanssen, J.L., Chuu, C.S., Raizen, M.G.: A high frequency optical trap for atoms using Hermite-Gaussian beams. Opt. Express. **13**, 2843 (2005)
75. Lafong, A., Hossack, W.J., Arlt, J., Nowakowski, T.J., Read, N.D.: Time-multiplexed Laguerre-Gaussian holographic optical tweezers for biological applications. Opt. Express. **14**, 3065 (2006)
76. Milne, G., Dholakia, K., McGloin, D., Volke-Sepulveda, K., Zemanek, P.: Transverse particle dynamics in a Bessel beam. Opt. Express. **15**, 13972 (2007)
77. Jia, B., Gan, X., Gu, M.: Strong tangential force within a small trapping volume under near-field Laguerre-Gaussian beam illumination. Opt. Express. **16**, 15191 (2008)

78. Meresman, H., Wills, J.B., Summers, M., McGloin, D., Reid, J.P.: Manipulation and characterisation of accumulation and coarse mode aerosol particles using a Bessel beam trap. Phys. Chem. Chem. Phys. **11**, 11333 (2009)
79. Ambrosio, L.A., Hernandez-Figueroa, H.E.: Gradient forces on double-negative particles in optical tweezers using Bessel beams in the ray optics regime. Opt. Express. **18**, 24287 (2010)
80. Guilford, W.H., Tournas, J.A., Dascalu, D., Watson, D.S.: Creating multiple time-shared laser traps with simultaneous displacement detection using digital signal processing hardware. Anal. Biochem. **326**, 153 (2004)
81. Dufresne, E.R., Spalding, G.C., Dearing, M.T., Sheets, S.A., Grier, D.G.: Computer-generated holographic optical tweezer arrays. Rev. Sci. Instrum. **72**, 1810 (2001)
82. Schmitz, C.H.J., Spatz, J.P., Curtis, J.E.: High-precision steering of multiple holographic optical traps. Opt. Express. **13**, 8678 (2005)
83. Grier, D.G., Roichman, Y.: Holographic optical trapping. Appl. Opt. **45**, 880 (2006)
84. Mejean, C.O., Schaefer, A.W., Millman, E.A., Forscher, P., Dufresne, E.R.: Multiplexed force measurements on live cells with holographic optical tweezers. Opt. Express. **17**, 6209 (2009)
85. Constable, A., Kim, J., Mervis, J., Zarinetchi, F., Prentiss, M.: Demonstration of a fiber-optical light-force trap. Opt. Lett. **18**, 1867 (1993)
86. Lyons, E.R., Sonek, G.J.: Confinement and bistability in a tapered hemispherically lensed optical fiber trap. Appl. Phys. Lett. **66**, 1584 (1995)
87. Sidick, E., Collins, S.D., Knoesen, A.: Trapping forces in a multiple-beam fiber-optic trap. Appl. Opt. **36**, 6423 (1997)
88. Collins, S.D., Baskin, R.J., Howitt, D.G.: Microinstrument gradient-force optical trap. Appl. Opt. **38**, 6068 (1999)
89. Renn, M.J., Pastel, R., Lewandowski, H.J.: Laser guidance and trapping of mesoscale particles in hollow-core optical fibers. Phys. Rev. Lett. **82**, 1574 (1999)
90. Guck, J., Ananthakrishnan, R., Mahmood, H., Moon, T.J., Cunningham, C.C., Käs, J.: The optical stretcher: a novel laser tool to micromanipulate cells. Biophys. J. **81**, 767 (2001)
91. Kreysing, M.K., Kießling, T., Fritsch, A., Dietrich, C., Guck, J.R., Käs, J.A.: The optical cell rotator. Opt. Express. **16**, 16984 (2008)
92. Kawata, S., Sugiura, T.: Movement of micrometer-sized particles in the evanescent field of a laser beam. Opt. Lett. **17**, 772 (1992)
93. Lutti, J., Langbein, W., Borri, P.: High q optical resonances of polystyrene microspheres in water controlled by optical tweezers. Appl. Phys. Lett. **91** (2007). https://doi.org/10.1063/1.2795332
94. Novotny, L., Bian, R.X., Xie, X.S.: Theory of nanometric optical tweezers. Phys. Rev. Lett. **79**, 645 (1997)
95. Grigorenko, A.N., Roberts, N.W., Dickinson, M.R., Zhang, Y.: Nanometric optical tweezers based on nanostructured substrates. Nat. Photonics. **2**, 365 (2008)
96. Righini, M., Volpe, G., Girard, C., Petrov, D., Quidant, R.: Surface plasmon optical tweezers: tunable optical manipulation in the femtonewton range. Phys. Rev. Lett. **100**, 186804 (2008)
97. Yang, A.H.J., Moore, S.D., Schmidt, B.S., Klug, M., Lipson, M., Erickson, D.: Optical manipulation of nanoparticles and biomolecules in sub-wavelength slot waveguides. Nature. **457**, 71 (2009)
98. Galajda, P., Ormos, P.: Complex micromachines produced and driven by light. Appl. Phys. Lett. **78**, 249 (2001)
99. Khan, M., Mohanty, S.K., Sood, A.K.: Optically-driven red blood cell rotor in linearly polarized laser tweezers. Pramana. **65**, 777 (2005)
100. Khan, M., Soni, H., Sood, A.K.: Optical tweezer for probing erythrocyte membrane deformability. Appl. Phys. Lett. **95** (2009). https://doi.org/10.1063/1.3272269
101. Bradac, C.: Nanoscale optical trapping: a review. Adv. Opt. Mater. **6** (2018). https://doi.org/10.1002/adom.201800005
102. Gittes, F., Schmidt, C.F.: Interference model for back-focal-plane displacement detection in optical tweezers. Opt. Lett. **23**, 7 (1998)

103. Svoboda, K., Block, S.M.: Biological applications of optical forces. Annu. Rev. Biophys. Biomol. Struct. **23**, 247 (1994)
104. Bohren, C.F., Huffman, D.R.: Absorption and Scattering of Light by Small Particles, 1st edn. Wiley (1998)
105. Loebich, O.: The optical properties of gold: a review of their technical utilisation. Gold Bull. **5**, 2 (1972)
106. Yakubovsky, D.I., Kirtaev, R.V., Stebunov, Y.S., Arsenin, A.V., Volkov, V.S.: Morphology and effective dielectric functions of ultra-thin gold films. J. Phys. Conf. Ser. **1092**, 012167 (2018)
107. Jiang, H.-R., Yoshinaga, N., Sano, M.: Active motion of a Janus particle by self-thermophoresis in a defocused laser beam. Phys. Rev. Lett. **105**, 268302 (2010)
108. Buttinoni, I., Volpe, G., Kümmel, F., Volpe, G., Bechinger, C.: Active Brownian motion tunable by light. J. Phys. Condens. Matter. **24**, 284129 (2012)
109. Lavergne, F.A., Wendehenne, H., Bauerle, T., Bechinger, C.: Group formation and cohesion of active particles with visual perception–dependent motility. Science. **364**, 70 (2019)
110. Paul, D., Chand, R., Kumar, G.V.P.: Optothermal evolution of active colloidal matter in a defocused laser trap. ACS Photonics. **9**, 3440 (2022)
111. Ross, D., Gaitan, M., Locascio, L.E.: Temperature measurement in microfluidic systems using a temperature-dependent fluorescent dye. Anal. Chem. **73**, 4117 (2001)
112. Duhr, S., Arduini, S., Braun, D.: Thermophoresis of DNA determined by microfluidic fluorescence. Eur. Phys. J. E. **15**, 277 (2004)
113. Natrajan, V.K., Christensen, K.T.: Fluorescent thermometry. In: Li, D. (ed.) Encyclopedia of Microfluidics and Nanofluidics, pp. 750–759. Springer, Boston, MA (2008)
114. Zhou, J., Yang, W., Yin, Y., Chen, S., Yan, B., Mu, J., Qi, X.: Nonlinear temperature calibration equation for rhodamine b in different solutions for wide-temperature-range applications. Appl. Opt. **58**, 1514 (2019)
115. Clapham, D.E.: Trp channels as cellular sensors. Nature. **426**, 517 (2003)
116. Okabe, K., Inada, N., Gota, C., Harada, Y., Funatsu, T., Uchiyama, S.: Intracellular temperature mapping with a fluorescent polymeric thermometer and fluorescence lifetime imaging microscopy. Nat. Commun. **3** (2012). https://doi.org/10.1038/ncomms1714
117. Gobin, A.M., Lee, M.H., Halas, N.J., James, W.D., Drezek, R.A., West, J.L.: Near-infrared resonant nanoshells for combined optical imaging and photothermal cancer therapy. Nano Lett. **7**, 1929 (2007)
118. Oyama, K., Mizuno, A., Shintani, S.A., Itoh, H., Serizawa, T., Fukuda, N., Suzuki, M., Ishiwata, S.: Microscopic heat pulses induce contraction of cardiomyocytes without calcium transients. Biochem. Biophys. Res. Commun. **417**, 607 (2012)
119. Oyama, K., Takabayashi, M., Takei, Y., Arai, S., Takeoka, S., Ishiwata, S., Suzuki, M.: Walking nanothermometers: spatiotemporal temperature measurement of transported acidic organelles in single living cells. Lab Chip. **12**, 1591 (2012)
120. Colombelli, J., Grill, S.W., Stelzer, E.H.K.: Ultraviolet diffraction limited nanosurgery of live biological tissues. Rev. Sci. Instrum. **75**, 472 (2004)
121. Berns, M., Rounds, D., Olson, R.: Effects of laser micro-irradiation on chromosomes. Exp. Cell Res. **56**, 292 (1969)
122. Berns, M.W., Olson, R.S., Rounds, D.E.: In vitro production of chromosomal lesions with an argon laser microbeam. Nature. **221**, 74 (1969)
123. Berns, M.W., Cheng, W.K., Floyd, A.D., Ohnuki, Y.: Chromosome lesions produced with an argon laser microbeam without dye sensitization. Science. **171**, 903 (1971)
124. Basehoar, G., Berns, M.W.: Cloning of rat kangaroo (ptk 2) cells following laser microirradiation of selected mitotic chromosomes. Science. **179**, 1333 (1973)
125. Berns, M.: A simple and versatile argon laser microbeam. Exp. Cell Res. **65**, 470 (1971)
126. Berns, M.W.: Biological Microirradiation Biological Techniques Series. Prentice-Hall, Englewood Cliffs, NJ (1974)

127. Berns, M.W., Aist, J., Edwards, J., Strahs, K., Girton, J., McNeill, P., Rattner, J.B., Kitzes, M., Hammer-Wilson, M., Liaw, L.-H., Siemens, A., Koonce, M., Peterson, S., Brenner, S., Burt, J., Walter, R., Bryant, P.J., van Dyk, D., Coulombe, J., Cahill, T., Berns, G.S.: Laser micro-surgery in cell and developmental biology. Science. **213**, 505 (1981)
128. McNeill, P.A., Berns, M.W.: Chromosome behavior after laser microirradiation of a single kinetochore in mitotic ptk2 cells. J. Cell Biol. **88**, 543 (1981)
129. Wang, H., Shi, L.Z., Wong, C.C.L., Han, X., Hwang, P.Y.-H., Truong, L.N., Zhu, Q., Shao, Z., Chen, D.J., Berns, M.W., Yates, J.R., Chen, L., Wu, X.: The interaction of ctip and nbs1 connects cdk and atm to regulate hr–mediated double-strand break repair. PLoS Genet. **9**, e1003277 (2013)
130. Truong, L.N., Li, Y., Shi, L.Z., Hwang, P.Y.-H., He, J., Wang, H., Razavian, N., Berns, M.W., Wu, X.: Microhomology-mediated end joining and homologous recombination share the initial end resection step to repair DNA double-strand breaks in mammalian cells. Proc. Natl. Acad. Sci. **110**, 7720 (2013)
131. Wang, H., Li, Y., Truong, L., Shi, L., Hwang, P.-H., He, J., Do, J., Cho, M., Li, H., Negrete, A., Shiloach, J., Berns, M., Shen, B., Chen, L., Wu, X.: CtIP maintains stability at common fragile sites and inverted repeats by end resection-independent endonuclease activity. Mol. Cell. **54**, 1012 (2014)
132. Srinivasan, R.: Kinetics of the ablative photodecomposition of organic polymers in the far ultraviolet 193 nm. J. Vac. Sci. Technol. B. **1**, 923 (1983)
133. Srinivasan, R., Braren, B.: Ultraviolet laser ablation of organic polymers. Chem. Rev. **89**, 1303 (1989)
134. Dupuy, C.G., Beach, D.B., Hurst, J.E., Jasinski, J.M.: Laser induced deposition of copper from (triethylphosphine)cyclopentadienylcopper(i). Chem. Mater. **1**, 16 (1989)
135. Misawa, H., Koshioka, M., Sasaki, K., Kitamura, N., Masuhara, H.: Laser trapping, spectroscopy, and ablation of a single latex particle in water. Chem. Lett. **19**, 1479 (1990)
136. Misawa, H., Kitamura, N., Masuhara, H.: Laser manipulation and ablation of a single microcapsule in water. J. Am. Chem. Soc. **113**, 7859 (1991)
137. Mason, T.G., Weitz, D.A.: Optical measurements of frequency-dependent linear viscoelastic moduli of complex fluids. Phys. Rev. Lett. **74**, 1250 (1995)
138. Mason, T.G., Ganesan, K., van Zanten, J.H., Wirtz, D., Kuo, S.C.: Particle tracking microrheology of complex fluids. Phys. Rev. Lett. **79**, 3282 (1997)
139. Addas, K.M., Schmidt, C.F., Tang, J.X.: Microrheology of solutions of semiflexible biopolymer filaments using laser tweezers interferometry. Phys. Rev. E. **70**, 021503 (2004)
140. Bishop, A.I., Nieminen, T.A., Heckenberg, N.R., Rubinsztein-Dunlop, H.: Optical microrheology using rotating laser-trapped particles. Phys. Rev. Lett. **92**, 198104 (2004)
141. Squires, T.M., Brady, J.F.: A simple paradigm for active and nonlinear microrheology. Phys. Fluids. **17** (2005). https://doi.org/10.1063/1.1960607
142. Wei, M.-T., Zaorski, A., Yalcin, H.C., Wang, J., Hallow, M., Ghadiali, S.N., Chiou, A., Ou-Yang, H.D.: A comparative study of living cell micromechanical properties by oscillatory optical tweezers. Opt. Express. **16**, 8594 (2008)
143. Wilson, L.G., Poon, W.C.K.: Small-world rheology: an introduction to probe-based active microrheology. Phys. Chem. Chem. Phys. **13**, 10617 (2011)
144. Neckernuss, T., Mertens, L.K., Martin, I., Paust, T., Beil, M., Marti, O.: Active microrheology with optical tweezers: a versatile tool to investigate anisotropies in intermediate filament networks. J. Phys. D. Appl. Phys. **49**, 045401 (2015)
145. Furst, E.M., Squires, T.M.: Active microrheology. In: Microrheology, pp. 279–301. Oxford University Press, Oxford (2017)
146. Robertson-Anderson, R.M.: Optical tweezers microrheology: from the basics to advanced techniques and applications. ACS Macro Lett. **7**, 968 (2018)
147. Khan, M., Regan, K., Robertson-Anderson, R.M.: Optical tweezers microrheology maps the dynamics of strain-induced local inhomogeneities in entangled polymers. Phys. Rev. Lett. **123**, 038001 (2019)

148. Squires, T.M., Mason, T.G.: Fluid mechanics of microrheology. Annu. Rev. Fluid Mech. **42**, 413 (2010)
149. Atakhorrami, M., Sulkowska, J.I., Addas, K.M., Koenderink, G.H., Tang, J.X., Levine, A.J., MacKintosh, F.C., Schmidt, C.F.: Correlated fluctuations of microparticles in viscoelastic solutions: quantitative measurement of material properties by microrheology in the presence of optical traps. Phys. Rev. E. **73**, 061501 (2006)
150. Squires, T.M.: Nonlinear microrheology bulk stresses versus direct interactions. Langmuir. **24**, 1147 (2007)
151. Helfer, E., Harlepp, S., Bourdieu, L., Robert, J., MacKintosh, F.C., Chatenay, D.: Microrheology of biopolymer-membrane complexes. Phys. Rev. Lett. **85**, 457 (2000)
152. Chapman, C.D., Robertson-Anderson, R.M.: Nonlinear microrheology reveals entanglement-driven molecular-level viscoelasticity of concentrated DNA. Phys. Rev. Lett. **113**, 098303 (2014)
153. Ginot, F., Caspers, J., Reinalter, L.F., Krishna Kumar, K., Krüger, M., Bechinger, C.: Recoil experiments determine the eigenmodes of viscoelastic fluids. New J. Phys. **24**, 123013 (2022)
154. Chapman, C.D., Lee, K., Henze, D., Smith, D.E., Robertson-Anderson, R.M.: Onset of non-continuum effects in microrheology of entangled polymer solutions. Macromolecules. **47**, 1181 (2014)
155. Knezevic, M., Podgurski, L.E.A., Stark, H.: Oscillatory active microrheology of active suspensions. Sci. Rep. **11**, 22706 (2021). https://doi.org/10.1038/s41598-021-02103-7
156. Paul, S., Kumar, R., Banerjee, A.: Two-point active microrheology in a viscous medium exploiting a motional resonance excited in dual-trap optical tweezers. Phys. Rev. E. **97**, 042606 (2018)
157. Baoukina, S., Marrink, S., Tieleman, D.: Molecular structure of membrane tethers. Biophys. J. **102**, 1866 (2012)
158. Dai, J., Sheetz, M.: Mechanical properties of neuronal growth cone membranes studied by tether formation with laser optical tweezers. Biophys. J. **68**, 988 (1995)
159. Hochmuth, F., Shao, J., Dai, J., Sheetz, M.: Deformation and flow of membrane into tethers extracted from neuronal growth cones. Biophys. J. **70**, 358 (1996)
160. Dai, J., Sheetz, M.P.: Membrane tether formation from blebbing cells. Biophys. J. **77**, 3363 (1999)
161. Pontes, B., Viana, N., Salgado, L., Farina, M., Neto, V., Nussenzveig, H.: Cell cytoskeleton and tether extraction. Biophys. J. **101**, 43 (2011)
162. Datar, A., Bornschlogl, T., Bassereau, P., Prost, J., Pullarkat, P.: Dynamics of membrane tethers reveal novel aspects of cytoskeleton-membrane interactions in axons. Biophys. J. **108**, 489 (2015)
163. Dao, M., Lim, C.T., Suresh, S.: Mechanics of the human red blood cell deformed by optical tweezers. J. Mech. Phys. Solids. **51**, 2259 (2003)
164. Henon, S., Lenormand, G., Richert, A., Gallet, F.: A new determination of the shear modulus of the human erythrocyte membrane using optical tweezers. Biophys. J. **76**, 1145 (1999)
165. Lim, C.T., Dao, M., Suresh, S., Sow, C.H., Chew, K.T.: Large deformation of living cells using laser traps. Acta Mater. **52**, 1837 (2004)
166. Dao, M., Li, J., Suresh, S.: Molecularly based analysis of deformation of spectrin network and human erythrocyte. Mater. Sci. Eng. C. **26**, 1232 (2006)
167. Yoon, Y.-Z., Kotar, J., Yoon, G., Cicuta, P.: The nonlinear mechanical response of the red blood cell. Phys. Biol. **5**, 036007 (2008)
168. Yoon, Y.Z., Kotar, J., Brown, A.T., Cicuta, P.: Red blood cell dynamics: from spontaneous fluctuations to non-linear response. Soft Matter. **7**, 2042 (2011)
169. Bronkhorst, P.J., Streekstra, G.J., Grimbergen, J., Nijhof, E.J., Sixma, J.J., Brakenhoff, G.J.: A new method to study shape recovery of red blood cells using multiple optical trapping. Biophys. J. **69**, 1666 (1995)

170. Mills, J.P., Diez-Silva, M., Quinn, D.J., Dao, M., Lang, M.J., Tan, K.S.W., Lim, C.T., Milon, G., David, P.H., Mercereau-Puijalon, O., Bonnefoy, S., Suresh, S.: Effect of plasmodial RESA protein on deformability of human red blood cells harboring Plasmodium falciparum. Proc. Natl. Acad. Sci. **104**, 9213 (2007)
171. Crick, A.J., Theron, M., Tiffert, T., Lew, V.L., Cicuta, P., Rayner, J.C.: Quantitation of malaria parasite-erythrocyte cell-cell interactions using optical tweezers. Biophys. J. **107**, 846 (2014)
172. Introini, V., Marin-Menendez, A., Nettesheim, G., Lin, Y.-C., Kariuki, S.N., Smith, A.L., Jean, L., Brewin, J.N., Rees, D.C., Cicuta, P., Rayner, J.C., Penman, B.S.: The erythrocyte membrane properties of beta thalassaemia heterozygotes and their consequences for Plasmodium falciparum invasion. Sci. Rep. **12** (2022). https://doi.org/10.1038/s41598-022-12060-4
173. Kariuki, S.N., Marin-Menendez, A., Introini, V., Ravenhill, B.J., Lin, Y.-C., Macharia, A., Makale, J., Tendwa, M., Nyamu, W., Kotar, J., Carrasquilla, M., Rowe, J.A., Rockett, K., Kwiatkowski, D., Weekes, M.P., Cicuta, P., Williams, T.N., Rayner, J.C.: Red blood cell tension protects against severe malaria in the Dantu blood group. Nature. **585**, 579 (2020)
174. Betz, T., Lenz, M., Joanny, J.-F., Sykes, C.: ATP-dependent mechanics of red blood cells. Proc. Natl. Acad. Sci. **106**, 15320 (2009)
175. Suresh, S.: Mechanical response of human red blood cells in health and disease: some structure-property-function relationships. J. Mater. Res. **21**, 1871 (2006)
176. Kim, J., Lee, H., Shin, S.: Advances in the measurement of red blood cell deformability: a brief review. J. Cell. Biotechnol. **1**, 63 (2015)
177. Zhu, R., Avsievich, T., Popov, A., Meglinski, I.: Optical tweezers in studies of red blood cells. Cells. **9**, 545 (2020)
178. Avsievich, T., Zhu, R., Popov, A., Bykov, A., Meglinski, I.: The advancement of blood cell research by optical tweezers. Rev. Phys. **5**, 100043 (2020)
179. Xie, Y., Liu, X.: Multifunctional manipulation of red blood cells using optical tweezers. J. Biophotonics. **15**, e202100315 (2021). https://doi.org/10.1002/jbio.202100315
180. Matthews, K., Lamoureux, E.S., Myrand-Lapierre, M.-E., Duffy, S.P., Ma, H.: Technologies for measuring red blood cell deformability. Lab Chip. **22**, 1254 (2022)
181. Sweeney, H.L., Holzbaur, E.L.: Motor proteins. Cold Spring Harb. Perspect. Biol. **10**, a021931 (2018)
182. Molloy, J.E., Padgett, M.J.: Lights, action: optical tweezers. Contemp. Phys. **43**, 241 (2002)
183. Moffitt, J.R., Chemla, Y.R., Smith, S.B., Bustamante, C.: Recent advances in optical tweezers. Annu. Rev. Biochem. **77**, 205 (2008)
184. Choudhary, D., Mossa, A., Jadhav, M., Cecconi, C.: Bio-molecular applications of recent developments in optical tweezers. Biomol. Ther. **9**, 23 (2019)
185. Wang, H., Guo, Y., Zou, R., Hu, H., Wang, Y., Wang, F., Ju, L.A.: Recent advances of optical tweezers–based dynamic force spectroscopy and mechanical measurement assays for live-cell mechanobiology. Front. Phys. **10** (2022). https://doi.org/10.3389/fphy.2022.771111
186. Veigel, C., Coluccio, L.M., Jontes, J.D., Sparrow, J.C., Milligan, R.A., Molloy, J.E.: The motor protein myosin-i produces its working stroke in two steps. Nature. **398**, 530 (1999)
187. Uemura, S., Higuchi, H., Olivares, A.O., De La Cruz, E.M., Ishiwata, S.: Mechanochemical coupling of two substeps in a single myosin v motor. Nat. Struct. Mol. Biol. **11**, 877 (2004)
188. Reinemann, D.N., Norris, S.R., Ohi, R., Lang, M.J.: Processive kinesin-14 hset exhibits directional flexibility depending on motor traffic. Curr. Biol. **28**, 2356 (2018)
189. Ariga, T., Tateishi, K., Tomishige, M., Mizuno, D.: Noise-induced acceleration of single molecule kinesin-1. Phys. Rev. Lett. **127**, 178101 (2021)
190. Kellermayer, M.S.Z., Smith, S.B., Granzier, H.L., Bustamante, C.: Folding-unfolding transitions in single titin molecules characterized with laser tweezers. Science. **276**, 1112 (1997)
191. Liphardt, J., Onoa, B., Smith, S.B., Tinoco, I., Bustamante, C.: Reversible unfolding of single RNA molecules by mechanical force. Science. **292**, 733 (2001)
192. La Porta, A., Wang, M.D.: Optical torque wrench: angular trapping, rotation, and torque detection of quartz microparticles. Phys. Rev. Lett. **92**, 190801 (2004)

193. Deufel, C., Forth, S., Simmons, C.R., Dejgosha, S., Wang, M.D.: Nanofabricated quartz cylinders for angular trapping: DNA supercoiling torque detection. Nat. Methods. **4**, 223 (2007)
194. Chen, Y., Ju, L., Rushdi, M., Ge, C., Zhu, C.: Receptor-mediated cell mechanosensing. Mol. Biol. Cell. **28**, 3134 (2017)
195. Jin, P., Jan, L.Y., Jan, Y.-N.: Mechanosensitive ion channels: structural features relevant to mechanotransduction mechanisms. Annu. Rev. Neurosci. **43**, 207 (2020)
196. Saotome, K., Murthy, S.E., Kefauver, J.M., Whitwam, T., Patapoutian, A., Ward, A.B.: Structure of the mechanically activated ion channel piezo1. Nature. **554**, 481 (2017)
197. Yang, T., Park, C., Rah, S.-H., Shon, M.J.: Nano-precision tweezers for mechanosensitive proteins and beyond. Mol. Cells. **45**, 16 (2022)
198. Zhu, C., Chen, Y., Ju, L.A.: Dynamic bonds and their roles in mechanosensing. Curr. Opin. Chem. Biol. **53**, 88 (2019)
199. Kim, S.T., Takeuchi, K., Sun, Z.-Y.J., Touma, M., Castro, C.E., Fahmy, A., Lang, M.J., Wagner, G., Reinherz, E.L.: The alpha beta t cell receptor is an anisotropic mechanosensor. J. Biol. Chem. **284**, 31028 (2009)
200. Feng, Y., Brazin, K.N., Kobayashi, E., Mallis, R.J., Reinherz, E.L., Lang, M.J.: Mechanosensing drives acuity of alpha beta t-cell recognition. Proc. Natl. Acad. Sci. **114**, E8204 (2017). https://doi.org/10.1073/pnas.1703559114
201. Das, R., Lin, L.-C., Catala-Castro, F., Malaiwong, N., Sanfeliu-Cerdan, N., Porta-de-la Riva, M., Pidde, A., Krieg, M.: An asymmetric mechanical code ciphers curvature-dependent proprioceptor activity. Sci. Adv. **7** (2021). https://doi.org/10.1126/sciadv.abg4617
202. Falleroni, F., Torre, V., Cojoc, D.: Cell mechanotransduction with piconewton forces applied by optical tweezers. Front. Cell. Neurosci. **12** (2018). https://doi.org/10.3389/fncel.2018.00130
203. Meloty-Kapella, L., Shergill, B., Kuon, J., Botvinick, E., Weinmaster, G.: Notch ligand endocytosis generates mechanical pulling force dependent on dynamin, epsins, and actin. Dev. Cell. **22**, 1299 (2012)
204. Shergill, B., Meloty-Kapella, L., Musse, A., Weinmaster, G., Botvinick, E.: Optical tweezers studies on notch: single-molecule interaction strength is independent of ligand endocytosis. Dev. Cell. **22**, 1313 (2012)
205. Stephenson, N.L., Avis, J.M.: Direct observation of proteolytic cleavage at the s2 site upon forced unfolding of the notch negative regulatory region. Proc. Natl. Acad. Sci. **109**, E2757 (2012). https://doi.org/10.1073/pnas.1205788109
206. Gordon, W., Zimmerman, B., He, L., Miles, L., Huang, J., Tiyanont, K., McArthur, D., Aster, J., Perrimon, N., Loparo, J., Blacklow, S.: Mechanical allostery: evidence for a force requirement in the proteolytic activation of notch. Dev. Cell. **33**, 729 (2015)
207. Ruggeri, Z.M., Mendolicchio, G.L.: Adhesion mechanisms in platelet function. Circ. Res. **100**, 1673 (2007)
208. Kim, J., Zhang, C.-Z., Zhang, X., Springer, T.A.: A mechanically stabilized receptor–ligand flex-bond important in the vasculature. Nature. **466**, 992 (2010)
209. Kim, J., Hudson, N.E., Springer, T.A.: Force-induced on-rate switching and modulation by mutations in gain-of-function von Willebrand diseases. Proc. Natl. Acad. Sci. **112**, 4648 (2015)
210. Mammen, M., Helmerson, K., Kishore, R., Choi, S.-K., Phillips, W.D., Whitesides, G.M.: Optically controlled collisions of biological objects to evaluate potent polyvalent inhibitors of virus-cell adhesion. Chem. Biol. **3**, 757 (1996)
211. Litvinov, R.I., Shuman, H., Bennett, J.S., Weisel, J.W.: Binding strength and activation state of single fibrinogen-integrin pairs on living cells. Proc. Natl. Acad. Sci. **99**, 7426 (2002)
212. Meier-Ruge, W., Bielser, W., Remy, E., Hillenkamp, F., Nitsche, R., Unsld, R.: The laser in the Lowry technique for microdissection of freeze-dried tissue slices. Histochem. J. **8**, 387 (1976)

213. Leitz, G., Lundberg, C., Fällman, E., Axner, O., Sellstedt, A.: Laser-based micromanipulation for separation and identification of individual Frankia vesicles. FEMS Microbiol. Lett. **224**, 97 (2003)
214. Kumar, S., Maxwell, I.Z., Heisterkamp, A., Polte, T.R., Lele, T.P., Salanga, M., Mazur, E., Ingber, D.E.: Viscoelastic retraction of single living stress fibers and its impact on cell shape, cytoskeletal organization, and extracellular matrix mechanics. Biophys. J. **90**, 3762 (2006)
215. Berns, M.W., Wright, W.H., Tromberg, B.J., Profeta, G.A., Andrews, J.J., Walter, R.J.: Use of a laser-induced optical force trap to study chromosome movement on the mitotic spindle. Proc. Natl. Acad. Sci. **86**, 4539 (1989)
216. Emmert-Buck, M.R., Bonner, R.F., Smith, P.D., Chuaqui, R.F., Zhuang, Z., Goldstein, S.R., Weiss, R.A., Liotta, L.A.: Laser capture microdissection. Science. **274**, 998 (1996)
217. Schütze, K., Becker, I., Becker, K.-F., Thalhammer, S., Stark, R., Heckl, W.M., Bohm, M., Posl, H.: Cut out or poke in—the key to the world of single genes: laser micromanipulation as a valuable tool on the look-out for the origin of disease. Genet. Anal. Biomol. Eng. **14**, 1 (1997)
218. Schiitze, K., Bohm, M., Thalhammer, S., Stark, R., Heckl, W.M., Posl, H.: Laser microablation for genetic analysis on a single cell basis. BIOforum. **20**, 82 (1997)
219. Schütze, K., Lahr, G.: Identification of expressed genes by laser-mediated manipulation of single cells. Nat. Biotechnol. **16**, 737 (1998)
220. Berns, M.W., Wright, W.H., Steubing, R.W.: Laser microbeam as a tool in cell biology. Int. Rev. Cytol. **129**, 1–44 (1991)
221. Tsukakoshi, M., Kurata, S., Nomiya, Y., Ikawa, Y., Kasuya, T.: A novel method of DNA transfection by laser microbeam cell surgery. Appl. Phys. B Photophys. Laser Chem. **35**, 135 (1984)
222. Kurata, S.-I., Tsukakoshi, M., Kasuya, T., Ikawa, Y.: The laser method for efficient introduction of foreign DNA into cultured cells. Exp. Cell Res. **162**, 372 (1986)
223. Tao, W., Wilkinson, J., Stanbridge, E.J., Berns, M.W.: Direct gene transfer into human cultured cells facilitated by laser micropuncture of the cell membrane. Proc. Natl. Acad. Sci. **84**, 4180 (1987)
224. Weber, G., Monajembashi, S., Greulich, K., Wolfrum, J.: Genetic manipulation of plant cells and organelles with a laser microbeam. In: Progress in Plant Protoplast Research, vol. 12, pp. 319–322. Springer, Dordrecht (1988)
225. Weber, G., Monajembashi, S., Greulich, K., Wolfrum, J.: Microperforation of plant tissue with a UV laser microbeam and injection of DNA into cells. Naturwissenschaften. **75**, 35 (1988)
226. Weber, G., Monajembashi, S., Wolfrum, J., Greulich, K.: A laser microbeam as a tool to introduce genes into cells and organelles of higher plants. Ber. Bunsenges. Phys. Chem. **93**, 252 (1989)
227. Weber, G., Monajembashi, S., Wolfrum, J., Greulich, K.-O.: Genetic changes induced in higher plant cells by a laser microbeam. Physiol. Plant. **79**, 190 (1990)
228. Stuhrmann, B., Jahnke, H.-G., Schmidt, M., Jahn, K., Betz, T., Müller, K., Rothermel, A., Kas, J., Robitzki, A.A.: Versatile optical manipulation system for inspection, laser processing, and isolation of individual living cells. Rev. Sci. Instrum. **77** (2006). https://doi.org/10.1063/1.2214961
229. Schopper, B., Ludwig, M., Edenfeld, J., Al-Hasani, S., Diedrich, K.: Possible applications of lasers in assisted reproductive technologies. Hum. Reprod. **14**, 186 (1999)
230. Clement-Sengewald, A., Schütze, K., Ashkin, A., Palma, G.A., Kerlen, G., Brem, G.: Fertilization of bovine oocytes induced solely with combined laser microbeam and optical tweezers. J. Assist. Reprod. Genet. **13**, 259 (1996)
231. Enginsu, M., Schütze, K., Bellanca, S., Pensis, M., Campo, R., Bassil, S., Donnez, J., Gordts, S.: Endocrinology: micromanipulation of mouse gametes with laser microbeam and optical tweezers. Hum. Reprod. **10**, 1761 (1995)
232. Helmerson, K., Kishore, R., Phillips, W.D., Weetall, H.H.: Optical tweezers-based immunosensor detects femtomolar concentrations of antigens. Clin. Chem. **43**, 379 (1997)

233. Townes-Anderson, E., Jules, R., Sherry, D., Lichtenberger, J., Hassanain, M.: Micromanipulation of retinal neurons by optical tweezers. Mol. Vis. **4**, 12 (1998)
234. Pine, J., Chow, G.: Moving live dissociated neurons with an optical tweezer. IEEE Trans. Biomed. Eng. **56**, 1184 (2009)
235. Becker, I., Becker, K., Rohrl, M., Minkus, G., Schutze, K., Hofler, H.: Single cell mutation analysis of tumors from stained histologic slides. Lab. Invest. **75**, 801 (1996)
236. Seeger, S., Monajembashi, S., Hutter, K.-J., Futterman, G., Wolfrum, J., Greulich, K.: Application of laser optical tweezers in immunology and molecular genetics. Cytometry. **12**, 497 (1991)
237. Higurashi, E., Ukita, H., Tanaka, H., Ohguchi, O.: Optically induced rotation of anisotropic micro-objects fabricated by surface micromachining. Appl. Phys. Lett. **64**, 2209 (1994)
238. Friese, M.E.J., Rubinsztein-Dunlop, H., Gold, J., Hagberg, P., Hanstorp, D.: Optically driven micromachine elements. Appl. Phys. Lett. **78**, 547 (2001)
239. Zong, Y., Liu, J., Liu, R., Guo, H., Yang, M., Li, Z., Chen, K.: An optically driven bistable Janus rotor with patterned metal coatings. ACS Nano. **9**, 10844 (2015)
240. Petkov, T., Yang, M., Ren, K., Pouligny, B., Loudet, J.-C.: Behaviors of ellipsoidal microparticles within a two-beam optical levitator. J. Quant. Spectrosc. Radiat. Transfer. **195**, 85 (2017)
241. Fernandez-Nieves, A., Cristobal, G., Garces-Chavez, V., Spalding, G., Dholakia, K., Weitz, D.: Optically anisotropic colloids of controllable shape. Adv. Mater. **17**, 680 (2005)
242. Stilgoe, A.B., Nieminen, T.A., Knoener, G., Heckenberg, N.R., Rubinsztein-Dunlop, H.: The effect of Mie resonances on trapping in optical tweezers. Opt. Express. **16**, 15039 (2008)
243. Tkachenko, G., Brasselet, E.: Helicity-dependent three-dimensional optical trapping of chiral microparticles. Nat. Commun. **5**, 4491 (2014). https://doi.org/10.1038/ncomms5491
244. Vovk, I.A., Baimuratov, A.S., Zhu, W., Shalkovskiy, A.G., Baranov, A.V., Fedorov, A.V., Rukhlenko, I.D.: Chiral nanoparticles in singular light fields. Sci. Rep. **7** (2017). https://doi.org/10.1038/srep45925
245. Dharmadhikari, J., Roy, S., Dharmadhikari, A., Sharma, S., Mathur, D.: Torque-generating malaria-infected red blood cells in an optical trap. Opt. Express. **12**, 1179 (2004)
246. Ghosh, A., Sinha, S., Dharmadhikari, J.A., Roy, S., Dharmadhikari, A.K., Samuel, J., Sharma, S., Mathur, D.: Euler buckling-induced folding and rotation of red blood cells in an optical trap. Phys. Biol. **3**, 67 (2006)
247. Bambardekar, K., Dharmadhikari, A.K., Dharmadhikari, J.A., Mathur, D., Sharma, S.: Measuring erythrocyte deformability with fluorescence, fluid forces, and optical trapping. J. Biomed. Opt. **13**, 1 (2008)
248. Allen, L., Beijersbergen, M.W., Spreeuw, R.J.C., Woerdman, J.P.: Orbital angular momentum of light and the transformation of Laguerre-Gaussian laser modes. Phys. Rev. A. **45**, 8185 (1992)
249. O'Neil, A.T., MacVicar, I., Allen, L., Padgett, M.J.: Intrinsic and extrinsic nature of the orbital angular momentum of a light beam. Phys. Rev. Lett. **88**, 053601 (2002)
250. Garces-Chavez, V., McGloin, D., Padgett, M.J., Dultz, W., Schmitzer, H., Dholakia, K.: Observation of the transfer of the local angular momentum density of a multiringed light beam to an optically trapped particle. Phys. Rev. Lett. **91**, 093602 (2003)
251. Parkin, S., Knoner, G., Nieminen, T.A., Heckenberg, N.R., Rubinsztein-Dunlop, H.: Measurement of the total optical angular momentum transfer in optical tweezers. Opt. Express. **14**, 6963 (2006)
252. Wilking, J.N., Mason, T.G.: Optically driven nonlinear microrheology of gelatin. Phys. Rev. E. **77**, 055101 (2008)
253. Faucheux, L.P., Bourdieu, L.S., Kaplan, P.D., Libchaber, A.J.: Optical thermal ratchet. Phys. Rev. Lett. **74**, 1504 (1995)
254. Bang, J., Pan, R., Hoang, T.M., Ahn, J., Jarzynski, C., Quan, H.T., Li, T.: Experimental realization of Feynman's ratchet. New J. Phys. **20**, 103032 (2018)

255. Arzola, A.V., Villasante-Barahona, M., Volke-Sepúlveda, K., Jakl, P., Zemanek, P.: Omnidirectional transport in fully reconfigurable two dimensional optical ratchets. Phys. Rev. Lett. **118**, 138002 (2017)
256. Lopez, B.J., Kuwada, N.J., Craig, E.M., Long, B.R., Linke, H.: Realization of a feedback controlled flashing ratchet. Phys. Rev. Lett. **101**, 220601 (2008)
257. Liesfeld, B., Nambiar, R., Meiners, J.C.: Particle transport in asymmetric scanning-line optical tweezers. Phys. Rev. E. **68**, 051907 (2003)
258. Liu, J., Guo, H.-L., Li, Z.-Y.: Self-propelled round-trip motion of janus particles in static line optical tweezers. Nanoscale. **8**, 19894 (2016)
259. Shen, Y., Weitz, D.A., Forde, N.R., Shayegan, M.: Line optical tweezers as controllable micromachines: techniques and emerging trends. Soft Matter. **18**, 5359 (2022)
260. Gahagan, K.T., Swartzlander, G.A.: Optical vortex trapping of particles. Opt. Lett. **21**, 827 (1996)
261. McCann, L.I., Dykman, M., Golding, B.: Thermally activated transitions in a bistable three-dimensional optical trap. Nature. **402**, 785 (1999)
262. Lutz, C., Kollmann, M., Bechinger, C.: Single-file diffusion of colloids in one-dimensional channels. Phys. Rev. Lett. **93**, 026001 (2004)
263. Mu, W., Liu, Z., Luan, L., Wang, G., Spalding, G.C., Ketterson, J.B.: Enhanced particle transport in an oscillating sinusoidal optical potential. New J. Phys. **11**, 103017 (2009)
264. Roichman, Y., Sun, B., Stolarski, A., Grier, D.G.: Influence of nonconservative optical forces on the dynamics of optically trapped colloidal spheres: the fountain of probability. Phys. Rev. Lett. **101**, 128301 (2008)
265. Sun, B., Lin, J., Darby, E., Grosberg, A.Y., Grier, D.G.: Brownian vortexes. Phys. Rev. E. **80**, 010401 (2009)
266. Niculescu, A.-G., Chircov, C., Bîrcă, A.C., Grumezescu, A.M.: Fabrication and applications of microfluidic devices: a review. Int J Mol Sci. **22**, 2011 (2021)
267. Leach, J., Mushfique, H., di Leonardo, R., Padgett, M., Cooper, J.: An optically driven pump for microfluidics. Lab Chip. **6**, 735 (2006)
268. Cran-McGreehin, S., Krauss, T.F., Dholakia, K.: Integrated monolithic optical manipulation. Lab Chip. **6**, 1122 (2006)
269. Enger, J., Goksor, M., Ramser, K., Hagberg, P., Hanstorp, D.: Optical tweezers applied to a microfluidic system. Lab Chip. **4**, 196 (2004)
270. Dholakia, K., Reece, P.: Optical micromanipulation takes hold. Nano Today. **1**, 18 (2006)
271. Monat, C., Domachuk, P., Grillet, C., Collins, M., Eggleton, B.J., Cronin-Golomb, M., Mutzenich, S., Mahmud, T., Rosengarten, G., Mitchell, A.: Optofluidics: a novel generation of reconfigurable and adaptive compact architectures. Microfluid. Nanofluid. **4**, 81 (2007)
272. Terray, A., Oakey, J., Marr, D.W.M.: Fabrication of linear colloidal structures for microfluidic applications. Appl. Phys. Lett. **81**, 1555 (2002)
273. Applegate, R.W., Squier, J., Vestad, T., Oakey, J., Marr, D.W.M.: Optical trapping, manipulation, and sorting of cells and colloids in microfluidic systems with diode laser bars. Opt. Express. **12**, 4390 (2004)
274. Paterson, L., Papagiakoumou, E., Milne, G., Garces-Chavez, V., Tatarkova, S.A., Sibbett, W., Gunn-Moore, F.J., Bryant, P.E., Riches, A.C., Dholakia, K.: Light-induced cell separation in a tailored optical landscape. Appl. Phys. Lett. **87**, 123901 (2005)
275. Wang, M.M., Tu, E., Raymond, D.E., Yang, J.M., Zhang, H., Hagen, N., Dees, B., Mercer, E.M., Forster, A.H., Kariv, I., Marchand, P.J., Butler, W.F.: Microfluidic sorting of mammalian cells by optical force switching. Nat. Biotechnol. **23**, 83 (2005)
276. Wang, X., Chen, S., Kong, M., Wang, Z., Costa, K.D., Li, R.A., Sun, D.: Enhanced cell sorting and manipulation with combined optical tweezer and microfluidic chip technologies. Lab Chip. **11**, 3656 (2011)
277. Paie, P., Zandrini, T., Vazquez, R., Osellame, R., Bragheri, F.: Particle manipulation by optical forces in microfluidic devices. Micromachines. **9**, 200 (2018)

278. Wang, Y., Shevate, S., Wintermantel, T.M., Morgado, M., Lochead, G., Whitlock, S.: Preparation of hundreds of microscopic atomic ensembles in optical tweezer arrays. npj Quant. Inf. **6**, 54 (2020)
279. Endres, M., Bernien, H., Keesling, A., Levine, H., Anschuetz, E.R., Krajenbrink, A., Senko, C., Vuletic, V., Greiner, M., Lukin, M.D.: Atom-by-atom assembly of defect-free one-dimensional cold atom arrays. Science. **354**, 1024 (2016)
280. Chapin, S.C., Germain, V., Dufresne, E.R.: Automated trapping, assembly, and sorting with holographic optical tweezers. Opt. Express. **14**, 13095 (2006)
281. Zemánek, P., Jonás, A., Liska, M.: Simplified description of optical forces acting on a nanoparticle in the Gaussian standing wave. J. Opt. Soc. Am. A. **19**, 1025 (2002)
282. MacDonald, M.P., Spalding, G.C., Dholakia, K.: Microfluidic sorting in an optical lattice. Nature. **426**, 421 (2003)
283. Mellor, C.D., Bain, C.D.: Array formation in evanescent waves. ChemPhysChem. **7**, 329 (2006)
284. Garces-Chavez, V., Quidant, R., Reece, P.J., Badenes, G., Torner, L., Dholakia, K.: Extended organization of colloidal microparticles by surface plasmon polariton excitation. Phys. Rev. B. **73**, 085417 (2006)

The manufacturer's authorised representative in the EU is Springer Nature Customer Service Centre GmbH, Europaplatz 3, 69115 Heidelberg, Germany. If you have any concerns regarding our products, please contact ProductSafety@springernature.com

Printed and bound by CPI Group (UK) Ltd, Croydon, CR0 4YY
01/10/2026
02225930-0015